Climate Change Management

Series editor

Walter Leal Filho, Hamburg, Germany

More information about this series at http://www.springer.com/series/8740

Fátima Alves · Walter Leal Filho
Ulisses Azeiteiro
Editors

Theory and Practice of Climate Adaptation

Springer

Editors
Fátima Alves
Department of Social Sciences and Management
Universidade Aberta
Lisbon
Portugal

and

Centre for Functional Ecology, Science for People and the Planet
University of Coimbra
Coimbra
Portugal

Walter Leal Filho
Faculty of Life Science
HAW Hamburg
Hamburg
Germany

Ulisses Azeiteiro
Department of Biology & CESAM
University of Aveiro
Aveiro
Portugal

ISSN 1610-2002 ISSN 1610-2010 (electronic)
Climate Change Management
ISBN 978-3-030-10283-8 ISBN 978-3-319-72874-2 (eBook)
https://doi.org/10.1007/978-3-319-72874-2

Printed on acid-free paper

This Springer imprint is published by the registered company Springer International Publishing AG part of Springer Nature
The registered company address is: Gewerbestrasse 11, 6330 Cham, Switzerland

Preface

Climate change is among the major challenges of modern times. As the Fifth Assessment Report (AR5) produced by the Intergovernmental Panel on Climate Change (IPCC) and the recommendations from the Paris Agreement have shown, it is important to strengthen the global response to the threat of climate change by keeping a global temperature rise this century well below 2 degrees Celsius above pre-industrial levels and to pursue efforts to limit the temperature increase even further to 1.5 °C.

The endorsement of the Paris Agreement, which was ratified by 172 Parties—out of the 197 Parties to the UN Framework Convention on Climate Change (UNFCCC)—means that there is a perceived need to strengthen the ability of countries—especially the developing ones—to deal with the impacts of climate change. To reach these ambitious goals, a number of measures are needed. For example,

- The need to ensure appropriate financial flows;
- The need for a new technology framework;
- The need for an enhanced capacity building framework;
- The need to continuously measure and monitor progress.

A further need, acknowledged by both industrialising and developing nations alike, is the requirement for the documentation and dissemination of experiences and good practice on climate change adaptation, which may serve as inspiration and motivation.

Consistent with the need to fill in the above gap, "the 2nd World Symposium on Climate Change Adaptation" was organised by the University of Coimbra (Portugal), the Research and Transfer Centre "Sustainable Development and Climate Change Management" of the Hamburg University of Applied Sciences (Germany) and the Baltic University Programme in Uppsala (Sweden), in cooperation with the

International Climate Change Information Programme (ICCIP) and the United Nations University initiative "Regional Centres of Expertise on Education for Sustainable Development" (RCE). The symposium was a truly interdisciplinary event, covering some of the key areas in the field of climate change adaptation. The event focused on "integrative approaches to implement climate change adaptation" and was meant to contribute to the further development of this fast-growing field.

This book titled "Theory and Practice of Climate Adaptation" is one of the outcomes of the event. Its aims are as follows:

i. To provide research institutions, universities, NGOs and enterprises all around the world with an opportunity to document their works in the field of climate change adaptation;
ii. To foster the exchange of information, ideas and experiences acquired in the execution of climate change adaptation projects, especially successful initiatives and good practice;
iii. To introduce methodological approaches and experiences deriving from case studies and projects, which aim to show how the principles of climate change adaptation may be implemented in practice.

Thirty-two double-blind peer-reviewed chapters from several geographic origins cover different subjects related to the theme of this book that is organised in five sections. Part I "Climate Change Governance and Policy" gathers 6 contributions about community resilience priorities, ecosystem-based adaptation and climate change adaptation policies; Part II "Climate Change Adaptation, Resilience and Disaster Risk Management" gathers 6 contributions and brings experiences from different continents and geographies (e.g. Mozambique in Africa, India in Asia and Brazil in South America); Part III "Social Perceptions and Socio-cultural Dimensions of Climate Change Impacts and Adaptation" gathers 7 contributions about community perceptions, attitudes, ability and willingness, environmental justice and climate change adaptation in the context of risk society, migrations, small island coastal communities and climate risk communication; Part IV "Food Security and Climate Change: Adapting Agriculture to Climate Change" gathers 8 contributions and brings experiences from Portugal and the Mediterranean, Malawi, Argentina and Brazil; and Part V "Technological approaches to Climate Change Adaptation/Innovative Approaches towards Low Carbon Economics" gathers 5 contributions about agriculture practices as initiatives for mitigation and adaptation to climate change, climate adaptation practices in building constructions, water-saving technologies to adapt to global changes and adaptation technologies in water sector, emission trading schemes and carbon markets in the NDCs.

Last but not least, a further aim of the book is to document and disseminate the wealth of experiences on climate change adaptation available today across the world.

We thank the authors for their willingness to share their knowledge, know-how and experiences, as well as the many peer reviewers, which have helped us to ensure the quality of the manuscripts.

Enjoy your reading!

Lisbon, Portugal — Fátima Alves
Hamburg, Germany — Walter Leal Filho
Aveiro, Portugal — Ulisses Azeiteiro
Summer 2018

Contents

Part I
Climate Change Governance and Policy

Decentralized Governance and Climate Change Adaptation: Working Locally to Address Community Resilience Priorities

Erin Martin, Christopher Perine, Veronique Lee and Jeff Ratcliffe

Abstract With climate change adaptation emerging as a top priority in many developing countries, national governments and international and local organizations seek the most effective policy and governance mechanisms through which to foster progress. One avenue of keen interest is the devolution of authority for adaptation planning and implementation to the local level. To date, little research is available on the link between local governance effectiveness and progress on adaptation. To help address that need, this paper explores governance and climate change adaptation dynamics at the local level in Mali, focusing on the commune, the smallest-scale elected body. The discussion centers on a legally-mandated local planning process, for which community consultation is required, as the entry point to analyze the effectiveness of local efforts to identify and implement climate change adaptation priorities. The paper provides the results of activities undertaken by the United States Agency for International Development (USAID), presenting findings from a qualitative study of five communes, together with a case study on Koro commune, which highlight lessons on integration of climate change adaption into commune level development plans. The discussion is intended to foster

This paper was made possible through the support of USAID but its contents are the sole responsibility of the authors and do not necessarily reflect the views of USAID or the United States Government.

E. Martin (✉) · C. Perine · V. Lee
Project staff of USAID Adaptation, Thought Leadership and Assessments (ATLAS),
1717 H Street NW, Washington, D.C 20001, USA
e-mail: eemartin08@yahoo.com

C. Perine
e-mail: cperine@chemonics.com

V. Lee
e-mail: vlee@chemonics.com

J. Ratcliffe
Project staff of USAID Climate Change Adaptation Activity,
Bamako, Mali
e-mail: jratcliffe@maliccaa.com

F. Alves et al. (eds.), *Theory and Practice of Climate Adaptation*,
Climate Change Management, https://doi.org/10.1007/978-3-319-72874-2_1

understanding of the opportunities and conditions that enable successful climate adaptation interventions at the local level, as well as the extent to which a decentralized system of government can confer an adaptive advantage. The key takeaway of the discussion is that even in the context of fundamental governance challenges, such as those Mali currently experiences, climate-resilient local planning is possible given adequate resources, technical expertise and commitment to engaging citizens in identifying priorities. This finding and others described here are relevant to national governments and other development actors investing in climate change adaptation at the local level.

Keywords Climate change adaptation · Decentralization · Mali
USAID · Resilience

Introduction

Among the poorest countries in the world, Mali and its people have endured a series of shocks over the last decade ranging from armed conflict and political violence to droughts that have added to the burden of food insecurity and poor health affecting much of the population. While certainly the sociopolitical situation results from a complex set of factors, climate-related stresses are increasingly viewed as a key driver; competition over natural resources, including water, farmland and rangeland, is common. Working within Mali's system of decentralized governance, USAID and other donors are looking for effective strategies to help local elected leaders and the people they represent in assessing future climate changes, mitigating their effects and adapting to potential outcomes.

The country's 17.6 million people, some 60% of whom live in rural areas, are exceedingly vulnerable to shocks and stressors (World Bank 2017). More than 80% rely on agriculture or livestock for their livelihoods, making them highly dependent on rainfall (USAID CDCS 2015). While typical weather conditions vary from year to year, overall annual rainfall declined by 12% between 1960 and 2009, and temperature rose 0.8 °C (USGS/USAID 2012). Prevailing scenarios for future climate suggest high variability in annual rainfall coupled with significant increases in warming, a projection that is raising alarm about permanent declines in rainfed farming in southern regions (Traore et al. 2017). Already, over the last six decades, the northern limit for rainfed millet and sorghum, key household subsistence crops, shifted southward by approximately 50 km (CGIAR 2014). This trend is likely to continue as temperature increases affect soil moisture (Lippman and Lewis 1998).

Another projection of serious concern is anticipated population growth (USAID CDCS 2015). With the third highest fertility rate in the world, the population of Mali is expected to double by 2035 (CIA Fact Book 2017).

Against this backdrop, climate change adaptation is seen to be critical in stemming potential climate-driven increases in household poverty, food insecurity and poor health outcomes. Interventions in the agriculture sector, such as new seed technologies, improved soil and water management, switching to climate-resilient crops, and diversification away from climate-dependent livelihoods, are not only viewed as a response strategy, but also, if managed well, a potentially game-changing opportunity to make development gains. Mali's significant potential to increase irrigated agriculture, livestock production and inland fisheries makes this scenario viable (USAID CDCS 2015).

Yet persistent conflict, including deliberate attacks on local elected and traditional leaders, has created an atmosphere of fear and hampered development progress in some areas (Reuters 2016). This politically-motivated violence lingers from the 2012 to 2013 armed conflict, during which Islamist rebels sought the secession of the Saharan and Sahelian zones that comprise northern Mali. A 2015 peace accord stalled in 2016 as rebels extended their operations south, launching dozens of attacks on Malian security forces, international peacekeepers and civilians (Human Rights Watch 2017).

Among the brutal incidents of violence were the summary executions of 27 men in Mopti, Ségou, Timbuktu and Kidal regions, most of them mayors, deputy mayors, village chiefs and teachers (Human Rights Watch 2017). Some victims were accused of collaborating with the government and French troops on counterterrorism operations; others were reportedly killed as punishment for alleged corruption. Still other killings of local authorities were attributed to intercommunal or interpersonal conflicts over land, water and community leadership (Human Rights Watch 2017).

Climate change has been cited as one of many factors driving the conflict. The Islamist groups behind the violence have allegedly leveraged community frustration over access to grazing land and water as a recruiting tool (Walch 2017).

The targeting of local government officials has had a chilling effect on governance at the commune level (Reuters 2016). As violence has spread from northern Mali into central and even some southern areas, some local government leaders have reportedly fled their communes, seeking to govern remotely from Mopti and Sevare cities (Human Rights Watch 2017).

Invested in the idea that effective local governance will help safeguard national unity, the national government is pressing forward with its 25-year-old decentralization effort (IMF 2015). Born of political compromise after a prior revolt in the North, this system of governance was enshrined in the Malian Constitution in 1992 and legislated in 1993. Its main objective is to create institutions of power that manage local-level problem-solving and development, engage the public in decision-making, and channel resources to address local priorities (IMF 2015).

Since 2012, a new series of laws and decrees has sought to reignite action on poverty reduction at the local level by increasing the transfer of resources from central to local governments, up to as much as one-third of central budget (IMF 2015). Political commitment to decentralization appears to be strong: the minister of decentralization in 2015 affirmed plans to increase revenue allocation from the central level to communes from 12 to 30% by 2020 (USAID CDCS 2015).

Per the decentralization law, communes are mandated to create a five-year development plan, known as the *Plan de Développement Economique, Social et*

Culturel, or PDESC. The process, which includes rounds of consultation with villagers, is meant to be a cornerstone of participatory development. It is the primary mechanism through which local government can influence and effect climate change adaptation (USAID ATLAS 2017a, b).

At the commune level, local governance has the potential to harness the traditions of shared decision-making and collective action and expand information-sharing across villages. Local governments can also work vertically to link up with external actors at both the national and international levels, potentially unlocking an inflow of resources and information (USAID ATLAS 2017a, b).

USAID, Mali's largest international donor, has supported decentralization efforts there since the 1980s (Lippman and Lewis 1998). Its 2015–2020 Country Development Cooperation Strategy (CDCS) responds to the socioeconomic and political conditions that helped spark the 2012–2013 conflict (and are continuing to fuel ongoing tensions), specifically through support for good governance and resilience-building. A centerpiece of the CDCS is assistance to increase local government effectiveness on service delivery, improve management and financial oversight, and strengthen coordination with regional and national government as well as civil society.

To understand the complex relationships between governance and local-level adaptation to climate change, USAID supported a 2017 study *Governance and Climate Change Adaptation: A Case Study on Mali*. Conducted by USAID's Adaptation, Thought Leadership and Assessments (ATLAS) activity, the study sought to ascertain whether Mali's decentralization system provided an advantage in helping households and communities adapt to climate change. In particular, the study considered:

- Characteristics of the predominant local livelihoods
- Local perceptions of climate variability and change with associated environmental and socioeconomic impacts
- Household and community strategies of response to climate and environmental stressors and shocks
- Experience of decentralization as it is understood and operates at local level
- Current or potential impacts of this local governance on climate change adaptation

The results, intended to inform future USAID programming in Mali, may also be applicable to adaptation activities in other countries.

Field research took place in three regions of southern and central Mali (Koulikoro, Mopti and Sikasso), specifically within five communes: Kléla, Konna, Koula, Minidian and Ourikela (Fig. 1). The University of Arizona and the Malian nongovernmental organization (NGO) Sahel Eco conducted the field work and analysis during June and July 2016. Within each commune, three villages were selected for public opinion research based on underlying agro-ecological and livelihood characteristics. In each of those villages, two focus group discussions were carried out, one with men and one with women. In addition, six to 10 household interviews were conducted.

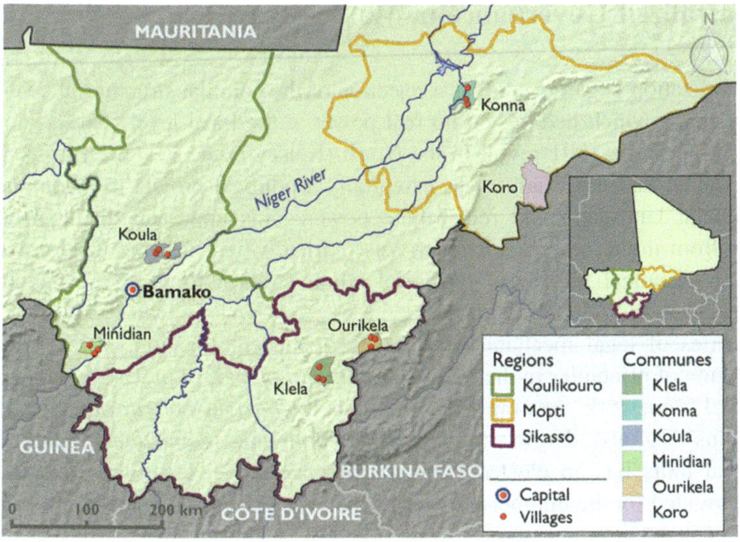

Fig. 1 Map of Mali showing communes discussed in this paper

At the commune level, key informant interviews were conducted with local elected officials; local government administrators; agriculture, livestock and forestry extension agents; school and health clinic staff; and local merchants. In all, the field team completed 193 interviews and more than 800 pages of responses.

The field work was complemented by a review of secondary literature, existing reports and a comparative analysis of the five communes' PDESC development plans.

Alongside the qualitative study, on-the-ground experience with development projects in Mali sheds light on local government progress on climate change adaptation. Among these projects is the five-year USAID Mali Climate Change and Adaptation Activity (MCCAA), which is active in the central region of Mopti, an area highly threatened by climate change. In 2016, MCCAA, working closely with the NGO *Groupe de recherches et d'applications techniques* (GRAT), collaborated with 25 communes in the region to help integrate climate change activities into their PDESCs. The development and adoption of these plans has presented an opportunity for close coordination and collaboration between MCCAA and other USAID and other donor-funded projects.

The ATLAS study and MCCAA's experience in Mopti, particularly in the commune of Kono, are discussed in the following sections.

Decentralized Governance in Mali

Decentralization began in Mali as a rejection of the colonial structure of centralized government—which had vested no real power at the local level—and as conciliation following the 1991 revolt in the North (Kelsey et al. 2011). From the beginning, Malian decentralization was essentially a political project to reconfigure the institutional landscape and redistribute power. This approach distinguished the system from initiatives elsewhere in West Africa, where decentralization was primarily a technical reform of territorial administration to meet the demands of donors (SNV and CEDELO 2004).

A series of legal mechanisms gave life to decentralization, starting with the articulation of principles in the 1992 Constitution. In 1993, *Loi 93-003* established a structural framework, creating territorial units (*collectivités territoriales*) consisting of regions (*régions*), districts (*cercles*), and communes (*communes*). Each of these territorial units has an elected council and governing executives. Legislation in 2012 provided for the number of regions to expand from its original eight to 11, but not all of these have been constituted (IMF 2015). Currently, Mali has 10 regions, 49 districts and 703 communes. Bamako, the capital, is its own district (Fig. 2).

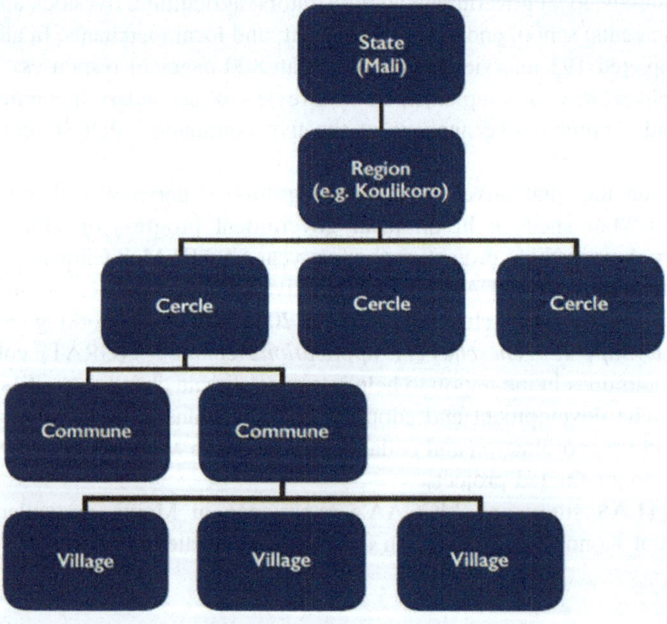

Fig. 2 Levels of government in Mali's decentralized system

At the commune level, decision-making is managed by the elected communal council (*conseil communal*), which chooses a mayor and deputy mayors from its membership. Traditional leaders, through a village chief and village development committees (*Comités Villageois de Développement*), mediate the distribution of customary land rights, settle conflicts related to land tenure and herder-farmer relations, and also interact with elected bodies.

In 1996, the Principal Decentralization Law (*Codes des Collectivités Territoriales*) established communes as administratively and fiscally autonomous subnational units. That same year, *Loi N° 96-050* shifted responsibilities for protecting natural resources and managing lands to local government, specifically designating responsibility for protecting the environment, managing public lands, formulating land use plans, protecting and developing government land, and organizing agroforestry production (Kelsey et al. 2011).

Today, political decentralization is more or less complete, but fiscal and administrative decentralization lag. Communes rely primarily on central government for funding but do not have input into the central planning or budgeting process. This creates inefficiencies and inaccuracies related to the amount of funds disbursed to them (IMF 2015).

Two particular challenges in implementing decentralization reform in Mali are the lack of human and financial resources and incidents of corruption. Local revenue generation, though promising, is not adequate to drive development. The central government continues to be the main funding source of the territorial units but decision-makers are reluctant to transfer financial resources to the communes until they can demonstrate a higher level of capacity to manage the money, including systems to combat and control corruption. This paradoxical situation serves to hinder the decentralization process (IMF 2015).

Overall, commune-level management and administration is hampered by a lack of local technical expertise, limited engagement of the public in decision-making and lack of funding (USAID ATLAS 2017a, b). On land and natural resource management particularly, communes are judged to be lacking in financial capacity to effectively manage the resources within their charge (Kelsey et al. 2011).

Climate Change Adaptation in Mali

Livelihoods Context in Mali

Across Mali, rainfall is by far the most decisive factor in determining differences between livelihood zones.

Northern Mali's Saharan and Sahelian zones are sparsely populated, home to just 10% of the country's population, despite accounting for a major share of its total landmass. There, pastoralism is the dominant livelihood. In central and southern Mali, where annual rainfall is comparatively higher, the majority of people depend on a mix of agriculture and livestock. Some areas receive enough rainfall to sustain cotton farming and irrigated rice production. Many other subsistence farmers plant basic grains. Across the country, seasonal and long-term migration are common, with adult males frequently moving to urban centers, neighboring countries or Europe in search of work (USAID FEWS NET 2010).

The areas of Mali covered by the ATLAS study, as well as the Koro commune, fall in the Center-eastern Millet and Livestock Livelihood Zone (USAID FEWS NET 2015). This area includes the center of Koulikoro and Ségou regions and the eastern part of Mopti. In most but not all areas of this livelihood zone, subsistence crop production in a normal year usually covers less than six months of annual food needs. Livestock rearing offers households a way to cover crop production deficits.

Typical livelihood activities include:

- Herding of cattle, sheep and goats, with frequent migration to seasonal pastures, even while maintaining some cultivated land
- Crop farming, relying on rain for cultivation of staples such as sorghum, millet, rice and maize
- Fishing, particularly along the Niger River and in seasonal wetlands
- Cotton farming
- Household production of peanuts, sesame and other vegetables on small irrigated plots, typically cultivated by women and sold in local markets
- Non-agricultural livelihoods including skilled labor (blacksmithing, leathercraft, masonry, carpentry, etc.) as well as handicraft production, processing of local tree products, gold mining and unskilled day labor

The main sources of income for poor households in the livelihood zone are livestock sales, migration, farm and non-farm labor, self-employment, and crop sales (USAID FEWS NET 2015). In-kind payments in the form of crops are a significant source of food. For the middle and better-off wealth groups, livestock are an important source of cash income and the principal form of savings/investment. Men will often migrate to the Niger Delta for temporary work during the rice harvest. Fattening and selling sheep is seen as a key income earning opportunity; these animals are usually kept apart from the main herd, tethered in the household's compound and fed livestock feed.

The months of July to September are the hardest, particularly for the poorest groups, whose harvests are likely to have run out by that time. Malaria incidence is at its height then. The very poor and poor receive some food in gifts from wealthier households but depend on the market for the majority of their food needs. They most commonly buy cheap staples such as millet and sorghum (USAID FEWS NET 2010 and 2015).

Local Perceptions of Climate Change

While rural Malians are generally unfamiliar with the scientific explanations, they have heard of climate change (Diiro et al. 2016). Given the country's variable climate, they are also accustomed to wide-ranging weather conditions. In the ATLAS field interviews across communes, local leaders and citizens—including farmers and herders—described a series of negative changes in climate and weather, all with a concurrent negative impact on livelihoods. Those perceptions included:

- Overall reduction in the amount of precipitation
- Delayed onset of and shorter rainy season
- Unusual dry spells during the rainy season
- Excessive temperatures including more days over 40 °C
- Increasing frequency of extreme winds

Although not all of these claims are explicitly supported by scientific evidence, the observations on reduced precipitation and increasing temperatures correlate with historical climate data. These data show a decrease of 12% from the mid-1960s to 2009 in average annual rainfall and a corresponding increase in average annual temperature of 0.8 °C (USGS/USAID 2012).

The shift in rainfall patterns was seen as a contributor to declines in crop production, particularly those crops with production cycles requiring sustained moisture. These include staples such as sorghum, millet, maize and rice, as well as the country's largest cash crop, cotton. Farmers also associated the excessive heat with reduced rainfall and intra-seasonal periods of drought. These dynamics have raised respondents' concerns over future household food security.

Herders particularly noted climate change as the cause of the reduced yield of natural rangelands and the disappearance of important grass species. They pointed to a relationship between climate change and human activity such as deforestation, saying the rapid destruction of trees for charcoal, fuel and agricultural land was responsible for negative changes in rainfall regimes and excessive temperatures, as well as an overall reduction of soil fertility.

Soil erosion was another concern. One causal factor cited was the increase in strong winds, including the seasonal Harmattan winds, which can be strong enough to destroy crops in the fields. Extreme storm events including flooding were also mentioned as contributing to soil erosion (USAID ATLAS 2017a, b).

While citing climate change as a factor in declining crop production and shrinking rangelands, study participants acknowledged that prevailing natural resource management practices played a major role in the changes they observed. These included the conversion of traditional rangelands to cropping areas, poor soil management and unsustainable deforestation (USAID ATLAS 2017a, b).

Despite the recognition of a changing environment, the long-term implications of climate change remain vague in the minds of local residents and even among technical specialists in local government. In many cases, the concept of climate change was introduced to local villages by NGO workers and some government service providers, but specific understanding of the science is uncommon. This dynamic appears to affect villagers' sense of urgency about the need for systematic, forward-looking climate adaptation strategies (USAID ATLAS 2017a, b).

Survey of Adaptive Practices Currently in Use

Since the devastating famines of the 1970s and 1980s, significant effort has gone into improving agricultural production in Mali. In addition, rural households are accustomed to variability and weather, and consequently have some ability—though limited—to cope. Understanding these strategies is critical to identifying adaptation opportunities (USAID ATLAS 2017a, b).

During the research, households cited a range of such strategies, including:

- Improved technology for inputs, such as better seeds and use of fertilizer
- Improved water and soil conservation techniques, including the introduction of micro-irrigation on small plots for vegetable production
- More confined management of herds
- Adoption of cash crops such as cotton and sesame
- Diversification into non-agricultural microenterprise activities such as small business, handicraft production, food processing, etc.
- Migration, both seasonal and semi-permanent

Across Mali, farmers are increasingly using drought-resistant or short-season seed varieties to adjust to changes in the rainy season (USAID ATLAS 2017a, b). These seeds, along with chemical fertilizers, have been introduced by NGOs, government technical staff and the parastatal cotton company, *Compagnie Malienne du Developpement Textiles*.

Also, water harvesting and soil management techniques are used to capture and maximize available moisture. These include rock and earthen bunds, live hedges, planting pits and other structures that channel water to specific sites and into the soil. To combat declining soil fertility, farmers commonly incorporate manure and compost into their fields.

Herders have adapted, too, reducing herd sizes and altering their composition toward small ruminants. Along with poorer-quality rangeland, they are also coping with the loss of traditional ranges near their homes, which, in many cases, have been converted to cropland. They and their herds are migrating to more distant pastures in the highlands and plateaus. In addition, many households have opted to put their cattle into the care of transhumant herding groups in exchange for milk and part of the calf production. In some cases, households are forced to purchase cattle feed from the market during the dry season to sustain their animals.

Intersection of Climate Adaptation and Governance in Local Planning

At the commune level, the PDESC, as the primary mechanism for development planning, is also the most important vehicle for systematic integration of climate change adaptation.

The development of the PDESC is guided by communal councils, which range in size from 11 to 45 members, depending on the population of the commune. These elected leaders are supported by technical advisors representing the sectors under the council's jurisdiction; for example: agricultural, health and forestry extension agents. The councils may also have support from NGO and community-based organizations such as local cooperatives and associations.

The development of a PDESC is meant to follow a standard set of procedures across all communes, beginning with the council's appointment and training of a field outreach team. The team is to visit every village in the commune to meet with the village development committees and conduct a public assembly to which all residents, including women and youth, are invited. After each visit, the team compiles an assessment of the expressed needs and priorities.

Next, these assessments are synthesized into a draft PDESC at a multiday workshop organized and managed by the council and technical staff. The draft plan is then presented to citizens at another series of public meetings in each village. Based on this public consultation, a committee finalizes the draft and sends it to the prefecture at the regional level. Once approved by the region, the plan is made public and then circulated among donors, NGOs and government agencies in an effort to attract financial and technical support.

To better understand the effectiveness of these processes in considering climate adaptation, the USAID ATLAS project's 2017 qualitative study analyzed the PDESCs from the five communes considered in the study. The discussion that follows is based on that review.

PDESC Content

Across the five communes, the PDESCs varied in quality and level of detail, but they nonetheless served as a starting point for setting the needs, priorities and resources for each commune.

The PDESCs included a description of the commune, an overview of priority needs and a set of proposed activities, organized by sector. Common priorities included:

- *Agriculture*: improvements to cropland including irrigation; farmer training and farmer organization

- *Livestock*: extended water sources for animals, new livestock corridors to reduce potential conflict with farmers and improved animal health
- *Aquaculture*: Increased aquaculture projects, including improvement of the local fleet in one PDESC
- *Water and sanitation*: Numerous projects including the installation of tube wells and latrines
- *Environment*: Reduced deforestation through the creation of forest reserves, reforestation and enforcement activities
- *Energy*: Improved energy sources; including the need for solar panels in one PDESC
- *Health and education*: Improved schools/classrooms and health centers
- *Infrastructure*: Construction, paving and/or rehabilitation of roads

Funding projections for the PDESC activities ranged from around US$500,000 to US$15 million.

Only one of the five PDESCs reviewed had a specific focus on climate change adaptation.

The PDESC and Climate Change Adaptation

The study found that the mere existence of governance institutions at the commune and village level confers an advantage in integrating climate adaptation in the sense that these structures (1) function as an intermediary or gatekeeper through which external resources and information on technology, government regulations and policies can flow, and (2) provide community-level structures that can facilitate public participation. The study concluded that localized governance in Mali influences climate change adaptation in the following ways:

- **Decentralized governance has likely played a positive role in the wide diffusion of adaptation technologies, through a combination of information-sharing across villages and with external actors who introduce the technologies**. A number of adaptation technologies are evident across villages and communes; some developed locally, some borrowed and improved, and some imported ready-made. For example, land management techniques, water harvesting, micro-irrigation and improved seed adoption have been integrated into most livelihood systems.
- **In some communes, local government action on natural resource management has favored climate change adaptation**. Even if compliance is uneven, government regulations on forest use—especially related to deforestation and charcoal production and land management related to bush fires—and collective livestock corridors are well-known at by local residents, elected officials and technical services. In some communes, the leadership has enforced the laws, created small reserves, promoted local reforestation, established

livestock corridors and instituted local rules for collective natural resource management. These positive steps are interpreted as examples of the vertical mechanisms by which decentralization can advance climate change adaptation.

- **In communes such as Konna, the PDESC has supported strategies for livelihood diversification**. For example, lands are set aside for female-managed garden plots, improved fishing infrastructure such as landing docks, and improvements in market facilities and transportation infrastructure.
- **Finally, effective commune governance, particularly manifested in the active leadership of the mayor and council, seems to mobilize more external resources, both governmental and nongovernmental**. With communal leaders acting as two-way brokers—horizontally across villages and vertically with government agencies, NGOs and donors—access to investment and technical resources appears to be greater. Examples from the study region include:
  - Catholic Relief Services' work to promote women's gardening using micro-irrigation in Konna and Kléla, and
  - Sahel Eco's introduction of agro-ecology interventions in Konna based on farmer-managed natural regeneration to improve soil fertility and increase agricultural production. This project includes work with women to develop their capacity for harvesting non-timber forest products.

These types of NGO-driven interventions have strong adaptation benefits. They seem to have a greater probability of successful adoption where local governance institutions are more effective. It was not possible to discern whether NGOs target communes that are known to perform more effectively or if the presence of an NGO enables improved governance.

While localized governance can play a positive role in climate change adaptation, the study noted several issues impeding the effectiveness of the PDESC process:

- Lack of technical capacity at the commune level. Most communes contracted specialized firms to prepare the document. This, in effect, isolates important parts of the process from the constituency, reducing participation.
- Of the five PDESCs reviewed, most appeared to reflect perfunctory preparation, with little evidence of a development vision, little assessment of the likely effectiveness of the proposed interventions, and no review of the outcomes of the previous PDESC.
- Most respondents interviewed in the commune said they had little or no knowledge of the PDESC or its content. None said they had read it. Even the government staff who provide technical services related to agriculture, livestock and natural resource management lacked familiarity with the plan.
- No local sources of revenue were readily available to fund projects in the PDESC, making it more of a wish list than an operational tool.

The ATLAS analysis found that decentralized governance can play a key role in facilitating problem-solving across villages and create important vertical linkages with external actors and resources. The realization of these benefits, however, is directly related to the quality of such governance.

Overall, evidence from the study indicates that most communes do not have the technical or financial resources to elaborate a development pathway sensitive to the increasing pressures of climate and environmental change. No widespread adaptation vision guides development planning; the PDESC tends to be much more pedestrian in its scope, focused on solving specific, concrete, short-term and immediate problems. Critically, the PDESCs are not typically informed by any climate analysis, such as a climate vulnerability assessment, which could help identify specific risks. Currently, while the interventions identified in a PDESC may address risks posed in whole or part by climate change, this outcome is incidental rather than evidence-based.

The local presence of an active NGO, however, helps overcome the major resource and technical constraints to development planning and can make the PDESC a true guiding document, rather than a wish list.

The next section, looking at experience in the commune of Koro, stands out an example of how leadership and cooperation can help communities develop a vision of local development that includes climate adaptation.

Case Study: Koro Commune

The inclusion of climate change considerations in governance systems is critical to facilitating the adoption of local solutions to climate impacts. In Mopti, a central region that serves as gateway to the north and south, the effects of climate change are being felt by every household. As part of their response, local governments have pledged to integrate activities for climate change adaptation activities in their PDESCs.

During 2016, USAID's Mali Climate Change Adaptation Activity partnered with the Malian NGO GRAT, to help communes in Mopti region create and validate communal climate change adaptation plans.

GRAT's work centered on a participatory and inclusive climate-proofing method adopted to the Mali context from an approach developed by the German national development agency GIZ (*Deutsche Gesellschaft für Internationale Zusammenarbeit*). The process centers on identifying risks and constraints associated with the biophysical and socioeconomic impacts of climate change, and using those conclusions to identify concrete steps to decrease vulnerability and foster resilience. Through a consultative process with commune authorities and key stakeholders, MCCAA and GRAT helped 25 communes develop and adopt climate-specific plans; these were integrated into the communes' PDESCs. MCCAA is continuing to play a facilitating role in sharing the plan with donors, including USAID, and other local development projects.

This section highlights the experience in one commune, Koro, which is notable for the technical strength of the climate-proofing process, the integration of the climate adaptation plan into the PDESC and immediate implementation efforts. Central to the success of this effort was the leadership of the mayor and other elected officials in engaging external supporters and leveraging funding to kick-start an initiative that is already delivering tangible benefits to the community.

About Koro

Koro commune, one of 16 communes in Koro district, has an ethnically diverse population of nearly 63,000 people living in 49 villages. Located near the Burkina Faso border, Koro's population is heavily dependent on agriculture.

While farmers in many areas of Mopti region do not produce enough food to feed themselves, Koro is a surplus-producing area. Predominant livelihoods include animal husbandry (cattle, goats, sheep, some camels) and rainfed cultivation of millet and sorghum, along with intercropping of cowpeas and groundnuts for consumption and sale. In addition, wealthier households cultivate some rice, and some women maintain small, hand-irrigated gardens of onions, garlic and tobacco. Soil infertility is recognized as a significant problem (Diiro et al. 2016). Use of drought-resistant seeds is relatively low: a 2014 survey of farmers in Mopti found that fewer than one-third planted drought-resistant seeds. While compost and mineral fertilizers are frequently used, chemical fertilizers are not commonly applied due to their cost and the distance of fertilizer sellers (Diiro et al. 2016). Hazards include droughts, floods, crop pests and animal diseases (USAID FEWS NET 2015).

Historically, the tourism industry, focused primarily on Dogon culture and architecture, generated significant local revenues, but this activity declined in the wake of the 2012–2013 armed conflict and the continuing violence in the central Mali. Between 2006 and 2016, the number of annual visitors to Mopti region declined from nearly 40,000 to fewer than 4,000, and tourist taxes dropped from more than CFA 16 million to CFA 282,000 (about US$27,000 to US$500) (Regional Tourism Directorate 2017). Many families who relied on tourism for income turned to agriculture, adding to natural resource pressures.

Climatically, Koro lies in the semi-arid Sahelian belt of Mali, where annual rainfall typically ranges from 600 to 800 mm. Insufficient rainfall is often cited as a concern and a contributing factor to a lack of grazing land.

Projections for the Sahelian climate in the coming decades for rainfall vary substantially, with some anticipating declines and others suggesting an increase. Most projections agree that rainfall is likely to become more variable in onset and distribution, and the rainy season is likely to be shorter (USAID Sahel Climate Risk Profile 2017a). More significantly, temperature is expected to rise at a faster rate than in other parts of Africa and the world, with an estimated increase of 4 °C by 2100 (USAID Sahel Climate Risk Profile 2017a).

Should such conditions prevail in Koro, the effects on people and livelihoods will be substantial. Crop yields are likely to decline significantly and livestock will become less productive. Those to face the greatest consequences will be low-income, subsistence farmers and their families, as well as the extension agents, traders and other service providers employed in the sector.

MCCAA and GRAT Support in Koro

Koro was one of the 25 communes that received assistance from MCCAA and GRAT in developing its climate adaptation plan. GRAT facilitated the creation of a climate proofing committee to lead the process and trained its participants on the climate proofing methodology. Based on the committee's assessment and consultations with community members, key adaptation priorities included: animal fattening projects, reforestation, soil and water conservation, and agroforestry.

These inputs were used to finalize the commune's climate adaptation plan. It was later adopted by the communal council and integrated into the commune's PDESC.

Plan Implementation and Donor Coordination

To immediately breathe life into the plan, the mayor of Koro and MCCAA leadership reached out to a local development initiative known as RIC4REC to seek their support. As part of its ongoing initiatives to build resilience to climate stresses, RIC4REC established an animal fattening project for women in Koro commune. Animal husbandry is widely recognized as a climate-friendly alternative to crop farming (Turner et al. 2014). The aim of the RIC4REC initiative is help poor households build climate resilience by improving their livelihood options. (RIC4REC project is funded by DFID, the British Department for International Development, through its BRACED project, Building Resilience to Climate Extremes and Disasters.)

Partnership with RIC4REC exemplifies the kind of dynamic opportunities available to help communes implement adaptation plans. It is also a reminder of the importance of close coordination and collaboration between donor-funded projects in Mopti region, and in Mali at large.

Conclusion: Applying Lessons Learned in Mali to Other Contexts

The ATLAS study and the Koro experience show that certain conditions are required to help communes facilitate the development of climate-sensitive PDESCs.

Challenging governance conditions, particularly the lack of capacity (and the lack of resources to engage a mechanism to develop that capacity) at the local level, can be addressed through a combination of strong local leadership and partnerships that help strengthen the technical expertise of mayors, councils and technical staff. Another necessary element is the political will to involve village households in identifying priorities. These conditions put local leaders and their constituents in a position to articulate a strong vision of resilience for their communities and then move forward with practical, achievable adaptation approaches.

Recommendations

Several lessons learned and recommendations may assist leaders in Mali as well as actors in other development contexts as they seek to support local government action on climate change adaptation. They include efforts to:

- **Encourage broader participation of citizens through increased public education and knowledge-sharing on climate change adaptation**: This includes basic information on climate change impacts, its importance to local planning, and opportunities to get involved. The goal might be participation in planning as well as long-term citizen engagement in a continuously iterative process in which the document is discussed and systematically monitored. In other words, the aim should be to make planning strategies "living documents" that are vigorously promoted and shared with the public in local languages.
- **Provide climate change training for local officials**: To lead and guide the local conversation on climate change adaptation, elected officials and technical specialists must understand why it is important and how long-term planning is needed. Training may be required on climate change impacts, best practices in adaptation options, the value of public participation, strategies for interacting with central government, and other context-specific information. This training should bring together elected and technical officials from various localities, with the idea of creating information-sharing networks to facilitate regular exchanges on progress toward implementation and best practices in adaptation.
- **Maximize the use of available resources by formalizing collaboration among donors**: In some contexts, a coordinated strategy or regular donor forum might be appropriate in helping donors, NGOs and government actors understand existing plans and programs, and cooperate on activities. The involvement of national government and nongovernmental actors will clearly enhance this conversation and add credibility to the process.
- **Refine the role of the regional or provincial level government**: In Mali, for example, the regions might be given a more explicit role as the liaison and coordinating body between central government and sub-national government (commune). Opening this channel may lead to stronger relationships between levels of government and more opportunities for communes to work with national and international NGOs on the implementation of climate change adaptation priorities.

- **Undertake policy and administrative reforms to strengthen the integration of climate change adaptation as an organizational theme**, including: mandating its consideration in planning processes and formalizing monitoring and evaluation of its implementation. Structures for monitoring and evaluation will enable systematic reviews and surveys of lessons learned, both of which can improve implementation and inform future plans.
- **Invest in increased use and broader dissemination of climate information to identify risks and utilize the information in PDESC development.** While it is not realistic for each commune to undertake a detailed, science-based climate change vulnerability assessment, it may be possible over time to undertake this work at the regional scale, or to complete qualitative vulnerability assessments at the commune level through community engagement. This latter, participatory approach could be an important precursor to the PDESC process.
- **Invest in financial planning and management capacity**: In Mali, the ability of local actors to implement their PDESC depends in large part on whether they have adequate financial support. A critical element of this is regular dialogue between local and central levels of government on budget priorities. Along with increased capacity building on financial management for local leaders, local and central actors should coordinate on integrated financial planning procedures and processes. In Mali, this would require the involvement of government financing agencies such as the *Agence Nationale d'Investissement pour les Collectivités Territoriales*, ANICT. In addition, to reinvigorate central government's interest and support for decentralization, new financial opportunities might be created, including a small grants program to fund PDESCs. This could be funded through international donor support to the national ministry responsible for overseeing local government, the *Ministère de l'Administration Territoriale et des Collectivités Locales* (Fig. 3).

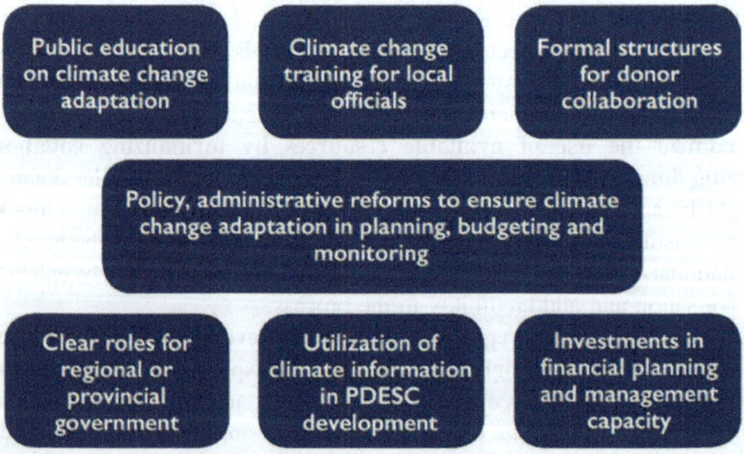

Fig. 3 Recommendations for strengthening action on climate adaptation at the local level in Mali. While specific to the Mali context, they may also apply in other countries with a decentralized system

A common thread of these lessons learned—relevant in Mali and many other developing countries—is that progress at the local level on climate change adaptation cannot be operationalized by local governments alone. Significant support is required from central government as well as donors. While political structures may be in place, the true consolidation of decentralization will occur when administrative and financial capacity are strong. This element of capacity is a key success factor in development planning generally and climate change adaptation specifically.

Limitations

Limitations on this research are in three areas:

- Little research has been done on the relationship between decentralization and climate change adaptation, particularly in the Malian context. The decision to undertake a qualitative study with data drawn from a combination of focus groups, household interviews, key informant interviews and a desk review was an effort to establish a foundational understanding of the topic. Future research in this area would greatly strengthen the available knowledge base.
- The small number of communes sampled. While research on the five communes was comprehensive and analysis of the findings was thorough, these communes represent just a small percentage of Mali's 703 total communes.
- The limitations of methodologies that rely on self-reporting data. Participants and respondents in focus groups, key informant interviews and household interviews provide data and information that is difficult to verify independently. Their responses are also subject to selective memories, confounding of memories and events, hypothesizing, exaggeration and other biases. In addition, the account of MCCAA's work in Koro relies on self-reporting by staff engaged in the program.

References

Consortium of International Agricultural Research Centers (CGIAR) Research Program on Climate Change, Agriculture and Food Security. (2014). *How can effective dialogue be established between researchers and policy makers on climate change adaptation in Mali?*. CGIAR, Washington, D.C.: Info Note.

Central Intelligence Agency (CIA). (2017). World fact book, Washington D.C.

Diarra, A., & Ag Anara, S. (2016). Mali's local elections marred by boycotts, kidnapping. Reuters World News, November 20, 2016.

Diiro, G., Petri, M., Birhanu, Z. B., Sinare, B., Dicko, M., & Traore, D., et al. (2016). *Gendered analysis of stakeholder perceptions of climate change, and the barriers to its adaptation in*

Mopti region in Mali. Research Report no. 68. Technical Report. ICRISAT, Patancheru, Telangana, India.

Direction Régional du tourisme et de l'hôtellerie in Mopti. (2017). Annual Report.

Human Rights Watch (HRW). (2017). *World Report 2017.* New York: HRW.

International Monetary Fund (IMF). (2015). *Technical assistance report—implementing fiscal decentralization.* Washington D.C.: IMF.

Kelsey, J.-C., Hughes, A. K., & Knox, A. (2011). *The challenge of decentralization in Mali: Lesson 2 on focus on land in Africa*, Brief. World Resources Institute in partnership with Landesa. Washington D.C.: Rural Development Institute.

Lippman, H., Lewis, B. (1998). *Democratic decentralization in Mali: a work in progress.* Washington D.C.: United States Agency for International Development.

SNV Netherlands Development Organisation and Centre d'Expertises pour la Decentralisation et le Developpement local. (2004). *Decentralization in Mali: putting policy into practice.* KIT development, policy and practice—series: Decentralization and local governance, Bulletin 362. Amsterdam: Royal Tropical Institute.

Traore, B., Descheemaeker, K., van Wijk, M. T., Corbeels, M., Supit, I., & Giller, K. E. (2017). Modelling cereal crops to assess future climate risk for family food self-sufficiency in southern Mali. *Field Crops Research, 201,* 133–145.

Turner, M., McPeak, J., & Ayantunde, A. (2014). The role of livestock mobility in the livelihood strategies of rural peoples in semi-arid West Africa. *Human Ecology, 42,* 231–247.

United States Agency for International Development (USAID). (2015). *Country development cooperation strategy mali forward.* Washington D.C.: USAID.

USAID Adaptation Thought Leadership and Assessments (ATLAS). (2017a). *Climate change risk profile West Africa/Sahel.* Washington D.C.: USAID ATLAS.

USAID Adaptation Thought Leadership and Assessments (ATLAS). (2017b). *Decentralized governance and climate change adaptation: A case study on Mali.* Washington D.C.: USAID ATLAS.

USAID Famine Early Warning Systems Network (FEWS NET). (2010). *Mali livelihoods profile.* Washington D.C.: USAID FEWS NET.

USAID Famine Early Warning Systems Network (FEWS NET). (2015). Mali livelihoods zone descriptions. Washington D.C.: USAID FEWS NET.

United States Geological Survey (USGS)/United States Agency for International Development (USAID). (2012). Climate trend analysis of Mali. Washington D.C.:USGS/USAID.

Walch, C. (2017). *Fertile ground? Climate change and jihadism in Mali.* New Security Beat Blog, The Wilson Center, May 2017, Washington D.C.

World Bank. (2017). Data bank for Mali. Washington D.C.:World Bank.

Using the Open Standards-Based Framework for Planning and Implementing Ecosystem-Based Adaptation Projects in the High Mountainous Regions of Central Asia

Paul Schumacher, Tobias Garstecki, Bunafsha Mislimshoeva, John Morrison, Benedikt Ibele, Corey Lesk, Salamat Dzhumabaeva, Umed Bulbulshoev and Shaun Martin

Abstract The ecosystems of the high mountain regions of Central Asia are rich in biodiversity and provide essential services, such as the regulation and provision of water for the population of the entire region. However, inappropriate land management of these fragile ecosystems combined with their vulnerability to climate change threats (e.g. melting glaciers, changes in water flow regime, droughts) decrease their resilience and thus, the ability to provide continued services to

P. Schumacher (✉)
Deutsche Gesellschaft für Internationale Zusammenarbeit (GIZ) GmbH, Country Office Kyrgyzstan, 22, Erkindik Boulevard, 720040 Bishkek, Kyrgyzstan
e-mail: paul.schumacher@giz.de

B. Ibele
Deutsche Gesellschaft für Internationale Zusammenarbeit (GIZ) GmbH, Ministry of Environment Protection and Agriculture, 6 Gulua St., 0114 Tbilisi, Georgia
e-mail: benedikt.ibele@giz.de

T. Garstecki
Independent Consultant, Feuerbachstraße 41, 14471 Potsdam, Germany
e-mail: garsteckit@gmail.com

B. Mislimshoeva
UNIQUE Forestry and Land Use GmbH, Schnewlinstr 10, 79098 Freiburg, Germany
e-mail: bunafsha.mislimshoeva@unique-landuse.de

J. Morrison · S. Martin
World Wildlife Fund US, 1250 24th Street, N.W., DC 20037 Washington, USA
e-mail: john.morrison@wwfus.org

S. Martin
e-mail: shaun.martin@wwfus.org

C. Lesk
Center for Climate Systems Research, Earth Institute, Columbia University, 116th and Broadway, New York, NY 10027, USA
e-mail: lesk@ldeo.columbia.edu

F. Alves et al. (eds.), *Theory and Practice of Climate Adaptation*,
Climate Change Management, https://doi.org/10.1007/978-3-319-72874-2_2

people. A BMUB (German Federal Ministry for the Environment, Nature Conservation, Building and Nuclear Safety) IKI (International Climate Initiative) funded and GIZ-led (Deutsche Gesellschaft für Internationale Zusammenarbeit GmbH) consortium explores the use of an Ecosystem-based Adaptation approach to help people adapt to the adverse impacts of climate change. It aims to strengthen the provision of ecosystem services and thereby enhance the livelihoods of the population depending on them. Although the application of potential ecosystem-based measures in Central Asia is not new, typically relevant climate risk information on people and ecosystems is not considered and thus, has a higher risk of introducing maladaptive interventions. This project used a modified form of the Open Standards for the Practice of Conservation to systematically develop and test an integrated planning framework that used climate risk information to identify key vulnerabilities of people and ecosystem services under several plausible climate change scenarios and developed potential adaptation options. The framework's guidance can contribute to a paradigm shift: moving away from business as usual approaches to climate informed adaptation processes. In addition, it constitutes a cornerstone for building a bridge between international climate finance mechanisms and climate-informed adaptation of local communities.

Keywords Climate change · Ecosystem-based Adaptation · Open Standards
High mountains · Central Asia

Introduction

The Central Asian countries of Kazakhstan, Kyrgyzstan, Tajikistan, Turkmenistan and Uzbekistan are presently facing major environmental problems that threaten the livelihoods of more than 60 million people and the unique ecosystems of the region. The main environmental challenges in the areas of water, energy, agriculture, transport and industry are becoming more complex with climate change—increasing temperature and changes in precipitation patterns (Reyer et al. 2015; WB 2015). The biodiverse mountainous areas of Central Asia (CA) are expected to experience locally amplified climate change and are already extensively affected by human activities (CEPF 2016; Reyer et al. 2015). These areas are therefore a priority for adaptation interventions.

S. Dzhumabaeva
Public Foundation CAMP Alatoo, 3, Ufimskiy Pereulok, 720040 Bishkek, Kyrgyzstan
e-mail: salamat@camp.kg

U. Bulbulshoev
Public Foundation CAMP Tabiat, 9 Sohibnazarova Street, 734002 Khorog, Tajikistan
e-mail: director@camptabiat.org

Ecosystem-based Adaptation (EbA) is a promising approach that has potential to address the complex socio-ecological problems facing CA. However, adaptation interventions in the region, that could be considered EbA, are often applied as stand-alone measures in single-sector adaptation strategies rather than being part of an overall integrated adaptation strategy that addresses multiple needs across sectors at the local, national or regional levels. In addition, practitioners often select EbA measures based on conservation priorities and do not consider climate risk affecting ecosystems meant to help people adapt to climate change. They therefore may actually be maladaptive in the long-term. These shortcomings are not CA specific, but rather common for EbA approaches. The authors state that communities will be able to manage climate risks by both, using and strengthening natural systems and succeed only, if a holistic approach to adaptation is considered, and if communities develop ownership as well as make use of tailored climate information for taking decisions.

In a project led by GIZ (Deutsche Gesellschaft für Internationale Zusammenarbeit GmbH) and commissioned by BMUB (German Federal Ministry for the Environment, Nature Conservation, Building and Nuclear Safety) IKI (International Climate Initiative), a consortium of international and local conservation, development and academic institutions tackles the above mentioned challenges by developing a systematic planning framework for elaborating an overall adaptation strategy together with local communities. The strategy considers EbA measures first, supplementing them with other adaptation and livelihood improvement measures that EbA alone cannot address. The local planning framework and lessons learned from pilots on the ground can support the development of national and regional adaptation strategies. The objective of the paper is to discuss the elaboration and implementation of the planning framework for an EbA project in the high mountainous regions of CA and to show how to initiate an adaptation process which goes beyond development project life spans. Furthermore, it presents the dissemination potential for climate adaptation investment projects.

Ecosystem-Based Adaptation—an Innovative Approach to Help People Adapt to Climate Change

EbA has received growing attention from conservation and development practitioners, policy-makers and donor agencies alike. While some institutions have developed their own working definitions of EbA, most of these are similar to the definition adopted by the Convention on Biological Biodiversity (CBD) in 2009 that states, "Ecosystem-based adaptation is the use of biodiversity and ecosystem services as part of an overall adaptation strategy to help people to adapt to the adverse effects of climate change."

In spite of this commonly agreed upon definition, the need to operationalize EbA has created a proliferation of tools and frameworks, on-the-ground activities and research effectiveness and cost-benefits that differ greatly in their understanding of what EbA and its intended purpose is. An IKI-supported inventory of methodologies has identified nearly 160 tools and guidance documents for EbA practitioners (Hicks et al. in prep). Furthermore, the confusion on the meaning of EbA hinders the integration of EbA into adaptation planning, as some organizations still see EbA as the adaptation of ecosystems to climate change rather than the use of ecosystems for human adaptation to climate change (Doswald et al. 2014). It is therefore useful to consider that the definition of EbA by the CBD (2009) comprises four major elements:

1. The use of biodiversity and ecosystem services (a tactic or sub-strategy)
2. as part of an overall strategy (the strategy)
3. to help people (the target)
4. adapt to the adverse effects of climate change (the goal)

There are many pitfalls for the would-be EbA practitioner if not all four elements of this definition are applied. If not seen as a component of an overall strategy, EbA moves from being a tactic embedded within a strategy to becoming the strategy itself. This can exclude other important tactics necessary to help people adapt to climate change. Additionally, different institutions can value different aspects of EbA depending on their missions. A conservation organization, for example, may (improperly) treat biodiversity and ecosystem services as the target, while a development organization may focus (too narrowly) on a tactic that uses man-made ecosystems like household gardens without considering the wide range of natural biodiversity available to help communities to adapt. Moreover, even the goal of adaptation itself is at risk of being lost when climate-related vulnerabilities are not considered, leading to the use of "business-as-usual" practices that may be in fact maladaptive in the long-term.

To help clarify targets, goals, strategies, and tactics among the diverse stakeholders involved in this project, we focused on the role of biodiversity and ecosystem services in climate change adaptation to help identify an array of tactics (or measures) that could be used to develop a comprehensive adaptation strategy to help the target communities. We categorized adaptation tactics into four broad categories (Table 1): ecosystem-centric, ecosystem-based, ecosystem-friendly and "other," each with varying combinations of targets and activities but all sharing the common goal of climate change adaptation.

It is important to note that each of these tactics often confers benefits to the other three and as such, it is usually not helpful to consider any set of activities in isolation when developing a full adaptation strategy.

Project Area and Its Climate Change Impacts

The elaboration and implementation of the EbA planning framework has been piloted in two watersheds in Tajikistan's and Kyrgyzstan's mountainous areas, one in the Naryn Province of Kyrgyzstan, and the other in the Bartang Valley of Tajikistan (Fig. 1).

Table 1 Categorized adaptation tactics

Overall adaptation strategy				
Goal	adapt to climate change			
Tactics	Ecosystem-centric	Ecosystem-based	Ecosystem-friendly	Other
Target	Biodiversity	People	People	People
Activities	Use climate-informed conservation activities (biological corridors, climate refugia, etc.)	Use biodiversity and ecosystem services (pastures for livestock, kitchen gardens, etc.)	Use activities that do little or no harm to biodiversity (rainwater harvesting, silos, improved irrigation)	Use activities that are unrelated to biodiversity (weather monitoring, early warning systems, improved governance)

The economy of both pilot sites is dominated by subsistence cropping and grazing livestock, both of which depend on rainfall and irrigation from glacier and snowmelt runoff (Table 2). The primary climate vulnerabilities in these areas therefore relate to changing quantity and timing of water provision.

Climate projections for the pilot areas, which were derived from a suite of 21 general circulation models (GCMs), indicate warming in all seasons over the coming decades, which is likely to be accompanied by later fall frost, earlier spring thaw, and increased intensity and frequency of summer heat extremes. As in much of the world, projected precipitation changes are highly uncertain at the pilot sites: the most likely range of future changes comprises slight decreases to substantial increases in total annual precipitation. Some degree of intensification of heavy rainfall is likely. Conditions are highly likely to become wetter during the winter and spring seasons, while rainfall may increase or decrease during the summer and autumn. The distribution of total annual precipitation across the seasons is therefore likely to shift, and year-to-year variability will continue to pose a challenge to livelihoods (Annex 1).

The combination of increased winter and spring precipitation with warming springs is likely to result in earlier and possibly more intense spring floods. Meanwhile, warmer summers will enhance evapotranspiration. A net summer drying of the landscape is probable, even if rainfall increases slightly, as summers are already dry and hot in the mountainous regions of CA. Regulation of streamflow, stabilization of stream banks, and water storage are therefore priority adaptation goals to prevent damaging floods and ensure availability of irrigation water for pastures and crops. Warmer seasonal average temperatures imply an increased risk of longer, hotter, and more frequent periods of extreme heat. Such events can reduce the productivity of crops and harm the health of people as well as livestock. Extreme heat impacts may also compound with projected changes in rainfall, and should be taken into consideration when designing EbA activities.

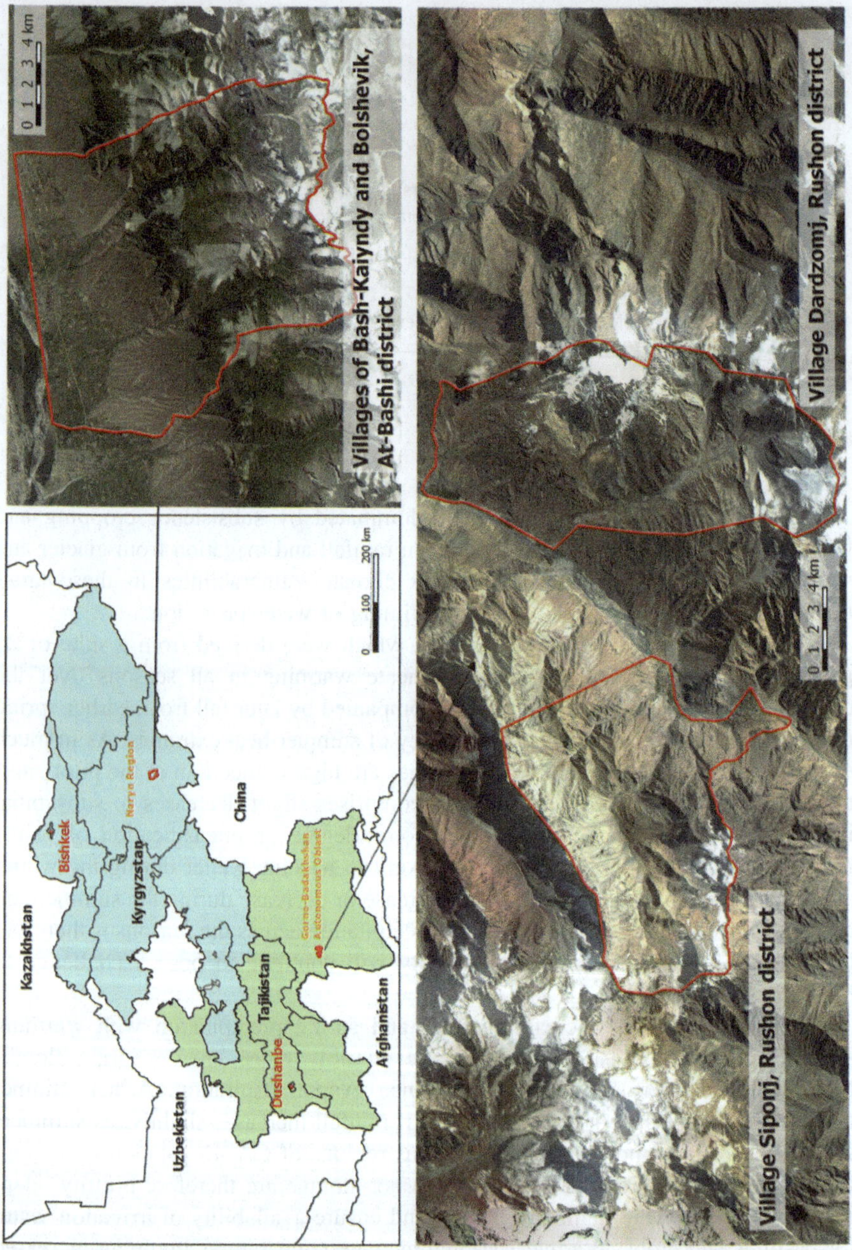

Fig. 1 Map of project areas (GADM 2015, Image ©2017 DigitalGlobe and ©2016 Google)

Table 2 Main characteristics of the pilot areas in Kyrgyzstan and Tajikistan

	Kyrgyzstan	Tajikistan
Location	Naryn province, Bash-Kaiyndy community, Bash-Kaiyndy and Bolshevik villages	Gorno Badakhshan Autonomous Oblast (GBAO), Rushan district, Bartang valley, Siponj and Darjomj villages
Coordinates	41° 09′24, 76″N 75° 54′06, 23″ E, Bash Kaiyndy 41° 09′48, 92″N 75° 58′48, 12″ E, Bolshevik	38° 03′45, 87″N 71° 52′19, 91″ E, Siponj 38° 07′33, 44″N 71° 56′26, 32″ E, Darjomj
Approx. area (sq. km)	197	113
Altitude (meter above sea level)	2100-4300 (At-Bashy River to mountain peak)	2170–5000 (Bartang River floodplain to mountain peak)
Population	Bash-Kaiyndy—approximately 4752 inhabitants, around 1087 households; Bolshevik—approximately 1000 inhabitants, 232 households	Siponj—approximately 328 inhabitants, around 97 households; Darjomj—approximately 233 inhabitants, around 52 households
Livelihood strategies	Subsistence agriculture, livestock and the cultivation of vegetables and fruits	Subsistence agriculture, livestock and the cultivation of vegetables and fruits
Main conventional threats to livelihood strategies	Pasture degradation due to overgrazing, Forest degradation due to overcutting and grazing pressure, poor water management	Overgrazing and overcutting in riparian forests, loss of fertile land due to overuse
Main climate change related threats	Shifting seasonal distribution of precipitation, increased intensity and frequency of summer heat extremes, more intense spring flooding	Shifting seasonal distribution of precipitation, increased intensity and frequency of summer heat extremes, more intense spring flooding, increase in intensity of natural hazards such as mudflows

Methodology

Using the Open Standards for EbA Planning and Implementation

The planning framework is based on the Open Standards (OS) for the Practice of Conservation (Conservation Measures Partnership (CMP) 2013). The OS are a common planning and adaptive management framework usually applied to biodiversity conservation projects and programs. As such, they enjoy a broad global constituency that is familiar with its process and terminology (Schwartz et al. 2012; Lamoreux et al. 2014). Efforts are underway to promote uptake in Central Asia, including translation of the OS into Russian. The general approach is a detailed version of the project cycle (Fig. 2), which is adapted to conservation and ecosystem management.

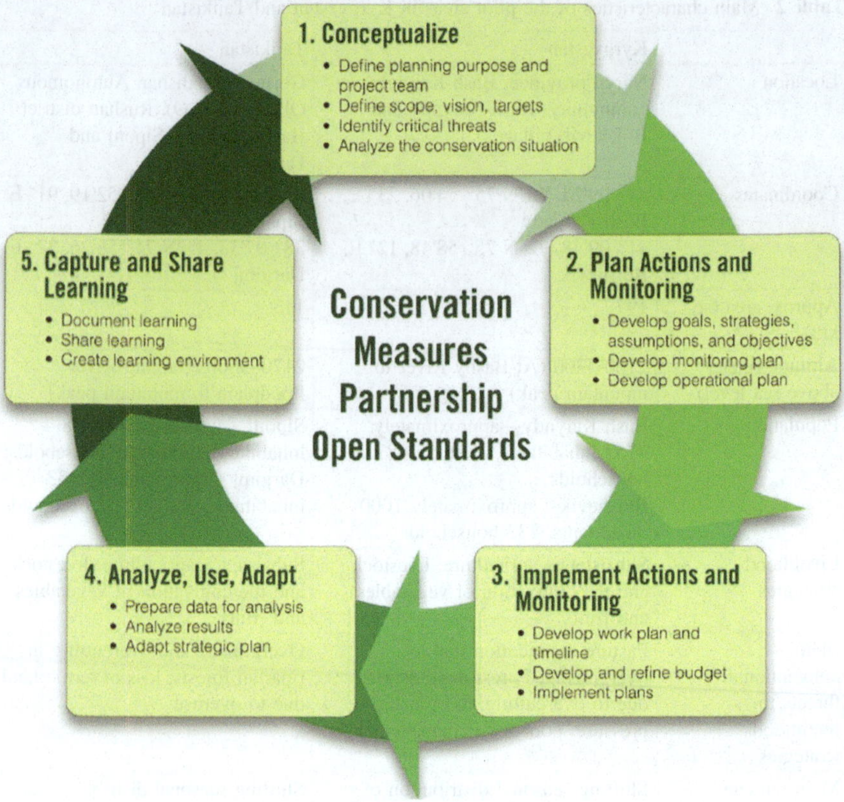

Fig. 2 Overall approach of the open standards for the practice of conservation (CMP 2013)

The EbA approach is different from typical conservation projects. Whereas the goal of a conservation project is to conserve populations or ecosystems for their own sake (at least), an EbA project is attempting to assist humans to adapt to climate change by safeguarding and enhancing those ecosystems and ecosystem services supporting human well-being, to make the human population more resilient to a changing climate. As such, the EbA approach is not primarily focused on conservation but more widely on interactions between ecosystems and society. EbA also requires a comprehensive understanding of the impacts of climate to both humans and the ecosystems they depend on.

To render application of the OS most useful in an EbA context, minor modifications to the general approach were made. These included: (1) the site selection procedure; (2) an emphasis on understanding the relationship at a site between ecosystems, ecosystem services, and people; (3) obtaining a thorough understanding of the level of uncertainty in local climate projections; (4) the development of climate scenarios that encompass that uncertainty; (5) obtaining a full understanding of direct and indirect climate impacts on ecosystems and humans (for a series of potential climate scenarios); (6) and the use of intervention points for EbA measures that differ from those generally used in conservation projects.

Results

The results of the project consist of a climate-informed Open Standards-based design, planning and management framework for local EbA measures, and of the lessons learned during the pilot application of this framework in Kyrgyzstan and Tajikistan. The steps of the "Climate-informed Open Standards Ecosystem-based Adaptation" framework that emerged are:

1. Define the social, thematic and geographic scope of the EbA planning process
2. Understand current and desired state of ecosystems[1]
3. Analyse conventional vulnerabilities of people and ecosystems (those not related to climate change)
4. Identify climate change-related threats
5. Develop a general model of the socioeconomic-ecological system (including climate impacts)
6. Re-evaluate project objectives
7. Identify and select strategies measures based on general model
8. Implement and monitor selected adaptation measures, learn, adapt

Step 1. Define the social, thematic and geographic scope of the EbA planning process

This step includes defining the community that the project aims to support and delineating the range of ecosystem services and location of the ecosystems it depends on, in order to set the geographic scope—at least initially. This step can involve extensive discussion about which human communities should be involved, and which stakeholders need to be consulted. It is quite common that the scope of the project—and consequently the stakeholders involved—will shift as the process proceeds, and new questions and information comes to the fore. Based on initial thinking about the social scope of the project, the project team analyzes what ecosystem services human communities are dependent on, and which ecosystems are providing those ecosystem services. The linkages between ecosystems, ecosystem services, and human well-being are represented graphically in a conceptual model (Fig. 7). The more specific the identified ecosystem services are, the better. The analysis should focus on ecosystem services relevant at the local level, rather than global ecosystem services that the project cannot effectively address.

[1]This could be widened to also assess current and desired state of ecosystem service provision and communities.

Step 2. Understand current and desired state of ecosystems
The second step is about understanding how the ecosystems provide critical ecosystem services to project human communities currently, and what the future condition ought to be to continue doing so. This desired future state provides a guide to what the project is trying to achieve to ensure stable or enhanced provision of ecosystem services in the future, and to make human communities more resilient. This step, which follows standard OS methodology (CMP 2013), involves determining the "key ecological attributes", along with associated measurable indicators and target values for each critical ecosystem—to help understand current and future conditions (Wulffraat and Morrison 2013; Parrish et al. 2003).

Step 3. Analyse conventional vulnerabilities of people and ecosystems (those not related to climate change)
This step is aimed at understanding conventional threats to critical ecosystems—i.e. those not related to climate change—with their respective contributing factors. It is completed based on general OS methodology (CMP 2013), and focuses mainly on the types of conventional threats to ecosystem integrity as listed in Salafsky et al. (2008).

Step 4. Identify climate change-related threats
Once an understanding of the current situation and the extent to which it is already affected by non-climate related anthropogenic factors is reached, the climate perspective is integrated. Together with the previous steps, this then leads to a full-fledged analysis of the current conventional and future climate change-related vulnerabilities of ecosystems, and the people who depend on them.

Step 4.1. Develop plausible climate scenarios for the project area
In cooperation with the Center for Climate Systems Research (CCSR) of Columbia University in New York, USA, we developed a method to derive plausible climate scenarios for use in EbA planning. Based on participatory compilation of a "local" climate calendar that reflects how the local ecosystems and native species, communities' livelihoods, and social and cultural events are tied to the calendar and associated climatic and seasonal events/shifts, the most relevant "seasons" in the project area for further analysis by climate scientists can be identified (Fig. 3).

A climate team then runs a full suite (20–30) of GCMs to approximately the year 2050. These models are downscaled to improve local relevance. The downscaled GCM outputs are analyzed. The climate team looks for parameters (e.g. degree of warming, amount or timing of precipitation) where the models all agree, as well as parameters where they differ. If all models agree on all parameters, this means that there is little uncertainty and that the future climate is clear (i.e., there is only one plausible future climate scenario). However, GCM outputs that differ for one or more parameter indicate uncertainty in future climate. For those outputs where there are differences, target communities are consulted to help decide which uncertain parameters would be most significant for human lives and activities there. No more than two key parameters are prioritized typically, to avoid an intractable multiplication of plausible future climate scenarios. These (two) selected parameters

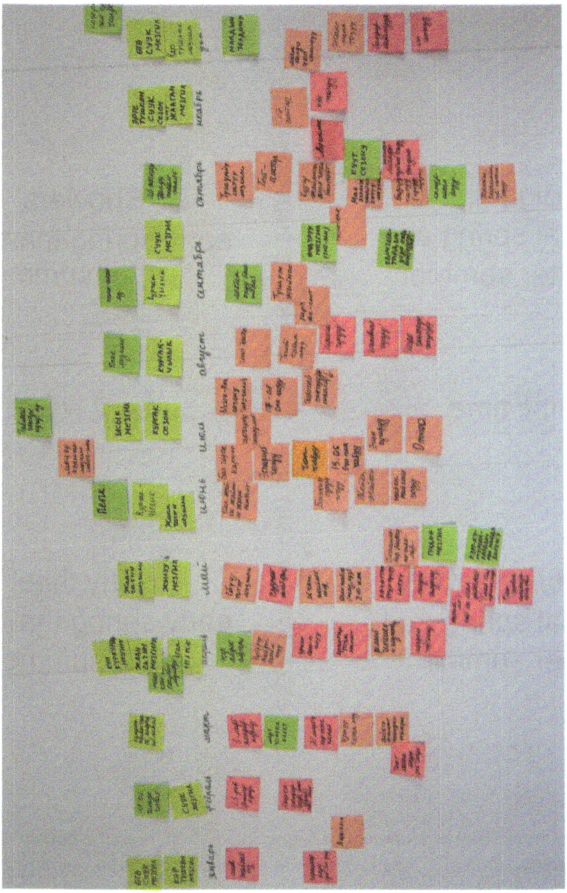

Fig. 3 Project example for participatory compilation of seasonal calendar. The colored sticky notes show livelihood activities by month. This information can then be used to identify seasons which are most relevant from an EbA perspective

become orthogonal axes in a quadrant graph. Each quadrant can be named (see Fig. 4 for project example, in which timing of spring and summer precipitation had been identified as the most uncertain parameters).

Step 4.2. Understand climate impacts to ecosystems and people

Step 4.1. produces up to four possible and relevant climate scenarios for the area of interest. The likely impacts of climate change for each of the developed scenarios are then determined in Step 4.2. The objective is to understand direct impacts of projected climate change on the relevant ecosystems (green), resulting impacts of climate change on humans and human systems (orange), and impacts of any human reactions to the climate changes on ecosystems (red). A common tool to harvest local knowledge about likely impacts is the "ecological drawing", a large drawing of the relevant ecosystems and human activities within the landscape, as well as the

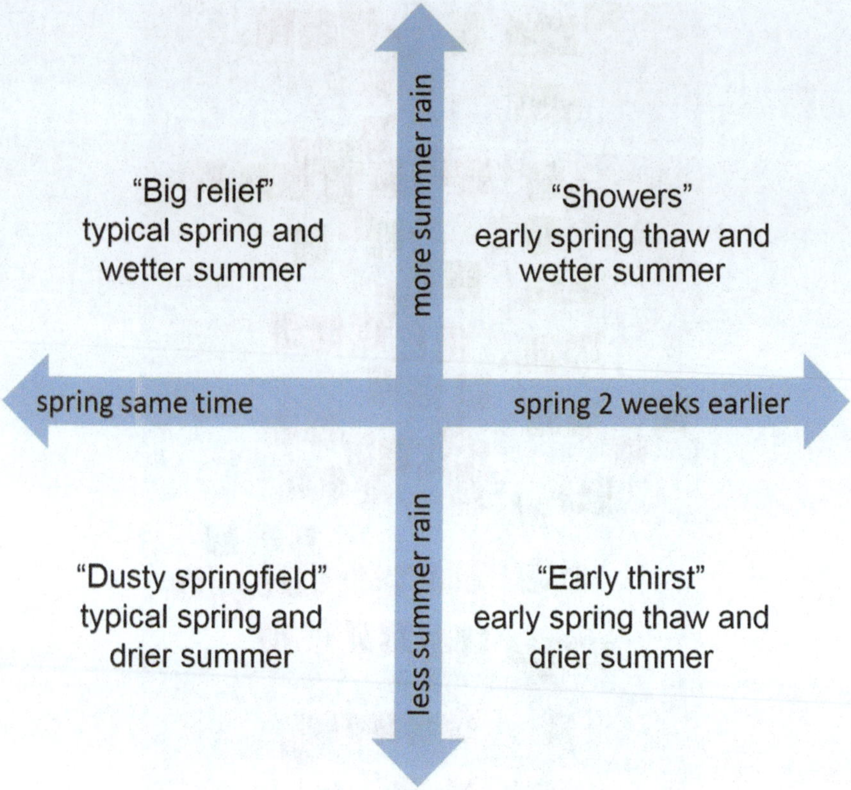

Fig. 4 Project example for development of four climate scenarios in a situation where timing of spring and summer precipitation had been identified as most uncertain parameters

ecosystem services connecting them. A separate ecological drawing for each climate scenario is prepared and the relevant expected climate impacts associated with the respective climate scenario are added to the drawing (see Fig. 5 for project example, for the scenarios identified in Fig. 4). The informed local expectations about impacts of projected change are collected in this way. This is because there is minimal published research on specific impacts under specific climate scenarios. Local people may not have experienced one of the scenarios entirely, but they have usually experienced each of the different elements of those scenarios in recent history, together with their respective impacts.

The developed information is then transferred into a table format and also summarized on a quadrant chart of the climate scenarios. Figure 6 shows the main impacts that were identified by the local communities based on the scenarios identified in Fig. 4 as well as using the ecological drawings as illustrated in Fig. 5.

Fig. 5 Project example for four ecological drawings for different climate scenarios as derived from Fig. 4. (*Photos* Shaun Martin)

Step 5. Develop a general model of the socioeconomic-ecological system (including climate impacts)

The entire socioeconomic-ecological system is summarized in a box-and-arrow diagram (Fig. 7). This represents—by a standard set of box symbols—the relevant ecosystems, the ecosystem services that accrue from them, and the elements of human wellbeing that are supported by those ecosystem services. The direct conventional threats to the ecosystems, and their ultimate drivers as well as the climate impacts—both observed and derived from future climate scenarios—are also included in the diagram as shown in Fig. 7. This includes how climate change directly affects ecosystems (e.g. loss of productivity of land due to increased drought), how climate impacts exacerbate existing conventional threats (e.g. more frequent droughts leading to more crop failures and therefore clearing forests for more extensive agriculture), and how climate impacts bring about new conventional threats (e.g. humans start impounding water to address increasing drought).

In order to understand the relative importance of both, conventional threats (Salafsky et al. 2008) and climate impacts to the ecosystems upon which humans depend in the project area, threats are then rated based on generic Open Standards protocols, using the criteria of scope (i.e., what proportion of the ecosystem is affected by the threat in question), severity (i.e., how severe is the threat to the

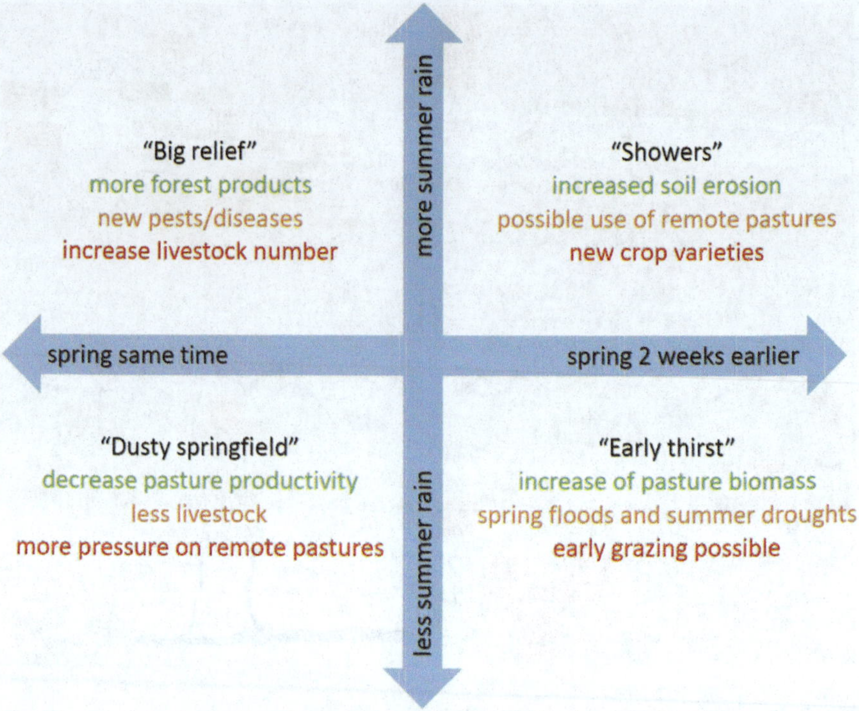

Fig. 6 Project example of the types of climate impacts identified in four climate scenarios as derived from Fig. 4, based on the methodology illustrated in Fig. 5

ecosystem in question), and irreversibility (i.e., how long or difficult will it be before the effects of the threat are no longer felt) (CMP 2013).

Step 6. Re-evaluate project objectives
Steps 1–5 clarify the range of conventional and climate threats which are degrading—or might degrade—relevant ecosystems. These outputs should then be re-considered. For example, will all of the identified ecosystems (and associated services) be present under all future climate scenarios? Are the goals which were developed earlier (desired future conditions) still attainable in all future climate scenarios? It is possible that, as a result of this consideration, the project team abandons goals associated with particular ecosystems and instead shifts to goals associated with whatever natural cover will still be present under different future climate scenarios. This would not be an easy decision, but it should nevertheless be discussed at this time.

Step 7. Brainstorm, select and test adaptation measures
Based on the information gathered in Steps 1–6 above, adaptation measures are identified, selected and prioritized together with local stakeholders, with the help of

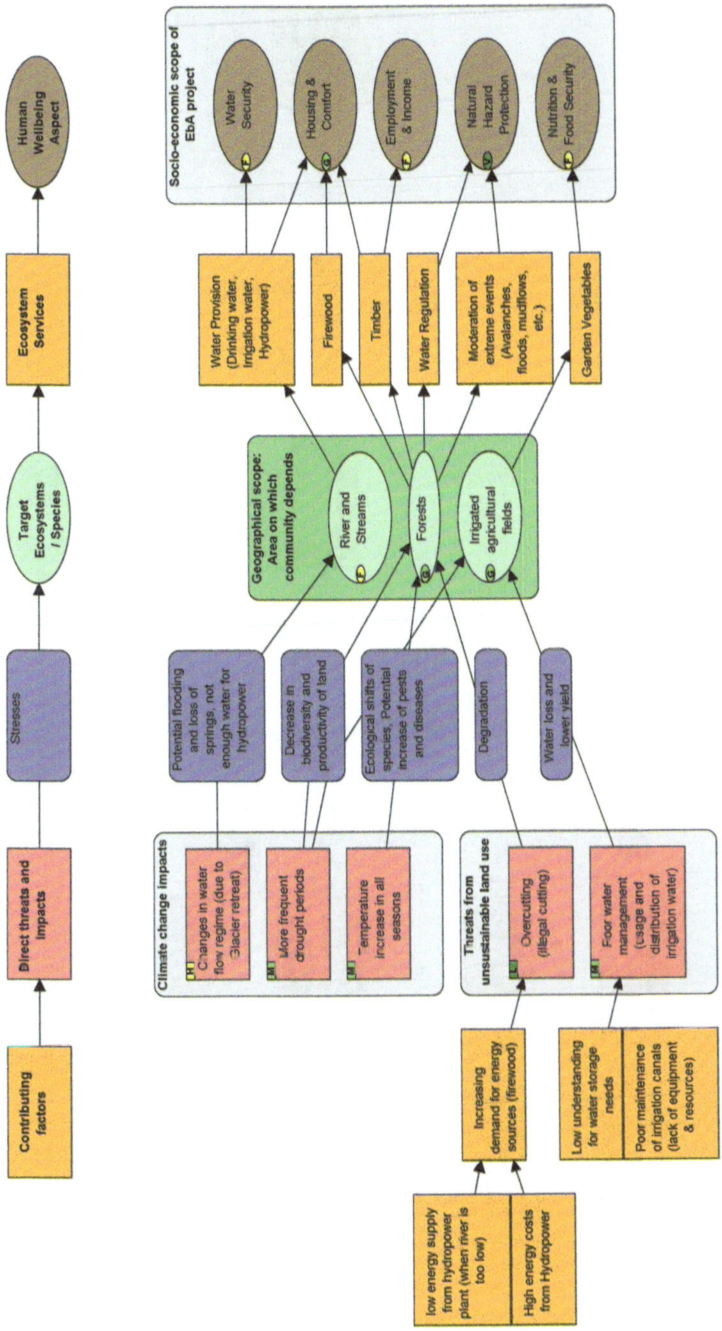

Fig. 7 Project example of a simplified conceptual model for including both conventional threats and climate change related impacts to ecosystems. The simplified diagram on top of the more detailed example contains explanations for the meaning of all boxes

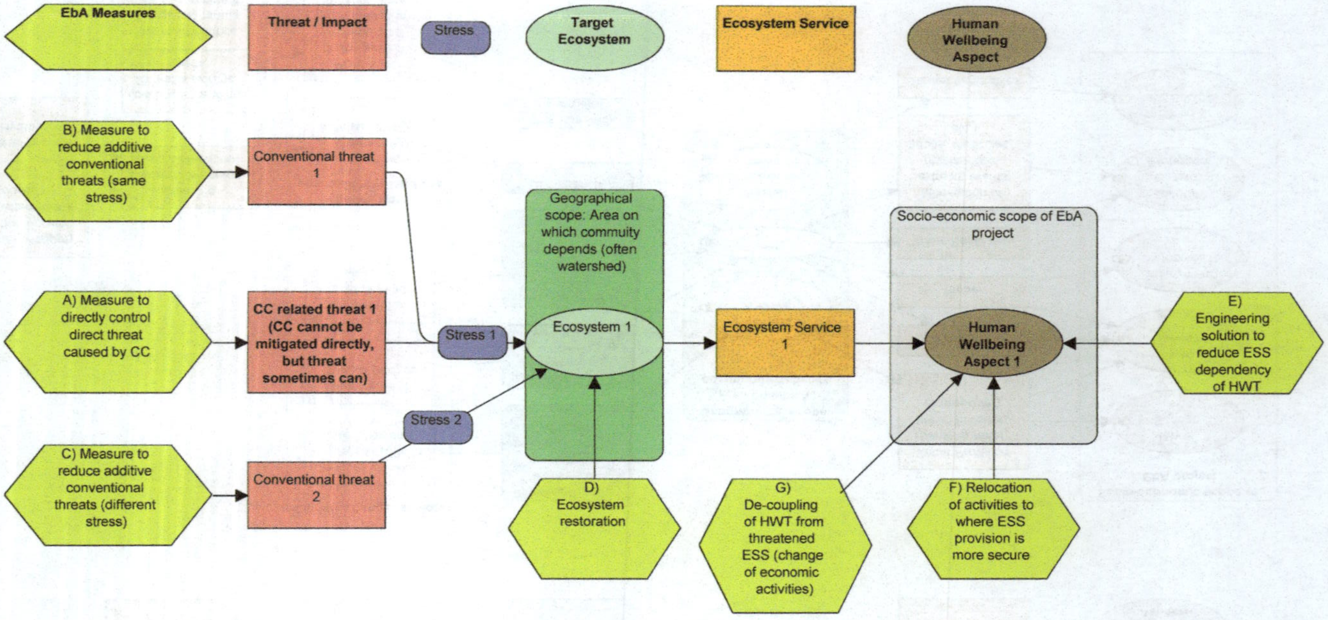

Fig. 8 Various potential intervention points for EbA measures. See text for further explanations

rating criteria and expert consultations. Prior to the implementation of selected measures, result chains for each measure are analysed to check feasibility and identify implicit assumptions.

Step 7.1. Identify and select measures based on general model
This step involves using the conceptual model (see Step 5), with the ecosystems, ecosystem services, human well-being, conventional and climate threats, and drivers, as a basis to identify intervention points and brainstorm potential measures to address the threats and vulnerabilities that have emerged. No idea should be dismissed at the brainstorming step, because it may require creative, not previously considered ideas to address some of the issues raised. This may also include types of measures and intervention points that are not usually used as part of conservation projects (Fig. 8).

Figure 8 shows a draft classification of potential measures for an overall EbA strategy[2]. The types of measures shown in this figure can be defined as follows:

(A) **Managing ecosystems to buffer impacts from climate change on other ecosystems**: It is impossible to affect the changing climate, but it is sometimes possible to manage ecosystems so that climate change impacts are reduced. For instance, sustainably managing a riparian forest or wetland might reduce flood impacts on agricultural ecosystems downstream.
(B) **Reduction of synergistic conventional threats**: If an ecosystem is stressed by both, human activities and climate change, then reducing threats resulting from human activities that add to climate change related threats can improve the viability of the ecosystem and hence ensure ecosystem services provision and contribute to EbA. This will be most effective where the conventional and climate change related threats add to produce the same stresses[3] in the ecosystems in question. E.g. where the capacity of a pasture ecosystem to provide biomass is threatened both by overgrazing and less precipitation, reduced stocking and introducing a better rotation system could result in a stabilized provision of biomass.
(C) As (B), but "conventional threat 2" (see Fig. 8) results in a different stress than the climate change related threat affecting the ecosystem. It appears likely that EbA measures addressing this type of threat will be less effective than those described under (A) (but this depends on the way in which key ecological attributes for each ecosystem are related). For the practice, this means that

[2]Only ecosystems are considered here although the same classification might apply to species, where these are relevant to EbA at all. The figure shows a simplified geographical scope with only one ecosystem affected by three threats (one CC related and two conventional), and providing one ESS supporting one HWT. Measures are linked to direct threats in this simplified scheme although in practice, they might well link to indirect factors that offer intervention points to address these.

[3]A stress is an impaired aspect of an ecosystem that results directly or indirectly from human activities (e.g. low population size; reduced river flows; increased sedimentation; lowered groundwater table level). It is generally equivalent to a degraded key ecological attribute (CMP 2013).

measures of type (B)/(C) should be chosen only after careful examination of the relationship between the conventional threat to be addressed and the climate change related threat that triggers the need for EbA.

(D) **Ecosystem restoration** to re-establish its functionality and hence ecosystem services provision. Usually only effective if it is likely that threat reduction measures [types (B) and (C)] have already sufficiently reduced the direct threats that originally caused the decline in ecosystem functionality. In practice, this means that type (C) typically should be applied in concert with (B) and (C), as well as sometimes (A). An example might be reforestation.

(E) **Engineering to reduce sensitivity**: There may sometimes be engineering solutions to gradually reduce dependency of socio-economic activities depending on ecosystem services that are declining because of climate change. Examples are energy-efficient ovens or more effective irrigation systems, drought-resistant crops where climate change results in reduced precipitation or runoff and hence water provision from ecosystem for agriculture, or where reduced hydrological buffering capacity of ecosystem changes seasonal patterns of water availability.

(F) **Small scale relocation**: Where climate change results in small-scale spatial shifts of ESS provision (e.g. where vegetation zones with their associated ESS shift horizontally, or in response to reduced water supply), EbA could consist of re-locating ecosystem services dependent economic activities in response. Similarly, where the capacity of ecosystems to protect people and activities against natural hazards is reduced by climate change in one place, measures could include relocating people or activities to where this protection is likely to persist. This type is likely to be more relevant over the long term.

(G) **De-coupling**: Where climate change reduces ecosystem functionality to such an extent that ecosystem services provision does not support existing economic activities anymore, the only solution from a livelihood and economic development perspective may be to replace these with other economic activities that do not depend on the eroded ecosystem services. An example might be the establishment of greenhouse gardening.

Step 7.2. Use climate scenarios as a filter to evaluate measures

The range of measures brainstormed in the previous step can be narrowed down using a number of prioritization criteria such as those suggested by CMP (2013). They can include likely impact, cost, technical or social feasibility, organizational niche, etc. An additional and particularly important criterion is whether each suggested measure will be feasible and sensible in all investigated climate scenarios. If a measure works in all of them, then it can be considered “climate-robust”. If this is not the case, further thought is needed, and the strategy is potentially not “climate-robust”.

Step 7.3. Develop detailed logic chains to show how measures will work
For those measures selected, a logical model diagram illustrating the hypothesized mechanisms by which the set of selected measures should work is then constructed, based on generic OS guidance. (CMP 2013; Margoluis et al. 2013). This includes setting threat reduction and intermediate objectives, against which monitoring can be conducted during implementation. Peer review of these logic chains is strongly encouraged in order to tease out hidden assumptions on the part of the team that developed the logic chain.

Steps 2–7 above represent a vulnerability assessment (VA) as described in the climate adaptation community. While many VAs are quite general and difficult to turn into action, this one evaluates the climate exposures to the relevant elements of ecosystems that support human well-being via ecosystem services. Each climate scenario spells out the specific exposure of the ecosystems to climate change. The sensitivity of ecosystems is considered explicitly when the impacts of climate change on the various scenarios are evaluated.

Step 8. Implement and monitor selected adaptation measures, learn, adapt
The OS for the Practice of Conservation provide a generic framework for the development of operational plans and monitoring plans, as well as for the integration of monitoring and learning into the implementation and—if necessary—adaptation of the identified measures (CMP 2013). The climate-informed OS based framework follows this guidance. The only exception is that Step 6—re-consideration of project goals and objectives—should be repeated at sensible intervals, based on the monitoring data gathered.

Discussion

The climate-informed OS proved to be a useful framework for the development and implementation of an EbA strategy in high-mountainous regions of Central Asia. In particular the consistent logical chain from climate stimuli, provided by locally adapted climate projections, to the adaptation measure identified, is considered a strength of the approach, clearly distinguishing it from 'business-as-usual-adaptation'. Furthermore, the framework is designed to be applied in a participatory manner, involving the local population in the whole process. This is considered a particular advantage for creating ownership of the adaptation process by those that need to adapt to the adverse effects of climate change on the ground.

Throughout the implementation of the project, it became evident that capacity development on multiple levels is a key success factor for EbA. First, the facilitators need to have a good understanding about EbA in general and the climate-informed OS for the development and implementation of an EbA strategy in particular. This is even more important given that the framework is new. The facilitators then are crucial for the development of the capacities for EbA of the local communities. However, not only the capacities of people should be supported to enable communities to adapt to climate change. The EbA strategies developed need to be systematically integrated into local development plans and the organisations in charge need to have the mandate and capacities necessary to implement and monitor the, then climate change informed, development plans. Last but not least, a common and comprehensive understanding of the EbA concept and framework is crucial for project staff too, prior to project implementation.

Traditional land use practices are challenged greatly by changes in climate and therefore it becomes necessary to transfer scientific knowledge about future climate change scenarios and their respective impacts on livelihoods and ecosystem services for use in adaptation planning. The modified, climate-informed OS framework uses scientific projections about climate change in a participatory scenario-planning approach for adaptation with the community stakeholders. However, this is only a first step to trigger and enable a comprehensive adaptation process. After the intervention, scientific information should be transferred to the communities so they can make use of it for land use and adaptation planning even beyond the lifetime of projects. Even more so, the scientific information about climate change impacts needs to be systematically mainstreamed into local development planning and monitored over time to confirm accuracy. This process has started in the current EbA project, but is not yet finalized.

Combining the efforts of capacity building, institutionalizing of the EbA approach and making scientific information available and accessible for local development planners paves the way to come from a project to a process, because adaptation is a process and not a project.

There were several important lessons learned from implementing the various steps of this EbA framework. When the scope is defined (Step 1), it is crucial to consider the ecosystems that are *most* important for the livelihoods of the target communities (rather than for conservation priorities), keeping in mind that ecosystems such as watersheds may extend beyond the defined project area. Some ecosystems, such as remote summer pastures in the case of this project, might seem quite disconnected from the target community when they are in fact an important source of livelihoods.

VAs are of little use for the development of an adaptation strategy if conducted before the needs of communities are fully understood and/or findings are not systematically integrated into the framework. In fact, the authors maintain that the climate-informed OS framework has a 'built-in' VA (Steps 1–5) and, if applied thoroughly, makes a separate VA unessential. Even more so, the framework systematically and participatory integrates scientific and locally adapted climate information into the planning process, which is considered a key factor for successful adaptation by the authors (Step 4). However, despite the importance of discussing climate scenarios with the target communities, there is the danger that climate change can be misunderstood as the source of all environmental problems. Therefore, this process must be facilitated with deliberate consideration of this situation. It is important that communities understand the linkages between unstainable natural resource management and the impacts of climate change and not treat either as sole sources of environmental degradation.

When defining the adaptation strategy (Step 7), it is close to impossible to rely solely on EbA measures. EbA works best when combined with other approaches to adaptation and can only work as part of an overall adaptation strategy. Whether the framework presented in this paper is used to define an overall adaptation strategy or define the sub-EbA strategy of an existing strategy depends on the scope of the application (Step 1). In the case presented in this paper, the framework was used to define an overall adaptation strategy for local communities in high mountainous regions in CA with a clear priority (but not limited to) measures using ecosystems and their goods and services. Finally, the consideration of locally existing knowledge on sustainable land use has to be integrated in the definition and implementation of adaptation measures.

The next step for this EbA project is to implement measures identified by communities in the adaptation strategy. Implementation must be accompanied by planned, systematic monitoring. Beyond the two pilot regions, the framework will be replicated in another region of Tajikistan and there, it will be contrasted to a land use planning approach that does not systematically use locally adapted climate information. This will provide valuable information on the added value of the framework for adaptation in CA. Then, the lessons learned and the framework itself will be systematically documented for replication and will be publicly available. For future projects and research, it will be necessary to test and discuss to which extent the elaborated framework is specific for the high mountainous regions of CA and what can be transferred to other ecoregions of the world.

The process of integrating the EbA approach into national adaptation strategies of the project countries as well as into programs and strategies of international partner organisations and accredited entities for climate funds implementation has been initiated. For this, the project develops guidelines for applying the EbA approach based on a brand key concept highlighting among others the values, benefits and essence of the approach in comparison to business as usual approaches. The process of mainstreaming and disseminating EbA is accompanied by capacity development of national experts and by this enhancing national ownership for project development processes for international climate funds.

Conclusion

In conclusion, the framework described in this paper presents an innovative solution for adaptation planning in high mountainous regions of Central Asia. While the elements of the framework are not new, it is the combination of those that trigger a paradigm shift in climate change adaptation. This paradigm shift consists of the systematic integration of localized climate information, by which adaptation planning directly responds to specific climate threats on the local level and the understanding of adaptation as a process rather than a project. The framework is complemented by a set of tools, ready to be taken up and applied by donors in Central Asia.

Annex 1: Temperature and Precipitation Projections for Pilot Areas in Tajikistan and Kyrgyzstan

The Center for Climate Systems Research conducted a detailed study on historical and projected temperature and precipitation for the pilot areas in Tajikistan and Kyrgyzstan. Below are tabulated summarized climate projections based on a suite of 21 climate models run under 2 RCP emissions scenarios. The results are presented as a most likely range of future climate change, reflecting the degree of uncertainty in the projections. The results within the most likely range, extending from the low estimate to the high estimate, can all be considered plausible futures.

Table X1: Temperature in Villages of Bash Kaiyndy

Projected annual and seasonal temperature in the villages in the early 21st century (2011–2040) and mid-21st century (2041–2070) in comparison to the 1980–2005 baseline. Projections show values that represent both the degree change in relation to the baseline (±°C) and estimated absolute future average temperature (°C) in each time slice. The table shows the most likely range of future climate changes. The low estimate in the range refers to the 25th percentile of model outcomes in greenhouse gas emissions scenario RCP 4.5, and the high estimate in the range refers to the 75th percentile of model outcomes in greenhouse gas emissions scenario RCP 8.5.

Villages	Baseline 1980–2005 (°C)	2011–2040 low estimate to high estimate	2041–2070 low estimate to high estimate
Annual	2.0	2.9 to 3.8 °C[a] (+0.9°C to +1.8 °C)[b]	3.7 to 6.0 °C (+1.7 to 3.9 °C)
Winter (Nov to Feb)	−12.3	−11.5 to −10.6 °C (+0.7°C to +1.7)	−10.6 to −8.7 °C (+1.6 to +3.6)
Spring (Mar to May)	4.6	5.4 to 6.4 °C (+0.8 to +1.8)	6.0 to 8.5 °C (+1.4 to +4.0)
Summer (Jun to Aug)	14.9	15.8 to 16.8 °C (+0.9 to +1.9)	16.7 to 19.1 °C (+1.7 to +4.1)
Autumn (Sept to Oct)	7.6	8.6 to 9.4 °C (+1.1 to +1.8)	9.6 to 11.8 °C (+2.1 to +4.3)

[a]Represents estimated absolute future average temperature for the time period
[b]Represents change in temperature from the baseline value

Table X2: Precipitation in Villages of Bash Kaiyndy

Projected annual and seasonal precipitation in the villages of the Bash Kaiyndy project site in Kyrgyzstan in the early 21st century (2011–2040) and mid-21st century (2041–2070) in comparison to the 1980–2005 baseline. Projections show values that represent both the percentage change in relation to the baseline (%) and estimated absolute future average precipitation totals (mm) in each time slice. The table shows the most likely range of future climate changes. The low estimate in the range refers to the 25th percentile of model outcomes in greenhouse gas emissions scenario RCP 4.5, and the high estimate in the range refers to the 75th percentile of model outcomes in greenhouse gas emissions scenario RCP 8.5.

Villages	Baseline 1980–2005 (mm)	2011–2040 low estimate to high estimate	2041–2070 low estimate to high estimate
Annual	295	274 to 354 mm[a] (−7 to +20%)[b]	275 to 366 mm (−7 to +24%)
Winter (Nov to Feb)	46	47 to 58 mm (+2 to +25%)	52 to 68 mm (+13 to +48%)
Spring (Mar to May)	105	105 to 126 mm (0 to +20%)	109 to 138 mm (+4 to +32%)
Summer (Jun to Aug)	116	100 to 139 mm (−14 to +20%)	93 to 129 mm (−20 to +11%)
Autumn (Sept to Oct)	28	22 to 31 mm (−20 to +12%)	21 to 30 mm (−26 to +7%)

[a]Represents estimated absolute future average precipitation for the time period
[b]Represents percentage change in precipitation from the baseline value

Table X3: Temperature in Darjomj and Siponj Villages

Projected annual and seasonal temperature in the Darjomj and Siponj villages in Tajikistan in the early 21st century (2011–2040) and mid-21st century (2041–2070) in comparison to the 1980–2005 baseline. Projections show values which represent both, the degree change in relation to the baseline (±°C), and estimated absolute future average absolute temperature (°C) in each timeslice. The table shows the most likely range of future climate changes. The low estimate in the range refers to the 25th percentile of model outcomes in greenhouse gas emissions scenario RCP 4.5, and the high estimate in the range refers to the 75th percentile of model outcomes in greenhouse gas emissions scenario RCP 8.5.

Tajikistan	Baseline 1980–2005 (°C)	2011–2040 Low estimate to high estimate	2041–2070 Low estimate to high estimate
Annual	−0.6	0.4 to 1.3 °C[a] (+0.9 to 1.9 °C)[b]	1.3 to 3.7 °C (+1.9 to +4.4 °C)
Winter (Nov to Feb)	−10.5	−9.6 to −8.6 °C[a] (+0.8°C to +1.8)[b]	−8.5 to −6.5 °C (+2.0 to +4.0)
Spring (Mar to May)	−1.6	−0.8 to 0.3 °C (+0.8 to +1.8)	0.0 to 2.6 °C (+1.5 to +4.2)
Summer (Jun to Aug)	10.4	11.5 to 12.2 °C (+1.1 to +1.8)	12.4 to 14.6 °C (+2.0 to +4.2)
Autumn (Sept to Oct)	4.3	5.4 to 6.4 °C (+1.2 to +2.1)	6.5 to 9.5 °C (+2.3 to +5.2)

[a]Represents estimated absolute future average temperature for the time period
[b]Represents change in temperature from the baseline value

Based on the projections, it appears likely that annual precipitation will increase overall, by as much as 20% in the 2011–2040 time slice, and by as much as 32% in the 2041–2070 time period.

Table X4: Precipitation in Darjomj and Siponj Villages in Tajikistan

Projected annual and seasonal precipitation in the Darjomj and Siponj villages in Tajikistan in the early 21st century (2011–2040) and mid-21st century (2041–2070) in comparison to the 1980–2005 baseline. Projections show values which represent both the percentage change in relation to the baseline (%), and estimated absolute future average precipitation totals (mm) in each timeslice. The table shows the most likely range of future climate changes. The low estimate in the range refers to the 25th percentile of model outcomes in greenhouse gas emissions scenario RCP 4.5, and the high estimate in the range refers to the 75th percentile of model outcomes in greenhouse gas emissions scenario RCP 8.5.

Tajikistan	Baseline 1980–2005 (mm)	2011–2040 low estimate to high estimate	2041–2070 low estimate to high estimate
Annual	601	582 mm to 719 mm[a] (−3% to +20%)[b]	591 to 796 mm (−2 to +32%)
Winter (Nov to Feb)	266	266 to 322 mm (0 to +21%)	276 to 373 mm (+4 to +21%)
Spring (Mar to May)	249	245 to 292 mm (−2 to +17%)	249 to 313 mm (0 to +26%)
Summer (Jun to Aug)	46	38 to 56 mm (−17 to +22%)	32 to 60 mm (−29 to +31%)

(continued)

(continued)

Tajikistan	Baseline 1980–2005 (mm)	2011–2040 low estimate to high estimate	2041–2070 low estimate to high estimate
Autumn (Sept to Oct)	40	33 to 49 mm (−17 to +23%)	34 to 50 mm (−15 to +25%)

[a]Represents estimated absolute future average precipitation for the time period
[b]Represents percentage change in precipitation from the baseline value

References

Conservation Measures Partnership (CMP). (2013). *Open standards for the practice of conservation.* Version3/April 2013. Conservation Measures Partnership (51 p). Downloaded on April 28, 2017 at http://cmp-openstandards.org/wp-content/uploads/2014/03/CMP-OS-V3-0-Final.pdf.

Convention on Biological Biodiversity (CBD). (2009). *Ecosystem-based adaptation.* Downloaded on April 28, 2017 at https://cmsdata.iucn.org/downloads/iucn_position_paper_eba_september_09.pdf.

Critical Ecosystem Partnership Fund (CEPF). (2016). Mountains of Central Asia. Accessed on April 20, 2017 at http://www.cepf.net/resources/hotspots/Europe-and-Central-Asia/Pages/Mountains-of-Central-Asia.aspx.

Doswald, N., Munroe, R., Roe, D., Giuliani, A., Castelli, I., Stephens, J., Möller, I., Spencer, T., Vira, B., Reid, H. (2014). Effectiveness of ecosystem-based approaches for adaptation: review of the evidence-base. *Climate and Development, 6*(2). dx.doi.org/https://doi.org/10.1080/17565529.2013.867247.

Global Administrative Areas (GADM). (2015). Boundaries without limits. Version is 2.8. Downloaded on April 24, 2017 at http://www.gadm.org/.

Hicks, C., Reid, H., Rivi, A. R., Bertram, M. (in prep). An inventory of tools and methodologies relevant for Ecosystem-based Adaptation (EbA) practitioners.

Lamoreux, J. A., Chatwin, M., Foster, C., Kakoyannis, C., Vynne, K., & Wolniakowski, C. et al. (2014). Overcoming the funder's dilemma. *Biological Conservation, 175*, 74–81. Accessed on April 28, 2017 at http://www.sciencedirect.com/science/article/pii/S0006320714001682.

Margoluis, R., Stem, C., Swaminathan, V., Brown, M., Johnson, A., & Placci, G., et al. (2013). Results chains: A tool for conservation action design, management, and evaluation. *Ecology and Society, 18*(3), 22. Accessed on April 28, 2017 at https://www.ecologyandsociety.org/vol18/iss3/art22/.

Parrish, J. D., Braun, D. P., & Unnasch R (2003). Are we conserving what we say we are? Measuring ecological integrity within protected areas. *BioScience, 53*(9), 851–860. Accessed on April 28, 2017 at http://www.wec.ufl.edu/academics/courses/wis4554/WebUpdate/ReadingsWIS5555/Indicators/Parrish_et_al03%20Bioscience53_850.pdf.

Reyer, C. P. O., Otto, I. M., Adams, S., Albrecht, T., Baarsch, F., Cartsburg, M., et al. (2015). Climate change impacts in Central Asia and their implications for development. *Regional Environmental Change.* https://doi.org/10.1007/s10113-015-0893-z.

Salafsky, N. D., Salzer, A. J., Stattersfield, C., Hilton-Taylor, R., Neugarten, S. H. M., & Butchart, B., et al. (2008). A standard lexicon for biodiversity conservation: unified classifications of threats and actions. *Conservation Biology, 22*(4), 897–911. Accessed on April 28, 2017 at http://www.fosonline.org/wordpress/wp-content/uploads/2010/11/Classification-of-threats-and-actions.pdf.

Schwartz, M. W., Deiner, K., Forrester, T., Grof-Tisza, P., Muir, M. J., Santos, M. J., et al. (2012). Perspectives on the open standards for the practice of conservation. *Biological Conservation, 155,* 169–177.

World Bank (WB). (2015). Climate adaptation and mitigation program for the aral sea basin (CAMP4ASB). Environmental Management Framework, Vol. III. http://documents.worldbank.org/curated/en/729711468228293250/pdf/SFG1018-V3-REVISED-EA-P151363-Box396251B-PUBLIC-Disclosed-5-6-2016.pdf.

Wulffraat, S., & Morrison, J. (2013). Measuring biological indicators for status assessment of the heart of Borneo. *Environmental Conservation, 40*(3), 277–286.

Understanding Climate Change Adaptation: The Role of Citizens' Perceptions and Appraisals About Extreme Weather Events

Samuel Domingos, Rui Gaspar, João Marôco and Rita Beja

Abstract Climate change is driving dramatic environmental changes and posing new demands to citizens, health authorities, and policy makers worldwide. This is due to an increased frequency, intensity, and duration of associated extreme weather events. Recent calls for better understanding of how citizens adapt to such demands and the role that psychological processes' play in that adaptation, have been put forward. We contributed in this regard by (1) applying the Biopsychosocial Model of Challenge and Threat (e.g. Blascovich 2008) to the study of human responses (psychological, physiological, and behavioural) to extreme weather events; (2) using it as the conceptual basis for a mixed methods study aimed at exploring citizens' perceptions, beliefs, and appraisals of the demands posed by such events and available resources to cope with them. Preliminary qualitative results are presented and potential implications for stakeholders and policy makers in the climate change domain are discussed. An example of how such conceptual and methodological approaches may contribute to developing evidence-based strategies for incrementing citizens' resilience and adaptation to

S. Domingos (✉) · J. Marôco · R. Beja
William James Center for Research, ISPA-Instituto Universitário,
Rua Jardim do Tabaco, 34, 1149-041 Lisbon, Portugal
e-mail: samuel.domingos@sapo.pt

J. Marôco
e-mail: jpmaroco@ispa.pt

R. Beja
e-mail: ritamb_14@hotmail.com

R. Gaspar
Universidade do Algarve, Departamento de Psicologia e Ciências da Educação,
Faculdade de Ciências Humanas e Sociais, Campus de Gambelas,
8005-139 Faro, Portugal
e-mail: rgaspar@fch.lisboa.ucp.pt

R. Gaspar
Catolica Research Centre for Psychological, Family and Social Wellbeing (CRC-W),
Universidade Católica Portuguesa, Universidade Católica Portuguesa,
Palma de Cima, 1649-023 Lisbon, Portugal

F. Alves et al. (eds.), *Theory and Practice of Climate Adaptation*,
Climate Change Management, https://doi.org/10.1007/978-3-319-72874-2_3

climate change, will be provided. This allow a better understanding of citizen appraisals and perceptions' role in shaping adaptive behaviour, in order to provide them with the necessary personal and social resources to cope with extreme weather events and increment future resilience.

Keywords Climate change · Extreme weather events · Adaptation Psychological processes · Coping

Introduction

Climate change is creating new challenges to citizens, health authorities, and policy makers worldwide, as a result from dramatic changes to the Earth's climate, biological, and human systems (IPCC 2014), and an increased frequency, intensity, and duration of extreme weather events (Lefevre et al. 2015; Santos 2006; Spence et al. 2009). The profound alterations in climate systems (e.g. floods, droughts, heatwaves, coastal erosion, sea level changes, riverbed changes, air quality), biological systems (e.g. terrestrial and marine ecosystems) and human systems (e.g. mortality, lower physical and mental health, food production breakdowns) translates into increased socioeconomic costs (IPCC 2014). Due to these multidimensional consequences, climate change is already seen as one of the major societal risks to our planet and its inhabitants (Ahmed and Long 2010; Spence et al. 2012; Swim et al. 2011). This makes the understanding and promotion of climate change adaptation a high priority worldwide.

Although climate changes naturally overtime, what is relatively recent in the planet's history is the role that human activity plays in that change. Climate has changed in many ways over the Earth's geological time, but for the first time, over the past century, human activity (e.g. fossil fuel consumption, agricultural malpractices, greenhouse gas emissions) has become a significant cause of climate change (Swim et al. 2011). This implies that climate systems and human systems now interact in new and often unpredictable ways, magnifying old problems, and creating new ones. Due to this interactive aspect, current research on climate change demands an integrative work between the physical and social sciences, in diverse and multidisciplinary fields. This has been the case in topics like, for example: (a) understanding the socio-physical processes underlying human activities that may cause climate change (e.g. physical, chemical and social changes that lead to climate change); (b) understanding how key areas for human sustainability (e.g. agriculture, water management, forests and rivers management, biodiversity and wildlife conservation, coastal management, urban sustainability) can be adapted to mitigate climate change demands; (c) understanding potential and observed impacts of climate change on human, social, natural, biological, hydrological, and physical systems; and more recently, (d) understanding the human dimensions of climate change (Casimiro et al. 2003; Chen et al. 1983; Gilman et al. 2007; Glick et al. 2009; Grothmann and Patt 2005; IPCC 2014; Swim et al. 2011).

As shown by Swim et al. (2011), climate and human systems can be integrated in a conceptual model which establishes connections among them. According to this, people (human systems) influence climate through activities (e.g. fossil fuel consumption) that may contribute to changes in climate systems (human contribution to climate change). In turn, changes in climate systems may have consequences over human systems by altering key aspects of the environment that support humans and other living beings (e.g. changing the frequency of storms and droughts, the availability of water, the viability of food crops). These changes may have, among others, socio-individual consequences (e.g. emotional responses, stress, mental and physical health deterioration, intergroup tensions, aggressiveness, threats to cultures and ways of life), that demand mitigation responses (e.g. implementation of policies that limit human contribution to climate change) and adaptation (e.g. cognitive and behavioural strategies implemented with the goal of mitigating direct impacts of climate change). Central to this process, is the role played by psychological factors that mediate the relationship between changes in human and climate systems and their socio-individual consequences. These include cognitive (e.g. human understandings, perceptions, beliefs, and appraisals of climate change and related events), affective (e.g. affective responses to climate change and related events), and motivational (e.g. motivation to mitigate, adapt, and implement protective behaviours against climate change and related events) processes. On one side, these processes may be influenced by and directly influence the way humans contribute to climate change (thus indirectly influencing climate systems). On another side, they may influence human's perceptions and beliefs about climate change risks and consequences, as well as mitigation and adaptation responses.

Despite the central role that psychological factors may have, these are still understudied in research. Indeed, there is evidence that people can reduce their susceptibility to climate change events and adapt overtime (Arbuthnott et al. 2016). However, it is still somewhat unclear how this process takes place and the role that psychological variables play in it. This need for clarification is highlighted by recent calls for more psychological research on climate change adaptation (APA 2011; Clayton et al. 2015). A better understanding of the role that psychological factors play in the global climate change process is therefore vital for promoting human adaptation to the demands posed by climate change. One example, could be understanding citizens' perceptions, beliefs, and appraisals about climate change and its associated extreme weather events, and how cognitive, affective, and motivational processes interact to shape these and the associated adaptation responses. This could enable, for example, identifying common misconceptions about causes and consequences of climate change and develop tailored communication strategies to target them. By better understanding the role that citizens' perceptions, beliefs, and appraisals have in shaping adaptive behaviour, we can provide citizens with the necessary personal and social resources to cope with climate change in general and with extreme weather events in particular. The ultimate goal in this regard is incrementing their resilience and adaptation to climate change. To achieve this, we not only need a better understanding of these processes through more research on the subject, but also, we need to develop models that enable us a better conceptualization of these. This is the focus of the next sections.

Understanding Psychology's Role in Promoting Adaptation to Climate Change

Psychological research can foster the understanding of human causes and consequences of climate change thus contributing to promote the continuous human process of adaptation to that phenomenon. These contributions may assume the form of new and original research or allow new insights from psychological science to research in other science fields (Bradley and Reser 2017). For example, what is the gain of having communication activities that increase citizens' knowledge of climate change and extreme weather events protective measures, if they lack the motivation to use such information in deciding whether to implement such protective behaviours? Failure to implement or maintain behaviour can be due to psychological processes (e.g. Schwarzer 1992), which should be identified but often are not. To face a multidimensional threat like climate change and promote adaptive behaviours and resilience, multidisciplinary work from various scientific domains is therefore essential.

When we consider the main human contributions for climate change (e.g. human behaviours that contribute to climate change) we acknowledge that these mainly involve a set of collective actions driven by individuals, groups, or organisations short-term benefits (e.g. goals). The latter actions degrade a long-term common good such as climate sustainability (Gifford 2008; Swim et al. 2011), with consequences (positive, negative, and neutral) to such individuals, groups, and organizations. These consequences in turn influence the way they think and feel about climate change. This makes clear that not everyone consciously engages in or supports the behaviours that contribute to climate change. Nevertheless, all without exception engage in them to some extent (whether consciously or unconsciously), and all without exception suffer its effects to some extent (whether consciously or unconsciously). This individual and social variability in experiencing and perceiving climate change and its consequences can lead, for example, to different attitudes, goals, beliefs, perceptions, and appraisals concerning the need to act, adapt, and mitigate its effects. Stakeholders, organizations, and individuals that work in the field of climate change adaptation and mitigation must therefore be prepared to face such variability in the way citizens experience, perceive, appraise (e.g. evaluate), respond, and adapt to climate change (Bradley and Reser 2017).

The differential effects that human systems can have on climate systems and vice versa, will be detailed next.

Human Systems Influence on Climate Systems

Regarding human contributions to climate change, we know that these result mainly from actions related to unsustainable high levels of consumption (e.g. fossil fuel consumption, mass consumption practices) that have underlying individual and collective goals which lead to specific behavioural patterns (Bradley and Reser 2017; Swim et al. 2011). Psychology can not only help promote a better understanding of such goals and associated perceptions, beliefs, and appraisals, but also of other psychological processes (cognitive, affective, motivational) that play a role in it. For example, it can help identifying implicit and explicit predictors of adaptive (individual and collective) behaviours to climate change.

This contribution can also translate into designing evidence-based intervention strategies aimed at restructuring goal prioritizations, perceptions, appraisals, and beliefs, as well as reinforcing existent or promoting new adaptive behaviours. Despite the above focus on global climate change adaptation, the same principles apply to specific events identified as direct or indirect causes or consequences of climate change (e.g. heatwaves, floods, energy consumption, recycling, etc.).

Climate Systems Influence on Human Systems

As important as an understanding of the influence of human systems on climate systems, is an understanding of the latter's influence on the former, from a psychological level of analysis. For example, as noted by Doherty and Clayton (2011), with the increase in extreme weather events occurrence due to climate change, citizens tend to become concerned and worry by anticipation which may lead to higher levels of stress and anxiety, gradually eroding the quality of life, creating new tensions, and threatening the mental health of the affected communities. This is even more problematic considering the effects on vulnerable population groups such as, for example, children and the elderly. Children in particular, are expected to experience higher levels of anxiety and frustration regarding climate change, because they find the magnitude of this threat confusing (Fritze et al. 2008) and present low confidence in their abilities to contribute to solving it (Connell et al. 1999; Özdem et al. 2014). In this regard, children's climate change coping strategies are identified as an important research area (Clayton et al. 2014; Ojala 2012). Another example of climate change effects over vulnerable population groups, concerns the increasing competition for scarce environmental resources, combined with pre-existing inequalities and disproportionalities among social groups, organizations, and nations. This may lead to an increase in interpersonal and intergroup conflicts, further increasing citizens stress and anxiety levels, contributing to the global degradation of the quality of life and mental health worldwide. Psychology is suited to understand and address these issues, and promote adaptation, as evidenced by its large body of research on intra and inter-group, and intra and inter-personal processes (Swim et al. 2011).

Another important aspect is describing, explaining, and informing citizens adaptive responses to climate change and related events. Adaptive responses also involve cognitive, affective, and behavioural dimensions that allow citizens, for example, to become more attentive to the issue, accept its reality and implications, act in ways that facilitate psychological adjustment, and adopt problem-solving and pro-environmental attitudes (Reser et al. 2012). Therefore, it is important to unveil the influence that these cognitive, affective, and behavioural dimensions may have on adaptive responses to climate change and related events.

Adaptation to climate change and related events extends beyond making physical and structural adjustments, also requiring a range of adjustments on individual coping strategies (e.g. cognitive and behavioural efforts made by citizens in order to adapt and cope with potentially adverse and stressful situations) and psychological processes (e.g. how people perceive and appraise the situation and the control they have over it, how people think about the situation and make sense of it, how people feel about the situation) that need to be addressed.

Adapting to and coping with climate change and its related events is a dynamic process comprised by the interaction of many intricate intrapsychic and social processes that influence citizens' reactions (behavioural and psychological), preparations, and adaptive responses to the adverse impacts of climate change (Swim et al. 2011). These social processes are also an important dimension. Indeed, individual adaptation processes are influenced by what other people do and think, by the media representations of climate change, and by formal and informal social discourse. In turn, this leads to socially shared constructions of climate change (e.g. the way society sees it), and associated perceptions, beliefs, norms (e.g. how to behave in a certain situation), and representations (e.g. sets of beliefs about causes, consequences, risks, and protective behaviours). All of this needs to be considered when promoting adaptation to climate change (Doherty and Clayton 2011).

Individual's Perceptions and Appraisals of Climate Change Related Events: The Case of Extreme Weather Events

Climate change effects, especially those in the form of extreme weather events, are particularly worrisome due to high and fast increases in mortality rates, and to the rapid degradation of human health and quality of life. Although research is being developed with the aim of finding ways to respond to these challenges, statistics regarding human adaptation do climate and environmental changes show that more work is still needed (e.g. Prüss-Ustün et al. 2016). To reduce such human susceptibility, it is important to capacitate individuals to better adapt to such events and train/prepare them to future events, by promoting resilience. This capacitation and training/preparation should draw on scientific evidence focused on the cognitive, affective, and motivational processes that influence how individuals perceive and appraise the events, the demands, the existent resources to cope with such demands, and how they respond to such events (expressed through cognitions, affect and/or behaviours). Moreover, such knowledge may not only enable mitigating both human consequences of climate change but also human contributions to climate change.

With the goal of achieving a better understanding of how individuals perceive, appraise, and respond to climate change related events, as is the case of extreme weather events, we adapted the theoretical approach put forward by Blascovich: The Biopsychosocial Model of Challenge and Threat (BPS Model; for a review see Blascovich 2008; Blascovich and Mendes 2001) to the study of human reactions (psychological, physiological, and behavioural) to extreme weather events. Drawing on this model it is possible, in our view, to explore citizens perceptions, beliefs, and appraisals about (1) extreme weather events situations, (2) the demands that are posed to them by such events, (3) their individual and social consequences, (4) the personal and social resources available to them to cope with those demands, and (5) the potential strategies that can be implemented to cope with these events, thus promoting adaption.

The BPS Model of Challenge and Threat has been extensively supported by research (e.g. Blascovich 2008; Blascovich and Mendes 2001, 2010; Blascovich and Tomaka 1996; Fonseca et al. 2014; Garcia-Marques et al. 2015; Gaspar et al. 2015; Tomaka et al. 1993; Tomaka et al. 1997). To our knowledge, this model has never been applied in the field of climate change adaptation and particularly to the appraisal of extreme weather events. It is our contention that its utility to this field derives from its integrative goal of understanding human's, by considering the joint role of biological, psychological, and social processes, while also considering the synergistic role of affective and socio-cognitive factors, separately considered in the past (Lazarus 1984; Zajonc 1984).

According to the BPS Model, when faced with a given potentially stress inducing situation that may require action—a "motivated performance situation"—individuals tend to evaluate (more or less consciously) the perceived demands that are posed by that situation and the perceived available resources (personal and social) they have to cope with such demands. For example, when faced with an extreme heat weather event, people may perceive "being more thirst than usual" as a demand, and "carry a bottle of water" as a resource to cope with that demand.

When individuals perceive they have sufficient personal and social resources (R) to meet the situationally imposed demands (D), the situation will be appraised as a challenge to cope with (R > D = Challenge). When resources are perceived as insufficient to cope with the demands, the situation will be appraised as a threat (R < D = Threat). These challenge or threat appraisals determine the strategies that are implemented to cope with such demands (Blascovich 2008; Blascovich and Mendes 2001, 2010; Gaspar et al. 2015), such as, for example, seeking or avoiding information about the event(s), among others.

This appraisal process is mediated by the synergistic influence of socio-cognitive (e.g. thoughts) and affective (e.g. emotions, feelings) factors. These may, in turn, be influenced by situational (specific characteristics of the context where the situation occurs; e.g. how "new" the situation is) and dispositional (specific characteristics of the individuals that are in the situation; e.g. personality traits and states) factors.

A heatwave, for example, represents a motivated performance situation, as it disrupts individuals' current goals (e.g. they may not be able to achieve the goal of going to work or do their everyday tasks). This deviation from what they consider to be a "normal" day or the "norm" (Gaspar et al. 2015), drives them to appraise (directly or indirectly) how much and what type of demands this new situation poses to them, and what are the available resources to cope with such demands. Depending on their evaluation of the situation/events as posing a challenge or a threat to cope with, they may respond in different ways (e.g. express fear if a threat is appraised vs. express trust in the authorities' protection, if a challenge is perceived). However, different people when faced with the same event may respond in different ways, depending on the influence of socio-cognitive and affective factors (e.g. individuals' who express negative affectivity more often, may tend to perceive events as more demanding).

Although the BPS Model is not primarily focused on adaptive processes, it does present an explanation of the process through which individuals appraise/evaluate potentially stressful situations/events, which determines the subsequent implementation of strategies to cope with such situations/events (Gaspar et al. 2015). Particularly, this model can be applied to study the influence that certain motivated performance situations/events (climate change in general, specific climate change related situations like extreme weather events) and their characteristics (e.g. how much the situation/event differentiates from the "norm"; geographical and seasonal weather variations; if its seen as more negative or positive) may have on how individuals perceive and appraise such events/situations.

In order to understand how individuals' (1) perceive extreme weather events, (2) the demands posed by potentially stressful situations such as extreme weather events, (3) the available resources to cope with such demands, (4) how they overall appraise these events (challenge or threat to cope with), and (5) study the influence that the way how the situation is framed (with a negative or positive affective valence; i.e. if they think about it in a more pleasant or unpleasant way; see Bruine de Bruin et al. 2016) may have on those perceptions and appraisals, we have developed a preliminary mixed methods quantitative-qualitative study.

Next, we will present this study's methodological approach and preliminary results, discussing them, with a focus on its potential implications for promoting citizen's adaptation to climate change. In our view, conceptual approaches like the BPS Model applied to climate change and extreme weather events, and research approaches like the one exemplified next, may provide a first step to develop evidence-based strategies for incrementing citizens' resilience and adaptation to climate change.

Study Methodology

A total of 159 (33 male; 126 female) participants with ages between 18 and 88 years old (*M* = 41.86; *SD* = 25.48) were recruited through a non-random sampling procedure and randomly assigned to one of three experimental conditions (53 participants per group/condition). The study followed a mixed methods approach based on the application of individual semi-structured interviews and a structured questionnaire.

The procedure partially replicates the work developed by Bruine de Bruin et al. (2016) and was aimed at identifying individual beliefs, perceptions, and appraisals about Extreme Heat Weather Events (EHWEs; commonly known as heatwaves).

Particularly, it was focused on individual's beliefs and perceptions of demands, resources, overall appraisals (challenge or threat), and the influence of affective valence framing of the situation based on the random allocation of participants to one of three experimental conditions: pleasant/unpleasant/no-frame. Thus, in this study, we requested participants to think about what it is for them a typical pleasant/unpleasant/no-frame EHWE, followed by a semi-structured qualitative interview and questionnaire.

The qualitative interview allowed exploring individual and social beliefs and perceptions about EHWEs evoked situations, its associated components (demands, resources, and consequences), protection behaviours, and norm deviation (e.g. perceptions about how the situation they thought about deviates from normal everyday situations, see Gaspar et al. 2015). The questionnaire allowed collecting appraisal, protective intention, and temperature estimation measures. The appraisal measures were adapted from the work developed by Tomaka et al. (1993). These measures requested participants to separately rate the level of demands posed by the EHWE they thought about, and the level of resources to cope with those demands. Participants estimations of the temperature that would best describe the EHWE they thought about were also collected, along with intentions to implement future protective behaviours against EHWEs. This measure was based on the work developed by Bruine de Bruin et al. (2016), and the Portuguese Directorate-General for Health recommendations for heatwave situations (DGS 2015). Participants age, gender, geographical location, marital status, and other socio-demographic variables were also collected in the end of the study.

Qualitative content analysis allowed exploring participants beliefs and perceptions about the evoked extreme heat weather events (e.g. how they thought about them, what were the demands and resources that first came to their minds when thinking about such events). Moreover, and because extreme heat weather events have a pleasant dimension (Bruine de Bruin et al. 2016; Lefevre et al. 2015) this allowed us to identify the effects of thinking more positively or negatively about them, and how can that link to different behaviours (e.g. exposure or avoidance).

Results and Discussion

The results reported here are preliminary and only qualitative differences in reported EHWEs situations, resources, and demands between the positive (participants asked to think positively about EHWEs) and negative (participants asked to think negatively about EHWEs) conditions are being considered. The only exception is the presentation of a quantitative result regarding differences in the perceived temperature between participants in the positive and negative conditions.

Preliminary results suggest qualitative differences between the situations, demands, and resources reported by the participants in the positively vs. negatively framed conditions. Regarding the type of situations reported by participants when requested to think about a typical situation of extreme heat, participants in the positive condition (requested to think about pleasant EHWEs) tended to evoke situations of exposure to extreme heat that involved pleasant outdoor activities as for example: go to a beach or pool, walking in the countryside, and engage in outdoor social activities with friends and family. Differently, participants in the negative condition (requested to think about unpleasant EHWEs), tended to evoke situations of avoidance and discomfort associated with exposure to extreme heat, as for example: stay at home with windows and shutters closed, stay isolated from others, avoid social contact, flee from the street to home or other cooler places near home. Participants estimations about the temperature (°C) of the extreme heat event they thought about were also explored, demonstrating significant differences between the positive and negative valence conditions (t (104) = 3.440; $p = 0.001$; $d = 0.67$; I.C. 95% =]1.355, 5.042[): the estimated temperature of an extreme heat situation was reported as significantly lower in the positive condition ($M = 35.18$ °C; $SD = 4.87$) than in the negative condition ($M = 38.38$ °C; $SD = 4.70$).

Another interesting result relates to the extent to which participants perceived that the evoked situation differed from normal everyday situations. In the positive condition, some participants perceived the situation as different from normal because they had the chance to do activities that they could not do on other days (e.g. go to a beach, walk in the countryside). In the negative condition, some participants perceived the situation as different from normal because it interfered negatively with their life (e.g. impossibility to do everyday tasks due to extreme heat weather).

Regarding perceived demands posed by EHWEs, preliminary results suggest that participants in the positive condition tended to report psychological demands (e.g. concentration difficulties, loss of patience, irritability, bad mood, concern, need for mental habituation) and material demands related to personal care concerns (e.g. concern about drinking more water, concerns about what clothing should be used, concerns about the need to wear sunscreen). Participants in the negative condition tended to report physical demands (e.g. indisposition, tiredness, breathing difficulty, dizziness, need for physical habituation, impossibility to perform tasks), which could also suggest higher levels of embodiment (e.g. feeling in the body what they are thinking about).

Regarding perceived resources to cope with the demands posed by EHWEs, preliminary results suggest that participants in the positive condition frequently reported resources that can be accessed outdoors, such as, for example, drinking water, going for a swim, use a hat or an umbrella, search for a shade, or apply sunscreen. However, this last reported resource represents a misconception of what could be an effective protection, given that sunscreen may protect against UV rays but not against heat. On the other hand, participants in the negative condition frequently reported resources associated with housing conditions (e.g. using fans or air conditioning, remain in colder areas of the household, take a bath, close windows and shutters).

Despite the preliminary nature of these results, they suggest interesting implications to the adaptation process associated to EHWEs. The first one relates to qualitative and quantitative differences in the types of EHWEs situations that came to the participant's minds. This result is in line with those reported by Bruine de Bruin et al. (2016), highlighting that the perceived need for risk protection may be influenced by the availability heuristic (e.g. the information most readily available in our minds has a higher potential of influencing our judgment and decision making) and the affect heuristic (e.g. the current affective state has a higher potential of influencing our judgment and decision making; for a review see Bruine de Bruin et al. 2016; Slovic et al. 2004; Tversky and Kahneman 1973). The preliminary results suggest that the manipulation (thinking positively vs. negatively about EHWEs) seem to make available different information in the participants' minds, which in turn originates perceptual differences about EHWEs situations, demands, and available resources. These may have implications not only in the way participants prepare themselves and cope with EHWEs (e.g. thinking positively or negatively about an EHWE can influence what information people retrieve from memory in that moment, as for example information on protective behaviours) but also in the way they behave and protect themselves.

Also, as thinking positively about EHWEs seemed to evoke an approach tendency towards extreme heat (approaching something "good") whereas thinking negatively seemed to evoke an avoidance tendency (avoiding something "bad") along with expressions of discomfort. This makes evident that the way in which authorities communicate about EHWEs (e.g. more positively or negatively) may determine how citizens appraise such events and consequently, their responses (behavioural and psychological). As showed in other psychological research about climate change, a higher psychological distance from risk was generally associated with lower levels of concern (e.g. Gifford 2011; Spence et al. 2012). If in fact thinking positively about EHWEs also makes people be less concerned about the associated risks (e.g. leading to more exposure) this may have implications in their behaviours and ultimately to their health. On the other hand, if thinking negatively about EHWEs also makes people more avoidant, this may also have implications in their health and quality of life (e.g. less social interaction). Thus, this is an important aspect that is worth of further research in the future.

The results also reveal existent misconceptions in participant's minds, as some resources reported in the positive condition may not be adequate to cope with EHWEs (e.g. sunscreen as a protective resource). Further analysis should allow the identification of other possible misconceptions.

Thinking positively or negatively about EHWEs may also link to specific thoughts about experiences and emotional responses to those events. Thus, in the same way that direct experience of negative extreme weather events seems to lead to an overall increase of emotional responses, greater perceived personal vulnerability, and risk perceptions (Demski et al. 2017), the mere act of thinking about an extreme weather event, without directly experiencing it, may activate similar or opposed responses, beliefs, and perceptions whether the person focus on the negative or positive aspects of it. This may have important implications in understanding how citizen's think and feel about these events, how they plan future behaviour, and ultimately how they behave.

Conclusions

We believe that the results presented here may have a practical and applied significance for entities and stakeholders working in the field of climate change adaptation and communication, especially when dealing with "pleasant hazards" (Bruine de Bruin et al. 2016; Lefevre et al. 2015) such as EHWEs or heatwaves. For example, when addressing the public, caution should be taken to ensure that recommendations that make salient negative aspects, do not also have the ironic effect of reminding people of pleasant activities that can actually increase their exposure to risks. These communication activities should rather focus on providing citizens with the necessary and adequate resources to cope with events (resources that citizens can really use, access, and relate with), while intervening on commonly identified misconceptions (e.g. sunscreen as a protection against heat). This also means that a collaboration between health authorities and media organizations should exist to standardize the information that is communicated, especially when dealing with events that, although negative in their consequences, may evoke positive affect (as it is the case of heatwaves).

Due to their preliminary nature caution should be taken when drawing conclusions regarding these results because these are still preliminary and not generalizable to every individual. In the same way, the magnitude and significance of the reported differences also needs to be further scrutinized to clarify their relevance.

Another limitation that needs to be considered is that the study focused on perceptions and imagined situations about EHWEs, and not the actual experience. Nevertheless, there is a clear link between thinking about and experiencing a situation, given that thinking about it may function as a cognitive simulation of the general experience people have when faced with such events.

Some limitations also arise from the sample that was collected for this study, and from the period in which it was collected. First, the sample is small and low in diversity (mostly comprised of young and older adults). Second, data was collected during the winter months, which raises the question if perceptions, beliefs, and appraisals are influenced by seasonality. Third, although geographical information about the participants was collected, the study was not designed to fully consider geographical differences. We intend to overcome these limitations in future studies by considering such sources of variability (e.g. temperature, geographical location, actual experience). Moreover, current results will be further scrutinized to evaluate possible quantitative differences between groups regarding demands vs. resources appraisals, and intentions to engage in protective behaviours. Comparisons with the control group will also be considered to characterize natural tendencies (without affective framing). By doing so, the study will allow a first step in the identification of the characteristics that are most and least commonly referred by the participants about EHWEs situations, demands, and resources to cope with those demands, while linking them to specific behavioural patterns.

Because climate change adaptation is a multidimensional and complex process, the goal of this paper was to allow a better understanding of how psychological processes can be conceptualized, analysed, and used to promote human adaptation to climate change and related extreme weather events. Given the above limitations, this work should therefore be seen as a first step in understanding the role that citizens' beliefs, perceptions, and appraisals play in climate change adaptation, exemplifying how this could be conceptualized and analysed. Future studies will help overcome limitations and clarify the link between thinking about and experiencing extreme weather events. Studies as the one presented here should further explore the data in order to identify the most frequently evoked demands and resources (e.g. those that are more readily cognitively accessible in any given situation) and the less frequently evoked demands and resources (e.g. those that are less readily cognitively accessible in any given situation). This second layer of analysis may provide clues about which resources and demands should be prioritized in the future when addressing the public, and what are the most common misconceptions that need to be clarified (Gaspar et al. 2016). Overall, this should provide entities and stakeholders involved in promoting adaptation to climate change with valuable tools to overcome citizens' misconceptions and promote effective adaptive behaviours.

Acknowledgements The first author's work is funded by the Foundation for Science and Technology (Fundação para a Ciência e a Tecnologia-FCT) with the grant PD/BD/128512/2017. William James Center for Research, ISPA-Instituto Universitário is supported by the FCT Grant No. UID/PSI/04810/2013.

Conflict of Interests The authors declare no conflicts of interest.

References

Ahmed, K., & Long, W. (2010). *Adaptation as a response to climate change: A literature review*. Munich Personal RePEc Archive. Retrieved from https://mpra.ub.uni-muenchen.de/45024/.

American Psychological Association (APA). (2011). *Resolution on affirming psychologists' role in addressing global climate change. American Psychological Association*. Retrieved from http://www.apa.org/about/policy/climate-change.aspx.

Arbuthnott, K., Hajat, S., Heaviside, C., & Vardoulakis, S. (2016). Changes in population susceptibility to heat and cold over time: Assessing adaptation to climate change. *Environmental Health, 15*(1), 33. https://doi.org/10.1186/s12940-016-0102-7.

Blascovich, J. (2008). Challenge, threat, and health. In J. Y. Shah & W. L. Gardener (Eds.), *Handbook of Motivation Science* (pp. 481–493). New York, USA: Guilford Press. ISBN:9781593855680.

Blascovich, J., & Mendes, W. B. (2001). Challenge and threat appraisals: the role of affective cues. In J. Forgas (Ed.), *Feeling and thinking: the role of affect in social cognition* (pp. 59–82). New York, USA: Cambridge University Press. ISBN: 9780521011891.

Blascovich, J. & Mendes, W. B. (2010). Social psychophysiology and embodiment. In S. T. Fiske, D. T. Gilbert, & G. Lindzey (Eds.), *The handbook of social psychology* (*5th Edn*., pp. 194–227). New York, USA: Wiley. ISBN: 9780470137482.

Blascovich, J., & Tomaka, J. (1996). The biopsychosocial model of arousal regulation. In M. P. Zanna (Ed.), *Advances in experimental social psychology* (vol. 28, pp. 1–51). San Diego, USA: Academic Press. ISBN: 9780120152285.

Bradley, G. L., & Reser, J. P. (2017). Adaptation processes in the context of climate change: a social and environmental psychology perspective. *Journal of Bioeconomics, 19*(1), 29–51. https://doi.org/10.1007/s10818-016-9231-x.

Bruine de Bruin, W., Lefevre, C. E., Taylor, A. L., Dessai, S., Fischhoff, B., & Kovats, S. (2016). Promoting protection against a threat that evokes positive affect: the case of heatwaves in the UK. *Journal of Experimental Psychology: Applied, 22*(3), 261–271. https://doi.org/10.1037/xap0000083.

Casimiro, E., Calheiros, J. M., & Dessai, S. (2003). Human health. In F. D. Santos, K. Forbes, & Moita, R. (Eds.), *Climate change in Portugal. Scenarios, impacts and adaptation measures—SIAM Project*. Lisboa, Portugal: Gradiva. ISBN: 972–662-843-1.

Chen, R. S., Boulding, E., & Schneider, S. H. (1983). *Social science research and climate change: an interdisciplinary appraisal*. Dordrecht, The Netherlands: Reidel. ISBN 978-94-009-7003-8.

Clayton, S., Devine-Wright, P., Stern, P. C., Whitmarsh, L., Carrico, A., Steg, L., et al. (2015). Psychological research and global climate change. *Nature Climate Change, 5*(7), 640–646. https://doi.org/10.1038/nclimate2622.

Clayton, S., Manning, C. M., & Hodge, C. (2014). *Beyond storms & droughts: the psychological impacts of climate change*. Washington, DC, USA: American Psychological Association and ecoAmerica.

Connell, S., Fien, J., Lee, J., Sykes, H., & Yencken, D. (1999). If it doesn't directly affect you, you don't think about it': a qualitative study of young people's environmental attitudes in two Australian cities. *Environmental Education Research, 5*(1), 95–113. https://doi.org/10.1080/1350462990050106.

Demski, C., Capstick, S., Pidgeon, N., Sposato, R. G., & Spence, A. (2017). Experience of extreme weather affects climate change mitigation and adaptation responses. *Climatic Change, 140*(2), 149–164. https://doi.org/10.1007/s10584-016-1837-4.

Direção-Geral da Saúde (DGS). (2015). *Plano de contingência para temperaturas extremas adversas - módulo calor*. Direção de Serviços de Prevenção da Doença e Promoção da Saúde, Divisão de Saúde Ambiental e Ocupacional. Lisboa, Portugal: Direção-Geral da Saúde.

Doherty, T. J., & Clayton, S. D. (2011). The psychological impacts of global climate change. *American Psychologist, 66*(4), 265–276. https://doi.org/10.1037/a0023141.

Fonseca, R., Blascovich, J., & Garcia-Marques, T. (2014). Challenge and threat motivation: effects on superficial and elaborative information processing. *Frontiers in Psychology, 5,* 1170. https://doi.org/10.3389/fpsyg.2014.01170.

Fritze, J. G., Blashki, G. A., Burke, S., & Wiseman, J. (2008). Hope, despair, and transformation: Climate change and the promotion of mental health and wellbeing. *International journal of mental health systems, 2,* 13. https://doi.org/10.1186/1752-4458-2-13.

Garcia-Marques, T., Fonseca, R., & Blascovich, J. (2015). Familiarity, challenge, and processing of persuasion messages. *Social Cognition, 33*(6), 585–604. https://doi.org/10.1521/soco.2015.33.6.585.

Gaspar, R., Barnett, J., & Seibt, B. (2015). Crisis as seen by the individual: the norm deviation approach. *Psyecology: Bilingual. Journal of Environmental Psychology, 6,* 103–135. https://doi.org/10.1080/21711976.2014.1002205.

Gaspar, R., Domingos, S., Diniz, A., & Falanga, R. (2016). Barriers to and facilitators of older adult's adherence to health recommendations: towards an engAging two-way health communication. In G. Graffigna (Ed.), *Promoting Patient Engagement and Participation for Effective Healthcare Reform* (pp. 58–82). Hershey, USA: IGI Global. https://doi.org/10.4018/978-1-4666-9992-2.ch004.

Gifford, R. (2008). Psychology's essential role in alleviating the impacts of climate change. *Canadian Psychology, 49*(4), 273–280. https://doi.org/10.1037/a0013234.

Gifford, R. (2011). The dragons of inaction: Psychological barriers that limit climate change mitigation and adaptation. *American Psychologist, 66*(4), 290–302. https://doi.org/10.1037/a0023566.

Gilman, N., Randall, D., & Schwartz, P. (2007). *Impacts of climate change: A system vulnerability approach to consider the potential impacts to 2050 of a mid-upper greenhouse gas emissions scenario.* Retrieved from goo.gl/oSC7dP.

Glick, P., Staudt, A., & Stein, B. (2009). *A new era for conservation: Review of climate change adaptation literature.* National Wildlife Federation. Retrieved from https://www.nwf.org/pdf/Reports/NWFClimateChangeAdaptationLiteratureReview.pdf.

Grothmann, T., & Patt, A. (2005). Adaptive capacity and human cognition. *Global Environmental Change, 15*(3), 199–213. https://doi.org/10.1016/j.gloenvcha.2005.01.002.

Intergovernmental Panel on Climate Change (IPCC). (2014). Climate change 2014: Synthesis report. In Core Writing Team, R. K. Pachauri & L. A. Meyer (Eds.), *Contribution of working groups i, ii and iii to the fifth assessment report of the intergovernmental panel on climate change.* Geneva, Switzerland: IPCC. ISBN 9789291691432.

Lazarus, R. S. (1984). On the primacy of cognition. *American Psychologist, 39*(2), 124–129. https://doi.org/10.1037/0003-066x.39.2.124.

Lefevre, C., Bruine de Bruin, W., Taylor, A., Dessai, S., Kovats, S., & Fischhoff, B. (2015). Heat protection behaviors and positive affect about heat during the 2013 heat wave in the United Kingdom. *Social Science and Medicine, 128,* 282–289. https://doi.org/10.1016/j.socscimed.2015.01.029.

Ojala, M. (2012). How do children cope with global climate change? Coping strategies, engagement, and well-being. *Journal of Environmental Psychology, 32*(3), 225–233. https://doi.org/10.1016/j.jenvp.2012.02.004.

Özdem, Y., Dal, B., Öztürk, N., Sönmez, D., & Alper, U. (2014). What is that thing called climate change? An investigation into the understanding of climate change by seventh-grade students. *International Research in Geographical and Environmental Education, 23*(4), 294–313. https://doi.org/10.1080/10382046.2014.946323.

Prüss-Ustün, A., Wolf, J., Corvalán, C., Bos, R., & Neira, M. (2016). *Preventing disease through healthy environments: A global assessment of the burden of disease from environmental risks.* Geneva, Switzerland: World Health Organization. ISBN 9789241565196.

Reser, J. P., Bradley, G. L., & Ellul, M. E. (2012). Coping with climate change: bringing psychological adaptation in from the cold. In B. Molinelli & V. Grimaldo (Eds.), *Handbook of the psychology of coping: Psychology of emotions, motivations, and actions* (pp. 1–34). New York, USA: Nova Science.

Santos, F. D. (2006). Problemática das alterações climáticas no início do século XXI. In F. D. Santos & P. Miranda (Eds.), *Alterações climáticas em Portugal: cenários, impactos e medidas de adaptação - Projeto SIAM II*. Lisboa, Portugal: Gradiva.

Schwarzer, R. (1992). Self-efficacy in the adoption and maintenance of health behaviors: Theoretical approaches and a new model. In R. Schwarzer (Ed.), *Self-efficacy: Thought control of action* (pp. 217–243). Washington, DC, USA: Hemisphere.

Slovic, P., Finucane, M. L., Peters, E., & MacGregor, D. G. (2004). Risk as analysis and risk as feelings: Some thoughts about affect, reason, risk, and rationality. *Risk Analysis, 24*(2), 311–322. https://doi.org/10.1111/j.0272-4332.2004.00433.x.

Spence, A., Pidgeon, N., & Uzzell, D. (2009). Climate change: Psychology's contribution. *The Psychologist, 21*, 108–111. ISSN: 09528229.

Spence, A., Poortinga, W., & Pidgeon, N. (2012). The psychological distance of climate change. *Risk Analysis, 32*(6), 957–972. ISSN 1539-6924.

Swim, J. K., Stern, P. C., Doherty, T. J., Clayton, S., Reser, J. P., Weber, E. U., et al. (2011). Psychology's contributions to understanding and addressing global climate change. *American Psychologist, 66*(4), 241–250. https://doi.org/10.1037/a0023220.

Tomaka, J., Blascovich, J., Kelsey, R. M., & Leitten, C. L. (1993). Subjective, physiological, and behavioral effects of threat and challenge appraisal. *Journal of Personality and Social Psychology, 65*(2), 248–260. https://doi.org/10.1037/0022-3514.65.2.248.

Tomaka, J., Blascovich, J., Kibler, J., & Ernst, J. M. (1997). Cognitive and physiological antecedents of threat and challenge appraisal. *Journal of Personality and Social Psychology, 73* (1), 63–72. https://doi.org/10.1037/0022-3514.73.1.63.

Tversky, A., & Kahneman, D. (1973). Availability: A heuristic for judging frequency and probability. *Cognitive Psychology, 5*(2), 207–232. https://doi.org/10.1016/0010-0285(73)90033-9.

Zajonc, R. B. (1984). On the primacy of affect. *American Psychologist, 39*(2), 117–123. https://doi.org/10.1037/0003-066x.39.2.117.

Diffusion of Climate Change Adaptation Policies Among Local Governments in Sub-Saharan Africa: Conceptual Review

Issah Justice Musah-Surugu, Albert Ahenkan, Justice Nyigmah Bawole and Samuel Antwi Darkwah

Abstract Many studies have confirmed the diffusion of climate change adaptation mainstreaming related policies and innovations across local governments (LGs). However, it is not clear under what conditions these policies get transferred, implemented or subsequently abandoned, especially in developing countries where LGs' capacity to mainstream adaptation barely exist. Based on retrospective literature analysis, this article adduces theoretical and empirical evidence to analyze the factors likely to influence the mainstreaming of adaptation into local governance in Sub-Saharan Africa (SSA). The outcome of the article is expected to help policy practitioners and academics to comprehend the preconditions that enhance or frustrate the adoption of adaptation policy innovations across LGs in SSA where adaptation remains critical developmental concerns. The preponderance of literature suggests that multiple factors, including those, outside the control of LGs affect their capacity and urgency to adopt adaptation policies. The chapter concludes that it is worthwhile to adopt multiple diffusion perfectives that reflect the complex web of factors conspicuously affecting the transfer of climate policy innovations at the micro level to explain the adoption of adaptation policies. The chapter argues that, by doing so, one is able to portray the nature of climate change policy diffusion in local governance in a more sophisticated, comprehensive and logical manner. The chapter therefore makes useful contributions to the emerging field of climate policy and provides useful lessons for practitioners on the factors that influence the adoption of adaptation policies.

Keywords Diffusion · Climate change · Policy · Mainstreaming Local government

---

I. J. Musah-Surugu (✉) · A. Ahenkan · J. N. Bawole
Department of Public Administration and Health Services Management, University of Ghana Business School, Legon, Accra P. O. Box LG 78 Ghana
e-mail: musah123@gmail.com; ijmusah_surugu@st.ug.edu.gh

S. A. Darkwah
Faculty of Regional Development and International Studies, Mendel University in Brno, Brno, Czech Republic

F. Alves et al. (eds.), *Theory and Practice of Climate Adaptation*, Climate Change Management, https://doi.org/10.1007/978-3-319-72874-2_4

Introduction

There is obviously less denial that the ravaging havocs of climate change are worsening developmental deficits that exist at sub-national level in Sub-Saharan Africa. As a result the idea of climate change adaptation policy mainstreaming (CCAPM), which aims to integrate adaptation needs into policy-making, budgeting and program implementation processes are gradually being diffused across local governments (LGs), where climate impacts are acute (Porter et al. 2015). Consequently, CCAPM is becoming a prominent concept in the sub-field of climate policy studies. Despite its' emerging prominence its' adoption across LGs has received insufficient attention in the emerging field of climate and policy studies (Adelle and Russel 2013). As a result evidence of why LGs adopt CCAPM practices remains largely unexplored. Meanwhile, numerous empirical evidences have confirmed that the adoption or spread of CCAPM strategies has not been isomorphic (Ayllet 2015; Rootes et al. 2012; Betsill and Bulkeley 2004) suggesting that, whiles some LGs are early adopters others are laggards. For example a study conducted by Porter et al. (2015) found that 82 LGs out of 116 sampled across England, Scotland and Wales made no demonstrable progress in integrating climate change into local governance despite significant improvement in LGs understanding of climate change. Similar previous studies on CCAPM have shown uneven levels of effort across LGs, suggesting significant variations in the pace of adoption (Ayllet 2015, Rootes et al. 2012; Betsill and Bulkeley 2004).

The limited research on the issue gives rise to the ancient policy diffusion question, "why do similar LGs act differently" in mainstreaming climate change into local governance. This question remains vital because it remains unclear why LGs' commitment to CCAPM varies granted that: (1) (inter)national climate policies places responsibility on all LG to act; (2) LGs are first of all levels of government which offers opportunity to set exemplary climate change leadership; (3) LGs are more closer to those vulnerable to climate change; (4) LGs appreciate the hazards of climate change better; (5) LGs possess vital local knowledge desired for moderating climate risk, etc. (Ayllet 2015; Rootes et al. 2012; Agrawal 2008; Bulkeley and Betsill 2013). Bridging the current research gap, this book chapter attempts to make a theoretical contribution by synthesizing the emerging climate policy literature to provide a possible causal explanatory framework on the diffusion of CCAPM practices in sub-Saharan Africa (SSA). This research is relevant because researchers cannot assume that the adoption of CCAPM practices essentially mimic the mechanisms through which other policies which has received much research attention get adopted by LGs. This concern was a vital element in the seminal works of Berry and Berry (1990) arguing that the nuances of policy adoption differ across issues, time and space. The examination of CCAPM among LGs in SSA will have three concerns. Firstly, the article would interrogate how contextual and agency factors affects LGs' willingness and ability to adopt CCAPM practices. Secondly, the article would examine how structural factors outside the domain of LG negotiate LGs' adoption of CCAPM innovations. The last concern reflects how biophysical and science-heaviness of climate change also influence LGs' responses to climate change. Diffusion and adoption is used interchangeable in this article. Following from Berry and Berry (1990) and Walker (1969) CCAPM practices adoption in this article refers to the broad commitment of LGs to integrate climate change adaption actions into local governance. The chapter is organized into four parts: introduction, literature review, methodology, discussion and conclusion.

Literature Review: The Search for Explanatory Framework

Contextual and Agency Issues (Bottom-up) Affecting Micro-Level Policy Adoption?

Public policy literature has fingered numerous bottom-up factors affecting potential adopters' capacity and motivation to embrace new policies or innovations (Stadelmann and Castro 2014). Broadly, these factors include taxonomy of political-economy issues internal to potential adopters environment. Nevertheless, previous studies that have attempted to examine how internal factors influence LGs' ability to adopt innovation were caught-up in the "everything matter trap" that has become notorious in policy research. Crudely put, they end up listing numerous repetitive phrases as internal determinant of policy adoption at sub-national level (Ryan 2015). For example, in a related study Aylett (2015) cited 24 internal factors influencing LGs' climate adoption behaviour. Indeed, many studies have cited the following: financial, legal, and political costs, local leadership, policy home, public support, policy compatibility, political incentives, clarity of regulation, perception, manpower, supportive collaborators, etc. as contextual and agency factors influencing LGs responses to new policies (Rootes et al. 2012; Betsill and Bulkeley 2004). Despites the lack of certainty over the significance of these numerous but often cited internal determinant of policy adoption by micro-level agencies research conducted by Rootes et al. (2012), Tanner and Allouche (2011), Ryan (2015) and Betsill and Bulkeley (2004) provides five all-embracing key pointers useful for undertaking a comprehensive analyses of how contextual and agency related factors influences climate change policy adoption at the **micro level** (see, Table 1). It appears that these factors as illustrated in Table 1 configures almost all the often cited bottom-up determinants of policy adoption across the literature.

Structural Factors (Top-Down) Affecting Micro-Level Policies Adoption?

Despite inconclusive arguments past researches have indicated that politico-economic factors emanating from international systems at times influence

Table 1 Contextual and agency factors affecting micro-level policy adoption

Local-political economy	Implication for climate change policy adoption	Means of verification (e.g.)
Domestic framework for framing and defining climate problem	Framing climate change issues as indispensible to local development priority enhances the possibilities of building stronger and lasting local political coalitions to support climate initiatives	– Extend of awareness – Who is saying it – How is it said – When is it said – Medium of saying it – Clarity of message
Domestic climate policy framework	Adoption of climate policy in the face of uncertainty relies upon the feature of climate statutes and levels of institutional engagement- how it can respond rapidly to a range of different scenarios	– Political feasibility – Divisibility – Communicability – Simplicity – Compatibility
Local capacity	The ability of municipal governments to make resilient decisions and implement across a range of responsibilities and services depends on existing endowment and resources	– Autonomy – Fiscal strength – Expertise – Access to climate information
Political institutions and incentives	Governance arrangements and actors climate perception can create incentives or impediments for political actors to promote certain ideologies	– Voter preferences – Symbolic benefit – Instrumental benefits
Interest groups and preferences	Policy transfer build on existing technical and implementation support services provided by NGO/ civil society sector, technical and academic institutions	– Supportive collaborators – Collaborators preference

Authors, compilation 2017

the adoption of specific policies at the local level (Massey et al. 2014; Powell and DiMaggio 2012; Elazar 1972). In fact, the extent to which this suggestion holds true to climate change policy adoption has barely been systematically tested or analyzed. However, there are arguments that the understanding and ability to manage climate risk at local level is also shaped largely by events and decisions that occur at the national and international level, given the global breadth of climate change (IPCC 2014; Olwig 2012; Wang 2012; Betsill and Bulkeley 2004). The common view is that formal decisions such as UNFCCC resolutions could create a discursive links between local climate change politics and global events (Olwig 2012). For example some scholars have opined that policy actors in Africa have not been reacting to climate risk per se but quite often to the principles that emerged from international negotiations (Weisser et al. 2014). Global events could also create a dominant logic requiring all levels of government to kowtow. As Powell and DiMaggio 2012 argue actors quest for legitimacy to either participate or access some incentives within a network of actors creates subtle coercing forcing participants to adhere to governing logic. As a result it is possible that local actors may be overtly or covertly forced or motivated to adopt a particular climate statutes in order to access some incentives or avert possible sanctions. On the wholes, previous researches including that of Dodman and Mitlin (2013), Powell and DiMaggio 2012, and Massey et al. (2014) has illustrated three major structural factors that provide some basis for analyzing how factors outside the domain of potential policy adopter influence their behaviour: (1) the structure for financing climate change; (2) inter-actors relationship that occurs at global climate policy events; (3) quest for legitimacy. Table 2 provides summary of these factors.

How Does Biophysical and Climate Science-Heaviness Influence Climate Policy Adoption?

Despite progress in climate modeling future expected impact and reasonable adaptation options remains science-heavy, something shrouded in irreducible uncertainties (IPCC 2014; Wang 2012; Dessai and Hulme 2004). Scholars argue that inherent climate uncertainty present multi-complex regime incapable of escaping from flecked opinion and unequal response capacity among similar policy

Table 2 Structural Factors (top-down) affecting micro-level policies adoption

International determinants	Implications for policy diffusion	Means of verification (e.g.)
Inter-actors relationship emanating from CC Conferences and networks	Multi-level communication and institutional articulation creates opportunities for learning, emulation and eventual adoption UNFCCC negotiations creates institutional and discursive links making the isolation of local responses to climate change from global discussions impossible	City-Networks Sister City relationships Climate change networks Agenda 21 signatory Participation in COP Signatory to COP decisions
Quest for legitimacy	When a new policy attains normative status, governments are induced to adopt it in the quest for legitimacy and so as not to appear 'deviant'	Sets of agreed values, principles and standards
CC financing architecture	The architectural make up of a climate change financing is gradually is creating isomorphism among institutions	– Sources of financing – % of external funds – External funding regulations

Authors, compilation 2017

Table 3 Biophysical and climate science-heaviness influence climate policy adoption

Climate change self-induced factors	Implications for policy diffusion	Means of verification (e.g.)
Biophysical sensitivity to climate risk	The severity of climate change risk is a function of geographical resources endowment	– Vulnerability – Exposure – Resilience
Local social co-benefits	Rationalist argue that the economic benefits of adopting a climate change policy (as well as the economic costs of not responding) is quintessential for adoption of related policies	– Welfare benefits – Governance – Socio-economic etc.
Science-heaviness of climate change	Debate, review and analysis of climate impact and desirable responses lacks consensus on conclusions which has perhaps created variegated political lines on institutional responses	– Communicability – Simplicity – Precision – Actionability

Authors, compilation 2017

actors. Specifically, uncertainties of climate change revolves around the precise future and current risk, cost-benefit, optimal solution etc. As a result crafting adaptation strategies remain a thorny area for both scientist and policy practitioners (Massey et al. 2014; Smith and Hitz 2002; Dessai and Hulme 2004). For example current understanding of natural and socio-economic processes that influence climate variability, resilience, exposure and adaptive capacities across multiple ecosystems is inadequate (IPCC 2014). Also, differences in biophysical endowment make some territories and sectors more expose to climate risk. Geographical sensitivity therefore creates spatio-temporal variability of institutional responses to climate change policies (Wang 2012). Clearly, the geographical diversity in resource endowments and climate science-heaviness create a new dimensional mechanization likely to influence its' policy adoption which are less treated within the traditional policy diffusion literature (Wang 2012; Wilson et al. 1999). The Table 3 below provides a summary presentation of these factors is likely to affect the diffusion of climate policies across governments.

Policy Diffusion Theories

Throughout this chapter of the book, there is a consistent argument that both context-agency and structural factors work out to influence the adoption of CCAPM practices within the boundaries of biophysical and climate science-heaviness. Introducing policy diffusion theories offers a stronger theoretical starting point to settle on how these factors work out to influence climate policy diffusion across micro-space. Policy diffusion research can be traced to the early 20th century when the French sociologist Gabriel Tarde asked why some new policy gets imitated and others ignored. Currently, there are two constellations of ideas explaining why policies get adopted: internal and external explanations. As argued sufficiently by scholars the internal determinants scholars justify policy diffusion as a function of institutions' internal characteristics (Berry and Berry 1992; Walker 1969; Mohr 1969; Gray 1973). They explained adoption as function of potential adopters' endowments. On the other hand those who view diffusion as a function of external forces indicate that factors outside potential adopter's immediate environment could also influence their behavior. For example isomorphic, regional diffusion and the leader-laggard models argue that neighboring agencies that have already adopted a particular policy may serve experimentation laboratory for potential adopters (Elazar 1972). Again, both the national interaction and vertical influence models are of the opnion that new policies are studied and communicated through peer relationships. Clearly, the existing theories on policy diffusion have illustrated the main mechanism influencing the transfer of new ideas at sub-national level. This chapter carefully adapts these views along those of the emerging climate policy literature to discuss a conceptual framework which explicate how adaptation policies get adopted at the local level.

Methodology

The paper adopts a retrospective literature analysis of theoretical and empirical studies mainly drawn from journal articles and scholarly books that are relevant to climate change and climate policy analysis. Following Yeboah-Assiamah et al. (2017) structured phrases and questions that are relevant to the issue under examination were keyed into search engines to help identify a large pool of literature. Significant search engines used are Wiley Online Library, Emerald Full Text, Palgrave Macmillan Journals, JSTOR, Sage journals and Sciencedirect. Key phrases such as "climate change and local level action", "local government responses to climate change", "climate change policy diffusion", "adoption of policy among local government", "barriers and motivation for adaptation", and theories of policy diffusion were used. Over the 175 theoretical and empirical articles that were retrieved they were screened through stratification according to: contextual and agency, structural and climate change science-heaviness factors that affect LGs policy adoption behaviour. All authors met to eliminate duplicates and made a shortlist of abstracts for detailed and systematic review and inclusion in the paper. The study finally agreed on some 65 relevant articles considered pertinent to our research questions. The evidence that emerged from the sampled literature was analyzed using content analysis and peered appraisals among the authors. The authors make no claim of exhaustive inclusion of all relevant literature. The argument put forward in this article, however, reflect largely how policies get diffused across local governments. In the final discussion we classified the content into themes that emerged in the content analysis and peer appraisals.

Discussion: What Determines the Adoption of CCAPM Practices Among LGs?

As clearly demonstrated by policy diffusion theories in previous section, policy diffusion discourses focus on the vertical and horizontal interplay of local and non-local political economy system. Climate change policy adoption analysis therefore cannot escape from drawing multiple evidence from both remote and immediate institutional factors (e.g. values, incentives, structures) that affect potential adopter's capacity and willingness to act. Consequently, any study aimed at assessing how climate change policy innovations get adopted across LGs would have to: (1) examine the context and agency related concerns that affect the capacity of LGs to take onboard resilience related strategies; (2) the structural factors which are outside the control of LGs but affect their capacity to respond and the urgency weight they allocate to adaptation concerns; (3) the biophysical and science-heaviness of climate change which influence, among other thing, LGs' risk exposure, local contestation over adaptation complexities and expected social co-benefit (see, Table 4).

Table 4 Determines of CCAPM adoption among local governments

Determinants of adoption	Explanations
Context and agency	Context and agency determinants postulate that the incidental internal characteristics of potential adopters are vital policy adoption explanatory variables. It simply argues that adaptation policy adoption depend largely on the unique individual features of LG such as: fiscal, manpower, jurisdictional, leadership capabilities; sectorial structures; institutional memory; public support; political incentives, clarity of regulation; supportive collaborators; etc.
Structural	Structure determinant on the other hand, focus on the macro and meso institutional contexts in which potential adopters and the new policy are embedded. Financial incentives, degree of subsidiarity, appetite for legitimacy, etc. which are considered outside the domain of potential adopters may coerce or incentivize them to adopt
Biophysical and science-heaviness	On the other hand the geographical resources endowment of potential adopters and climate uncertainties may influence its sensitivity to climate havoc and debates over responses respectively

Adapted from Oteman et al. (2014) and Ryan (2015)

Determinants of LGs' Adoption of Policies: Analytical Framework

From the review above, the paper provide a graphical view of how the three determinants of policy diffusion underpin and influence the embracement of adaptation policy by micro-institutions with empirical evidence largely from the literature. Largely drawing from the works of the traditional diffusion literature (Walker 1969); Gray (1973); Mohr (1969); Berry and Berry (1992); Oteman et al. (2014) and the growing climate policy literature (Ryan (2015), the framework below explicate how adaptation policies get adopted at the local level. From the framework (Fig. 1), LGs' adoption of CCAPM practices is a function of contextual-agency factors, structural and biophysical and climate science-heaviness. The main object here is to adduce evidence from the literature from different context within SSA to drum home how these three determinant works out to influence the adoption of CCAPM practice among LGs. As clearly indicated on Fig. 1 both contextual-agency and structural factors influence the capacity of local institutional to adopt climate innovations. At the heart of climate change discourse is the urgency (speed/pace) of response. The Fig. 1 further confirms that structural and biophysical and climate science heaviness influence local institutions pace of response to climate innovations. Its therefore shows that incentives and sanctions emanating from super-level government as well as geographical sensitivity to climate risk determines the speed with which local agencies adopt climate related policies.

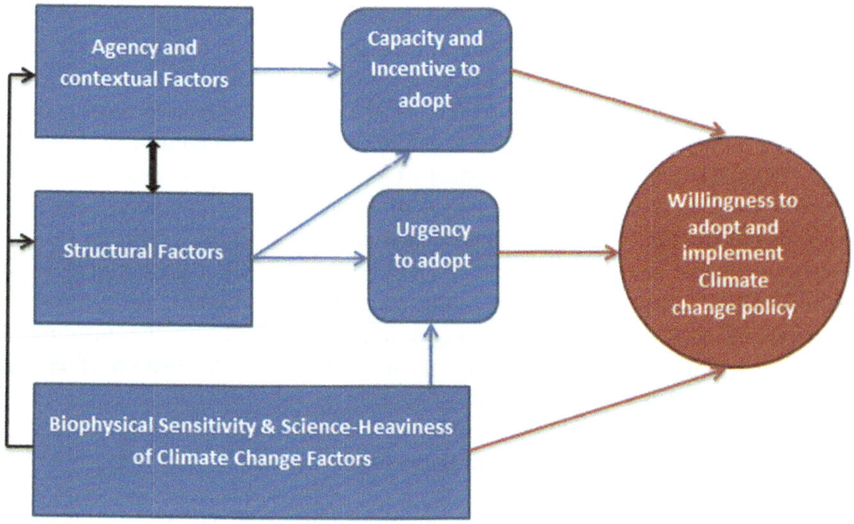

Fig. 1 Key factors likely to influence LGs' adoption of Climate Policy. *Source* Author's conceptualization 2017

Contextual and Agency Factors Inflowing LGs Ability to Adopt Climate Policy

Climate Change Issues Should Be Framed as Development Priority

Many studies have concluded that LGs' adoption of adaptation focus policies depends largely on the extent to which it fit with the LGs' broader sectorial goals (Betsill and Bulkeley 2004; de Oliveira 2009). The proposition is that for climate change policy to be embraced at the local level considerable effort must be made at carving climate mainstreaming as a surest way to achieving local developmental needs. There are empirical studies to suggest that LGs in SSA are saddled with numerous but equally competitive local needs, against limited fiscal capacity (see, Bawole 2017). As a result, if climate change is not well framed in a way that would be seen as indispensable developmental concern it might be swept under the carpet of local policy agenda setting (see, Carter and Jacobs 2014; Down 1972). This claim is confirmed by a study conducted by Ryan (2015) arguing that poor framing of climate risk weight of urgency local actors alots to it against other local needs. Another study conducted by Bulkeley and Betsil (2004) found that despite adaptation been a development challenge in most developing countries it still remains marginal issues due to its poor framing.

Adoption Is a Function of Existing Institutional Capacity

The review further confirms that in some cases adaptation is well framed nonetheless it barely gets adopted. The major factors that apparently influence adoption that came up strongly in the review is institutional resource endowment of potential adopters. According to Wang (2012) the capacity to adopt a policy refers to the ability to plan, stimulate and organize resources vital for accommodating new statutes or strategies. A careful review of the literature point to three critical resources influencing LGs' capacity to adopt innovations: statutory; institutional; organizational resource-husbandry capacity (Ayllet 2015; Pasquini et al. 2013). The statutory (legal) capacity refers to degree of autonomy LGs have over adaptation domains (Ryan 2015). Evidence shows that most LGs in SSA operate on subsidiarity principle and therefore remains subservient to central government weakening their authority over areas affecting adaptation governance. As a result the incentive to adopt climate policy might not exist. Similarly, the institutional capacity of LGs (size, form, and internal arrangement) is a function of climate policy adoption (Ayllet 2015). A study conducted by Adu-Boateng (2015) in Ghana shows that the size (Metropolitan, Municipal, District), material resources capacity (old or newly created), extent of departmentalization influence their capacity to take climate action.

Furthermore, the incremental cost of integrating climate change into local development plans, the skills set requires to act, access to climatic information, etc. constitute organizational resource-set that influence LGs' adoption decision.

Existing research on climate change policy adoption have consistently put forward sub-national institution that has weak fiscal capacity and poor manpower are likely to delay action on climate change (Adu-Boateng 2015). In Sub-Saharan Africa financial resources is in fact important determinant of policy adoption at the local level because documented evidence further reveals that LGs in the region are grappling with dwindling resources to meet mounting traditional responsibilities (Jibao and Prichard 2016). Empirical studies conducted by Roberts (2010) in the region suggested that limited funds and manpower affected the implementation of adaptation strategy lunched by Ethiekwini municipality in South Africa in 2006. Flowing from similar perspective Ryan (2015) argues local agencies adoption of adaptation strategies in developing countries is a function of it organizational resource-set. Similar study conducted by Amankwah-Amoah and Sarpong (2016) also confirmed that limited organizational resources-set affected the evolution and scaling-up of resilience action (adoption of solar PV) in Ghana between the years 1980 and 2010. The evidence confirms the widely reported failure of decentralization to offer the needed means for undertaken LGs fidicial responsibilities in Ghana, Mali, Gambia, Burkina Faso, Cameroon, Guinea, Malawi, and Niger (Bawole 2017; Ribot 2003).

There Should Be Attractive Political Market for Climate Change Policy Adoption

Existing literature provides list of political market building block and incentives that influence actors and institutions ability and willingness to adopt new policies. In typical LGs setting these building blocks may include the extent of executive and fiscal autonomy, policy entrepreneurs, number of veto players, checks and balances, citizens' expectations, bureaucracy, etc. The literature clearly shows the nature of these building block can frustrate or encourage the adoption of innovations (North 1990). The new institutional literature have argued extensively that the number of veto player on policy adoption, decision clearance points, incentives, and the characteristics of a policy in question affects the extent to which it gets diffused across scale (Ostrom 2007). For example the number of veto players within LG who must decide whether to adopt climate policy or not will determine the extent of contestation over local climate change governance.

The paper therefore argue that in LGs where there are more veto players involved in agreeing on the adoption of resilience policy, the less likely it would happen. Despite delays that the numerous veto players may unleash on possible adoption, it is possible for new policies to be adopted if there are clear incentives for the actors. This hypothesis has been texted in several environmental policy adoption studies across developing countries, for example by Wood (2015) and Amankwah-Amoah and Sarpong (2016). They established that veto players within LG are rational actors who weigh the incentives related to climate change policy adoption to determine their reaction. This suggest that if the veto players (actors) within LGs see potential advantage in mainstreaming climate change in local development plans they are more likely to adopt it. Invariably, the relative instrumental and symbolic benefit of climate policy is *sine qua non* for adoption. The incentive can be in the form of possible electoral gains, access to extra external resilience based funding, institutional performance improvement, etc. (Obradovich and Zimmerman 2016). Thus, if climate change policy adoption provides opportunities for accessing external funding opportunities LGs are more likely to integrate climate concerns into local governance (Samuels 2000). Similarly, if the adoption of climate policy is a performance appraisal metrics then LGs are covertly woe to adopt since it would help their score on LGs performance rating league.

Another important argument advance in this paper is that the adoption of climate change policy depends on the perceived implication on voters' behaviour and political power (Elster 1986). As clearly argued by rationalist in LGs where climate change is seen by voters as a development challenges or "need base" issue political office holders have no reason whatsoever not to integrate climate concerns into local development plans (Obradovich and Zimmerman 2016). A study by Obradovich and Zimmerman (2016) shows that due to non-rewarding nature of climate change action Malawian politician is not likely to giver urgent attention to climate change. There are also empirical discourses that provide evidence that the nature of institutional bureaucracy could affect policy adoption behaviour. In many LGs where various units and departments are highly sectorised it's often difficult to plan and adopt climate change policies that have a boundary spanning character. LGs that have flexible bureaucratic system and allow consultation and collaboration among its separate department have the right political market (bureaucracy) favorable for adoption of climate change policies.

Adoptable Statutes Must Be Unambiguous and in Sync with Sector Wide Policies

At the local level, there are array of agencies that are supposed to be coordinated and consulted over the adoption of climate change policies. The success of such coordination depends on the characteristics of the policy in question and the potential adopter: perceived advantage, communicability, divisibility, simplicity and compatibility. It's been argued that climate change policies could be diffused at the local level if existing institutions, sectors, and the climate change policy itself are flexible to enhance the integration without huge alteration in institutionalized norms. For example many previous researches have strongly put forward that the compatibility of new policies with potential adopter's institutionalized structures, norms, values and goals enhances the adoption processes (Aubert and Hamel 2001; Fennell and Warnecke 1988). Alternatively, policy innovations that are seen as complex and extremely difficult to align with existing institutional logic are barely adopted (Denis et al. 2002; Grilli and Lomas 1994; Marshall 1990). Perceived complexity of new policy can therefore delay it adoption in SSA countries where LGs are highly sectorised. Reports from Ghanaian LGs show that the existence of numerous autonomous institutions operating within similar local government whose functional areas affect adaptation, perhaps, slow the pace for embracing innovations meant for resolving complex-crises that requires the coming together of all the numerous institutions in some way (Adu-Boateng 2015). In such LGs climate change policies are supposed be calved in a manner that is compatible with overarching local strategic plans of the numerous autonomous entities in an attempt to resolve bureaucratic competition and institutional conflict which manifest within local administrative system (Adu-Boateng 2015).

Adaptation Policies Must Be Attractive to Interest Groups

Local governments that host large number of enthusiastic environmental interest groups are more likely to adopt environment related innovation. The argument is that environmental interest groups have the vibrancy to exert necessary pressure on LG to integrate or adopt environmental policy innovations. As the literature clearly suggests, environmentally conscious actors within LG are able to speed up policy adoption at all levels of government in developing countries given their social legitimacy, resource powers, agenda setting skills, mass mobilization and campaigns proficiency (Zahran et al. 2008; Vachon and Menz 2006). Along the same line of analysis de Oliveira (2009) put forward that civil society groups operating within LG can serve as a conduit for policy diffusion due to their specialized skills in organizing civil protest, seeking legal remedies, providing vital information and lobbying LG on environmentally important issues. Empirical evidence from SSA underscores the relative strength of civil society organisations in policy transfer across local territories.

For example, Nelson and Agrawal (2008) explained that NGOs in Tanzania were most visible in the diffusion of land reforms innovations across municipalities through their series of campaigns. These assertions have also been confirmed in a study conducted by Shar et al. (2011). Again, in Ghana the Association of Communities Affected by Mining (WACAM) which is a community based NGOs have successfully lead many micro-level mining laws reforms and compensation systems that has significantly improves local people's rights significantly (www.modernghana 2016). This paper therefore concludes that LGs adoption of climate policy directly correlates with the existence and preferences of other local development actors such as NGOs.

Structural Factors Affecting LGs' Adoption of Climate Policy (Top-Down in Nature)

The Nature of Global Climate Change Financing Architecture

The top-down architecture of financing climate change globally makes donors countries, and international financial institutions with financial leverage, to a large extent, influential in dictating what happens at both national and local level

(Smith et al. 2014). This assertion resonates well with the dominant hypothesis of the international diffusion literature that suggest that some external policy actors use their resources powers (control of the purse) to coerce (either subtle or overt) smaller government through conditionalities (Shipan and Volden 2008). What this means is that if a LG in developing countries intends to access funds from the current global climate change financing architecture then it must meet the conditionalities of those providing the funds. Alternatively, LG that see the opportunity to grab some financial resource from the global funding regime would also have to respond to their conditionalities which may include among other things the integration of resilience into local governance. Both coercion and grabbing incentive mechanisms have been confirmed in diffusion related studies in developing countries. For Instance research evidence on community-based wildlife sector reforms in developing countries shows that local institutions are often motivated to reform their activities because of demands from foreign donors and international conservation organizations that provide funding supports for these institutions (Baldus et al. 2003; Goldman 2003). The subordination position of most SSA LGs in the hierarchy of government and their fiscal weakness which have been discuss in previous sections, is a significant factor that put them at the mercy of international inducements.

Is Legitimacy Over Adaptation Domain Relevant to Actors?

Actors have often sought legitimacy within their network by pursuing commonly agreed practices, both informally and informally. It is on this basis that the international diffusion literature argues that the quest for legitimacy within the network leads to transfer of common values across territories (Powell and DiMaggio 2012). It appears that inherent interaction and communication across network of policy actors creates awareness on innovation, offer learning opportunity, emulations and subtle coercion influencing members to shift towards a dominant logic (Powell and DiMaggio 2012). The literature clearly shows that networks enhance diffusion of ideas because of the trust, autonomous norms and social capital that are apparent in most network (Coleman 2000). Using 15 years newspapers articles on climate change from three countries (1996–2010), Schäfer et al. (2014) concluded that international climate summits has stronger impacts on its policy diffusion. Other empirical evidence shows that many sustainability policies have been easily diffused among over 1000 LGs that belongs to the International Council for Local Environmental Initiatives (ICLEI) (ICLEI 2002). It is also evident across various researches that the participation in networks such as the World Mayors Council on Climate Change (WMCCC), C40 cities, International Solar Cities Initiative (ISCI), Sustainable Cities International Network (SCI), e Global Legislators' Organisation (Globe) etc. is more likely to influence the diffusion of climate policy innovation across participants (Bulkeley 2010). Experience from Ghana's field studies confirms that the constant participation of Accra Metropolitan Assembly in international climate change events has had a significant influence on its budgetary allocation to adaptation related intervention and the establishment of a resilience desk.

Bio-Physical and Science-Heaviness of Climate Change Factors

Science-Heaviness of Climate Change to Local Actors

Climate change remains science heavy that its exact future impacts are not known. Given the lack of certainty about climate impact policy actors are left in elution and future solution are modeled on mere assumptions without realistic facts (Dessai and Hulme 2004). This situation has given room to many skeptisim and contestation as to whether LGs that are confronted with numerous developmental needs should also consider responding to the incremental cost and investment needs of something that cannot be scientifically measured. As clearly hinted, the science-heaviness of climate change has made adaptation very fluid concept creating huge contestation over what constitute adaptation (Smith and Hitz 2002). Many local adaptation domains has failed to incorporate climate change because LG actors are unable to have a common ground on what exactly should be considered adaptation. Given the fact that uncertainties over public issues remains a key sources of erratic patterns of decision making on wicked problems (Koppenjan and Klijn 2004), the article concludes that the science heaviness of climate change influence LGs ability and willingness to adopt its innovations.

How Severe Is Climate Change Risk Perceived- Geographical Location of LG

Climate policy theorist has argued that the perceived severity of climate related risk, sensitivity, exposure and vulnerability of LG to ongoing and future impacts affect its responses to climate policy (Agrawal 2008; Wang 2012). At the local level, there is apparent huge diversity on factors mediating climate risk, sensitivity and vulnerability. This diversity has been cited as been responsible for the heterogeneous climate change response at the local level. This implies that if climate change is not perceived by LGs actors as risk to local development its related policy innovation barely gets adopted, it won't simply get through the muddy agenda table (Adu-Boateng 2015; Ayllet 2015). As a result LGs that are more vulnerable to climate change risk due to its physical characteristics (for example closer to the sea, economic resources are climate sensitive etc.) are more likely to adopt climate policy innovation. Observation from ongoing study on the political economy of climate change policy across selected LGs in Ghana shows that although national directive enjoins all Ghanaian LGs to integrate sustainable adaptation strategies in district development plans, some districts tend to be more serious than others because of variations in exposure and sensitivity to climate risk. The field observation further reveals that in areas where climate change has a profound impact, particularly on the livelihood of the local people, climate change policies is framed as a 'need-based' issue and consequently integrated into development activities, by LG. In such LG, it was noted that climate change is mainstreamed into the district's budget, with clearly located funds and accompanied interventions.

Are There Any Local Social Co-Benefits

Again, public policy discussions have held that positive externality that emanates from adopting a new policy is vital determinant of adoption. The climate policy literature is in consonance with this hypothesis and therefore upholds that if climate change policy has clear and unambiguous social co-benefits to potential adopters, then it is easier to be adopted (Dirksen et al. 1996; and Rogers 2003). In Sub-Saharan Africa where LGs are plagued with competitive needs, the adoption of CCAPM practices would depend on the predictable non-climatic social benefits related to such policy. The argument suggests that such local social co-benefits could gunner the needed bureaucratic, political and public support for adoption (see, Obradovich and Zimmerman 2016). For example, if local people become aware that planting of trees along streets, in schools and at all public places could simultaneously reduce storm havoc and also increase air quality they are likely to support the adoption of such climate related policy. Also the symbolic benefits of adopting a new policy to elected and appointed LG staffs would influence their willingness to provide the needed leadership for adoption. Social scientist suggests that the extent to which the adoption of climate change could affect electoral victory (as discussed previously under incentives) would also affects the behaviour of elected officials: the more secure they are if adopted the more likely for them to provided needed resources for the adoption of policies that are popular with the electorate (Walker 1969; Berry and Berry 1992).

Coincidently, research evidence in Ghana shows that the tendency of Ghanaian political office holders to allocate public resources to certain public needs could be determined by the desire to reach out to as many potential voters as possible in order to win and maintain political power (Abdulai 2012). Therefore both the instrumental and symbolic benefits that accrues from adopting climate change policy may influence it adoption by LG.

Policy and Governance Suggestions

Achieving substantial progress in climate change management reignites the discussion and scholarship on the diffusion of public policies aimed at resolving complex and wicked public issues such as climate change. Though this study attempted to contribute to ongoing discourse on climate change management it suffers from the exclusive use of secondary information. Future studies should therefore build upon this study by testing the claims made here through primary data. Notwithstanding, this paper has successfully discussed three major factors that simultaneously work out to influence the diffusion of climate change policies at the sub-national level. The analysis so far points to several potential generalizations about the factors that may influence LGs' adoption of CCAPM in SSA. Drawing on various theories and empirical studies across different fields "salad" of factors were succinctly configured into three major determinants of climate change policy adoption. Consequently, the article concludes that LGs' adoption of CCAPM practices is a function of: contextual and agency, structural and Science-heaviness factors. As clearly presented in Fig. 1, the contextual and agency factors indicates that factors leading to the adoption of CCAPM practices at the local level include social, economic and political resources that influence LGs adaptive capacity. Similarly, the structural factors argue that factors external to local government but capable of motivating or frustrating LGs adaptive capacity determine LGs pace of adoption. In contrast the bio-physical and science-heaviness of climate change factor posits that climate science-heaviness, uncertainty, severity of climate change risk, and the expected social co-benefit of climate change policy affects LGs reaction towards climate policy adoption. The article therefore put forward that to present a theory that seeks to offer superior explanation for the adoption of climate change adaptation mainstreaming innovations at the local level in developing countries requires a holistic knowledge reflecting the context and the capacity of the agency concerned, structural, and biophysical and science-heaviness factors that influence LGs adaptive capacity and urgency to act. By doing so, one is able to portray the nature of climate change policy diffusion in local governance in a more sophisticated, comprehensive and logical manner.

Given the fact that internal capacity concerns, external inducements, uncertainties, skepticisms and contestation surrounding climate change affects LGs ability and willingness to adopt climate change policy it is would be methodologically inappropriate to use a single theory (either internal determinants or diffusion theories) to study how climate related policy gets adopted. The paper thus, proposes an embracement of multiple diffusion perspectives for studying climate change policy diffusion which resonates well with the central concerns within the existing policy literature. The analysis thus far confirms that actors within LG whose daily interaction affects policy diffusion are not passive recipients of new ideas but rather rational actors often confronted with limited resources, enormous local priorities demanding equal attention and are also subjected to organizational logic (norms), which exposes them to both internal and external inducements (Lindblom 1959). Thus, events taking place within the global system and at the local level affect how CCAPM would get diffused across LG. Also, LGs' policy actors subjects new policies to intense experiment, evaluation, definitions, reframing, avoidance, or acceptance through negotiations and dialogue with others from both remote and immediate environment before final action. Flowing from the above, there is enough theoretical evidence to agree that there are diverse set of factors that could be found on the causal pathway of climate change policy diffusion at the sub-national level. However, the significance of each factor to influence the adoption of climate change policy may vary across various LGs. The apparent diversity of factors, perhaps, explains the reported evidence of non-isomorphic LGs' responses to climate change concerns across the literature.

References

Abdulai, A.-G. (2012). *State elites and the politics of regional inequality in Ghana* (Doctoral thesis). University of Manchester.

Adelle, C., & Russel, D. (2013). Climate policy integration: A case of déjà vu? *Environmental Policy and Governance, 23*(1), 1–12.

Adu-Boateng, A. (2015). Barriers to climate change policy responses for urban areas: A study of Tamale Metropolitan Assembly, Ghana. *Current Opinion in Environmental Sustainability, 13,* 49–57.

Agrawal, A. (2008). The role of local institutions in adaptation to climate change. Retrieved February 21, 2017 from http://www.icarus.info/wp-content/uploads/2009/11/agrawal-adaptation-institutions-livelihoods.pdf

Amankwah-Amoah, J., & Sarpong, D. (2016). Historical pathways to a green economy: The evolution and scaling-up of solar PV in Ghana, 1980–2010. *Technological Forecasting and Social Change, 102,* 90–101.

Aubert, B. A., & Hamel, G. (2001). Adoption of smart cards in the medical sector:: the Canadian experience. *Social Science & Medicine, 53*(7), 879–894.

Aylett, A. (2015). Institutionalizing the urban governance of climate change adaptation: Results of an international survey. *Urban Climate, 14,* 4–16.

Baldus, R., Kibonde, B., Siege, L. (2003). Seeking conservation partnerships in the Selous game reserve, *Tanzania. Parks, 13*(1), 50–61.

Bawole, J. N. (2017). Pro-poor decentralization in Ghana: Exploring the facilitators and the limitations. *Administrative Theory & Praxis, 39*(2), 122–139.

Berry, F., & Berry, W. (1990). State lottery adoptions as policy innovations: An event history analysis. *American Political Science Review, 84*(2), 395–415.

Berry, F. S., & Berry, W. D. (1992). Tax innovation in the states: Capitalizing on political opportunity. *American Journal of Political Science*, 715–742.

Betsill, M. M., & Bulkeley, H. (2004). Transnational networks and global environmental governance: The cities for climate protection program. *International Studies Quarterly, 48*(2), 471–493.

Bulkeley, H. (2010). Cities and the governing of climate change. *Annual Review of Environment and Resources,* 35.

Bulkeley, H., & Betsill, M. M. (2013). Revisiting the urban politics of climate change. *Environmental Politics, 22,* 136–154.

Carter, N., & Jacobs, M. (2014). Explaining radical policy change: The case of climate change and energy policy under the British labour government 2006–10. *Public Administration, 92*(1), 125–141.

Coleman, J. S. (2000). *Social capital in the creation of human capital. In Knowledge and social capital* (pp. 17-41).

de Oliveira, J. A. P. (2009). The implementation of climate change related policies at the subnational level: An analysis of three countries. *Habitat International, 33*(3), 253–259.

Denis, J. L., Hébert, Y., Langley, A., Lozeau, D., & Trottier, L. H. (2002). Explaining diffusion patterns for complex health care innovations. *Health care management review, 27*(3), 60–73.

Dessai, S., & Hulme, M. (2004). Does climate adaptation policy need probabilities?. *Climate policy, 4*(2), 107–128.

Dirksen, C. D., Ament, A. H., & Go, P. M. (1996). Diffusion of six surgical endoscopic procedures in the Netherlands. Stimulating and restraining factors. *Health policy, 37*(2), 91–104.
Dodman, D., & Mitlin, D. (2013). Challenges for community-based adaptation: Discovering the potential for transformation. *Journal of International Development, 25*(5), 640–659.
Downs, A. (1972). The issue-attention cycle and the political economy of improving our environment. *The political economy of environmental control* (pp. 9–34). Berkeley: University of California Press.
Elazar, D. J. (1972). *American federalism: A view from the states*. Crowell.
Elster, J. (1986). *Rational choice*. Oxford: Basil Blackwell.
Fennell, M. L., & Warnecke, R. B. (1988). *The diffusion of medical innovations: An applied network analysis*. New York: Plenum.
Gray, V. (1973). Innovation in the states: A diffusion study. *American Political Science Review, 67*(04), 1174–1185.
Grilli, R., & Lomas, J. (1994). Evaluating the message: The relationship between compliance rate and the subject of a practice guideline. *Medical care*, 202–213.
Goldman, M. (2003). Partitioned nature, privileged knowledge: Community-based conservation in Tanzania. *Development and change, 34*(5), 833–862.
https://www.modernghana.com/news/190303/a-decade-of-community-based-mining-advocacy-in-ghana-waca.html. Accessed on December 28, 2016
IPCC (2014). Summary for policy makers. *Climate change: synthesis report*. Retrieved February 21, 2017 from https://www.ipcc.ch/pdf/assessment-report/ar5/wg3/ipcc_wg3_ar5_summary-for-policymakers.pdf
ICLEI. (2002). What is the cities for climate protection campaign (CCP)? http://www.iclei.org/us/usccp.html.
Jibao, S., & Prichard, W. (2016). Rebuilding local government finances after conflict: Lessons from a property tax reform programme in post-conflict Sierra Leone. *The Journal of Development Studies,* 1–17.
Koppenjan, J. F. M., & Klijn, E. H. (2004). *Managing uncertainties in networks: A network approach to problem solving and decision making*. Psychology Press.
Lindblom, C. E. (1959). The science of "muddling through". *Public Administration Review,* 79–88.
Massey, E., Biesbroek, R., Huitema, D., & Jordan, A. (2014). Climate policy innovation: The adoption and diffusion of adaptation policies across Europe. *Global Environmental Change, 29,* 434–443.
Marshall, J. G. (1990). Diffusion of innovation theory and end-user searching. *Library and Information Science Research, 6*(1), 55–69.
Mohr, L. B. (1969). Determinants of innovation in organizations. *American Political Science Review, 63*(01), 111–126.
Nelson, F., & Agrawal, A. (2008). Patronage or participation? Community-based natural resource management reform in sub-Saharan Africa. *Development and Change, 39*(4), 557–585.
North, D. (1990). Institutions, economic theory and economic performance. *Institutions, institutional change and economic performance.* New York: Cambridge University Press.
Obradovich, N., & Zimmerman, B. (2016). African voters indicate lack of support for climate change policies. *Environmental Science & Policy, 66,* 292–298.
Olwig, M. F. (2012). Multi-sited resilience: The mutual construction of "local" and "global" understandings and practices of adaptation and innovation. *Applied Geography, 33,* 112–118.
Ostrom, E. (2007). Institutional rational choice: An assessment of the institutional analysis and development framework. Retrieved February 20, 2017 from https://digitalcommons.usu.edu/unf_research/60/

Oteman, M., Wiering, M., & Helderman, J. K. (2014). The institutional space of community initiatives for renewable energy: A comparative case study of the Netherlands, Germany and Denmark. *Energy, sustainability and society, 4*(1), 11.
Pasquini, L., Cowling, R. M., & Ziervogel, G. (2013). Facing the heat: Barriers to mainstreaming climate change adaptation in local government in the Western Cape Province, South Africa. *Habitat International, 40*, 225–232.
Porter, J. J., Demeritt, D., & Dessai, S. (2015). The right stuff? Informing adaptation to climate change in British local government. *Global Environmental Change, 35*, 411–422.
Powell, W. W. & DiMaggio, P. J. (Eds.). (2012). *The new institutionalism in organizational analysis*. University of Chicago Press.
Ribot, J. C. (2003). *Democratic decentralization of natural resources. Beyond Structural Adjustment The Institutional Context of African Development* (pp. 159–182). US: Palgrave Macmillan.
Roberts, D. (2010). Prioritizing climate change adaptation and local level resilience in Durban, South Africa. *Environment and Urbanization, 22*(2), 397–413.
Rogers, E. (2003). *Diffusion of innovations* (5th ed.). New York: Free Press.
Rootes, C., Zito, A., & Barry, J. (2012). Climate change, national politics and grassroots action: An introduction. *Environmental Politics, 21*(5), 677–690.
Ryan, D. (2015). From commitment to action: A literature review on climate policy implementation at city level. *Climatic Change, 131*(4), 519–529.
Samuels, D. (2000). Reinventing local government? Municipalities and intergovernmental relations in democratic Brazil. *Democratic Brazil: actors, institutions, and processes*, 77–100.
Schäfer, M. S., Ivanova, A., & Schmidt, A. (2014). What drives media attention for climate change? Explaining issue attention in Australian, German and Indian print media from 1996 to 2010. *International Communication Gazette, 76*(2), 152–176.
Shipan, C. R., & Volden, C. (2008). The mechanisms of policy diffusion. *American journal of political science, 52*(4), 840–857.
Smith, J. B., & Hitz, S. (2002). Estimating global impacts from climate change: Background paper for the OECD. WPGSP project "*Benefits of climate policy*", Stratus Consulting Inc., Boulder (CO).
Smith, B., Brown, D. & Dodman, D. (2014). *Reconfiguring urban adaptation finance*. IIED Working Paper.
Stadelmann, M., & Castro, P. (2014). Climate policy innovation in the South-Domestic and international determinants of renewable energy policies in developing and emerging countries. *Global Environmental Change, 29*, 413–423.
Tanner, T., & Allouche, J. (2011). Towards a new political economy of climate change and development. *IDS Bulletin, 42*(3), 1–14.
Vachon, S., & Menz, F. C. (2006). The role of social, political, and economic interests in promoting state green electricity policies. *Environmental Science & Policy, 9*(7–8), 652–662.
Walker, J. L. (1969). The diffusion of innovations among the American states. *American Political Science Review, 63*(03), 880–899.
Wang, R. (2012). Leaders, followers, and laggards: Adoption of the US conference of mayors climate protection agreement in California. *Environment and Planning C: Government and Policy, 30*(6), 1116–1128.
Weisser, F., Bollig, M., Doevenspeck, M., & Müller-Mahn, D. (2014). Translating the 'adaptation to climate change'paradigm: The politics of a travelling idea in Africa. *The Geographical Journal, 180*(2), 111–119.
Wilson, A. L., Ramamurthy, K., & Nystrom, P. C. (1999). A multi-attribute measure for innovation adoption: The context of imaging technology. *IEEE transactions on engineering management, 46*(3), 311–321.

Wood, A. (2015). The politics of policy circulation: Unpacking the relationship between South African and South American cities in the adoption of bus rapid transit. *Antipode, 47*(4), 1062–1079.
Yeboah-Assiamah, E., Muller, K., & Domfeh, K. A. (2017). Institutional assessment in natural resource governance: *A conceptual overview. Forest Policy and Economics, 74*, 1–12.
Zahran, S., Brody, S., Vedlitz, A., Grover, H., & Miller, C. (2008). Vulnerability and capacity: Explaining local commitment to climate-change policy. *Environ Planning C, Government Policy, 26*(3), 544.

The Politics and Governance of Negative Emissions Technologies

Klaus Radunsky

Abstract The starting point of this paper is the Paris Agreement under the UNFCCC agreed in 2015 and its global temperature goal. It highlights that the current pathway of decarbonization would result in a sharp temperature increase by 2100 (see Sect. 1) that would result in unacceptable risks of climate change (see Sect. 2). Those risks cannot be properly addressed by adaptation activities as long as there are no reliable expectations of the climate changes expected over the lifetime of current projects. The paper informs about such recently published projections that suggest that global surface temperatures will increase by approximately 5 °C (9 °F) over pre-industrial temperatures by the year 2100 and discusses options to bridge the gap to the goals specified in the Paris Agreement. In this context also two main geoengineering options carbon dioxide removal (CDR) and solar radiative management (SRM) and their associated risks are considered. At the center of the paper two proposals are made on a limited use of geoengineering that should finally help meet the goals of the Paris Agreement without increasing risks of geoengineering unduly, provided that decarbonization is significantly accelerated beyond the plans included in current National Determined Contributions. One conclusion is that geo-engineering can only provide a small contribution to meeting the goals of the Paris Agreement and that the main contribution needs to come from enhanced mitigation action. The paper also highlights how recommendations of the Financial Stability Board could contribute to result in the necessary shift of investments to accelerate GHG emission reduction and informs about a recent initiative to establish the necessary governance framework to manage geoengineering.

Keywords Paris agreement · Climate change risks · Geo-engineering
Climate action

K. Radunsky (✉)
Umweltbundesamt, Spittelauer Lände 5, 1090 Vienna, Austria
e-mail: klaus.radunsky@umweltbundesamt.at

F. Alves et al. (eds.), *Theory and Practice of Climate Adaptation*,
Climate Change Management, https://doi.org/10.1007/978-3-319-72874-2_5

Paris Agreement and Its Temperature Goal

The 2015 Paris Agreement of the 21st Conference of the United Nations Framework Convention on Climate Change agreed an ambitious goal: "Holding the increase in the global average temperature to well below 2 °C above pre-industrial levels and to pursue efforts to limit the temperature increase to 1.5 °C." This represents a huge policy challenge.

Several studies came to the conclusion that both, the 1.5 °C as well as even the 2 °C global warming limits are already out of reach without using controversial geoengineering technology such as negative emissions technology[1] (Joeri et al. 2016).

With extreme policy measures (to curb GHG emissions), a 2.5 °C limit is in reach (Nordhaus 2017).

A roadmap for rapid decarbonization has been proposed (Johan et al. 2017), the goal being to cut global CO_2 emissions by 50% every 10 years. Net-zero emissions should be achieved globally around mid-century. Such roadmap is seen as a planning instrument, linking shorter-term targets to longer-term goals. Given that global GHG emissions have peaked at best such roadmap would require a huge shift in investments as well as behavior in general.

A recent study on investments in the oil and gas industry (Mackenzie 2017) predicts demand for renewables would grow faster than oil in the next two decades: the analysts forecast annual growth rates of 6% for wind and 11% for solar, compared with 0.5% for oil demand. European oil and gas companies embrace renewables much more eagerly than US rivals such as Exxon and Chevron because the US firms have lower cost oil and gas production. Wood Mackenzie say returns of investments for renewables are about half those of oil and gas production, but the long-life of cash flow from assets such as windfarms could help firms support their dividends. According to this study wind and solar are poised to radically reshape energy markets and the growth opportunity in renewables cannot be ignored. Though the timing of the energy transition is highly uncertain, oil and gas companies need to build expertise, balance and optionality to hedge against any future erosion of the upstream value proposition. Thus the Majors are leading the way in shaping strategies as they look to capture a piece of the action in this fast growing market.

However, as long as regulations do not prevent decision making incoherent with the Paris Agreement temperature limits it seems justified assuming a global increase of temperature of about 5 °C above pre-industrial levels (or about 4 °C above current global average temperature level), e.g. in the context of climate proofing of long-term investments in infrastructure. This is reflected in proceedings prepared recently by the Institution of Civil Engineers (ICE 2017).

This assumption is justified insofar as the current plans related to reduction of GHG emissions as submitted in Intended Nationally Determined Contributions

[1]This reference includes references of several other studies.

(INDCs) imply a median warming of 2.6–3.1 °C by 2100 compared to the pre-industrial global temperature according to the already mentioned study in Nature (Joeri et al. 2016).

It is also important to note that the projections in the New Policies Scenario of the International Energy Agency (IEA) signal to policy-makers and other stakeholders only the direction in which today's policy ambitions are likely to take the energy sector. This does not, however, make this scenario a forecast—a point that needs constantly to be kept in mind. Those scenarios provide only a well-founded basis for expectations about the future and thereby also serve as an invitation for improvement: if the outcomes described are sub-optimal or, even, unacceptable, then policies and other conditions and factors need to change (Mackenzie 2017).

This helps to explain the significant differences between IEA scenarios and those required to meet the 2 °C goal and scenarios by others, e.g. Bloomberg. According to new analysis from (Bloomberg 2017) New Energy Finance (BNEF) renewables are to capture three-fourths of the $10 trillion the world will invest. Solar and wind will make up nearly a half of installed capacity and over a third of global power generation by 2040. That's a four-fold jump in wind capacity and a 14-fold jump in solar from today.

A recent development in the US demonstrates the significance of decisions on mitigation now for the temperature change by 2100: Many prominent U.S. corporations, opinion leaders and NGOs have announced their support of the Climate Leadership Council (CLC) "carbon dividends" proposal. The proposal includes a tax on carbon dioxide emissions, the return of tax revenues to all Americans in the form of monthly dividend payments and the rollback of some climate regulations, among other elements. This proposal easily would satisfy CLC's stated objective "to exceed the emissions reductions of current regulations." Thus, alongside well-designed complementary policies, an increasing carbon tax starting at $40 per ton could be the centrepiece of the U.S. strategy to achieve its 2025 emissions target and then rapidly transition to a low carbon energy system (Kaufmann 2017).

Potential Impacts at Expected Temperature Changes

This section summaries some findings as identified in the AR5 from the IPCC (2014a).

In the Fifth Assessment Report of the IPCC Working Group II used five integrative reasons for concern (RFCs) as a framework for summarizing key climate change risks across sectors and regions. Given the expected changes in temperature by 2100 the focus of this chapter is on the high end of the temperature changes assessed (for a more complete assessment see http://www.ipcc.ch/pdf/assessment-report/ar5/wg2/ar5_wgII_spm_en.pdf).

- Unique and threatened systems: Many species and systems with limited adaptive capacity are subject to very high risks with additional warming of 2 °C above global average temperature relative to 1986–2005, particularly Arctic-sea-ice and coral-reef systems.
- Extreme weather events: Climate-change-related risks from extreme events, such as heat waves, extreme precipitation, and coastal flooding, are already moderate and high with 1 °C additional warming. Those risks will increase further at higher temperatures.
- Global aggregate impacts: Risks of global aggregate impacts are moderate for additional warming between 1 and 2 °C, reflecting impacts to both Earth's biodiversity and the overall global economy. Extensive biodiversity loss with associated loss of ecosystem goods and services results in high risks around 3 °C additional warming. Aggregate economic damages accelerate with increasing temperature, but few quantitative estimates have been completed for additional warming around 3 °C or above.
- Large-scale singular events: With increasing warming, some physical systems or ecosystems may be at risk of abrupt and irreversible changes. Risks associated with such tipping points become moderate between 0 and 1 °C additional warming, and increase disproportionately as temperature increases between 1 and 2 °C additional warming and become high above 3 °C, due to the potential for a large and irreversible sea level rise from ice sheet loss. E.g. near-complete loss of the Greenland ice sheet would ultimately be contributing up to 7 m of global mean sea level rise.

The right part of the Fig. 1 below reflects graphically the climate change risks described above, based on a figure prepared by Working Group 2 for the Fifth Assessment Report (WG2 2014).

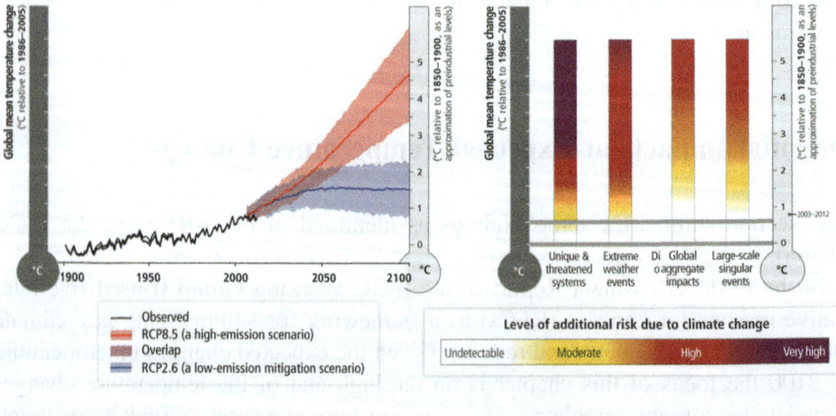

Fig. 1 Level of risks of climate change and their relationship to emission pathways

The following eight risks of climate change have been identified by AR5 WGII Report with high confidence:

(i) Risk of death, injury, ill-health, or disrupted livelihoods in low-lying coastal zones and small island developing states and other small islands, due to storm surges, coastal flooding, and sea level rise.
(ii) Risk of severe ill-health and disrupted livelihoods for large urban populations due to inland flooding in some regions.
(iii) Systemic risks due to extreme weather events leading to breakdown of infrastructure networks and critical services such as electricity, water supply, and health and emergency services.
(iv) Risk of mortality and morbidity during periods of extreme heat, particularly for vulnerable urban populations and those working outdoors in urban or rural areas.
(v) Risk of food insecurity and the breakdown of food systems linked to warming, drought, flooding, and precipitation variability and extremes, particularly for poorer populations in urban and rural settings.
(vi) Risk of loss of rural livelihoods and income due to insufficient access to drinking and irrigation water and reduced agricultural productivity, particularly for farmers and pastoralists with minimal capital in semi-arid regions.
(vii) Risk of loss of marine and coastal ecosystems, biodiversity, and the ecosystem goods, functions, and services they provide for coastal livelihoods, especially for fishing communities in the tropics and the Arctic.
(viii) Risk of loss of terrestrial and inland water ecosystems, biodiversity, and the ecosystem goods, functions, and services they provide for livelihoods.

Given the significant negative impacts identified above that are expected by scientists and economists by the change in the average global temperature that has to be assumed by 2100 and given the indicated limitations to avoid dangerous interference with the climate system by mitigation action only, some scientists and others are considering increasingly drastic action to reduce climate change risks by geoengineering.

Geoengineering Options

Geoengineering refers to a broad set of methods and technologies operating at a large scale that aim to deliberately alter the climate system in order to alleviate the impacts of climate change (IPCC 2014b). Most methods seek to either reduce the amount of absorbed solar energy in the climate system (solar radiative management, SRM) or increase the removal of carbon dioxide (CO_2) from the atmosphere by sinks to alter climate (carbon dioxide removal, CDR).

Such CO_2 removal by technical means, e.g. in the form of BECCS (Bio-energy with carbon capture and storage[2]) has been considered in all emission reduction scenarios considered in the Fifth Assessment Report of the Intergovernmental Panel on Climate Change that allow meeting the 2 °C global warming limit if global GHG emissions peak later than 2010.

Risks Associated with Geoengineering

"The best way to think of the options ahead is offering a balance of risks" (Porter 2017). What are those risks of geo-engineering?

In particular the following risks of SRM seem to be very relevant (Civil Society Briefing 2017):

- **No turning back**. SRM may be a one-way street. Because SRM masks the actual warming in the atmosphere, if it was terminated, a sudden jump in warming would occur, which would be much more difficult for ecosystems to adapt to and for society to tackle than gradual warming. The risk that SRM is terminated is so significant because of the very long time period that CO_2, the main greenhouse gas, will remain in the atmosphere, once emitted. According to the AR5 of the IPCC even after 1000 years of emissions there will still be an amount in the range of 40% in the atmosphere—and the further decay will be very slow, as the main decrease in the concentration happens right after the point in time of the emissions (Stocker 2013)[3].
- **Not addressing root causes**. SRM will not address the problem of atmospheric GHGs, which will continue increasing. Nor does SRM address ocean acidification (also caused by atmospheric buildup of carbon dioxide). Indeed, ocean acidification could be worsened by some SRM techniques.
- **SRM is the perfect excuse for inaction**. SRM, and geoengineering more broadly, is a "perfect excuse" for climate deniers and governments seeking to avoid the political costs of carbon emission reductions. For those looking to stall meaningful climate action the active development of tools and experiments will be presented as preferred pathway to address climate change and as an argument to ease restrictions on high carbon emitting industries. This line of argument was

[2]BECCS is a greenhouse gas mitigation technology which produces negative carbon dioxide emissions by combining bioenergy (energy from biomass) use with geologic carbon capture and storage.

[3]The removal process of CO_2 is as follows: The uptake of anthropogenic CO_2 by the ocean is primarily a response to increasing CO_2 in the atmosphere. Excess atmospheric CO_2 absorbed by the surface ocean or transported to the ocean through aquatic systems (e.g., rivers, groundwaters) gets buried in coastal sediments or transported to deep waters where it is stored for decades to centuries. The deep ocean carbon can dissolve ocean carbonate sediments to store excess CO_2 on time scales of centuries to millennia. (Stocker 2013)

already put forward by conservative think tanks in the United States such as the American Enterprise Institute.

SRM is already under a moratorium as these serious risks and unresolved issues justify banning SRM. As a precautionary measure, the 193 countries to the UN Convention on Biological Diversity (CBD)[4] established a de facto moratorium against most forms of geo-engineering including all forms of SRM.

With respect to carbon capture and storage (CCS) the following risks have been identified in an Article (Foyster 2017), building on several extensive studies:

- It is far too costly;
- Concerns around leakage;
- Increased risk of earthquakes if injecting carbon dioxide puts too much pressure in the underground reservoir.

The author: for all of these reasons, any form of CCS should be viewed with great skepticism.

A Working Paper of the Stockholm Environment Institute (SEI 2016) came to similar conclusions:

- Negative emission options will ultimately not prove feasible;
- Their large scale deployment involves unacceptable ecological and social impacts;
- Negative emissions activities prove less effective than hoped, either because they are subsequently reversed by human or natural forces, or because climate change impacts prove irreversible[5].

The Working Paper by the Stockholm Environment Institute has been based on the assessment of four main land-based negative emissions options in light of those risks: ecosystem restoration, mosaic-landscape restoration[6], reforestation and BECCS.

---

[4]It should be noted that the US is not a Party to the CBD.

[5]The explanation is as follows for the risk that climate impacts occurring during the period of concentration overshoot may prove irreversible: It is known that, for a given amount of total cumulative emissions, peak warming is higher for a pathway that overshoots before negative emissions begin to reduce concentrations. The peak warming is driven by time-integrated radiative forcing, and is a function of maximum cumulative emissions (before negative emissions start), rather than total cumulative emissions (including negative emissions). The higher peak warming causes greater climate impacts, and "increases the likelihood of crossing thresholds for 'dangerous' warming". Of particular concern is the potential to pass thresholds relating to sea ice, glaciers, ice sheets and permafrost, which can themselves create a positive feedback that causes additional warming (for example, through albedo effects or methane emissions). Literature: (Stocker 2013).

[6]Mosaic-type landscape restoration accommodates multiple land uses, such as agriculture, protected reserves, managed plantations and agroforestry systems (SEI 2016).

Similar assessments can be found in two other studies:

- Boysen et al. 2017 address two major issues: First, they calculate the extent of terrestrial carbon dioxide removal (tCDR) required to "repair" delayed or insufficient emissions reduction policies unable to prevent global mean temperature rise of 2.5 °C or even 4.5 °C above pre-industrial level. The results show that those tCDR measures are unable to counteract "business-as-usual" emissions without eliminating virtually all natural ecosystems. Even if considerable (Representative Concentration Pathway 4.5 [RCP4.5]) emissions reductions are assumed, tCDR with 50% storage efficiency requires >1.1 Gha of the most productive agricultural areas or the elimination of >50% of natural forests. In addition, >100 MtN/year fertilizers would be needed to remove the roughly 320 GtC foreseen in these scenarios. Such interventions would severely compromise food production and/or biosphere functioning. Second, they reanalyze the requirements for achieving the 160–190 GtC tCDR that would complement strong mitigation action (RCP2.6) in order to avoid 2 °C overshoot anytime. They find that a combination of high irrigation water input and/or more efficient conversion to stored carbon is necessary. In the face of severe trade-offs with society and the biosphere, they conclude that large-scale tCDR is not a viable alternative to aggressive emissions reduction. However, they argue that tCDR might serve as a valuable "supporting actor" for strong mitigation if sustainable schemes are established immediately.
- Stanford News Service (2017): Assuming easy carbon removal from the atmosphere is a high-stakes gamble, Stanford scientists say. The researchers (Chris Field, Katharine Mach and Devon Ryan) don't reject carbon capture, instead arguing that there are important near-term opportunities for carbon removal at modest scale, often with other benefits for nature and people, and critical needs now for developing the technologies of the future. But heavy reliance on biomass energy with carbon capture and storage could require tremendous land areas. For example, relying on the technology to achieve a temperature increase of 2 °C or less could require an amount of productive land equivalent to about 25–80% of total global cropland, up to about 8 percent of all of the land on Earth. The full study has the title "Rightsizing carbon dioxide removal".[7]

For the above reasons CDR is not, and never can be, a substitute for dramatic cuts in greenhouse gas emissions. But nor can it be dismissed out of hand. CO_2 persists in the atmosphere for centuries and millennia (see also footnote 3); it's almost certain we'll need to draw down some of it over in this century in order to avoid dangerous interference with the climate system. While CCS is still a nascent technology, it's more advanced than some other techniques to do the same job. The US National Academy of Sciences calls for a 'rapid expansion and scale up' of demonstration projects of CDR (News 2015).

[7]Science 19 May 2017: Vol. 356, Issue 6339, pp. 706–707.

The significant risks associated with additional GHG emissions highlighted in Sect. 2 in combination with the given limitations in accelerating the speed of reducing GHG emissions to an extent that would allow meeting the temperature goals of the Paris Agreement raise the question to which extent geo-engineering could offer some potential to reduce climate change risks without enhancing other risks significantly. The following Sect. 6 offers some first thoughts how this could be achieved without violating the precautionary principle that has been the basis for the moratorium under the CBD. But before moving to those suggestions some consideration of the carbon budget in Sect. 5.

Carbon Budget

The average global surface temperature rise has an almost linear relationship with the cumulative emissions of carbon dioxide (CO_2). This useful relationship has resulted in the concept of a remaining global "CO_2 budget" (the cumulative amount of CO_2 emitted over a given timeframe) that can be associated with a probability of remaining below a chosen temperature target.

According to a recent study of the IEA (2017), it was estimated that the CO_2 budget between 2015 and 2100 is 880 Gigatonnes (Gt) meeting the temperature goal with a 66% probability of keeping the average global surface temperature rise throughout the 21st century to below 2 °C. This is without any temporary overshoot and assuming non-CO_2 emissions originating from non-energy sectors similar to the scenarios from the Fifth Assessment Report of the Intergovernmental Panel on Climate Change (IPCC) and lies towards the middle of the 590–1 240 Gt CO_2 range from a study discussing CO_2 budgets commensurate with a 66% chance of staying below 2 °C.[8]

Proposals to Reduce Some Risks of Geo-Engineering

The first proposal is related to CCS but adds only risks to a quite limited extent, compared to the significant risks identified for CCS above in Sect. 4. It has the following three elements:

- **Enhance carbon sequestration by regrowth of degraded forests but do not use BECCS**. The Stockholm Environment Institute (SEI) recently concluded that allowing degraded forests areas to regrow could pull up to additional 330 billion tonnes (or Gt) of CO_2 out of the atmosphere over the 21st century (SEI 2016). This approach avoids the significant social and technological risks of BECCS. This amount of CO_2 corresponds to about 37% of the CO_2 emissions

[8]For comparison: the current global CO_2 emissions are in the range of 50 Gt per year.

coherent with the 2 °C goal. A side event at the UNFCCC session in May 2017 offered some insights on such approach.[9]

- **Do not consider to deploy other CDR technologies in the near term**. This helps to avoid other, more experimental technologies that would be available later and are too risky.
- **Agree first on a Global goal based on "zero fossil carbon"**. Any strategy based on future carbon removals leaves society at risk of insufficient decarbonization by counting on negative emissions that may not materialize. Agreement on a global goal based on "zero fossil carbon" as a first step does not pose that risk. Such step sends an unambiguous signal regarding the rate at which carbon emissions must be ceased. It offers no promise of future absolution based on negative emissions from still unproven land-based options, but neither does it exclude the use of them, should they be developed and proven effective to enhance the benefits of emission reductions.

The above described first proposal still requires a prompt, rapid and dramatic transformation of the economy and in particular of the energy system to shift away from fossil fuels and minimize land use-related emissions, albeit at a lower, more realistic speed compared to a scenario without using that option.

This approach should be feasible—last but not least because of the economic viability of renewables in particular in developing countries (Roberts 2017)—and it would help countries meet their sustainable development goals and it avoids unacceptable social and ecological impacts. However, there remains the risk that the actual decarbonization rate together with the limited sequestration potential of the regrowth of degraded land (geo-engineering option 1) will still result in a temperature change that becomes unmanageable due to its significant impacts (see Sect. 2 and Fig. 2).

Some scientists claim that SRM could be our last, best hope to deal with such climate change. How to limit the associated risks identified above? The second proposal tries to provide an answer to which extent SRM could help to manage climate change risks without adding high additional other risks.

Thus the second proposal relates to SRM but adds also only risks to a quite limited extent, compared to the huge risks identified for SRM above in Sect. 4. It has the following three elements:

- Agree on the development and deployment of technologies to mineralize CO_2. Atmospheric CO_2 can be sequestered by injecting it into basaltic rocks, providing a potentially valuable way to undo some of the damage done by fossil fuel burning. Matter et al. (2016) injected CO_2 into wells in Iceland that pass through basaltic lavas and hyaloclastites at depths between 400 and 800 m.

[9]This side event (http://enb.iisd.org/climate/sb46/enbots/17may.htm) concentrated on the contribution of forest landscape restoration (FLR) to the achievement of Nationally Determined Contributions (NDCs). Panelists shared views on the design and successful implementation of FLR policies and practices, and discussed how FLR contributes to reducing emissions and enhancing carbon stocks in forest landscapes. https://www.bonnchallenge.org/content/challenge

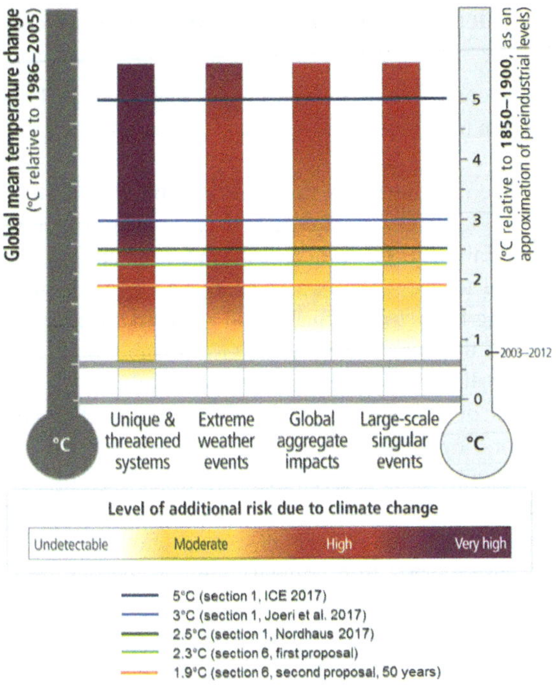

Fig. 2 Potential of feasible options to reduce climate change risks

Most of the injected CO_2 was mineralized in less than 2 years. Carbonate minerals are stable, so this approach should avoid the risk of carbon leakage as this proposal avoids the risk of reversal by human and/or natural forces. This would avoid also legal and other barriers experienced with CCS.

- In order to avoid further warming, allow to deploy SRM equivalent to an agreed amount of CO_2 sequestered in a specified period of time (e.g. 10 years).
- Repeat the above steps until the average global temperature is within the limits specified by the Paris Agreement and avoiding dangerous interference with the climate system.

This second proposal assumes temporary over-shooting of global average temperature above the temperature goal of the Paris Agreement but limits the use of SRM to the amount of CO_2 sequestered in a specified, but reasonable limited time period. This helps to control the "no turning back risk" which has been identified above. This approach also helps to reduce the risk of irreversible climate change impacts that increase with duration and extent of temperature overshoot. The proposal also allows for stable carbon storage avoiding the risk that negative emissions are subsequently reversed by human or natural forces by using mineralization. This expands the potential for geoengineering limited finally by the willingness to pay—see Sect. 7.

Implications of Limiting the Risks of Geo-Engineering

What could be the unknown but definitely limited potential to mineralize CO_2? This section tries to provide a first answer based on expert judgment of the willingness to pay.

The assumption is that carbon capture and mineralization has costs of at least 100 US$/t CO_2 sequestered. If the amount of money spent per year is limited to 1000 billion US$—which is 10 times the amount of money industrialized countries agreed to spent to address adaptation and mitigation under the Paris Agreement starting from 2020—this would allow to sequester 10 billion t CO_2/year—or about 1/5 of the current yearly GHG emissions or 100 billion t CO_2 in 10 years. This would limit the corresponding SRM activity to about 0.1 °C or at best to a few tenths of a degree C in the 21st century. The amount will depend on two key factors: by when such technology is mature enough for deployment at industrial scale and the costs. For comparison: warming by 2100 is currently expected to be at least around a factor of 10 higher.

Such amount of carbon sequestration is insofar not completely unreasonable because experts from industry are considering the need to compensate for process related CO_2 emissions in order to become carbon neutral as envisaged in the Paris Agreement. The scenarios assume carbon sequestration between 5 and 15 Gt C per year—what is a similar order of magnitude as the above assumption. And the necessary infrastructure is not only in the planning phase but started to become reality in the United Arab Emirates as reported recently (Bonn 2017) in a side event to the Bonn Climate Change Conference May 2017.

The two scenarios of the use of geo-engineering described above show that deployment of geoengineering options, including both, SRM and CDR, that avoid increasing risks in an unmanageable manner do not allow for the continued use of fossil fuels but can reduce climate change risks only to a quite limited extent. However, a smart combination of the available options could help meeting the goals of the Paris Agreement which are very likely out of reach with mitigation action only. Figure 2 below gives an indication to which extent the climate change risks (see Sect. 2) could be reduced by the approaches described above in Sects. 6 and 7.

Figure 2 visualizes that without rapid decarbonization which is able to reduce global average temperature by 2100 by about 2.5 °C the regions with low resilience to the impacts of climate change would probably be left—migration being the last resort for many (Wendle 2017). Geoengineering could contribute around 0.7 °C and thus much less than mitigation to lower the temperature increase in this century.

Main Risks of Climate Change and the Approach of the Financial Stability Board

The Financial Stability Board (FSB) is an international body that monitors and makes recommendations about the global financial system. At the Pittsburgh Summit, the Heads of State and Government of the Group of Twenty (G20) endorsed the FSB's original Charter of 25 September 2009 which set out i.a. the FSB's objectives and mandate. The FSB promotes international financial stability; it does so by coordinating national financial authorities and international standard-setting bodies as they work toward developing strong regulatory, supervisory and other financial sector policies. It fosters a level playing field by encouraging coherent implementation of these policies across sectors and jurisdictions.

The FSB, working through its members, seeks to strengthen financial systems and increase the stability of international financial markets.

The Task Force of the FSB on Climate-related Financial Disclosures prepared draft recommendations for helping businesses disclose climate-related financial risks and opportunities within the context of their existing disclosure requirements (FSB 2016).

The FSB has highlighted, that without effective disclosure of climate change risks, the financial impacts of climate change may not be correctly priced—and as the costs eventually become clearer, the potential for rapid adjustments could have destabilizing effects on markets.

The Task Force's draft recommendations aim to begin fixing this problem. What gets measured better gets managed better. Widespread adoption of the recommendations help ensure that climate-related financial issues are routinely considered in business and investment decisions and encourage an effective dialogue between companies and banks, insurers and investors. That will lead to smarter, more efficient allocation of capital, and—probably most importantly—speed the transition to a low-carbon economy.

The draft recommendations are structured around four thematic areas that represent core elements of how organizations operate: governance, strategy, risk management, and metrics and targets. The four overarching recommendations are supported by recommended disclosures that build out the framework with information that will help investors and others understand how reporting organizations think about and assess climate-related risks and opportunities.

The Task Force divided climate-related risks into two major categories: (1) risks related to the *transition* to a lower-carbon economy and (2) risks related to the *physical* impacts of climate change.

Transition Risks

Transitioning to a lower-carbon economy may entail extensive policy, legal, technology, and market changes to address mitigation and adaptation requirements related to climate change. Depending on the nature, speed, and focus of these changes, transition risks may pose varying levels of financial and reputational risk to organizations.

Physical Risks

Physical risks resulting from climate change can be event driven (acute) or longer-term shifts (chronic) in climate patterns. Physical risks may have financial implications for organizations, such as direct damage to assets and indirect impacts from supply chain disruption. Organizations' financial performance may also be affected by changes in water availability, sourcing, and quality; food security; and extreme temperature changes impacting organizations' premises, operations, supply chain, transport needs, and employee safety.

Based on many responses to the draft report the Task Force on Climate-related Financial Disclosures (TCFD) will publish the final report on 29 June 2017 and the report will be presented at the G20 Summit in July. It can be expected that this G20 meeting will endorse the recommendations.

The implementation of the recommendations will probably further shift investments from fossil fuels to renewables and enhancement of energy efficiency and will thus contribute meeting the temperature goals agreed in the Paris Agreement.

The final report including the "Recommendations of the Task Force on Climate-related Financial Disclosures" has been made public on 29 June 2017.[10]

For the time being the FSB did not consider risks from geo-engineering. This reflects that it is not to be expected that such activities will be deployed at scale in the near or medium term.

The Carnegie Climate Geoengineering Governance Initiative (C2G2)

The last section highlights some recent developments to catalyze the creation of effective governance for climate geoengineering technologies. The Carnegie Climate Geoengineering Governance Initiative (C2G2 2017) started its public work in January 2017 based on the following principles:

Impartiality

- We take an impartial approach: we are not for or against the research, testing or potential use of climate geoengineering technologies. That is a choice for society to make.

[10]Assessable at: http://www.fsb.org/wp-content/uploads/P290617-5.pdf.

- Hence our focus is on catalyzing transparent, inclusive policy discussions at the international, national and sub-national level on the risks posed by these technologies and on how to govern them.

Risk Management Approach

- The world faces an increasingly precarious future due to the growing impacts of climate change, coupled with other environmental, resource scarcity, equity and demographic challenges. Any response to climate change must be seen through a risk management lens.
- Inaction is itself a high-risk option. The world is currently headed toward a more than 3 °C temperature rise by the end of this century, which would be catastrophic for the survival and well-being of all species, including our own.

Higher uncertainty means larger climate risks and more need to cut carbon emissions. This the key message of an recent article (Lewandowsky et al. 2017).

Recently (June 2017), the setting up of an Advisory Group of leading international experts to guide and support the work of the initiative has been announced.[11]

Conclusions and Limitations of the Paper

The Paris Agreement has raised great expectations that countries will start to address climate change risks by actions on the ground. However, the dire reality after the recent decision of the US government to withdraw from the Paris Agreement did the Institution of Civil Engineers[12] let come to the recommendation that the world should prepare for a global warming of up to +5 °C above pre-industrial level despite the climate actions in the planning phase which should avoid such warming. [13]

[11]http://www.prweb.com/releases/2017/06/prweb14435483.htm.

[12]The Institution of Civil Engineers (ICE) is an independent professional association for civil engineers and a charitable body which exists to deliver benefits to the public. Based in London, ICE has nearly 89,000 members, of whom three quarters are located in the United Kingdom, while the rest are located in more than 150 countries around the world. ICE supports the civil engineering profession by offering professional qualification, promoting education, maintaining professional ethics, and liaising with industry, academia and government.

[13]The exact wording is as follows: While it is appropriate to utilise RCP 8·5 as a realistic 'nomitigation' scenario, a significant global societal change in the utilisation and generation of power may bring real emissions more in line with the best-case scenario. In fact, the economics of renewable energy such as wind and solar power are decreasing dramatically and now are close to, or less expensive than, that of more traditional fossil fuels. Despite these positive factors, until the implementation of robust international agreements to reduce emissions is successfully completed, RCP 8·5 should be considered the standard.

There is no doubt that climate change risks will further increase with further temperature increase but should we really climate proof against the worst case scenario? This paper tries to identify those climate actions that could be realized if we manage well and make the right decisions now.

It should be possible to limit warming to a level of about 2.5 °C by rapid deployment of mitigation actions based on already deployable technologies. To reduce warming further, use of geo-engineering options should be further explored, e.g. by research. This article includes two proposals related to geo-engineering, one related to CDR and the other including also to SRM. Those proposals could allow still meeting the 2 °C temperature goal of the Paris Agreement in the 21st century despite the current slow speed of decarbonization without increasing unduly other risks linked to geo-engineering. However, it might only be justified to design new infrastructure for less than 2.5–3 °C after the first global stocktake had triggered much more aggressive mitigation action. And only after solving the technical, economical, legal and other challenges before countries are ready to deploy geo-engineering at scale it would be justified to plan for a T-increase of 2° or even less.

The author of this paper hopes that the two proposals outlined in Sects. 6 and 7 inform the discussion on managing the risks of climate change, strengthen the efforts to speed up mitigation efforts at all levels and contribute to further efforts to limit the risks of geoengineering, e.g. with respect to research.

The sooner there is clarity about the long-term global temperature change the less will be the risk of maladaptation. True leadership could thus help avoid stranded investments and would help people to develop a clear and bright vision for our future.

References

Bloomberg. (2017). *New energy outlook 2017*. Available at https://about.bnef.com/new-energy-outlook/. Assessed 3 June 2017.

Bonn. (2017, May). *Global status of carbon capture and storage and the progress made over the past two decades*. Presented by the Global Carbon Capture and Storage (CCS) Institute. Available at: http://enb.iisd.org/climate/sb46/enbots/13may.html. Assessed at 20 June 2017.

Boysen, L. R., Lucht, W., Gerten, D., Heck, V., Lenton, T. M., & Schellnhuber, H. J. (2017). The limits to global-warming mitigation by terrestrial carbon removal. *Earth's Future, 5*, 463–474. https://doi.org/10.1002/2016ef000469. Available at: http://onlinelibrary.wiley.com/doi/10.1002/2016EF000469/full. Assessed at 8 June 2017.

C2G2. (2017). Carnegie Climate Geoengineering Governance Initiative (C2G2). Available at: https://www.carnegiecouncil.org/programs/ccgg. Assessed 14 June 2017.

Civil Society Briefing. (2017). *What is wrong with Solar Radiation Management*? 24 Mar 2017. Available at: http://etcgroup.org/sites/www.etcgroup.org/files/files/etc_briefing_why_srm_experiments_are_bad_idea_2.pdf. Assessed at 10 June 2017.

Foyster, G. (2017). *Climate pipe dreams—Eureka Street* (Vol. 27, No. 6). 30 Mar 2017. Available at: http://www.eurekastreet.com.au/article.aspx?aeid=51021. Assessed at: 10 June 2017.

FSB. (2016). *Recommendations of the task force on climate-related financial disclosures*. Available at: https://www.fsb-tcfd.org/publications/recommendations-report/ Assessed 15 December 2016.

ICE, Institution of Civil Engineers. (2017). *In Proceedings Briefing: Future Climate Projections Allow Engineering Planning*. Available at http://www.icevirtuallibrary.com/doi/pdf/10.1680/jfoen.17.00002. Assessed 2 June 2017.

IEA. (2017). *IEA study perspectives for the energy transition*. Available at: https://www.energiewende2017.com/wp-content/uploads/2017/03/Perspectives-for-the-Energy-Transition_WEB.pdf. Assessed at 19 June 2017.

IPCC. (2014a). Summary for policymakers. In: Climate change 2014: Impacts, adaptation, and vulnerability. Part A: global and sectoral aspects. In C. B. Field, V. R. Barros, D. J. Dokken, K. J. Mach, M. D. Mastrandrea, T. E. Bilir, et al. (Eds.), *Contribution of Working Group II to the Fifth Assessment Report of the Intergovernmental Panel on Climate*. Cambridge University Press, Cambridge, United Kingdom and New York, NY, USA, pp. 1–32. Available at: http://www.ipcc.ch/pdf/assessment-report/ar5/wg2/ar5_wgII_spm_en.pdf. Assessed at 20 June 2017.

IPCC. (2014b). Climate change 2014: Synthesis Report. In: Core Writing Team, R. K. Pachauri & L. A. Meyer (Eds.), *Contribution of Working Groups I, II and III to the Fifth Assessment Report of the Intergovernmental Panel on Climate Change*. IPCC, Geneva, Switzerland, 151 pp. Available at: https://www.ipcc.ch/report/ar5/syr/. Assessed at 20 June 2017.

Joeri, R., den Elzen, M., Höhne, N., Fransen, T., Fekete, H., Winkler, H., et al. (2016). Paris Agreement climate proposals need a boost to keep warming well below 2 °C. *Nature, 534*, 631–639. https://doi.org/10.1038/nature18307. Available at https://www.nature.com/nature/journal/v534/n7609/full/nature18307.html. Assessed 30 June 2016.

Kaufmann, N. (2017). *Republican-proposed "carbon dividend" is a great sign of progress*. 10 Feb 2017. Blog Post Available at: http://www.wri.org/blog/2017/02/republican-proposed-carbon-dividend-great-sign-progress. Assessed 3 June 2017.

Lewandowsky, S., Freeman M. C., & Mann, M. E. (2017). *Harnessing the uncertainty monster: Putting quantitative constraints on the intergenerational social discount rate*. Amsterdam: Elsevier. 25 April 2017. Available at: https://s100.copyright.com/AppDispatchServlet?publisherName=ELS&contentID=S0921818116302958&orderBeanReset=true. Assessed at 29 June 2017.

Mackenzie. (2017). *Report: Could renewables be the Majors' next big thing*? Available at: https://www.woodmac.com/reports/upstream-oil-and-gas-could-renewables-be-the-majors-next-big-thing-46827370. Assessed 1 June 2017.

Mackenzie, K. (2017). *Why IEA scenarios should be treated with extreme caution*. Financial Times, 24 May 2017. Available at: https://ftalphaville.ft.com/2017/05/24/2189189/guest-post-why-iea-scenarios-should-be-treated-with-extreme-caution/. Assessed 29 May 2017.

Matter, J.M., et al. (2016), Rapid carbon mineralization for permanent disposal of anthropogenic carbon dioxide emissions. *Science*, 10 Jun 2016: 352(6291), 1312–1314; Available at: https://doi.org/10.1126/science.aad8132; checked at 16 Feb 2018

News from the National Academies. (2015). Available at http://www8.nationalacademies.org/onpinews/newsitem.aspx?RecordID=02102015. Assessed at 20 June 2017.

Porter, E. (2017). *To curb global warming, science fiction may become fact*. Economic Scene, 4 April 2017. Available at: https://www.nytimes.com/2017/04/04/business/economy/geoengineering-climate-change.html. Assessed at 20 June 2017.

Roberts, D. (2017). *2 remarkable facts that illustrate solar power's declining cost*. Vox 3 February 2017. Available at: https://www.vox.com/science-and-health/2016/12/22/14022114/solar-year-two-remarkable-facts. Assessed 29 June 2017.

Rockström, J., Gaffney, O., Rogelj, J., Meinshausen, M., Nakicenovic, N., Schellnhuber, H. J. (2017). *Science, 355* (6331) 1269–1271, 23 March 2017. https://doi.org/10.1126/science.aah3443. Available at https://de.scribd.com/document/343117244/A-roadmap-for-rapid-decarbonization. Assessed 7 April 2017.

SEI. (2016). The risks of relying on tomorrow's 'negative emissions' to guide today's mitigation action. In: *SEI Working Paper*, August 2016. Available at: https://www.sei-international.org/

mediamanager/documents/Publications/Climate/SEI-WP-2016-08-Negative-emissions.pdf. Assessed at 10 June 2017.

Stanford News Service. (2017). *Assuming easy carbon removal from the atmosphere is a high-stakes gamble, Stanford scientists say*. Stanford News Service, 18 May 2017. Available at: http://news.stanford.edu/press-releases/2017/05/18/carbon-removal-tgh-stakes-gamble/. Assessed at 8 June 2017.

Stocker, T. F., Qin, D., Plattner, G.-K., Alexander, L. V., Allen, S. K., Bindoff, N. L., et al. (2013). Technical summary. In T. F. Stocker, D. Qin, G.-K. Plattner, M. Tignor, S.K. Allen, J. Boschung, et al. (Eds.), *Climate change 2013: The physical science basis. Contribution of Working Group I to the Fifth Assessment Report of the Intergovernmental Panel on Climate Change*. Cambridge University Press, Cambridge, United Kingdom and New York, NY, USA, Thematic Focus Element 7 Carbon Cycle Perturbation and Uncertainties, page 96. Available at: https://www.ipcc.ch/pdf/assessment-report/ar5/wg1/WG1AR5_TS_FINAL.pdf. Assessed at 20 June 2017.

Wendle, J. (2017). *The Ominous Story of Syria's climate refugees*. Scientific American March 2017. Available at: https://www.scientificamerican.com/article/ominous-story-of-syria-climate-refugees/. Assessed 29 June 2017.

WG2. (2014). In C. B Field, V. R. Barros, D. J. Dokken, K. J. Mach, M. D. Mastrandrea, T. E. Bilir (Eds.) *Assessment box SPM.1 figure 1 from IPCC, 2014: Climate change 2014: impacts, adaptation, and vulnerability. Part A: Global and sectoral aspects. Contribution of working group II to the fifth assessment report of the intergovernmental panel on climate change*. Cambridge University Press, Cambridge, United Kingdom and New York, NY, USA. Available at: https://www.ipcc.ch/report/graphics/index.php?t=Assessment%20Reports&r=AR5%20-%20WG2. Assessed 27 June 2017.

William, N. (2017). *Global 2C warming limit not feasible, warns top economist*. Available at: http://www.climatechangenews.com/2017/01/04/global-2c-warming-limit-not-feasible-warns-top-economist/. Assessed 10 January 2017.

Vulnerability Here, There, and Everywhere: What Happened to Ghana's Decentralized Climate Change Adaptation Policy?

Issah Justice Musah-Surugu, Albert Ahenkan, Justice Nyigmah Bawole and Antwi Samuel Darkwah

Abstract Across all the Bretton Wood institutions, decentralization has been touted as a fulcrum of good governance. This idea has had a sweeping effect across many areas of governance including climate change adaptation. However, the emerging climate change policy literature have had less focus on how decentralization can enhance adaptation governance at the local level. In Ghana, local governments have been given adaptation responsibility, through recently passed national climate change policies (NCCPs). This chapter of the book draws on experiences from the implementation of the NCCPs at the district level in Ghana which is perceived as a luminary of decentralization in Sub-Saharan Africa. The chapter specifically assesses the extent to which decentralization of NCCPs has impacted adaptation governance at the local level. The chapter therefore aims at fostering understanding of the nuances of implementing decentralized adaptation governance in developing countries among scholars and policy practitioners. It concludes that though decentralized adaptation governance in Ghana increases the institutional space for community participation in adaptation governance it is falling far short of creating a the management regime capable of building requisite adaptive capacity as envisage by the NCCPs at the local level. Although it is extremely premature to draw reliable conclusions, the chapter identified some positive trends amidst challenges.

Keywords Climate change · Adaptation · Decentralization · Local government Governance

---

I. J. Musah-Surugu (✉) · A. Ahenkan · J. N. Bawole
Department of Public Administration and Health Services Management, University of Ghana Business School, Legon, P. O. Box LG 78, Accra, Ghana
e-mail: musah123@gmail.com; ijmusah_surugu@st.ug.edu.gh

I. J. Musah-Surugu · A. S. Darkwah
Faculty of Regional Development and International Studies, Mendel University in Brno, Brno, Czech Republic

F. Alves et al. (eds.), *Theory and Practice of Climate Adaptation*,
Climate Change Management, https://doi.org/10.1007/978-3-319-72874-2_6

Introduction

Decentralization—a vehicle of New Public Management has been vigorously promoted by Bretton Wood institutions a fulcrum of `good governance across many developing countries. In Ghana, decentralization has involved the creation of 216 local governments (LGs). These LGs (known as districts assemblies) have since become the de facto implementers of national policies at the local level including climate change adaptation (CCA) related governance. Yet, the emerging climate change policy discourses largely ignore how existing decentralised governance space impact adaptation governance (Brockhaus and Kambiré 2009). This book chapter, therefore, examines: (1) the extent to which adaptation governance have been decentralized within Ghana's existing decentralization infrastructure; and (2) the extent to which the decentralization of adaptation policy implementation has also impacted adaptation governance at the sub-national level in Ghana. The chapter presents the results of qualitative research conducted across six selected districts in Ghana. These districts include two LGs each from Greater Accra, Northern Region and Ashanti region. The chapter aims to contribute to the emerging climate change policy literature.

The chapter commences with a review of Ghana's vulnerability to climate change and decentralization of adaptation governance in Ghana. It proceeds to examine the progress of decentralized CCA governance from a variety of perspectives—deconcentrated, devolved and autonomous local institutions accounts. It then moves on to examine how the decentralization of adaptation management in Ghana has provided local institutional space for participatory and pluralistic adaptation governance. The final part of the chapter provides conclusions and policy suggestion for enhancing decentralized adaptation governance. Adaptation governance in this study refers to a set of regulatory processes and mechanisms through which LGs actors influence pre-emptive measure (soft and hard or flow and stock) designed to systematically address climate related risk during and/or before climate hazard occurs. As well articulated in the literature "hard/stock" measures includes construction of irrigation, sea walls, and dykes, land use changes, etc. whiles 'soft/flow' measure includes climate induced fiscal realignments, information sharing, capacity building, insurance, etc. (Musah-Surugu et al. 2018; Kumamoto and Mills 2012). The study aligned its strategy and methods within the qualitative research approach which is relevant for conducting in-depth studies about the context and perspectives of participants (Creswell 2014). The qualitative research approach was considered appropriate for portraying the extent to which adaptation governance in Ghana have been decentralized and how the decentralization of adaptation governance has impacted CCA management at the micro level. Data used in this study include interviews and focus group discussions (FGDs conducted across six LGs in Ghana.

Multiple sampling techniques—purposive and cluster sampling were used to select 138 participants for both interviews and FGD. A purposive sampling technique was utilized to select 54 respondents (6 staff of deconcentrated agencies, 8 NGOs staff, 30 LG officials, and 10 opinion and quasi-community groups' leaders) and interviewed them. Using a combination of purposive and cluster sampling techniques, the remaining 84 are local citizens who by the nature of their occupation and livelihoods experience the impacts of climate were also selected. Out of the 84 respondents 30 face to face in-depth interviews was conducted. The remaining 54 participated in FGD which was made up of 6 different focus groups of 9 people each. The principal researcher facilitated each of the 6-FGDs and assisted by data collection assistant. The FGDs were very interactive and engaging; respondents were allowed to interact and argue among themselves interference from the researcher. Both interviews and FGDs lasted between 35 min and an hour. The data was collected between March and November 2016. Throughout the entire period of the data collection the researchers remained open to the information gathered from the field. Collected data was subsequently transcribed and read over several times to find common themes aiding the analysis and discussion. The multiple sources of data enhanced triangulation ensuring confluence of evidence that bleeds reliability and validity of the data obtained (Creswell).

Literature Review: Finding an Analytical Framework

Ghana's Vulnerability to Climate Change

Ghana, a former colony of the British, lies on the south-central coast of West Africa. The climate of Ghana is largely tropical. The country's climate which is strongly influenced by the West African Monsoon has witnessed dramatic changes over the past century (McSweeney et al. 2008). Ghana's climatic variability which is more than likely to continue for the next century would have unprecedented socio-ecological, economic and political ramification. Scientific accounts have established that climate change would significantly aggravate water stress, reduce food security, increase impacts from extreme weather events, displace many people (due to floods and sea level rise), and potentially increase the transmission of vector-borne diseases in Ghana (Eguavoen 2012; Codjoe et al. 2012; Agyemang-Bonsu et al. 2008; Mendelsohn et al. 2006; Dietz et al. 2003). These assertions are underscored by recent phenomenon of extreme droughts and floods with consequential loss of life and property, reduction in the gross domestic product, the crisis in power generation from the hydroelectric dam and spatial mobility of vulnerable groups. Whiles the impact of climate change are unequivocal and convoluted across space Ghana's' adaptive capacity remains extremely marginal (Owusu et al. 2015). Table 1 below provides the summary of Ghana's climate risk exposure and consequent adaptation needs. The complexity and ubiquitous nature of climate impact make a good case for participatory governance globally. Consequently, the next section discusses the institutional and development context within Ghana's participatory adaption governance unfolds.

Table 1 Climate change risk and adaptation needs in Ghana

Ecological zone	Climate risk	Risk level	Adaptation needs
Coastal savanna	Sea level rise	High	Value-added climate change based capacity building Systematic climate-resilient sectorial planning Protection of coastal infrastructure and eco-resources Enhanced alternative livelihood system Expand access to health and education Improved migrants and hosts communities dialogue
	Out migration	Medium	
	Weak livelihood support	Medium	
	See erosion	Extreme	
High forest	Erratic rainfall	Medium	Enhanced collaborative natural resources governance Sustainable forestry and on-farm conservation Access to alternative biofuel sources Augmented alternative livelihood system Sophisticated climate resistant agriculture Proactive response and prevention of bush fires
	Late state of rains	High	
	Early seasation rate of rains	High	
	Dry spell	Low	
Transitional	Low rainfall	High	Sufficient social support systems Early warning mechanism and public education Rapid response and disaster management Financial aid and insurance schemes Demand driven agriculture system State of the art alternative livelihood system Improve local governance across all economic sectors
	Rainfall extreme	High	
	Crop failure	High	
	Reduced minor seasons rains	Low	
	Mid-season brakes	High	
	Shortening of planting season	Medium	
Guinea and Sudan savannah	Long dry spell	High	Robust remittance flows infrastructure State of the art alternative livelihood Demand driven agriculture system Supportive ecosystem-based adaptation Strengthened disease surveillance and response systems Water storage and conservation systems Environmental sanitation and hygiene education
	Frequent flooding	High	
	Out migration	Extreme	
	Erratic rainfall	Medium	
	Rising temperature	High	

Sources Adapted from NCCP-2013; NCCAS-2010; Ghana 3rd National communication to UNFCCC-2014; Agyemang-Bonsu et al. (2008)

Ghana's Decentralized System: A Vehicle for Governing Adaptation at the Micro Level

At the heart of adaptation governance is the institutional and development context within which adaptation policy is to be condoned and implemented. It is, therefore, logical to place the above synthesized climate change risk and needs of Ghana within such context. Ghana's local development planning system is anchored on New Public Management paradigm- decentralization and participatory governance. Just as espoused by it devotees, decentralization of local development planning in Ghana is theoretically expected to improve local governance (Bawole 2017; Yeboah-Assiamah 2016; Ayee 1996). The current form of decentralization in Ghana has been implemented since 1988, with the aim of encouraging popular participation in local level decision making. Local governments are therefore created as the modicum of decentralization and are given responsibility for the overall development of their territory (Ayee 1996). It is then not far from this reason that the two primary instruments directly regulating adaptation management (National Climate Change Policy—NCCP and National CCA Strategy—NCCAS) in Ghana were prepared and designed within the context of national decentralized governance system. In effect, therefore, the 216 metropolitan, municipal, and district assemblies in Ghana should be responsible for the implementation of CCA actions aim at reducing institutional and individuals' exposure to climate havocs.

Legal Frameworks of decentralization (Constitution of Ghana and the Local Government Act of 1993, Act 462, NCCP-2013 and NCCAS-2010) all aim, among others, to achieve Ghana's sustainable development objective by clearly defining pathways for dealing with the challenges of climate change at the micro level. The NCCP and NCCAS share the responsibility of adaptation governance among state institutions with LGs been the lead actors. Nevertheless, there is an obvious top-down implementation structure where Ministries, Departments and Agencies (MDAs) at the national level are responsible for planning, monitoring and evaluation of adaptation initiatives while LGs' institutions execute. The lead agency for the implementation of the decentralized national adaptation governance is, therefore, the Ministry of Environment, Science and Technology assisted by the National Climate Change Committee.

Box 1:
The Functions of National Climate Change Committee:

1. The day-to-day management of the National Adaptation Strategy
2. The supervision of programmes/projects
3. The preparation of guidelines for programmes/projects
4. Monitoring and evaluation of programmes/projects and the
5. preparation of project reports

6. Support for the local institutions in capacity building and participation
7. Establish policy guidelines for programmes/projects operations
8. Approve programme/projects and recommend them for funding
9. Recommend impact evaluation studies to determine whether the
10. Specific interventions are producing their intended effects on the poor
11. and vulnerable groups; and
12. Review progress reports.

Sources Adapted from NCCP-2013 and NCCAS-2010

Even though the legal regime of Ghana's decentralization avers that DAs are established on the concept of devolution, anchored on subsidiarity principles, power relation are still skewed in favour of central level institutions (Yeboah-Assiamah 2016; Ayee 1996). In effect LGs' institutions are given guidelines from the National Climate Change Committee concerning the preparation of CCA initiatives. Similarly, at the local level, the Town/Area councils and unit committees are supposed to prepare adaptation plans and submit them to the District Assemblies for incorporation into the District Plans. District Environmental Management Committee (DEMC) is supposed to be strengthened by central government to take care of the implementation of adaptation actions (Porter and Young 1998). The policy envisaged participatory adaptation governance within a local space where LGs are assisted by other decentralized departments, Non-governmental organizations (NGOs), traditional authorities and the private sector in the preparation of detailed adaptation action plans and their implementation. The policy further requires that adaptations needs are carefully budgeted for and incorporate in the district development plans.

Narratives so far legitimise questions on the extent to which adaptation governance has actually been decentralized and impact it has made on adaptation governance at the local level. Constellation of reasons has been provided for decentralizing adaptation policy implementation. Major among these are: most resources that influence vulnerability, exposure, and sensitivity to climate change are local; LGs are intermediary for non-local agencies and local citizens; LGs are the first port of call for individuals and institutions looking for information on environmental related issues; and most importantly, local organizations set many strategic and long-term environmental policies that have implications for adaptation (Mearns and Norton 2010; Agrawal 2008; Ribot and Larson 2005). In indeed, the political and economic context of decentralization inevitably affects the implementation of climate change policies and the potential for real improvement in local people's adaptive capacity. Although Ghana has been presented as a role model of decentralization for the past three decades, it has not escaped some inherent challenges. A wealth of literature from Ghana confirms acute reluctance of the central government to allow genuine autonomy to decentralized institutions to manage their affairs and resources (Ayee 1996; Yeboah-Assiamah 2016; Bawole 2017).

Indeed, in many developing countries, a rhetorical political commitment, limited fiscal autonomy, poor local organizational resource and other assemblages of hurdles suffocates the usefulness of decentralization. These reported constraints are clearly evident in Ghana's efforts at decentralisation CCA administration. Previous researchers on adaptation related governance have confirmed that LGs in SSA are barely able to undertake their fiducial disaster management and environmental change responsibility due to weak resources husbandry, legal and institutional capacity (Brockhaus and Kambiré 2009; Porter and Young 1998).

Discussion of Data from the Field

Extent to Which Adaptation Governance Have Been Decentralized in Ghana

The structure of Ghana adaptation policies implementation reflects a fused decentralized system—an arrangement in which institutions extending from the central government (deconcentrated arm) and locally based institutions (devolved arm) are expected to be actively involved in the governance of adaptation at the sub-national level. The metropolitan, municipal and district (now after referred to as district assemblies or LGs) are the devolved institutions that perform deliberative, legislative and executive function that has a trickle-down effect on local resilience, sensitivity and exposure to climate variability. Similarly, extended line ministries such as specialised agencies are deconcentrated unit that is theoretically expected to collaborate with the devolved arm and other autonomous local agencies to perform locally based functions which could have both direct and indirect implication for adaptation.

The institutional arrangement for the governance of adaptation at the local level in Ghana hence mirrors the global view that robust adaptation response is that which is conceived, deliberated and organized around a network of interested stakeholders that engenders collective decision and responsibility capable of building adaptive capacity. It is on this basis that this part of the study interrogates the extent to which needful agencies are actually involved in the management of CCA at the sub-national level. Below, therefore, proceeds with the involvement of deconcentrated, devolved, and autonomous local agencies in CCA management respectively. They are the main developmental agencies at the local level in Ghana and have also been tasked by the various NCCPs to integrate CCA into their decision making.

Public Agencies and Specialized Bodies (Deconcentrated Arms) and Adaption Governance

The field data indicated that at the local level various departments of ministries and agencies which deal with some aspects of natural and built environment undertake actions that have both direct and indirect implication for building local adaptive capacity. Notable among these are the National Disaster Management Organisation (NADMO), Ministry of Food and Agriculture (MOFA), Ministry of health, Ministry of Environment, Science, Technology and Innovation (MESTI), the Ministry of Lands and Natural Resources, Ministry of Local Government and Rural Development (MLGRD), Environmental Protection Agency (EPA) and other research institutions. For example, it was evident across sampled LGs that NADMO and MOFA directorates have been in the forefront of creating awareness on impacts of extreme climate change impacts and develop mechanisms for adaptation. Similarly, in some of the LG it was noted that the department of the MOFA was exclusively implementing agricultural projects aimed at improving local farmers' productivity (adaptive capacity) through the provision of tolerable climate crops.

In spite of the need for interaction and cooperation among local development actors as envisioned by climate change related policies in Ghana in the governance of adaptation, due to its boundary spanning nature, many deconcentrated agencies operated independently of the various LGs. The above confirms the widely reported institutional silos in local government system that conspicuously works against effective local self-governance (Adu-Boateng 2015). Most of the adaptation programs implemented by deconcentrated agencies were less informed by the district-wide development plan—situation that reduces resources efficiency. Deconcentrated agencies rather take directives from upper ministries and hence are not professionally or legal confined to consult LGs that are legally mandated to initiate a district development plan and oversee its execution. An officer of a deconcentrated agency thus explains:

> Though the LG service act 2003 (Act 656) and subsequent Legislative instruments call for a transfer of some functions from central ministries into decentralized department of DAs, this has fallen flat on its face. We still take instruction from ministries and scarcely consult with the DAs on some of the things we do. Our action plans mimic that of the ministries'.

Local Government (Devolved Arm) and Adaptation Governance

At the lowest tier of government in Ghana, DAs serve as the melting pot for the implementation of adaptation programs. In fact both NCCP and NCCAS have christened them as the most crucial unit for building local resilience. Since the adoption of NCCAS and NCCP, all 216 DAs have started in earnest, implementing various adaptation oriented actions, albeit it's sporadic, non-isomorphic, and uncoordinated approaches. Typical areas considered as priority areas of various

DAs includes disaster risk reduction covering bushfire and flooding, livelihood diversification and economic empowerment, gender mainstreaming, awareness creation, agricultural productivity improvement, healthcare and land use management. Study participant from a DAs remarks:

> We undertake the following activities as a response to expected and ongoing impacts of climate change: managing waste to create wealth for local people, soil conservation, awareness creation, bush fire controls, eco-tourism development, land use management, afforestation, etc.

The integration of climate change into local governance are also anchored on assessment of specific sectors sensitivity to climate risk, availability of organizational resource-set such as human resources, climate data, political and public support, as well as other incentives that could be derived from central government and donors. A district planning officer indicates thus:

> We want to receive auxiliary funding supports such as the District Development Facility and those that could come from donors for incorporating responsible environmental strategies into local governance.

The field data confirms that DAs response to climate change varies. Whiles some DAs weigh adaptation as local development priority other consider it as a peripheral matter. Those who rate adaptation as development priority have CCA action plans, functional DEMC, earmarked resources, performance measures, etc. in place. This finding is also consistent with reported cases of variation in the implementation of adaptation policies across states and local agencies (Musah-Surugu et al. 2018; Yaro 2010; Wurtenberger et al. 2011). Many factors including lack of clarity over what constitutes adaptation, amount to be allocated to adaptation, bureau responsible and climate-science heaviness are more than the reasons for the apparent variations in commitment. The significance of this finding is that many factors affecting effective functioning of decentralization system in general affect that of CCA management.

Discussion with study respondents also confirms the views across the literature that most local government undertakes reactionary response to climate change risk. It appeared strongly across all the DAs studied that most of their responses to climate change were not planned. A situation they blamed on politics of local development, LGs subsidiarity to the central government, limited institutional capacity due to weak decentralization policy where responsibility is not accompanied with necessary means. For example, tentatively each DA is supposed to have a functional DEMC with an overall monitoring and coordination of environmental programs responsibility. It was startling, however, that in some district these bodies did not exist. In DAs where it existed, it was barely functioning due to the poor remuneration of the body. Another reason for their ineffectiveness is the lack of clear cut terms of reference for the committee in the LGs' Act, and thus no budget is made for its operations.

Other Locally Based Agencies (Autonomous Arm) and Adaptation Governance

There was evidence of local autonomous institutions such as NGOs seriously engaged in CCA governance or activities that simultaneous reflect the reduction of local exposure to climate change impacts. Specific projects and programs autonomous local level development partners undertook jointly with DAs includes but not limited to the promotion of non-formal environmental education, afforestation and agroforestry training, management of woodlands, watersheds and wildlife. NGOs and Epistemic community (e.g., Hunger Project—THP, Regional Institute of Population Studies -RIPS, German development Agency -GIZ) organized community based CCA durbars, developed actions and provides resources for implementations. Documentary evidence from THP shows over 4000 people on average received training between 2010 and 2013 on food security related issues. Interview with representative from RIPS coastal adaptation and resilience project also indicates that over 10,000 people who are most often affected by coastal erosion, floods, storms, surges, etc. have been trained on adaptation related options across hundreds of communities.

These field reports supports finding of Yaro et al. (2015) that NGOs are leading crusaders of CCA at the local level in Ghana. They use adverts, radio talks, community durbars, school seminars and climate competitions to set agenda for climate change within a decentralized adaptation governance space. A beneficiary of NGOs' adaptation program confirms this:

> GIZ various radio talk show and farm visits are gradually changing the face of farming in our community. Today many people in this community are not just aware of rain fall variability but are much better position to manage it menace.

According to the NGOs awareness creation and community dialogue on climate change is the first port of call for preventing maladaptation. An interviewer from GIZ buttresses this point as follows:

> We see awareness creation as an essential starting point for a successful implementation of CCA program. When a treatment group is not aware of its problems, it becomes extremely cumbersome for the group to accept the treatment and follow its course. This is why advocacy and awareness creation on ongoing impact of anthropogenic climate change is a critical component of our strategy to build local adaptive capacity.

An official of THP explained:

> Some of our epicenters have community demonstration farms, where villagers learn composting, intercropping and other methods, like drip irrigation, to improve farm output, restore soil fertility and make the best use of scarce resources.
>
> I can harvest three times more than I used to harvest after receiving training and drought resistant seeds from GIZ's AAESCC project. I think our weather information we are receiving from the granular weather station is very beneficial for planting at the right time.

Indeed, evidence abound that NGOs broaden local participation in governance processes that affect adaptation given their vibrancy in exerting the necessary pressure on development stakeholders, mobilize sufficient resource for reforms, and mobilize political support at the local level (Zahran et al. 2008; Vachon and Menz 2006). The relevance of autonomous local agencies in the face of decentralized adaptation governance have led to the conclusion that they are the complimentary arbiter for mobiling efficient bottom-up adaptation actions (Yaro et al. 2015).

How Has Decentralization of Adaptation Governance Impacted Local Adaptive Capacity?

This part of the study presents evidence from the field on the significance of decentralizating adaptation management for building local adaptation capacity.

Institutional Flexibility and Effective Coordination: Implication for Building Adaptive Capacity

Historical studies of resilience and societal collapse have suggested that adaptive capacity is a function of flexibility (see, Zheng et al. 2014; Butzer and Endfield 2012). Evidence from the various districts suggests that decentralization has enabled LGs to have the needed flexibility to integrate specific adaptation metrics into local development plans depending on the expected climate risk. The most visible flexibility cited by respondents was the autonomy LGs have over local development plans and decision making. Respondents concur that their ability to decide on what is needed, what should be done, who should be affected etc. at any material moment, albeit central government interference, provides a significant opportunity for reasonable adjustment necessary for building local adaptive capacity. They were further convinced that without decentralization it would be quite difficult to have the necessary flexibility to negotiate how resources should be used in respect of building the capacity of vulnerable local people. Another important reported issue that emerged from the interviews is that decentralization has enhanced the coordination of local development projects reducing overlapping and duplication of activities. According to the fields' data decentralization is an opportunities for ensuring that initiatives build on each other and make use of insights gained from earlier efforts.

Decentralization Permits Quicker and Effective Response Against Climatic Extremes

The pace of government response to climate change influence local sensitivity to climate impacts. As a result, effective adaptation decision is a function of institutional response pace. Relying on local institutions, structures, traditional knowledge and local experience with environmental change decentralization of adaptation governance can offer a just-in-time service for local people. In most of the focus group discussions it emerged that LGs' institutions were the first port of call during catastrophic climatic events. Some respondents were of the view that the pace with which local institutions respond to their needs, though could be improved, is better compared to central government responses. An interviewee from a DA reported thus:

> In 2015 when the marauding floods in Accra affected many lives, homes and properties Accra Metropolitan Assembly was, and NADMO was the first local agencies to respond with immediate intervention. In the past, the DA has been involved in emergency stabilization and recovery. Central government most often focuses on long-term measure needed to prevent the occurrence of such situations.

Resource Leveraging: Decentralization Allows Willing but Incapable LG to Access NGOs Support

NGOs have been seen as both instrumental and political tool for development in the era of collaborative public management. LGs that are willing to undertake adaptation action but lack the capacity have utilized the resource fields of autonomous local institutions as clearly envisioned by Ghana's NCCP and NCCAS. According to autonomous local development actors interviewed, the existence of decentralized structures in Ghana provides them with the opportunity to identify and cooperate with institutions that have an institutional memory for managing climate change related issues in the past. LGs officials were also of the opinion that they have gained much support from NGOs in penetration major difficult adaptation related fields and helping to change the narratives for building resilience. Similar responses dominated local adaptation project beneficiaries. Most beneficiaries agreed that most of the support they have received in the past that contributed to building their adaptive capacity came from NGOs with the help of LGs staffs. This narrative resonates with the literature which argues that exchange of resources among state and non-state actors represent a valuable leveraging mechanism for undertaking a participatory adaptation action (Agrawal 2008). Climate change remains a complex multidimensional problem that challenges the resource capacity of single development actor. The nature, depth, severity of climate impacts and cost remains highly uncertain and hence requires all-encompassing efforts. As clearly observed, the apparent decentralization of adaptation governance in Ghana allows all willing local actors to participate in the course of finding optimal solution.

Selective Planning: Exploring Diversity for Adaptive Capacity Building

Again, evidence from the field indicates that decentralised adaptation governance allows selective planning and implementation of the program taken into account spatial vulnerability. Generally, is has been established that the effect of climate variability is not uniformly felt across different spatial and societal scales, albeit its global nature. The differential impacts of climate change have been attributed to the varying adaptive capacity, sensitivity and vulnerability of different systems to the change. Respondent noted that decentralized NCCP allows LGs to undertake development actions based on the territorial needs/vulnerability.

Participatory Mapping: A Vehicle for Building Consensus, Legitimacy and Reduction in Areas of Tension

Evidence from the field demonstrated that (perhaps unconsciously) decentralization had encouraged participation of the governed into local adaptation based governance. This according to respondents helps in developing more legitimised adaptation projects that utilise bottom-up information. It was noted that through decentralization local adaptation decisions takes onboard "indigenous ecological knowledge", and "lived-experience" of local people that are relevant for adapting to the vagaries of climate. Although respondents largely differed on the extent to which decentralization take on "cultural knowledge", there was some consensus that decentralization has empowered local people to make input into local governance that affects adaptation to climate change . This was illustrated in this study on several occasions. For example, in one DANGO had partnered with the DA to initiative and developed community driven adaptation program which allows community members to use historical matrices, pair-wise ranking and cross impact analysis to access their vulnerability. The process allowed the use of long-standing local expertise to identify and prioritise adaptation strategies. According to one LG's official suggestions from the people were finally configured into resilience based programs that were later rolled out. A district planning officer confirms that decentralization enhance local participation and reported thus:

> Before the preparation of our district development plans, we are required by statute to involve all beneficiaries in order to develop a medium development plan that reflects the real needs of the people.

This position was corroborated by a district finance and budgeting officer when he reported:

> Our team from the planning department and relevant units just finalised local level consultation exercise across all zonal councils. The exercise was to allow the district apprise itself with the changing needs of the people and what the people feel can be done to improve their conditions.

These findings support that of Glazebrook (2011) who narrated that the agricultural and adaptation expertise of women, their sensitivity to local conditions and long-standing knowledge-practices about growing food in marginal conditions offer significant contributions to adaptation efforts. Despite reported cases of improved participation ostensibly emanating from decentralization, critics are of the view that instead of inviting an 'alternative view' into local adaptation decision-making, actors use political consideration to distill local views and therefore constraints the extent to which local participation can enhance adaptation (Nadasdy 2005). The view of a community member confirms this as she indicated that there are instances where the view of local people only remains at the level of consultation and does not translate into decision making. Similar views indicated that in most cases only a handful of influential community members are consulted, with those highly vulnerable (women, children) neglected. Despite the reported limitation, the findings confirm that through decentralization, the tendency to invite and utilise locally relevant knowledge into local adaptation governance is improved, at least some stakeholders get the opportunity to be part of the processes of finding adaptable solutions.

Determining Flows of External Support to Different Social Groups: Target Mapping for Capacity Building

In the adaptation literature, there is a consensus that local vulnerability is differentiated across social groups. Whiles top-down oriented social protection programs do not always lead to inclusion, equity and local ownership, decentralized adaptation governance envisages pluralistic adaptation governance where LGs would be better placed to determine the flow of external support to different social groups based on needs and urgency. Respondents' views clearly confirm that the decentralization system has critically contributed in undertaking target mapping that positively discriminates to ensure that those that are highly vulnerable are given special attention albeit some challenges. The position of and NGO official supports this view, and he argued thus:

> Sensitivity to climate change varies across different social groups based on location, economic sector, income and political clout. To build local capacity in the face of dwindling LGs' resource basket requires target mapping based on those who voice must be heard first – high risk but low adaptive ability. The decentralization of NCCP empowers us in some ways to help direct resources to those areas and people who are in dire need.

Decentralized Structures as Vehicles for Building Local Adaptive Capacity

It has have been argued throughout policy implementation discourses that the vehicle through which such interventions are carried out influence its effectiveness. In this study, respondents corroborated this long held view. They explained that decentralized structures at the local levels help local development actors to effectively implement programs that sought to improve local adaptive capacity to climate change. For example in some of the districts, it was found out that top-down pro-poor policy interventions such as Ghana Social Opportunities Project (GSOP) and the Ghana Environmental Management Program (GEMP) which was being implemented by the MEST through district assemblies help in creating climate change awareness, support farmers with improved farm technologies and also create direct jobs for other rural folks. As argued by Bawole (2017) the success of central government program such as the GEMP would have been least if the decentralized local institutions were not readily available and accessible. NADMO official supported this position when he argued thus:

> When the national adaptation policy and NCCP were passed in 2010 and 2013 respectively some offices within the various LG entities had the functional domains to readily help in the implementation of those goals envisioned in these policies. The existing structures of LGs' entities help in reducing the initial cost setting up a new institution to carry out NCCP goals. Similarly, it also allows our offices to undertake incremental actions that build on formal actions taken to reduce vulnerable people sensitivity to the scourge of climate change.

The GEMP example is one of local development interventions that simultaneously sought to reduce local vulnerability to the catastrophic impacts of climate change at the local level. Indeed, the study participant narrated several other central government, donor agencies and NGOs community led interventions that were specifically aimed at building local adaptive capacity or has indirect barring on local resilience. In their opinion, such intervention would have at best delayed or at worst have limited coverage impact without the existence of LG entities, which are creations of the nation's decentralization policy. The study confirmed that the decentralization of climate change policy implementation in Ghana allowed LGs entities to easily consider the integration of climate risk reduction, adaptive capacity building in many social protection programs such as the Livelihood Empowerment Against Poverty, the Ghana School Feeding Programme, and the Ghana Social Opportunities Project. Previous research confirms that such programs are directly impacting the lives of poorer people which could intern build their resilience against climatic shocks (Bawole 2017).

Availability and Access to Capable Caretakers of Non-state Actor Adaptation Projects

Evidence from the field illustrated that decentralization facilitates easy access to legitimate, representative and capable caretakers of adaptation projects undertaken by non-state agency. This, according to respondent enhances project lifespan and offers an opportunity for building cumulative efforts towards resilience. NGO-led adaptation project beneficiary position supports this claim, and he argued thus:

> NGOs that come to this community to help us often have limited number of years to implement intervention and exist the communities. LGs' institutions represent local identity which can take over both financial and regulatory responsibility needed to sustain such interventions.

According to respondents without the decentralised system adaptation project undertaken by non-state agencies would have limited revolving impact on adaptation. Some projects that were identified in the study which LGs have become caretakers after the exist of those who original built it include the Fish smoking at Usher towns, the drains project in Abogbloshie, and the Local Granula weather station at Gbogdaa that were all built by NGOs. An officer of other NGOs argued thus:

> Decentralization facilitates the identification of near-right local agencies to partner to increase the legitimacy of plan programs and subsequently handover the management of same after we exist. Currently, all our ongoing adaptation projects are done in conjunction with decentralized local structures such as zonal councils, NADMO units, district information department, etc. As we speak the Accra Metropolitan Development Agency has taken over the management of the fish shed that we constructed for the coastal fisher folks.

Conclusion and Policy Suggestion

Through qualitative research approach this chapter of the book specifically assessed the extent to which adaptation governance in Ghana have been decentralized and how it has/is impacting adaptation governance at the local level. Since generalization of results, conclusions and recommendation is not an expected output of qualitative research this study is accordingly limited to transfer its conclusion to many other areas, except those that are of similar characteristic. As reported so far, it clear that as Ghana's decentralization system requires, there are multiples of institutions involved in the governance of adaptation at the sub-national level. There is evidence of line ministries, departments and agencies involved in the implementation of climate change policy at the local level. Similarly, autonomous institutions such as NGOs are actively involved in adaptation services delivery. In all DAs are the de facto policy homes for the implementation of adaptation policies. Through local development plans, budgetary realignment and project implementation adaptation metrics are made integral part of local affairs. Despite the inability to conclude firmly that decentralization is the surest way to improve local adaptation governance based on this single study, evidence from Ghana thus provides good lessons. First, it was clear that decentralized CCA administration aid in generating good contact with those living with climate problems and further recognises the importance of their long-standing knowledge-practices (cultural, ecological knowledge) within the overall CCA administration. Second, it also helps in strengthening the democratic process of initiating and implementing actions that aim at building the capacity of those vulnerable to environmental changes. Third, it acknowledges the inadequacy of one-fit-all approaches to adaptation management. In effect, it takes into consideration spatial variability in adaptation planning—accounting for a wider range of local-level priorities. Fourth, it reflects a growing recognition of the importance of local initiatives, self-reliance, and participation in the development and implementation of resilience strategies.

However, some critical issues need to be considered if the marginal benefits of decentralizing CCA are to be enriched and sustained. Ideally, when responsibilities are transferred from central to LGs, the former is expected to contribute directly (e.g., technical support) to build the capacity of the later to efficiently function. Quite often, the central government has failed to do so. In this study, this concern came out strongly. It was noted that creation of a legal framework for transferring adaptation responsibility from the centre to periphery does not necessary leads to efficient public administration of CCA. Respondents cited numerous institutional and cultural challenges perilously hanging on the neck of Ghana's decentralization system which simultaneously affects the implementation of decentralized CCA governance. Major among these challenges include limited technical staff, funding, coordination, political manoeuvring and interest, embedded social norms as well as the complex nature of climate change. For example, as clearly reported by a respondent from NADMO significant funds for adaptation support services at the district level must come from the DAs, which is already grappling with limited resources against other equally important development priorities. Similarly, it was noted that the supposed DEMC is barely operational in most DA. In DA where they existed, their decisions remain on the books of the DA without actual implementation given the political consideration of many decision that is taken at the DA level. The study noted that Ghana's experience in decentralized adaptation management is by no means unique in SSA. In Kenya, for example, similar difficulties regarding the shortage of technical staff and funds at the district level have been reported (Poole and Leakey 1996).

To conclude, the Ghanaian government is committed to popular participation in public governance, but does not seem to have fully recognized the enormous difficulties which attend such approach, particularly in contestable fields like CCA. Adaptation is a highly contentious activity which inevitably involves conflict and negotiation between a wide range of actors with differing agendas and access to power. The paper proposes the following to enhance the governance of complex public issues such as climate change: (1) it is essential for local agencies to know much more about what CCA means for them in concrete and tangible terms. Particularly, they need to be provided with information on the nature of the risks they face and the urgency with which they must responds. Also, information on CCA needs must be framed in ways that make sense to local decision-makers. Again, the way out for LGs to improve their resource basket to respond to both incremental cost and investment needs of CCA they must think outside the box. They should, for example, look at their comparative advantage on the whole range of internal revenue menu and focus on where they have a competitive advantage. In addition to the above, there must be proper adherence to financial management regulations to ensure that LGs limited resources are well utilised. This recommendation is in line with the Auditor General reports that continue to depict all forms of financial malfeasance in Ghana's LGs system. At the central government level, it is also worth establishing Municipal Financial Authority which would give DAs the opportunity to raise funds from the capital market for productive investment which marginal benefits, perhaps, would improve local adaptive capacity.

References

Adu-Boateng, A. (2015). Barriers to climate change policy responses for urban areas: A study of Tamale Metropolitan Assembly, Ghana. *Current Opinion in Environmental Sustainability, 13*, 49–57.

Agrawal, A. (2008). Local Institution and climate change adaptation. *Social Development Notes: Community Driven Development, The Social Dimensions of climate change, 113*, 1–8.

Agyemang-Bonsu, W.K., Minia, Z., Dontwi, J., Dontwi, I.K., Buabeng, S. N., Baffoe-Bonnie, B., & Frimpong, E.B. (2008). Ghana climate change impacts, vulnerability and adaptation assessments. *Accra: Environmental Protection Agency.*

Ayee, J. R. (1996). The measurement of decentralization: The Ghanaian experience, 1988–92. *African Affairs, 95*(378), 31–50.

Bawole, J. N. (2017). Pro-poor decentralization in Ghana: Exploring the facilitators and the limitations. *Administrative Theory & Praxis, 39*(2), 122–139.

Brockhaus, M., & Kambiré, H. (2009). Decentralization: A window of opportunity for successful adaptation to climate change? In W. N. Adger, I. Lorenzoni, & K. L. O'Brien (Eds.), *Adapting to climate change: Thresholds, values, governance* (pp. 399–416). Cambridge: Cambridge University Press.

Butzer, K. W., & Endfield, G. H. (2012). Critical perspectives on historical collapse. *Proceedings of the National Academy of Sciences, 109*(10), 3628–3631.

Creswell, J. W. (2013). *Qualitative enquiry and research design: choosing among five traditions*. London: Sage Publications.

Codjoe, S.N.A., Atidoh, L.K., & Burkett, V. (2012). Gender and occupational perspectives on adaptation to climate extremes in the Afram Plains of Ghana. *Climatic Change, 110*(1–2), 431–454.

Dietz, T., Ostrom, E., & Stern, P. C. (2003). The struggle to govern the commons. *Science, 302*, 1907–1912.

Eguavoen, I. (2012). *Blessing and destruction: Climate change and trajectories of blame in Northern Ghana = Bénédiction et destruction* (No. 99). ZEF Working Paper Series.

Glazebrook, T. (2011). Women and climate change: A case-study from Northeast Ghana. *Hypatia, 26*(4), 762–782.

Kumamoto, M., & Mills, A. (2012). What African countries perceive to be adaptation priorities: results from 20 countries in the Africa adaptation programme. *Climate and Development, 4*(4), 265–274.

McSweeney, C., New, M., & Lizcano, G. (2008). UNDP climate change country profile: Ghana. *Oxford Report*. http://ncsp.undp.org/document/undpclimate-change-country-profile-11. visited, 11, 2014.

Mendelsohn, R., Dinar, A., & Williams, L. (2006). The distributional impact of climate change on rich and poor countries. *Environment and Development Economics, 11*(2), 159–178.

Musah-Surugu, J.I., Owusu, K., Yankson, P.W.K., & Ayisi, E.K. (2018). Mainstreaming climate change into local governance: financing and budgetary compliance in selected local governments in Ghana. *Development in Practice, 28*(1), 65–80.

Menz, F. C., & Vachon, S. (2006). The effectiveness of different policy regimes for promoting wind power: Experiences from the states. *Energy policy, 34*(14), 1786–1796.

Nadasdy, P. (2005). The anti-politics of TEK: The institutionalization of co-management discourse and practice. *Anthropologica, 47*(2), 215–232.

Owusu, K., Obour, P. B., & Asare-Baffour, S. (2015). Climate variability and climate change impacts on smallholder farmers in the Akuapem North District, Ghana. *Handbook of Climate Change Adaptation* (pp. 1791–1806). Berlin Heidelberg: Springer.

Poole, J.H., & Leakey, R.E. (1996). Kenyan case study. Lutz E. and Caldecott J.(eds.), 54–63.

Porter, G., & Young, E. (1998). Decentralized environmental management and popular participation in coastal Ghana. *Journal of International Development, 10*(4), 515–526.

Ribot, J. C., & Larson, A. M. (2005). *Decentralization of natural resources: experiences in Africa, Asia and Latin America*. London: Frank Cass.

World Bank, Mearns, R., & Norton, A. (2010). *Social dimensions of climate change*. Washington, DC: World Bank.

Würtenberger, L., Bunzeck, I.G., & van Tilburg, X. (2011). Initiatives related to climate change in Ghana

Yaro, J.A. (2010). The social dimensions of adaptation to climate change in Ghana. *World Bank Discussion Paper, 15*, 88.

Yaro, J. A., Teye, J., & Bawakyillenuo, S. (2015). Local institutions and adaptive capacity to climate change/variability in the northern savannah of Ghana. *Climate and Development, 7*(3), 235–245.

Yeboah-Assiamah, E. (2016). Power to the people! How far has the power gone to the people? A qualitative assessment of decentralization practice in Ghana. *Journal of Asian and African Studies, 51*(6), 683–699.

Zahran, S., Grover, H., Brody, S. D., & Vedlitz, A. (2008). Risk, stress, and capacity: Explaining metropolitan commitment to climate protection. *Urban Affairs Review, 43*(4), 447–474.

Zheng, J., Xiao, L., Fang, X., Hao, Z., Ge, Q., & Li, B. (2014). How climate change impacted the collapse of the Ming dynasty. *Climatic Change, 127*(2), 169–182.

Part II
Climate Change Adaptation, Resilience and Disaster Risk Management

Effect of a Drought on Cork Growth Along the Production Cycle

Carla Leite, Vanda Oliveira, Alexandra Lauw and Helena Pereira

Abstract Cork oak (*Quercus suber* L.) grows in the western Mediterranean region for which the most recent climatic scenarios predict higher temperatures and lower precipitation than usual values. Cork, the tree's outerbark, is obtained under a sustainable management system and has a considerable economic importance for forest producers and industry. Cork's specific set of properties allows multiple usages, from cork stoppers to insulating materials. This paper presents the first results of a dendroclimatological exploratory study about the effect of a severe drought in different moments of the cork 9-year production cycle, e.g. beginning, middle or end of the cycle. The results showed that the response of the phellogen (cork cambium) to the severe drought of 2004–2006 is independent of its age. In a mitigating strategy for the impact of the forthcoming more frequent drought events, and since cork growth decreases due to the reduction of water availability, forest managers should extend cork growth cycles and/or water cork oak stands. This is a way to ensure the production of cork with enough thickness to produce stoppers, thereby contributing to the overall sustainability of the cork sector in a climate change context.

Keywords Drought · Cork oak · Mediterranean region · Climate change

Introduction

Climate change is one of our most challenging issues in the 21st century[1] either regarding mitigation or adaptation strategies. The long term maintenance of forests is greatly determined by the adaptive capacity of species[2] under its two

[1]Stott et al. (2016).
[2]Braeuning et al. (2017).

C. Leite (✉) · V. Oliveira · A. Lauw · H. Pereira
Centro de Estudos Florestais, Instituto Superior de Agronomia,
Universidade de Lisboa, Tapada da Ajuda, 1349-017 Lisbon, Portugal
e-mail: cleite@isa.ulisboa.pt

F. Alves et al. (eds.), *Theory and Practice of Climate Adaptation*,
Climate Change Management, https://doi.org/10.1007/978-3-319-72874-2_7

components: the intrinsic adaptive capacity of trees and forests, and the socio-economic factors determining the capability to perform planned adaptation measures.[3] Although tree species can adjust to new environmental conditions, there is not much expertise about the processes involved.[4] In the meanwhile, forest managers do need to adapt their practices to the new reality in order to satisfy the needs of all their stakeholders, from the public in general to a rather competitive and demanding forest industry.

All climatic scenarios for Mediterranean ecosystems predict increasing temperatures, water deficits and periodicity, intensity and duration of extreme events, namely severe droughts, heat waves, heavy rainfall and fewer cold days.[5,6] Tree growth is constrained by water availability e.g. when evaporation is greater than soil moisture uptake, stomata close to reduce water loss decreasing carbon uptake and growth. For the Mediterranean basin, due to summer water deficits, it is expectable to have productivity losses and changes in species distribution.[7,8,9]

Cork oak forests cover an area of approximately 2.2 million ha and produce annually up to 200 thousand tons of cork that supply an important industry. [10] Portugal is the main producer of raw cork and cork products namely cork stoppers for the wine industry and boards.[11] The production of cork is based on the periodical removal of the cork layer from the stem. In the main producing regions, the period between consecutive cork removals is usually 9 years, enabling to have a cork plank with the desirable thickness (>24 mm) for the production of cork stoppers (the cork product with the major added value that supports all the cork segment).

The cork oak is ecologically plastic, adapting physiology and phenology to a changing environment, namely to summer drought and higher temperatures. This species vegetates well with mean annual precipitations of 600–800 mm, with 500 mm considered to be the minimum annual precipitation for a balanced tree development.[12] and [13] Furthermore it is sensitive to the amount and the moment when rains occur, namely precipitation in late spring has more influence on cork oak growth than the total annual precipitation, while it recovers quickly after a year of extreme dryness.[14]

---

[3]Lindner et al. (2010).

[4]See footnote 3.

[5]IPCC (2014).

[6]See footnote 3.

[7]Santos and Miranda (2006).

[8]Gea-Izquierdo et al. (2013).

[9]Piayda et al. (2014).

[10]APCOR (2016).

[11]Pereira (2007).

[12]Natividade (1950).

[13]See footnote 11.

[14]Besson et al. (2014)

In what relates to cork growth, results showed that rainfall, summer drought and temperature are determining factors; in particular, water availability is the most limiting factor for growth. [15,16,17] In fact, the rain period from May to September strongly influences activity of the phellogen (the cork cambium).[18] Cork growth is also extremely affected by drought conditions in short time scales (from 2 to 11 months) namely regarding spring precipitation. Nevertheless, cork growth rapidly recovers when drought conditions end.[19]

Similar to wood, it is possible to observe annual growth rings in cork as they are marked by the presence of a layer of darker latecork cells at the end of the growth season; the width of a cork ring is largely determined by the rate and duration of the meristematic division of the phellogen[20] which is seasonal with the environmental conditions of light and temperature as the main drivers for begin and end of dormancy.[21]

In spite of all the research performed so far about cork growth and its relation with climate (e.g. temperature and precipitation), we identified a gap of knowledge about the effect on cork growth of drought occurring in different moments of the 9-year cork growth cycle e.g. beginning, middle and end. Furthermore, as pointed out by[22] the present challenge is to translate the ecophysiology and forest ecology scientific research into forest management prescriptions useful for forest owners and managers.

This paper is an effort to overcome these shortcomings. With cork samples collected in three sites, located in one of the largest continuous forest areas of cork oak, we compared cork growth during the 2004–2006 drought that[23] considered as the driest event in the Iberian Peninsula over the last 140 years. The decrease of cork growth and also the recovery after the drought were analyzed. The complete growth (8 years) was also analyzed because the production of cork stoppers requires to have planks with more than 24 mm thickness. The sites were selected in order to have this drought event occurring in different moments along the growth cycle (see Table 1), allowing to study its effect according to the age of the phellogen in 2005 (the year in which the growth is negatively affected by the drought). Moreover, the results obtained are converted in forest management advices in order to adapt or mitigate the effects of the forthcoming more frequent droughts in the Mediterranean basin.

---

[15]Cherubini et al. (2003)

[16]Caritat et al. (2000)

[17]Oliveira et al. (2016)

[18]Pizzurro et al. (2010).

[19]See footnote 17.

[20]See footnote 11.

[21]Evert (2006).

[22]See footnote 2

[23]García-Herrera et al. (2007).

Table 1 Number of samples collected in each site, number of samples measured in this research, stripping year, civil years of the chronology (8 years) and age of the phellogen in 2005

	Number of samples collected	Number of samples measured	Stripping year	Civil years of the chronology	Age of the phellogen in 2005 (years)
Site 1	37	32	2007	1999–2006	7
Site 2	84	60	2009	2001–2008	5
Site 3	92	57	2012	2004–2011	2

Methods

Study Area

The cork samples used in this research were collected in three *montado* estates located in central-west of Portugal, in the Coruche municipality. The sites are set in the watershed of the Tagus river, in one of the main production areas of cork oak, where the Mediterranean climate prevails. In all the sites forest management is characterized by low tree density, with the trees being exploited for cork production and, in site 1, the herbaceous layer is used for extensive grazing.

Data Acquisition

A set of 224 cork samples was collected at breast height (1.30 m) during the stripping season of adult productive cork oak trees. The trees were randomly selected in the sites and a proportion of the total trees under striping was sampled. Table 1 shows some characteristics of the 3 sites sampled. All cork samples presented 8 complete years of growth with the previous debarking occurring 9 years before (the stripping is performed when phellogen is active thereby interrupting the cork growth in that year).

After debarking, the samples ($\pm 15 \times 15$ cm^2) were boiled in water for 1 h at atmospheric pressure and left to air-dry until equilibrium. For a more accurate observation of the growth rings the transversal sections were polished and two opposite radial strips with approximately 1 cm thickness were cut, fixed on a microscope slide (Fig. 1a) and digitalized in a laser based fluorescence imager (Fluoro Image Analyzer FLA -5100, Fujifilm, Life Science USA, Stamford, CT 06092).[24] Cork ring thickness was measured with AnalySIS® image processing software (Analysis Soft Imaging System GmbH, Münster, Germany, version 3.1) and along two transects in the radial direction with an accuracy of 0.01 mm. For each tree the media of the two measurements was determined.

[24] Surový et al. (2009).

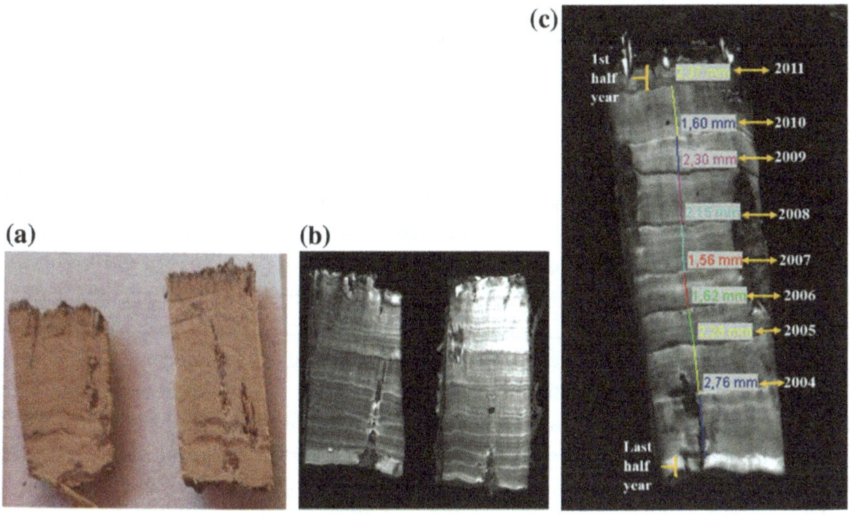

Fig. 1 **a** Slides with cork samples **b** example of a sample in which the delimitation of the 8 complete growth years is not possible **c** example of a sample from site 3 measured and dated

The initial and final half rings corresponding to the years of the present and last strippings were not considered for analysis as they are not complete growth years. As recognized earlier by several authors,[25,26] and[27] the delimitation of annual rings was not unequivocal in all the samples (Fig. 1b), and only in 70% (149) of the samples cork rings width were measured and cork rings dated (Fig. 1c).

Data Analysis

In order to take a brief characterization of our raw data a boxplot for each site was performed, allowing us to compare the sites and evaluate the respective variability.

As usually done in dendroclimatic studies,[28] each cork-growth curve was plotted and standardized in order to remove systematic changes of growth associated with phellogen age and maximize the inter-annual fluctuations due to weather. Due to the low age of the phellogen and in a similar approach as performed by[29] and[30] we used a negative exponential curve in the standardization process (detrend) through

[25]Caritat et al. (1996).
[26]See footnote 24.
[27]See footnote 17.
[28]Fritts (1976).
[29]See footnote 17.
[30]Costa et al. (2001).

the dendrochronology program library (dplr) of the R software[31] and[32]. This procedure fits the following model:

$$G_t = ae^{-bt} + k$$

where G_t is the growth trend and is estimated as a function of time t, being a, b and k, the coefficients. If that nonlinear model could not be fitted, then a standard linear model was adjusted $G_t = b_0 + b_1 t$ where b_0 is the intercept and b_1 the slope.[33]

After this procedure, for each tree, RWI—ring width indexes (or Cork Growth Indexes) were calculated dividing the real cork growth (R_t) by the expected growth at time t (G_t).

We initially performed the Shapiro-Wilk Normality Test (shapiro.test) in R, and then, as it failed at a significance level of 0.05, a non-parametric test—the Kruskal-Wallis rank sum test (kruskal.test) was performed for the RWI of 2005 (year which growth is affected by the drought) and for the difference between RWI of 2004 and 2005 ($RWI_{2005}–RWI_{2004}$) to analyze the effect of the drought in reducing RWI and between RWI of 2005 and 2006 ($RWI_{2006}–RWI_{2005}$) to evaluate the phellogen recovery.

Furthermore, as the total growth of the cork determines its industrial utilization, we also tested if the total growth was different according to the moment in which the drought occurred. The test was performed with the same methodology as before.

Results

The cork growth of the 8 complete years in each site is presented in Fig. 2. The minimum total growth is very similar in all the sites (approximately 12 mm) but the media and the third quartile of site 3 are lower than those of the other sites.

The mean series of annual cork growth indices suggest one strong climatic signal, and 2005 can be considered a negative pointer year in all the chronologies. In general, the cork ring indices from the same year have a similar behavior when compared to the preceding/next year, in all the mean chronologies (Fig. 3).

The Shapiro-Wilk Normality Test performed for the RWI for the year 2005 gave a p-value below 0.05 (3.71×10^{-6}—see Table 2) and we performed a non-parametric approach through the Kruskal-Wallis rank sum test. We concluded that the 2005-RWI is not different between all the sites as the p-value is under the significance level of 5% (Table 2).

We also compared $RWI_{2005}–RWI_{2004}$ and $RWI_{2006}–RWI_{2005}$ for the 3 sites. The normality was again rejected and the results from the Kruskal-Wallis rank sum test

[31]Bunn (2008).

[32]R Core Team (2017).

[33]See fotenote 31.

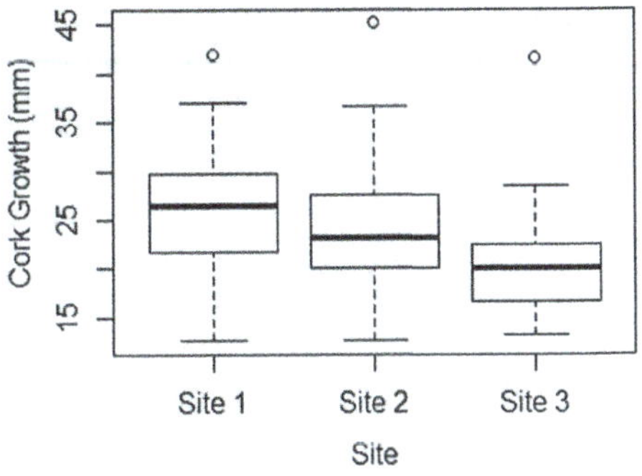

Fig. 2 Boxplot for the cork growth of 8 complete years of the 3 sites considered in this research

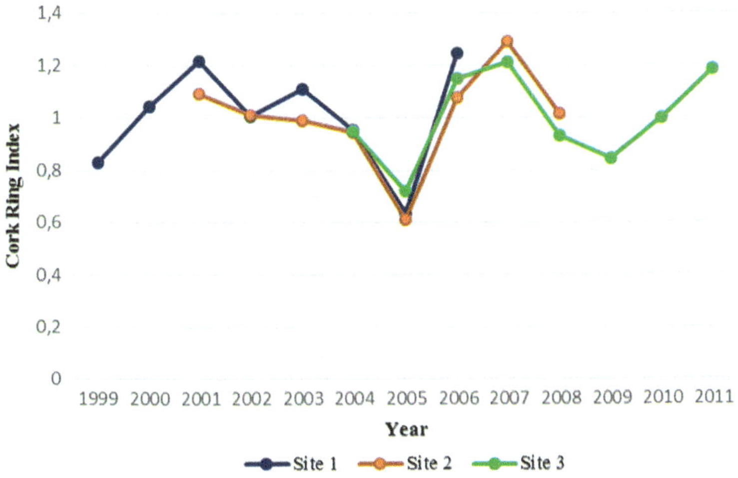

Fig. 3 Mean cork growth chronologies for each site

suggest that the phellogen behavior is the same (not different) in the 3 sites i.e. independent of its age. However the total cork growth (of the 8 years) showed differences between the sites (Table 2).

As there were differences between the sites, we calculated the average growth for each site: site 1 = 25.57 mm; site 2 = 24.18 mm and site 3 = 20.08 mm. By performing the Wilcoxon test (non-parametric) we concluded (with a significance level of 5%) that site 3 has less total growth than the other 2 sites, which are not different from each other.

Table 2 Results (p-values) of the Shapiro-Wilk Normality Test and Kruskal-Wallis rank sum test performed for the Cork Ring Index for the year 2005, for the differences of the indexes for the years 2005–2004 and 2006–2005 and for the growth of the 8 complete years

	Shapiro-Wilk normality test	Kruskal-Wallis rank sum test
RWI_{2005}	3.71×10^{-6}	0.0681
$RWI_{2005}–RWI_{2004}$	2.69×10^{-6}	0.5932
$RWI_{2006}–RWI_{2005}$	0.004727	0.1972
Total growth	0.000188	5.19×10^{-6}

Discussion

The values obtained for the cork growth of 8 complete years in each site are below the average values found by[34] in similar locations. The differences of the growth of the 8 complete years obtained for one of the sites are difficult to explain as several factors affect growth, namely climate, age of the trees, the site, competition and other disturbances.[35]

Analyzing the cork-growth chronology, the results show that in all the sites, the 2004–2006 drought negatively affected cork growth and that 2005 can be considered a negative pointer year in all the chronologies. This effect of drought on cork growth is consistent with results obtained by several authors.[36,37] The trees recovered well when precipitation returned to values similar to the long term mean, which is also in accordance of the findings of several authors[38] and.[39]

Moreover, our results indicate that the age of the phellogen has no influence on the intensity of growth decline in the drought pointer year, i.e. the ring width index was similar for all the sites regardless of its occurrence during the cork growth cycle. Also there was no effect of the age of the phellogen neither on the decreasing of growth due to the drought nor on the recovery after the drought.

In practical terms however, and considering the absolute values of cork ring in the successive years along the growth cycle, it is expected to have a stronger impact on the total 8-year cork growth if the drought event occurs in the beginning of the cycle.

[34]Costa et al. (2016).

[35]Cook and Kairiukstis (1989).

[36]See footnote 17.

[37]See footnote 34.

[38]See footnote 17.

[39]See footnote 14.

Conclusion

The results of this exploratory study suggest that the age of the phellogen has no effect on the response of the tree to a drought event and on its recovery. Consequently, forest managers shall adapt their management practices to drought events independently of the moment of its occurrence along the cork growing cycle.

The results once again confirmed that drought decreases cork growth, which may hinder the possibility of producing, in the usual 9-year cycle, a cork thick enough for the manufacture of cork stoppers, especially if occurring in the first part of the growth cycle. This is a strong signal for forest managers to consider the feasibility to water the cork oak trees and/or to enlarge the cork growing cycles of cork.

Nevertheless, the results obtained should be confirmed with a larger sample and number of chronologies in order to analyze the severe drought of 2004–2006 with samples for more sites as well as to evaluate the effects of other severe droughts.

Acknowledgements The research was carried out under the framework of Centro de Estudos Florestais, a research unit funded by Fundação para a Ciência e a Tecnologia, Portugal (UID/AGR/00239/2013). Carla Leite acknowledges a Ph.D grant by FCT under the SUSFOR doctoral program (PD/BD/113937/2015).

The authors acknowledge the collaboration of Associação dos Produtores Florestais do Concelho de Coruche e Limítrofes (APFC) in material supply.

References

APCOR. (2016). *Anuário de cortiça 2016 APCOR*. site accessed 9th Mar 2017. http://www.apcor.pt/wp-content/uploads/2016/09/Boletim-estatistico-2016.pdf.

Besson, C., Lobo-do-Vale, R., Rodrigues, M., Almeida, P., Herd, A., Grant, O., et al. (2014). Cork oak physiological responses to manipulated water availability in a Mediterranean woodland. *Agricultural and Forest Meteorology, 184*, 230–242. https://doi.org/10.1016/j.agrformet.2013.10.004.

Braeuning, A., Bolte, A., Nabais, C., Rossi, S., & Sass-Klaassen, U. (2017). Studying tree responses to extreme events. *Frontiers in Plant Science, 8*, 506. https://doi.org/10.3389/fpls.2017.00506.

Bunn, A. (2008). A dendrochronology program library in R (dplR). *Dendrochronologia, 26*(2), 115–124. https://doi.org/10.1016/j.dendro.2008.01.002.

Caritat, A., Gutiérrez, E., & Molinas, M. (2000). Influence of weather on cork-ring width. *Tree Physiology, 20*(13), 893–900. https://doi.org/10.1093/treephys/20.13.893.

Caritat, A., Molinas, M., & Gutierrez, E. (1996). Annual cork-ring width variability of *Quercus suber L.* in relation to temperature and precipitation (Extremadura, southwestern Spain). *Forest Ecology and Management, 86*(1), 113–120. https://doi.org/10.1016/S0378-1127(96)03787-5.

Cherubini, P., Gartner, B., Tognetti, R., Bräker, O., Schoch, W., & Innes, J. L. (2003). Identification, measurement and interpretation of tree rings in woody species from Mediterranean climates. *Biological Reviews, 78*(1), 119–148. https://doi.org/10.1017/S1464793102006000

Cook, E., & Kairiukstis, L. (1989). *Methods of dendrochronology: Applications in the environmental sciences*. Dordrech: Springer Science & Business Media. ISBN 978-90-481-4060-2.

Costa, A., Barbosa, I., Roussado, C., Graça, J., & Spiecker, H. (2016). Climate response of cork growth in the Mediterranean oak (*Quercus suber L.*) woodlands of southwestern Portugal. *Dendrochronologia, 38,* 72–81. https://doi.org/10.1016/j.dendro.2016.03.007.

Costa, A., Pereira, H., & Oliveira, A. (2001). A dendroclimatological approach to diameter growth in adult cork-oak trees under production. *Trees, 15*(7), 438–443. https://doi.org/10.1007/s004680100119.

Evert, R. F. (2006). *Esau's plant anatomy: Meristems, cells, and tissues of the plant body: Their structure, function, and development*. London: Wiley. ISBN 0-470-04737-2.

Fritts, H. (1976). *Tree rings and climate* (1st ed.). London: Academic Press. ISBN 978-0-323-14528-2.

García-Herrera, R., Hernández, E., Barriopedro, D., Paredes, D., Trigo, R., Trigo, I., et al. (2007). The outstanding 2004/05 drought in the Iberian Peninsula: Associated atmospheric circulation. *Journal of Hydrometeorology, 8*(3), 483–498. https://doi.org/10.1175/JHM578.1.

Gea-Izquierdo, G., Fernández-de-Uña, L., & Cañellas, I. (2013). Growth projections reveal local vulnerability of Mediterranean oaks with rising temperatures. *Forest Ecology and Management, 305,* 282–293. https://doi.org/10.1016/j.foreco.2013.05.058.

IPCC. (2014). In C. B. Field, V. R. Barros, D. J. Dokken, K. J. Mach, M. D. Mastrandrea, T. E. Bilir, et al. (Eds.) *Climate Change 2014: Impacts, Adaptation, and Vulnerability. Part A: Global and Sectoral Aspects. Contribution of Working Group II to the Fifth Assessment Report of the Intergovernmental Panel on Climate Change*. Cambridge University Press, Cambridge, United Kingdom and New York, NY, USA, 1132 pp.

Lindner, M., Maroschek, M., Netherer, S., Kremer, A., Barbati, A., Garcia-Gonzalo, J., et al. (2010). Climate change impacts, adaptive capacity, and vulnerability of European forest ecosystems. *Forest Ecology and Management, 259,* 698–709. https://doi.org/10.1016/j.foreco.2009.09.02.

Natividade, J. (1950). *Subericultura*. Lisboa: Ministério da Economia–Direcção Geral dos Serviços Florestais e Aquícolas.

Oliveira, V., Lauw, A., & Pereira, H. (2016). Sensitivity of cork growth to drought events: Insights from a 24-year chronology. *Climatic Change, 137*(1), 261–274. https://doi.org/10.1007/s10584-016-1680-7.

Pereira, H. (2007). *Cork: Biology, production and uses*. Amsterdam: Elsevier. ISBN 0-08-047686-4.

Piayda, A., Dubbert, M., Rebmann, C., Kolle, O., Silva, F., Correia, A., et al. (2014). Drought impact on carbon and water cycling in a Mediterranean *Quercus suber L.* woodland during the extreme drought event in 2012. *Biogeosciences, 11*(24), 7159–7178. https://doi.org/10.5194/bg-11-7159-2014.

Pizzurro, G., Maetzke, F., & Veca, D. (2010). Differences of raw cork quality in productive cork oak woods in Sicily in relation to stand density. *Forest Ecology and Management, 260*(5), 923–929. https://doi.org/10.1016/j.foreco.2010.06.013.

R Core Team. (2017). *R: A language and environment for statistical computing*. R Foundation for Statistical Computing. Retrieved from https://www.R-project.org/.

Santos, F. D., & Miranda, P. (2006). Alterações climáticas em Portugal. *Cenários, Impactos E Medidas de Adaptação–Projecto SIAM II*. ISBN: 989-616-081-3.

Stott, P., Christidis, N., Otto, F., Sun, Y., Vanderlinden, J. P., van Oldenborgh, G., et al. (2016). Attribution of extreme weather and climate-related events. *Wiley Interdisciplinary Reviews: Climate Change, 7*(1), 23–41. https://doi.org/10.1002/wcc.380.

Surový, P., Olbrich, A., Polle, A., Ribeiro, N. A., Sloboda, B., & Langenfeld-Heyser, R. (2009). A new method for measurement of annual growth rings in cork by means of autofluorescence. *Trees, 23*(6), 1237. https://doi.org/10.1007/s00468-009-0363-7.

Ozone Layer Depletion, Climate Change, Risks and Adaptation

Aliaksandr Krasouski and Siarhei Zenchanka

Abstract The recent years' publications have focused on the influence of the ozone layer depletion (OLD) to the climate change (CC). Some authors suggested that the reasons for it are: (a) the ozone absorbs the solar radiation, and (b) ozone is a greenhouse gas. Hence, the variations of ozone concentration in the atmosphere and climate change are closely related. The human risks associated with ozone layer depletion are skin cancer, damage to the eyes, local and systemic immunosuppression. Some of environmental risks associated with ozone layer depletion are a changing of climatic regions; changing of land productivity, impact on plants and animals. The aims of the article are the investigation of mutual influence of OLD and CC, analysis of the regional OLD and associated risks. Presented results show that there is one more reason for the OLD influence on CC concerning with the change of the tropopause height. In addition, positive anomalies of ozone can be just as serious preconditions for climate risks, as well as negative anomalies.

Keywords Ozone layer · Climate change · Risk · Adaptation
The Republic of Belarus

Introduction

The Vienna Convention for the Protection of the Ozone Layer (Convention 1985) and its Montreal Protocol (1987) on Substances that Deplete the Ozone Layer are the first treaties of any kind to achieve universal ratification. These conventions

A. Krasouski
Belarussian State University, National Ozone Monitoring Research and Education Center, 4, Nezavisimosti Avenue, 220030 Minsk, Belarus
e-mail: krasovsky@bsu.by

S. Zenchanka (✉)
Minsk Branch of Plekhanov Russian University of Economics, 127-2, Mayakovskogo Str, 220028 Minsk, Belarus
e-mail: zench@tut.by

F. Alves et al. (eds.), *Theory and Practice of Climate Adaptation*,
Climate Change Management, https://doi.org/10.1007/978-3-319-72874-2_8

defined main task directed on the reducing of human activity on ozone layer depletion (OLD). Currently, the Convention and the Protocol have enabled these reductions of more than 97% of all global consumption of controlled ozone-depleting substances.

Human influence on the climate system is clear, and recent anthropogenic emissions of greenhouse gases are the highest in history. Last year's climate changes have had widespread impacts on human and natural systems. It is extremely likely that more than half of the observed increasing in the global average surface temperature from 1951 to 2010 was caused by the anthropogenic increasing in GHG concentrations and other anthropogenic impacts together (IPCC 2014). The United Nation Framework Convention on Climate Change (UNFCCC 1992) and the Kyoto Protocol (1998) defined the measures for decreasing human influence on climate.

Nevertheless, both the ozone layer depletion and the climate change continue to have a negative impact on the environment and human beings. This requires adaptation to these processes.

The aims of the article are the investigation of mutual influence of OLD and CC, analysis of the regional OLD and associated risks.

Methods

The research was guided by the next questions:

- What is ozone layer depletion and how it can be measured?
- What is a climate change and what is a tendency of this changing?
- How OLD and CC correlate?
- Is it possible to adapt to OLD and CC?

To answer these questions the analysis of scientific reports, programs and articles were considered. The article presents the results of an investigation concerning with ozone concentration measurement.

Ozone Layer Depletion and Ozone Monitoring

In the atmosphere O_3 is presented in low concentration ($\sim$0.6 ppm), but it plays an important role by efficiently screening out harmful radiations. There are three wavelength regions of UV radiation: shorter wavelengths ranging from 100 to 280 nm (UV-C), the middle wavelength ranging from 280 to 315 nm (UV-B) and long wavelength ranging from 315 to 400 nm (UV-A). UV-C is completely absorbed by the ozone layer, no more than 5% of UV-B reaches the earth's surface, and nearly 95% of UV-A is able to penetrate the atmospheric layers

(Aggarwal et al. 2013). Accordingly, UV radiations affect all the elements of the biome including plants, pathogens, herbivores, carnivores and microorganisms.

Ozone occurs at three levels: the stratospheric ozone, the tropospheric ozone and surface ozone. Most of the ozone resides in the stratospheric ozone layer and a small percentage (about 5–10%) is produced in troposphere giving a contribution to the greenhouse effect (WMO 2014).

Most of the ozone was found in the stratosphere between 19 and 30 km up in the Earth's atmosphere and it forms so-called "ozone layer". This layer plays an important role in providing support the life on the Earth. Moreover, it is an essential factor in biological, environmental and health global phenomena, because without the ozone layer, a lot of ultraviolet (UV) radiation from the Sun reaches the Earth's surface, causing untold damage to most living species (Sivasakthivel and Reddy 2011).

To measure the total amount of ozone in an overhead column of the atmosphere the Dobson and Brewer spectrometers are used. The most common unit for stratospheric ozone measurement is the Dobson Unit (DU). Normal ozone concentration is about 300–350 DU. Term "ozone hole" should be applied when this concentration falls below 200 DU (Handbook 2008).

In their review Sivasakthivel and Reddy (2011) pointed that ozone-hole "now occurs every springtime above Antarctica and to a lesser extent in the Arctic, where special meteorological conditions and very low air temperatures accelerate and enhance the destruction of ozone loss by man-made ozone depleting chemicals". Firstly the ozone-hole over the Antarctic has been noticed since 1970s and during the last decade the Arctic ozone-hole was discovered.

After the adoption of the Montreal Protocol, the use of halons and chlorofluorocarbons has been discontinued and the size of the Antarctic ozone hole has stabilized with no significant trend since 1998. However, these compounds, due to their long lifetime, will remain in the atmosphere for many decades. More than enough chlorine and bromine present in the atmosphere to cause complete destruction of ozone at certain altitudes in the Antarctic from August to December, hence the size of the ozone hole from one year to the next is mostly governed by the meteorological conditions (WMO 2017).

To measure ozone concentration a global network of total-ozone observing stations was established in 1957 as part of the International Geophysical Year. Now, this network includes about 100 sites located around the world. Data from the network are essential for understanding the effects of different ozone-depleting substances on the global ozone layer. The accuracy of their measurements is maintained by regular instrument calibrations and inter-comparisons. Ground based instruments with excellent long-term stability and accuracy are now routinely used to help calibrate space-based observations of total ozone (Hegglin et al. 2015).

Climate Change

Climate Change is one of the most serious problems over the world. NASA's Goddard Institute for Space Studies presented an analysis of global temperature, which shows that February 2017 was the second-warmest February in 137 years of modern record keeping (https://data.giss.nasa.gov/gistemp/news/20170315/). February 2017 temperature was 1.1 °C warmer than the mean February temperature from 1951 to 1980. Accordingly the two top February temperature anomalies have occurred during the past two years. Cook et al. (2016) in their analyses of article concerning to the anthropogenic global warming (AGW) draw a conclusion that "the level of scientific agreement on AGW is overwhelmingly high because the supporting evidence is overwhelmingly strong".

Last WMO (2017) report confirms that 2016 was the warmest year on record: a remarkable 1.1 °C above the pre-industrial period, which is 0.06 °C above the previous record set in 2015. This increase in global temperature is consistent with other changes in the climate system. Moreover a globally averaged sea-surface temperature was also the warmest on record; global sea levels continued to rise; and Arctic sea-ice extent was well below average for most of the year.

The achieved consensus among scientists allowed politicians to adopt the Paris Agreement (2015), which set ambitious goals on climate change for the world community, including (1) a long-term goal of keeping the increase in global average temperature to well below 2 °C above pre-industrial levels; (2) an aim to limit the temperature increase to 1.5 °C, as this would significantly reduce the risks and impacts of climate change; and (3) the need for global emissions to peak as soon as possible, followed by a rapid decline—recognizing that this will take longer for developing countries (UNEP 2016). Consequently the entry into force of the Paris Agreement under the UN Framework Convention on Climate Change (UNFCCC) on 4 November 2016 represents a historic landmark. It is vital that its implementation becomes a reality and that the Agreement guides the global community in addressing climate change by curbing greenhouse gases, fostering climate resilience and mainstreaming climate adaptation into national development policies (WMO 2017).

At the same time, it should be noted that there are deniers of AGW which connect climate change with natural causes. Over the years, the deniers have employed a wide range of arguments against taking action on climate change, some of which contradict each other. For example, they have claimed that (http://www.davidsuzuki.org/issues/climate-change/science/climate-change-basics/climate-change-deniers/)

- Climate change is not occurring;
- The global climate is actually getting colder;
- The global climate is getting warmer, but not because of human activities;
- The global climate is getting warmer, in part because of human activities, but this will create greater benefits than costs;

- The global climate is getting warmer, in part because of human activities, but the impacts are not sufficient to require any policy response.

Nevertheless, WMO (2017) Statement underline that with carbon dioxide reaching a record annual average concentration of 400 parts per million (ppm) in the atmosphere, the influence of human activities on the climate system has become more and more evident. The Emissions Gap Report (UNEP 2016) pointed that "global greenhouse gas emissions continue to grow, and while the indications are encouraging that the growth rate of global carbon dioxide emissions from fossil fuel use and industry is slowing, it is still too early to say whether this is likely to be permanent". The powerful 2015/2016 El Niño played an important role in the year's climate and confirmed that, when natural variability interacts with anthropogenic climate change, the impacts on human societies and the natural environment can be severe (WMO 2017).

Anthropogenic greenhouse gas emissions have increased since the pre-industrial era, driven largely by economic and population growth, and are now higher than ever. Accordingly, this has led to atmospheric concentrations of carbon dioxide, methane and nitrous oxide that are unprecedented in at least the last 800,000 years. Their effects, together with those of other anthropogenic drivers, have been detected throughout the climate system and are extremely likely to have been the dominant cause of the observed warming since the mid-20th century (IPCC 2014).

Ozone Layer Depletion and Climate Change

Much Ozone Depleting Substances (ODS) are also potential greenhouse gases. Their phase out has therefore had a significant positive impact on climate change prevention. Therefore, by reducing emissions of ozone depleting substances, the Montreal Protocol has protected both the ozone layer and the climate at the same time. However, some of the ODS substitutes can equally contribute to climate change. The EEA (2015a) has recently published a report "Protecting the ozone layer while also preventing climate change" and presented a web page (http://www.eea.europa.eu/data-and-maps/daviz/consumption-of-controlled-ozone-depleting-substances-2#tab-chart_1) which provides more information on the links between ODS substitutes and climate change. The reduction of ODS emissions is not a uniformly positive story, it has indirectly led to new problems. In many sectors of economics, such as refrigeration and air conditioning applications, the fluorinated gases (F-gases), which include hydrofluorocarbons (HFCs), perfluorocarbons (PFCs) and sulphurhexafluoride (SF6), have been introduced as substitutes for ODS. These gases do not deplete the ozone layer, but they are greenhouse gases (EEA 2015b). In addition, some F-gases have a greenhouse effect that is up to 23,000 times more powerful than the same amount of carbon dioxide.

The new research has highlighted a key role that ozone, a major component of the stratosphere, plays in how climate change occurs, and the possible implications for predictions of global warming. Changes in ozone are often either not included, or are included a very simplified manner, in current climate models. This is due to the complexity and the sheer computing power it takes to calculate these changes, an important deficiency in some studies (Nowack et al. 2015).

Risks and Adaptation

Human activities have significant effects on the environment and vice versa. Further warming and long-lasting changes in all components of the climate system are defined by continued emission of greenhouse gases. Climate change limiting would require substantial and sustained reductions in greenhouse gas emissions, which, together with an adaptation, can limit risks connecting with climate change. The introduction of new chemicals into the market can lead to a change of the ozone layer. Reduced ozone level as a result of ozone depletion mean less protection from UV-B radiation at the Earth's surface.

The main effects of OLD on human life are (Aggarwal et al. 2013; Sivasakthivel and Reddy 2011):

- The eye diseases, skin cancer and infectious diseases;
- The changing of physiological and developmental processes in plants under the influence of UV-B radiation;
- The dependence of productivity of aquatic systems under increased levels of UV exposure;
- The influence of UV exposure on terrestrial and aquatic bio-geo-chemical cycles;
- Higher photo dissociation rates of key trace gases that control the chemical reactivity of the troposphere during the reduction of stratospheric ozone and increased penetration of UV-B radiation;
- Damage of mechanical properties of materials under the influence of UV radiation;
- The increasing in ground-level ultraviolet radiation with the depletion of the ozone layer.

In recent decades, changes in climate have caused impacts on natural and human systems on all continents and across the oceans. Impacts are due to observed climate change, irrespective of its cause, indicating the sensitivity of natural and human systems to changing climate. There are risks connected with climate change:

- Growth of temperature. Earth's temperatures in 2016 were the hottest ever recorded (WMO 2017). The global temperatures were strongly influenced by strong El Niño conditions that (NOAA 2015);
- Drought, storms, and floods. The biggest cause of droughts and floods around the world is the shifting of climate patterns between El Niño and La Niña events

(https://royalsociety.org/topics-policy/projects/climate-change-evidence-causes/question-13/);
- Sea-level rising due warming of ocean waters, melting the glaciers leads to flooding low-lying areas and islands, threaten dense coastal populations, erode shorelines, damage property and destroy ecosystems;
- Changing of landscapes and wildlife habitat. Experts predict that one-fourth of the Earth's species will be headed for extinction by 2050 if the warming trend continues at its current rate;
- Impact on the Economy. It is hard to predict, but many key economic sectors—from fishing to energy and water utilities—will feel long-term impacts of climate change.

The early effects of climate change and stratospheric ozone depletion on peoples' health in Europe were considered in the report of WMO (2000).

It is possible to define five main areas where influences by CC and OLD are most remarkable:

- Human health;
- Changing of biodiversity;
- Productivity of agriculture;
- Energy and fuel;
- Economy.

Such risks require special adaptation measures which include:

- Adaptation strategies in agriculture (Clements et al. 2011; Tubiello 2012);
- Fulfillment of the Montreal Protocol (1987) and its Amendment demands;
- Fulfillment of Kyoto Protocol (1998) demands;
- Development of special measures for fulfillment of Paris Agreement (2015).

Ozone Layer Depletion and Climate Change: Experience from Belarus

National Ozone Monitoring Research and Education Center (NOMREC) developed own unit for monitoring of total ozone, which is included in the Global network of total-ozone observing stations.

A double diffraction spectrophotometer PION-UV and its polar modification PION-UV-II, developed at NOMREC BSU, are able to record irradiance spectra (IS) in a wavelength range of 290–450 nm with a spectral resolution of 0.8 nm. In addition to a working channel, devices are equipped with a special "reference" channel to monitor the state of the clouds. In automated mode, the instruments measure spectra with an interval of 5 min during daylight hours (http://ozone.bsu.by/instr_3_pion-f.html

Total ozone instruments were calibrated using the WMO regional standard (Dobson N108 spectrophotometer) in Voeikov Main Geophysical Observatory (MGO), St. Petersburg, Russia. Based on the measurement results, UV index as well as the dose rates of basic biological effects of UV radiation (erythema, DNA damage, skin cancer, cataract, etc.) can be automatically determined.

The analysis of long-term changes in the state of the ozone layer over the territory of Belarus has been in progress based on the results of ground-based monitoring and satellite data of total ozone in the vertical atmospheric column (TO) for the city of Minsk (53.83 °N, 27.47 °E). Series of average daily, monthly and annual TO values for the periods of 1979–1992 and 1997–2016 have been analyzed consequently (Report 2017).

During the period from 1979 to the end of the 90s, trend magnitude was ~1.5 units/year. Since 2000, the negative trend has remained but its value has decreased, and for the period of 2000–2016 it amounts to ~0.3 units/year. Registered variations of the average annual TO values are presented in Fig. 1.

Analysis of long-term trends of average monthly TO values for the considered time periods shown that in 1980s and 1990s, negative trend TO has occurred in some way in all seasons, but the strongest decrease inTO was observed in winter and early spring. (Krasouski and Turyshev 2011). After 2000, the magnitude of the negative trend for all months except June and September has decreased significantly. In January and to a lesser extent in February, a positive trend of average monthly TO values has been observed.

Joint analysis of variations in average monthly TO values and variations of general circulation processes in the troposphere of the Northern Hemisphere have revealed their relationship (wherein variations in tropospheric general circulation are analyzed on the basis of a circulation link between high latitudes and middle

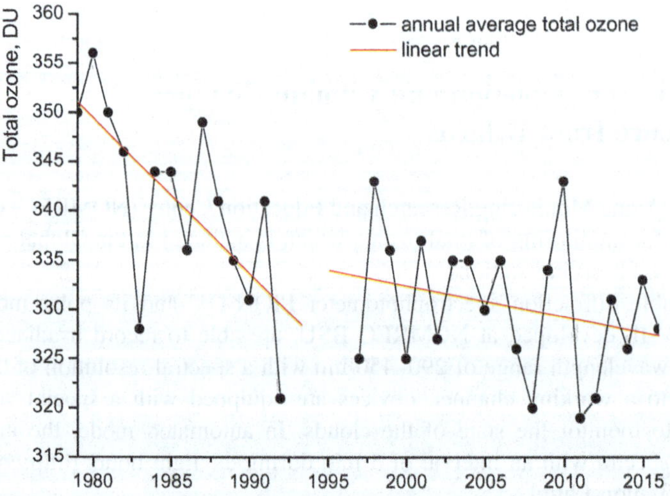

Fig. 1 The average value of TO and linear trend for the periods of 1979–1992 and 2000–2016

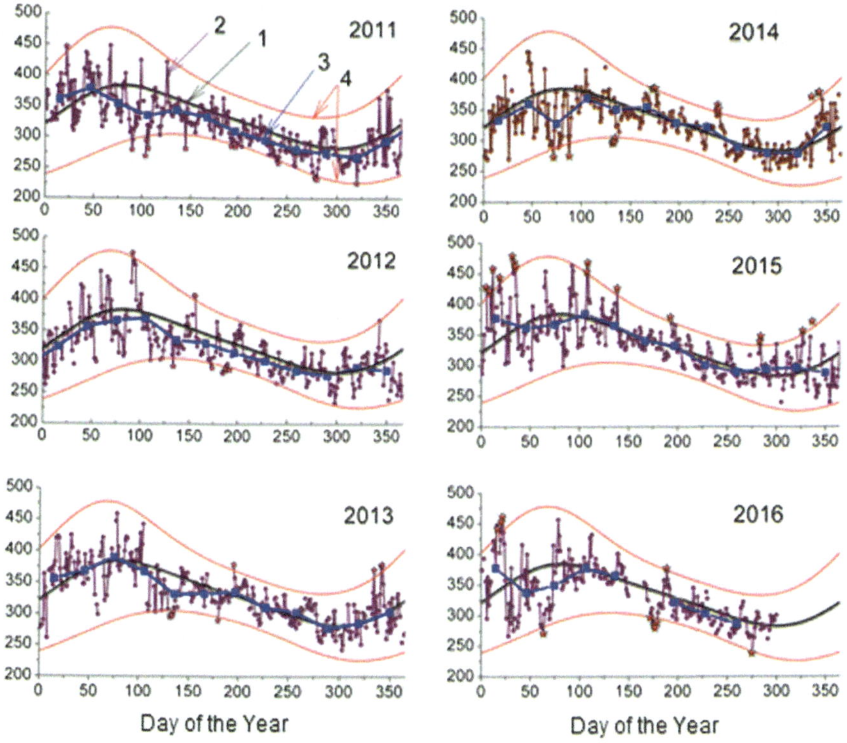

Fig. 2 The TO average daily values over the territory of Belarus in 2011–2016. 1—the annual variation of the climatic norm, 2—observations data, 3—monthly average values, 4—95% confidence interval

latitudes expressed by the amount of Arctic intrusions and their area, thereby allowing to identify repetitive circulation patterns in the Northern Hemisphere) (Report 2017).

Figure 2 presents results of TO observations in 2011–2016 compared to the TO climate norm and its monthly average values and the most pronounced ozone anomalies are marked. These are the ones falling beyond the 95% confidence interval of the variation of the daily average values that have been used to determine the climatic norm.

Apparently from the Fig. 2, since 2013 the number of the positive anomalies registered over the territory of Republic of Belarus grows and especially much they were in 2015. The TO average monthly values approach climatic norm which experiences slow growth now. Results shown on Figs. 1 and 2 were obtained by A. Krasovski with co-workers and were reported on Ozone Research Managers Meeting in Geneva, March 28–30, 2017.

Adaptation to Ozone Layer Depletion and Climate Change: Results from Belarus

Adaptation can reduce the risks of climate change impacts, but there are limits to its effectiveness, especially with greater magnitudes and rates of climate change. Taking a long term perspective, in the context of sustainable development, increases the likelihood that more immediate adaptation actions will also enhance future options and preparedness. Governance is prominent among the proposed avenues for reducing vulnerability to climate change in fisheries and aquaculture. The generation of new knowledge and information about the impacts of climate change on ecosystems is also fundamental. Without a fuller understanding of the functioning of ecosystems and of the uncertainty inherent in current climate models, optimal adaptation strategies are likely to be more difficult to design (FAO 2014).

Belarus, as a party to the Vienna Convention and Montreal Protocol, UNFCCC and Kyoto Protocol, has created and is developing a regulatory, legal and institutional framework for dealing with issues related to ozone layer depletion and climate change.

The basic principles of protection of the ozone layer and adaptation to its changes are defined in the Law (2001) of the Republic of Belarus "On the Protection of the Ozone Layer":

- State regulation in the field of protection of the ozone layer;
- Prevention of harmful effects on the ozone layer, harm to human life and health, the environment;
- Reducing the consumption of ozone-depleting substances and reducing (stopping) their use in accordance with international treaties of the Republic of Belarus.

These measures will demand advanced scientific and industrial developments:

- Introduction of ozone-safe technologies, as well as recycling (recycling) processes, recovery, neutralization and disposal of ozone-depleting substances;
- Continuous monitoring of the ozone layer;
- Providing access in accordance with legislative acts to environmental information in the field of ozone layer protection, including information on the state of the ozone layer, its impacts and measures for its protection;
- Responsibility for violation of the legislation on protection of the ozone layer.

The measures for people protection from UV irradiance were proposed in special Memo (2007) which covers basic measures for child protection, people working in open areas and tourist.

Scientific research, development of programs and preparation of recommendations for adaptation to climate change, as well as international cooperation are conducted in all sectors of the economy and spheres of people's lives. Moreover, protecting the environment and adapting the economy to climate change can not only reduce losses from its adverse effects, but also bring economic benefits to the country, at least in the near future.

"The State Program of Measures to Mitigate Climate Change for 2013–2020" (Program 2013) notes that the most vulnerable to climate change in Belarus are such sectors as agriculture, forestry and water management, energy, construction and the social sphere. The Communication (2015) points that the most dangerous can be an increase in the likelihood of low yields as a result of increased frequency and repeatability of droughts in region, a shift in growth phases and inadequate water availability in the spring. In forestry there is a shift in the range of some forest vegetation species, the growth of diseases and fires, the change in the hydrological regime in several regions. Implementation of measures for water supply requires a long time, therefore large water management measures should be planned with a lead time of about 25 years, and putting them into operation should exceed water requirements by 10–15 years.

The main activities for adaptation to climate change will include:

- Development and improvement of the system of criteria and conditions for the country's climate security;
- Assessment of the vulnerability of individual regions to climate change;
- Development of sectoral strategies for adaptation to climate change by the republican government bodies, implementation of these strategies;
- Stimulation of activities related to the implementation of agricultural activities to adapt to climate change;
- Minimizing of consequences of dangerous hydrometeorological phenomena caused by climate change, including the development of methods for calculating risks and assessing damage, as well as scenarios for adaptation to such phenomena;
- Minimizing the risk of reducing agricultural production, including a decrease in the productivity of farm animals, crop yields and gross crop yields;
- Introduction into production of thermophilic species and varieties of agricultural crops with the expansion of their sown areas;
- Assessment of the impact of climate change and potential threats to the biodiversity of natural ecosystems of the Republic of Belarus with the development of measures for their conservation.

Conclusion

Joining the UNFCCC and the Vienna Convention and their protocols the Republic of Belarus is doing a great job of meeting their requirements and adapting the economy to new environment and business conditions.

The official documents identify the most possible risks of ozone layer depletion and climate change and measures of mitigation and adaptation to them. The National Ozone Monitoring Research and Education Center, which is a part of the

Global network of total-ozone observing stations, constantly monitoring the content of total and surface ozone. The results of this monitoring show the effect of total ozone concentration on climate change.

Risks concerned with ozone layer depletion and climate change cause similar consequences for well-being, biodiversity changing, agriculture productivity, energy and fuel consumption and economy. Hence, adaptation measures should also be similar.

For Belarusian condition adaptation measures presented in the Law of the Republic of Belarus "On the Protection of the Ozone Layer" and in "The State Program of Measures to Mitigate Climate Change for 2013–2020". Most significant adaptation measures are:

(a) prevention of harmful effects on the ozone layer, harm to human life and health, the environment;
(b) reducing the consumption of ozone-depleting substances and reducing (stopping) their use in accordance with international treaties of the Republic of Belarus;
(c) minimizing of consequences of dangerous hydrometeorological phenomena caused by climate change, including the development of methods for calculating risks and assessing damage, as well as scenarios for adaptation to such phenomena; and
(d) minimizing the risk of reducing agricultural production, including a decrease in the productivity of farm animals, crop yields and gross crop yields.

The 10th Ozone Research Managers (ORM-10) meeting concluded that integrated earth science observation systems are essential to ensure that action to protect ozone layer also benefits the climate, given the complex and evolving interaction between the ozone layer and the climate system (http://web.unep.org/newscentre/increased-research-and-observation-crucial-efforts-continue-protect-ozone-layer-and-climate).

References

Aggarwal, A., Kumari, R., Mehla, N., Deepali, Singh, R. P., Bhatnagar, S., et al. (2013). Depletion of the ozone layer and its consequences: A review. *American Journal of Plant Sciences, 2013* (4), 1990–1997. Accessed 15th Mar 2017. http://file.scirp.org/pdf/AJPS_2013100915534704.pdf/.

Clements, R., Haggar, J., Quezada, A., & Torres, J. (2011). *Technologies for climate change adaptation—Agriculture sector*. In: X. Zhu (Ed.), UNEP Risø Centre, Roskilde.

Communication. (2015). *6th National communication of the Republic of Belarus in accordance with obligations under the UNFCCC*. The Ministry of Natural Resources and Environmental Protection. Scientific Research Center "Ecology". Minsk (in Russian).

Convention. (1985). *The Vienna convention for the protection of the ozone layer*. Accessed 25th Mar 2017. http://www.unep.org/ozone/.

Cook, J., Oreskes, N., Doran, P. T., Anderegg, W. R. L., Verheggen, B., Maibach, E. W., et.al. (2016). Consensus on consensus: A synthesis of consensus estimates on human-caused global warming. *Environmental Research Letters, 11*, 48002. Accessed 25th Mar 2017. http://iopscience.iop.org/article/10.1088/1748-9326/11/4/048002/pdf.

EEA. (2015a). *Protecting the ozone layer while also preventing climate change.* European Environment Agency. Kopenhagen, 2015.

EEA. (2015b). *Europe continues to phase out ozone layer-damaging chemicals.* European Environment Agency. Kopenhagen, 2015.

FAO. (2014). *The state offood and agriculture. Innovation in family farming.* Food and agriculture organization of the United Nations. Rome, 2014.

Handbook. (2008). *Operation handbook—Ozone observation with a Dobson spectrophotometer.* WMO. Global atmosphere watch. GAW No. 183. WMO/TD-No. 1469. 93 P.

Hegglin, M. I., Fahey, D. W., McFarland, M., Montzka, S. A., & Nash, E. R. (2015). *Twenty questions and answers about the ozone layer: 2014 update, scientific assessment of ozone depletion: 2014*, 84 pp., World Meteorological Organization, Geneva.

IPCC. (2014). Climate Change 2014: Synthesis Report. In Core Writing Team, R. K. Pachauri. & L.A. Meyer (Eds.), *Contribution of Working Groups I, II and III to the Fifth Assessment Report of the Intergovernmental Panel on Climate Change* (151 p). IPCC, Geneva, Switzerland.

Krasouski, A., & Turyshev, L. (2011). Interaction of "ozone-climate". *Science and innovations, 4*, 14–16. (In Russian).

Kyoto Protocol. (1998). *Kyoto Protocol to the United Nations Framework Convention on Climate Change*, United Nations, 1998. Accessed 25th Mar 2017. http://unfccc.int/resource/docs/convkp/kpeng.pdf.

Law. (2001). The Republic of Belarus of November 12, 2001 N 56-Z "*On Protection of the Ozone Layer*". Accessed 28th Mar 2017, www.minpriroda.gov.by/uploads/folderForLinks/zakon-ozon.docx"(in Russian).

Memo. (2007). *Safe use of ultraviolet radiation. Reminder for ecologist.* The Ministry of Natural Resources and Environmental Protection. The National Ozone Monitoring Research and Education Center. Minsk (in Russian), 2007.

Montreal Protocol. (1987). *European Commission, Luxembourg: Office for Official Publications of the European Communities.* Accessed 25th Mar 2017. https://treaties.un.org/doc/Publication/UNTS/Volume%201522/volume-1522-I-26369-English.pdf.

NOAA. (2015). *NOAA National Centers for Environmental Information, State of the Climate: Global Analysis for Annual 2015*, published online Jan 2016. Retrieved on 2 Apr 2017 from https://www.ncdc.noaa.gov/sotc/global/201513.

Nowack, P., et al. (2015). A large ozone-circulation feedback and its implications for global warming assessments. *Nature Climate Change, 5*, 41–45.

Paris Agreement. (2015). Accessed 25th Mar 2017. https://unfccc.int/files/meetings/paris_nov_2015/application/pdf/paris_agreement_english.pdf.

Program. (2013). *The state program of measures to mitigate climate change for 2013–2020.* Adopted by Decree of the Council of Ministers of the Republics of Belarus 21 June 2013, No 510 (in Russian).

Report. (2017). *National Report of the Republic of Belarus for the 10th WMO*, UNEP Ozone Research Managers Meeting Geneva, 28–30 Mar 2017.

Sivasakthivel, T., & Reddy, K. K. S. K. (2011). Ozone layer depletion and its effects: A review. *International Journal of Environmental Science and Development, 2*(1), 30–37.

Tubiello, F. (2012). Climate change adaptation and mitigation: challenges and opportunities in the food sector. Natural Resources Management and Environment Department, FAO, Rome. In *Prepared for the High-Level Conference on World Food Security: the Challenges of Climate Change and Bioenergy*, Rome, 3–5 June 2008.

UNEP. (2016). *The Emissions Gap Report 2016.* United Nations Environment Programme (UNEP), Nairobi.

UNFCCC. (1992). *The United Nation Framework Convention on Climate Change.* United Nations, 1992, FCCC/INFORMAL/84 GE.05-62220 (E) 200705.

WHO. (2000). In S. Kovats, B. Menne, A. McMichael, R. Bertollini, & C. Soskolne (Eds.), *Climate change and stratospheric ozone depletion. Early effects on our health in Europe*. WHO Regional Publications, European Series, No. 88. Denmark.
WMO. (2017). *WMO statement on the state of the global climate in 2016*. WMO—No. 1189. World Meteorological Organization, 2017.
WMO (World Meteorological Organization). (2014). *Assessment for decision-makers: Scientific assessment of ozone depletion: 2014* (Global Ozone Research and Monitoring Project—Report No. 56), 88 pp. Geneva, Switzerland.

Embedding Adaptation into Development Planning and Decision Making Process at the Municipal Levels in Mozambique

Luís Artur, Casimiro António, Olanda Bata, Felisberto Afonso and Gilberto Muai

Abstract Climate change has become one of the greatest challenges of the 21st century, particularly for growing cities. In places like Mozambique, the vulnerability of the cities (municipalities) is further compounded by socio-economic constraints that limit the ability of local and national governments to respond to extreme climatic events. All this means that climate change adaptation needs to be understood within a broader socio-economic and environment context. From this, we argue that adaptation interventions must be part of long lasting development plan and not a project per se. In so doing, the starting point should be to set a climate resilient development agenda that adaptation interventions will feed in. Currently indeed, most of the adaptation interventions are stand-alone initiatives based, in most of the cases, in assumptions made mainly by the international donors, NGOs and consultants that what is known globally around climate change apply to local context and can easily be replicated. Hence, pre-cooked ill locally researched interventions are proposed and implemented with very limited impacts. This trend is making adaptation less effective in developing countries despite a growing discourse of climate change adaptation. Against this backdrop, we present a different pathway. This paper addresses interventions by the Coastal City Adaptation Program (CCAP) in two municipalities in Mozambique which seek to apply action-research and use local participation (including local knowledge) to feed adaption interventions in a long lasting development agenda. The paper argues that this approach is far more effective for climate resilience because it responds to local development agenda, allows the prioritization of locally grounded interventions which are supported by a pool of local actors and, because it is locally grounded it allows the embedding of adaptation agenda in the local institution's everyday planning practices making it sustainable over time.

L. Artur
Eduardo Mondlane University, Maputo, Mozambique
e-mail: lartur@uem.mz

C. António (✉) · O. Bata · F. Afonso · G. Muai
USAID—Coastal City Adaptation Project Implemented by Chemonics International Inc, Armando Tivane Av 196, Maputo, Mozambique
e-mail: mobata@ccap-mz.org

F. Alves et al. (eds.), *Theory and Practice of Climate Adaptation*,
Climate Change Management, https://doi.org/10.1007/978-3-319-72874-2_9

Keywords Climate change · Adaptation · Risk · Vulnerability
Resilience

Introduction

Climate change is challenging the ways societies perceive and act upon nature and plan development. Global warming enhanced by the Green House Gases (GHG) emitted through unsustainable development model based on massive consumption of fossil fuels and environment degradation (mainly deforestation) led to a number of unforeseen consequences. The number of 'natural' disasters has more than quadrupled over the past 30 years; from an average of 90 disasters/year in the 1970s, it has now expanded to 450 disasters/year (Webster et al. 2009:5). The number of people affected by disasters has increased three times jumping from about 55 millions/year to about 262 million/year (UNDP 2009:30). In 2015 alone a total of 574 disasters were reported claiming lives of 32,550 people and losses estimated at USD 70.3 billion. To deal with rising number of disasters resulting from climate change and increased vulnerability the UN's annual appeal for international aid rose from USD 3.7 billion in 2004 to USD 20.1 billion in 2016 (IFRC 2016:12–13). If humanity fails to take urgent and concerted actions to address climate change an increased number of people will be affected and die from climate change related disasters and economic losses will continue to rise. NEF and BCAS (2002:2) estimates that annual losses due to climate change will surpass USD 300 billion over the current century while Stern (2006) estimated annual losses around 5% of the global GDP.[1]

Climate change does not only affect human beings; it also affects ecosystems which humans depend on for making a living. It is estimated that between 20 and 30% of flora and fauna species are likely to disappear due to climate change impacts (Ypersele 2008). Mayhew et al. (2007) suggest that global temperatures predicted for the coming centuries may trigger a new 'mass extinction event', where over 50% of animal and plant species would be wiped out.

Although climate change is a global threat (and opportunity), its impacts are disproportionally hitting harder on the poor. Due to higher vulnerability and limited capacity to anticipate, respond and cope with climate change impacts there are 13 times more deaths due to climate change related disasters in poor countries compared to the richer ones (IFRC 2004:164). And about 98% of these affected by climate related disasters live in developing countries (UNDP 2009:8).

Climate change poses a particular threat to cities and urban areas. Cities are home to more than half of the world's population and much of the world's industry.

[1]According to the World Bank figures, Global GDP was estimated at USD 51.31 trillion in 2006 and USD 74.15 trillion by 2015.

By 2050, more than 70% of the population—6.4 billion people—is projected to live in urban areas (OECD 2014). Climate change is already affecting cities and their residents, especially the poor, and more severe impacts are expected as climate extremes and variability increase. Increased numbers of heat waves affects lives and the urban economy which has to consume more energy and other resources; more frequent and intense droughts and inland floods compromises water and food supplies leading to market disruption; and for coastal cities, enhanced sea level rise and storm surges affects inhabitants and essential infrastructure, property, and ecosystems. These challenges highlight the need for cities to rethink how assets are deployed and people protected; how infrastructure investments are prioritized, and how climate will affect long-term growth and development plans (UCCRN 2011).

Building cities that are green, inclusive and sustainable should be the foundation of any local and national climate change agenda. This requires better management of cities, mobilization of a global array of stakeholders, additional financing, and strengthened partnerships, as well as specific sector policy reforms such as urban transport policies, sustainable city planning, and enhancing city resilience and energy efficiency (World Bank 2010a).

It is within the context of expanding climate threats and the need to respond to climate change impacts and opportunities in urban contexts that USAID designed and implemented the CCAP. As outlined above climate change threaten and will increasingly threaten the lives and livelihoods of urban settlers. CCAP is being implemented in five coastal cities in Mozambique. But for this paper we will focus on the first two namely Pemba and Quelimane cities. The interventions are based on the understanding that effective and long lasting adaptation interventions require the engagement of local actors and authorities and must be part of a long lasting development agenda. By fostering the understanding of climate change adaptation and implementing adaptation interventions aligned with local development agenda, CCAP's goal is to help the city's planners and respective residents to increase their understanding of the socio-economic impacts of climate change and engage in ways to deal with. This will help city's councils to improve their ability to deliver resilient services while local citizens will be better prepared to prevent and respond to climate risks. All this will foster resilient and long lasting development.

Structure of the paper

This paper is structured into 5 sections. The next Section "The Context", introduces the general context—Mozambique and then the two cities. Section "Conceptual Framework" provides the main conceptual frameworks on analyzing and building urban resilience. In so doing it ends up by providing the framework that CCAP used in its interventions. Section "Methods" presents key methodological approaches used. Section "Results" presents the stepwise approach used by CCAP to inform its interventions and, Section "Conclusions and Way Forward" provides main conclusions.

The Context

The General Context: Mozambique

Located in the east coast of Southern Africa, Mozambique may be best known for humanitarian crises. The country ranks third in global weather-related damage following Bangladesh and Ethiopia (Buys et al. 2007:38) and ranks 8th in global vulnerability (UNU-EHS 2011). On average Mozambique is affected by 1.17 disasters of great magnitude per year (INGC et al. 2003:7), which accounts for nearly one percent of its annual GDP.[2] As much as 25% of the population (about six million people) is at risk from natural hazards, and economic analysis suggests that annual GDP growth may fall between 4 and 14% in the decade 2040–2050 if climate change adaptation measures are not implemented (World Bank 2010b) (Fig. 1).

Although the majority of the population still live in rural context (about 70%), urbanization is growing fast in Mozambique and it is expected that by 2040 about 40% of the population will be living in cities (AEO 2016). Countries' economy is driven mainly through cities and urban contexts. Rural agriculture accounts for about 23% of the national GDP while urban industries and services accounts for the remaining 77% of the national GDP (GoM 2014). Hence, climate change impacts on urban contexts challenges the overwhelming pie of the national GDP.

Mozambique has already set a climate change agenda. In 2012 the government approved the National Climate Change Adaptation and Mitigation Strategy which sets a number of adaptation interventions around eight (8) areas namely: (1) disaster reduction; (2) water management; (3) agriculture and food security; (4) Social protection; (5) Health; (6) Biodiversity; (7) Forestry and (8) Infrastructure. There is also a dedicated national department and budget line on climate change. From 2014, the government started a national agenda on local adaptation on rural areas and a number of local climate change adaptation plans have been produced. However, there has been very little in the urban contexts. None of the cities had so far a coherent adaptation agenda and 'business as usual approach' still prevail in most of the urban planning and delivery of services.

CCAP is a 5-year's program designed to increase municipal understanding and application of urban adaptation solutions to boost local development. Initially designed for 2 cities (Quelimane and Pemba) CCAP has now expanded to additional 3 municipalities (Nacala, Ilha de Moçambique and Mocimboa da Praia). The program partners with cities municipal governments, the Mozambique Institute for Disaster Management—as well as, civil society, and universities. To create a high level of ownership among local governments and citizens, CCAP takes a stepwise approach to resilience by using interventions that match local capacity and needs. That is, interventions are preceded by a participatory inclusive process of consultation and decision making to feed into an overall development agenda.

[2]In 2015 national GDP was about USD 15 billion.

Fig. 1 Mozambique location

The Local Context: Pemba and Quelimane Cities

Pemba Municipality is located in the north of the country, and is the capital city of Cabo-Delgado province. The municipality measures 98 km^2 and has approximately 200,000 inhabitants. Pemba is considered one of the best bays of the world and has world class beaches and islands. Pemba community is composed mainly of three ethnic groups, namely the Macuas, the Makonde and Kimwani, coming from scattered parts of Cabo Delgado Province. The official language is Portuguese and

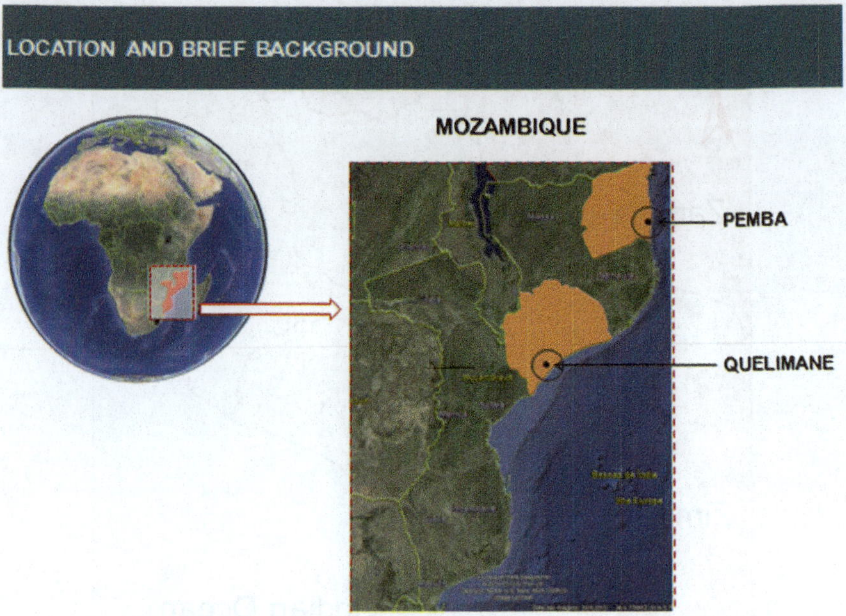

Fig. 2 Location of Pemba and Quelimane cities. *Source* Authors based on google maps

the Muslim and Catholic religions represent the most predominant religious beliefs in the community. Traditionally Pemba citizens depended upon on fishing, agriculture and informal trade for making a living. Over the past five[3] years indeed, Pemba has witnessed a *boom* on tourism and service sectors mainly to feed the demand produced by the recently discovered gas deposits in the nearby regions. Attracted by the huge natural gas reserves in the Rovuma Basin in northern Mozambique, the American company Anadarko Petroleum Corporation opened an office in the country in 2006. In addition to this, three other companies operating in the area of hydrocarbons are: Italy's ENI, the Norwegian Statoil, and Petronas of Malaysia (Fig. 2).

Anadarko and ENI stand out in terms of the volume of investment. The two companies recently announced the largest investment ever in Mozambique: 50 billion US dollars for the construction of up to ten plants of liquefied natural gas (LNG), a megaproject only surpassed by other similar in Qatar.

Pemba became one of the preferred destinations from people coming both from abroad and from other regions of the country in search of new employment opportunities, particularly for those who have high technical or academic qualifications and proficiency in foreign languages. There is therefore, seemingly a bright

[3]There has been indeed a slowdown in 2015/2016 due to political turmoil and national financial crisis explained later in the paper.

future for Pemba despite still the prevailing high rates of poverty (more than 50% of the population lives below the national poverty line) and weak and aging infrastructure.

Quelimane Municipality is the capital city of Zambezia province. It expands around 117 km^2 and has an estimated population of about 250,000 inhabitants. Population growth and city planning are some of the critical issues affecting the city. By 2007 the population was estimated at around 150,000 inhabitants and in only 10 years it has increased by nearly 66% an annual growth rate of about 6%. Contrary to Pemba where there is a visible economic *boom*, this population growth in Quelimane is not being followed by corresponding physical, urban planning and new livelihood options. People rely on petty trade, agriculture and fishing for livelihood and have been putting much pressure on local natural ecosystems such as mangroves for making a living. As a result, many of the areas on the periphery are poorly built and protected, not serviced and where they exist, the lack of maintenance has rendered these systems inoperative. This combination of loss of urban structure, expansion into areas unsuitable for construction such as on mangroves, and that are flood prone, as well as the very low level and declining supply of services, has created an urban environment, which by any urban standards is rated as of very low quality. By 2007, 83.5% of the population lived in huts, 50.2% had no basic sanitation and only 5.3% had access to pumped potable water in their houses (INE 2007).

Conceptual Framework

Over the past 20 years, there has been a growing body of knowledge on climate change impacts and ways forward in dealing with. Developing adaptive capacity has become the cornerstone and a consensual approach built over a period of time in different global agendas. The Paris agreement signed in 2015, which puts an end to years of global negotiations, emphasizes, in article 7, the need to establish the global goal on adaptation to enhance adaptive capacity, strengthen resilience and reducing vulnerability to climate change. Adaptive capacity is defined by IPCC (2007) as the ability or potential of a system to respond successfully to climate variability and change, and includes adjustments in both behavior and in resources and technologies. The presence of adaptive capacity has been shown to be a necessary condition for the design and implementation of effective adaptation strategies so as to reduce the likelihood and the magnitude of harmful outcomes resulting from climate change (Brooks et al. 2005). Adaptive capacity also enables sectors and institutions to take advantage of opportunities or benefits from climate change, such as a longer growing season or increased potential for tourism.

Increasing adaptive capacity which leads to resilience means also reducing the vulnerability; for this endeavor a number of frameworks have been proposed by different scholars. DFID (1999) sustains that adaptation and vulnerability reduction requires enhancing people's access to resources/capitals (Human, social, financial, physical and natural) and the reshaping of structures and processes that leads to increased vulnerability. Wisner et al. (2004) advocate that the root causes of vulnerability lies on global, national and local power relations embraced by political ideologies and economic systems. IPCC (2012) argues that vulnerability reduction encompasses three major issues: reducing exposure by, for instance, a good land use planning; reducing sensitivity by providing robust livelihoods and infrastructures, and increasing the adaptive capacity which is provided by re-thinking the ways people and society as a whole plan and develop their economies across scales and sectors.

For cities adaptation there are a number of guidelines and broad range of tools. In this section we just refer to some of them that have had conditioned a number of interventions. At 2011 Conference of Parties-held in Durban, South Africa, participants approved what has become known as the Durban Adaptation Charter (DAC). DAC proposes 10 principles for a resilient city namely: (1) mainstreaming adaptation, (2) understanding climate risks, (3) preparing city adaptation strategies, (4) aligning adaptation with mitigation, (5) meet needs of communities, (6) considering the role of ecosystems (7) access to funding, (8) designing a M&E, (9) good governance and (10) developing partnerships.

In 2011 the World Bank has prepared a guideline to help cities adapt to climate change. According to the guideline, the starting point is the assessment of the specific climate change impacts at the city level; based on that the city must develop adaptation plans, policies and actions which are thereafter implemented, monitored and evaluated (World Bank 2011).

In 2011 as well, the government of South Africa published a guideline for cities' adaptation encompassing six steps: (1) create an adaptation coordination committee; (2) Assess current climate trends and future projection for the city; (3) undertake a climate vulnerability and assessment for the city; (4) undertake an assessment of adaptation options; (5) develop a city adaptation plan and (6) implement, monitor, and adapt the interventions on ongoing basis (GoSA 2011).

In 2012 the German Technical Assistance (GIZ) published a manual called *'let's respond-integrating change into municipal planning'* which recommends five steps: prepare the process; analyze climate change trends and impacts; design a strategy to deal with CC; develop specific projects and then, integrate the project in the annual planning cycle (GIZ 2012).

From the literature review presented above it is our understanding that city's adaptation planning process could be summarized into four major dimensions:

Assess: this dimension looks at climate vulnerability and capacity analysis in order to understand local context and climate change impacts and opportunities;

Design: this represents the dimension where, through participatory processes, a strategy and locally context-based adaptation measures are proposed, discussed and agreed;

Implement: in this, the proposed measures are put in practice aiming to increase individual and institutional adaptive capacity;

Evaluate: this represents the dimension where the implemented adaptation interventions are evaluated in terms of their contributions to increasing resilience.

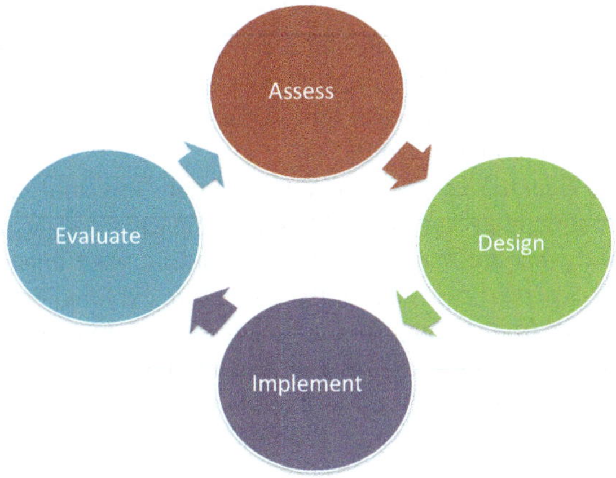

Fig. 3 The conceptual framework for climate change adaptation interventions. *Source* Authors

Note that resilience is a process and it may require a long-term time spam to achieve which routine short term evaluations may not be able to capture.

CCAP used this four dimension stepwise approach (Fig. 3) in designing its interventions in targeted cities in Mozambique.

Methods

CCAP used a multi-stakeholder platform for climate vulnerability and capacity assessment. This is based on action research methodology which is a research process used as a mean to help solve societal problems by applying relevant scientific and local knowledge. Reason and Bradbury (2008) argue that action research is a pathway for providing power to people to act in order to bring changes through conscious reflection on personal and/or collective experiences. The envisaged action research methodology involved a number of different methods and processes as follows:

During the course of the program climate assessment used both, scientific global models and local knowledge. For historical climate information CCAP used data available for two stations, Pemba (Station ID: CD000034, Lat = 12.98°S; Long = 40.53°E) and Quelimane (Station ID: ZB004001, Lat = 17.88°S; Long = 36.88°E). For climate modelling CCAP used 11 Global Circulation Models (GCMs). All 11 GCMs (MIROC-ESM, CNRM-CM5, CanESM2, FGOALS-s2, BNU-ESM, MIROC5, GFDL-ESM2G, MIROC-ESM-CHEM, GFDL-ESM2 M, MRI-CGCM3 and BCC-CSM1-1) were used in the IPCC 5th assessment report and forced with the Representative Concentration Pathways 8.5 W/m^2 (RCP.8.5) a method suggested by Moss et al. (2010).

To capture local knowledge the project used the participatory vulnerability and capacity assessment toolkit developed by CARE which encompasses:

Trends analysis: with this tool we discussed local perception on socio-economic and climate trends over time. This was useful for instance in validating the scientific data on climate change.

Vulnerability and capacity matrix: this tool allows the discussion on how communities and institutions are impacted and respond to climate related events and what are their perceptions regarding what must be done to build resilience;

Seasonal calendar: with this tool we explored how changes are experienced over a year from January to December in order to capture critical moments that require special attention by the government and partners like CCAP;

Venn diagram: this tool tends to capture existing institutions and actors in the area, what are their interventions and, the relationships between them;

Participatory mapping: this tool was used to capture local perception regarding resources distribution around the cities, the exposure, sensitivity and the capacity of infrastructures and business to stand under climate threats in each single neighbourhood *"bairro"*.

In order to run this exercise, municipal technical staff backed by university and CCAP experts run Focus Group Discussions and semi-structured interviews in the neighbourhood and institutions. After this, joint meeting with the representatives of the neighbourhoods and institutions were held to discuss the findings, and plan the climate resilient development agenda.

Based on exposure, sensitivity and adaptive capacity information—collected through the methods presented above, GIS tool was used to develop vulnerability and adaptive capacity maps that were later on discussed and agreed by the local stakeholders. For institutional capacity analysis another global tool was also used: the Local Government Self-Assessment Tool (LGSAT) by which local government actors assess themselves their own capacity. This capacity assessment is based on scorecards around 10 essentials for making resilient cities proposed by the United Nations Secretariat for Disaster Reduction (UNSDR) namely:

1. Engage, share understanding, and coordinate
2. Create financing and incentives
3. Identify and understand perils, probabilities and impacts
4. Make critical infrastructure disaster resilient
5. Make education and healthcare infrastructure disaster resilient
6. Apply risk-aware planning, land use and building codes
7. Build public awareness and capacity
8. Enhance and protect ecosystem services
9. Create warning systems and rehearse preparedness
10. Learn and build back better.

Results

Climate Change and Impacts in Pemba and Quelimane Cities

The climate in both Pemba and Quelimane is characterized by two seasons, one cold and dry and the other one hot and wet. The hot and wet season in both locations generally extends from October to April. The annual precipitation in Pemba ranges between 800 and 1000 mm while in Quelimane it ranges between 1000 and 1200 mm. The annual mean temperature in Pemba varies between 22 and 27 °C, with the highest temperatures on average of 30 °C being recorded in December. On the other hand, the annual mean temperature in Quelimane varies between 21 and 28 °C with the highest temperatures on average of 31 °C being recorded as well in December.

Temperature and precipitation trends have shifted indeed in both locations. Annual average maximum air temperature (Tmax) ***in Pemba*** has increased by 0.21 °C per decade, and minimum temperature (Tmin) shows an increasing trend in 0.55 °C per decade (Fig. 4).

Climate projections suggest that this trend on temperature is likely to continue over the coming decades. INGC (2009) suggest that the average maximum temperatures may shift from around 30 °C in 1980s to about 35 °C by 2100 (Fig. 4 left) while the average minimum temperatures may shift from about 19 °C to about 23 °C (Fig. 5 right).

For ***Quelimane*** city historical data shows that, annual maximum temperatures have increased by 1.73 °C during the 33 year period at an average rate of 0.52 °C per decade, while minimum temperature has increased by 1.16 °C between 1981 and 2013 at an average rate of 0.35 per decade (Fig. 6).

Climate projections for Quelimane city show also that both, maximum and minimum temperatures will increase over the coming decades. Average maximum is expected to shift from current 31 °C to about 33 °C by 2055 while average minimum may shift from 21 °C to about 23 °C (Fig. 7).

Apart from temperature increases, precipitation has been irregular although it is expected to slightly increase over the coming decades both—in Pemba and Quelimane. Figure 8 shows that the number of consecutive dry days is likely to increase on an average of 5 days in Quelimane and 2 days in Pemba over the period 2036–2065 (Fig. 8 left) while the total number of wet days is likely to increase by 40 days in Pemba and 20 days in Quelimane over the same period (Fig. 8 right).

During the discussions with the citizens in both municipalities, as part of the assessment, people have concurred that both cities are hotter now and precipitation has been irregular. Extreme events have increased with flooding/inundation becoming more frequent over the past 10 years compared to previous period. Sea level rise have also been recorded with a number of infrastructures being eroded and reallocated. City flooding, seal level rise, cyclone and environmental pollution

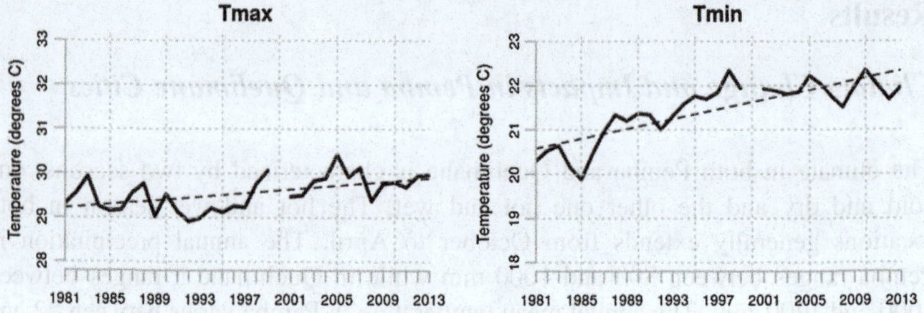

Fig. 4 Observed temperatures in Pemba. *Source CCAP* (2014)

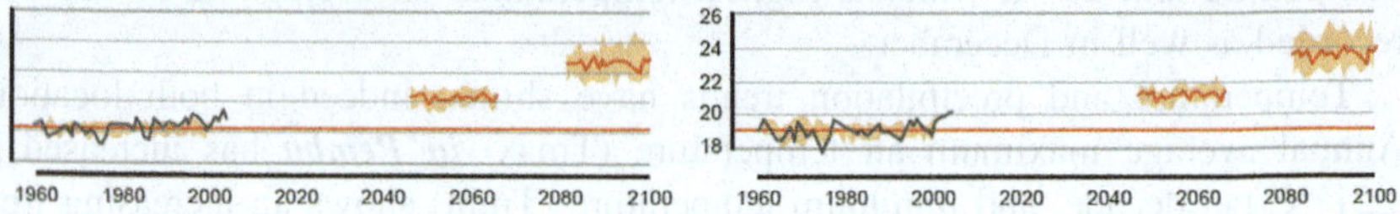

Fig. 5 Projected temperatures in Northern Mozambique (including Pemba). *Source* INGC (2009)

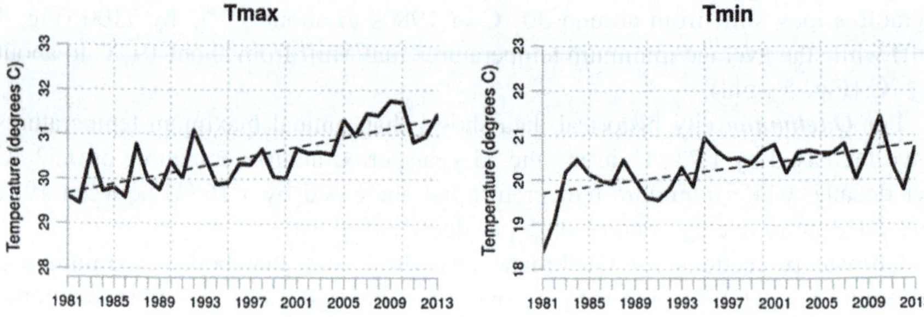

Fig. 6 Observed maximum and minimum temperatures for Quelimane city. *Source* CCAP (2014)

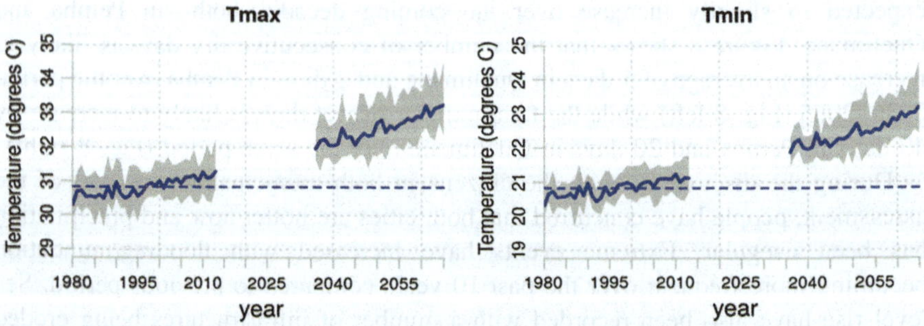

Fig. 7 Projected temperatures for Quelimane city. *Source* CCAP (2014)

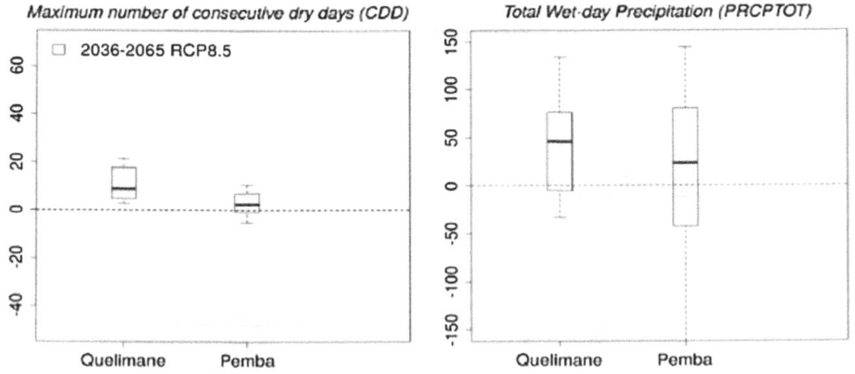

Fig. 8 Expected rainfall patterns for Quelimane and Pemba cities. *Source* CCAP (2014)

have been mentioned as major climate related hazards impacting the cities. Although no records have been made by the city councils on the socio-economic impacts—due mainly to lack of database and institutional capacity to collect and process information, all actors have referenced that more people are being affected by climate risks and economic losses tend to increase.

Vulnerability and Capacity Assessment

The vulnerability assessment has shown both cities to be highly vulnerable to climate extremes. Vulnerability map produced by the program for ***the Pemba city*** suggests that about 1/3 (34.7%) of the city is moderate to highly vulnerable to climate risks (yellow to red colors, Fig. 9). This is due to the fact that about ¼ of the city area is just about 50 meters above sea level and the major investments and densely populated areas are along the city's coastline.

Similarly, ***Quelimane*** is also a very vulnerable city. Almost the entire city is below the sea level and the city is surrounded by rivers and ocean. Over the course of years, as mentioned earlier, weak city structural planning and expanding poverty led to massive environmental degradation and increased exposure to inundation, cyclone and heat waves. The Fig. 10 shows the Quelimane vulnerability map where about 85% of the city is moderate to highly vulnerable and only about 1/10th (around the city center/business district) could not be classified as vulnerable to climate risks.

While this is the case for the vulnerability issue, analysis of the adaptive capacity shows that only about 1/10th (green color) of both cities (especially around the main towns) could be labeled as resilient to climate risks (Fig. 11).

Institutional capacity assessment for dealing with climate change has shown very limited capacity of the cities to deal with climate issues. Pemba municipality has a

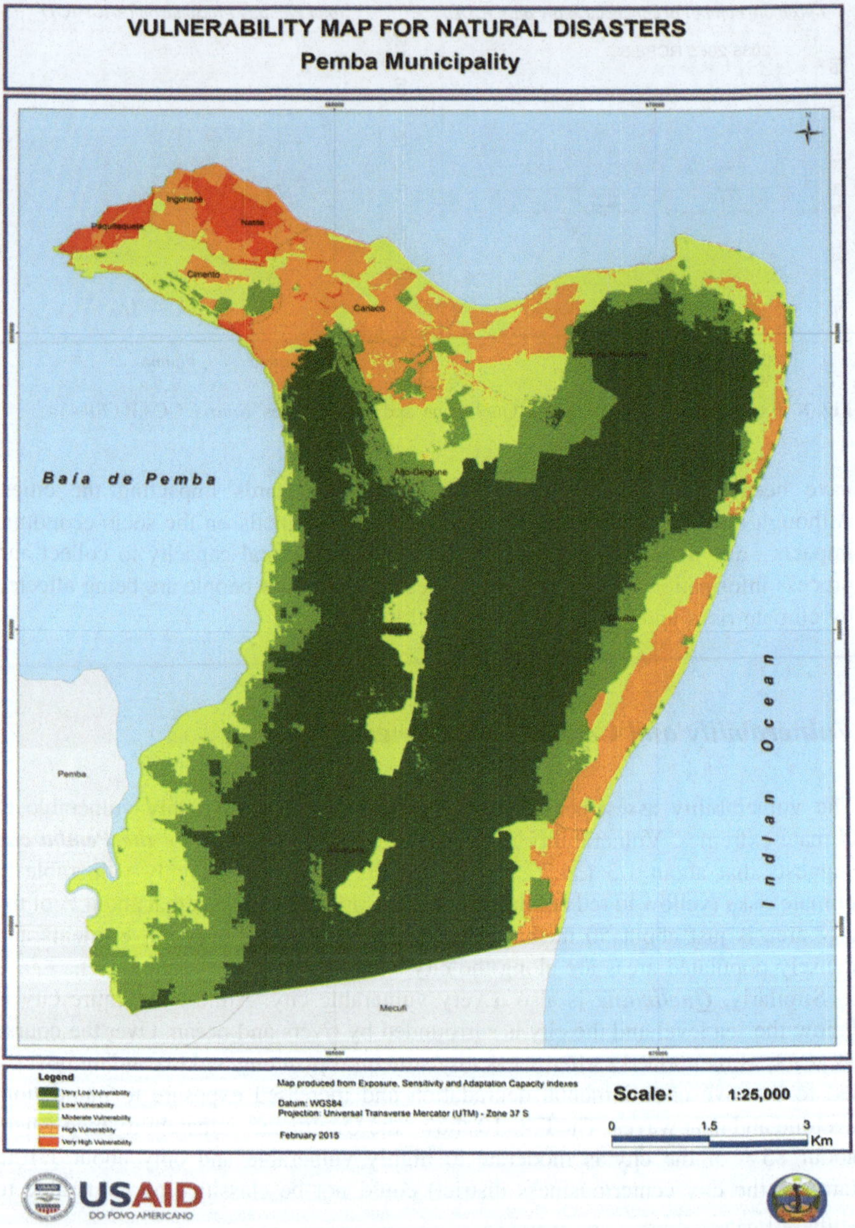

Fig. 9 Vulnerability map for Pemba city. *Source* CCAP (2014)

written development agenda but it has not incorporated the issue of climate change on it. Quelimane has no a written development agenda and annual plans and budgeting has virtually been without taking the issue of climate change into

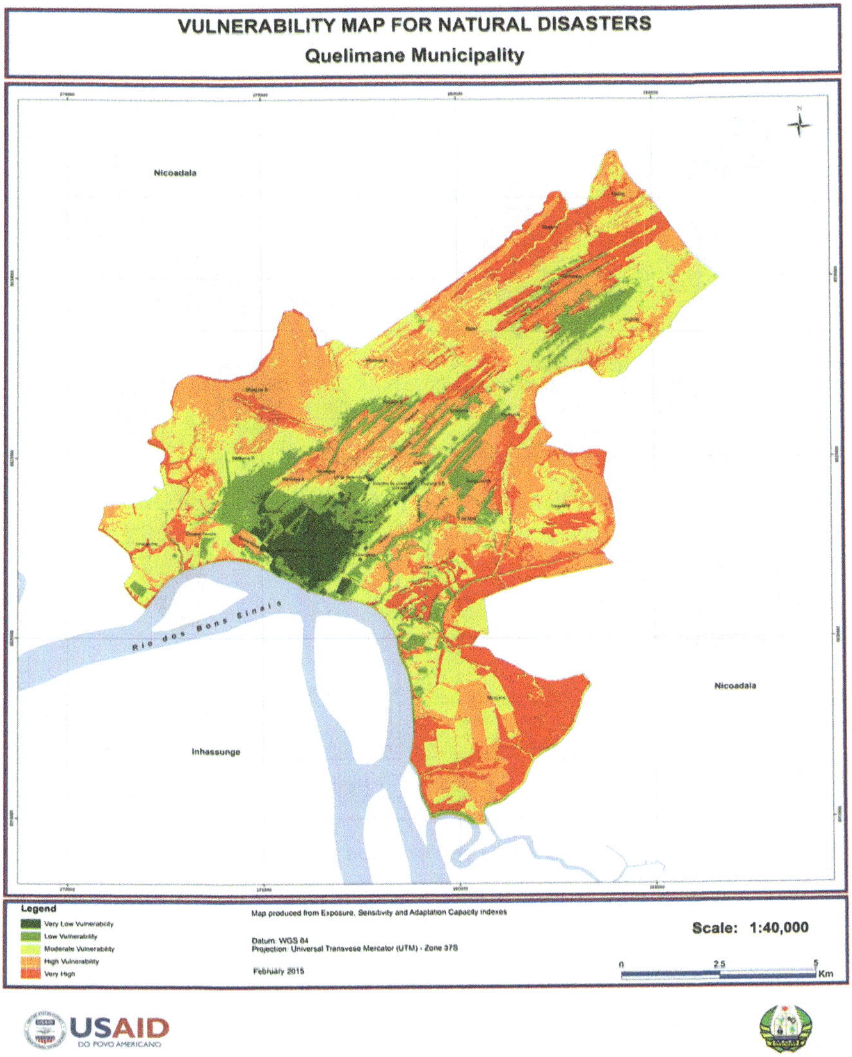

Fig. 10 Climate vulnerability for Quelimane city. *Source* CCAP (2014)

consideration. The overall score of the 10 essentials for a resilient city was 1.8 out of a maximum of 5 points for Pemba (Fig. 12) and 2 for Quelimane (Fig. 13). The lowest scores for both cities were on coordination (essential 1), finance and incentives (essential 2), understanding of CC and impacts (essential 3); risk awareness planning (essential 6) and design and implementation of warning systems (essential 9).

Overall the assessment has shown that climate change is happening in both cities. Indeed, both cities are vulnerable to climate risks and have limited adaptive

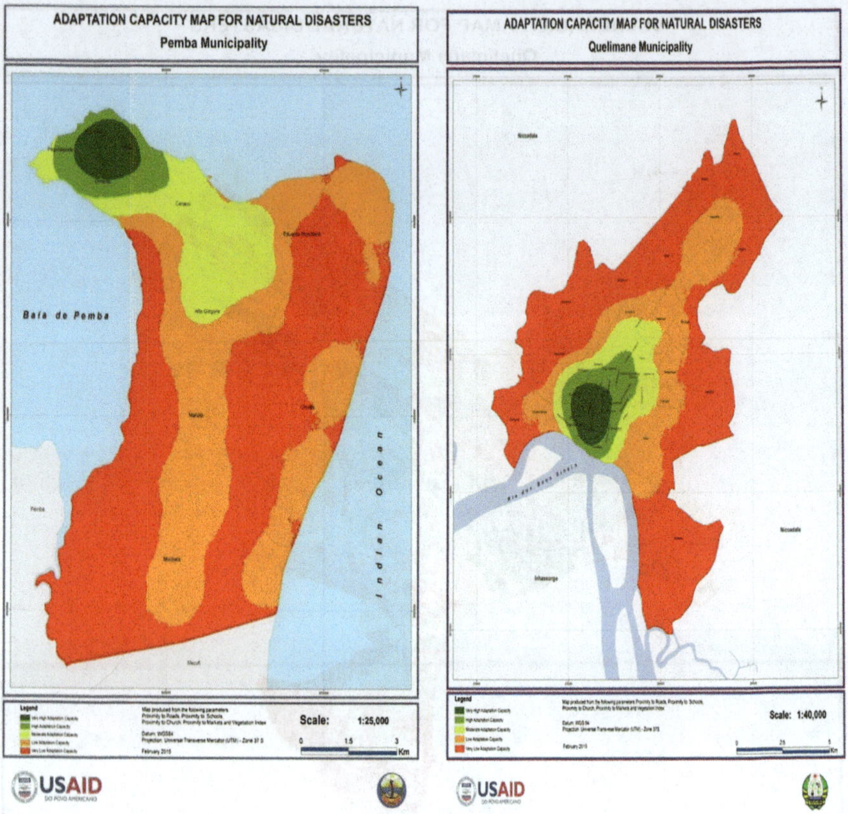

Fig. 11 Adaptive capacity of the Pemba city (left) and Quelimane city (right). *Source* CCAP (2014)

capacity to deal with. This was the background that informed the interventions planned in both cities which are discussed in the next section.

Designing Interventions: Setting a Climate Resilient Development Agenda Through the Local Adaptation Plans

The assessment presented above showed that none of the cities had a coherent climate resilient development agenda. To start interventions, CCAP envisaged a broad consultation process to design a climate resilient development agenda. The consultation process included all the neighborhoods (*bairros*) of the cities and to

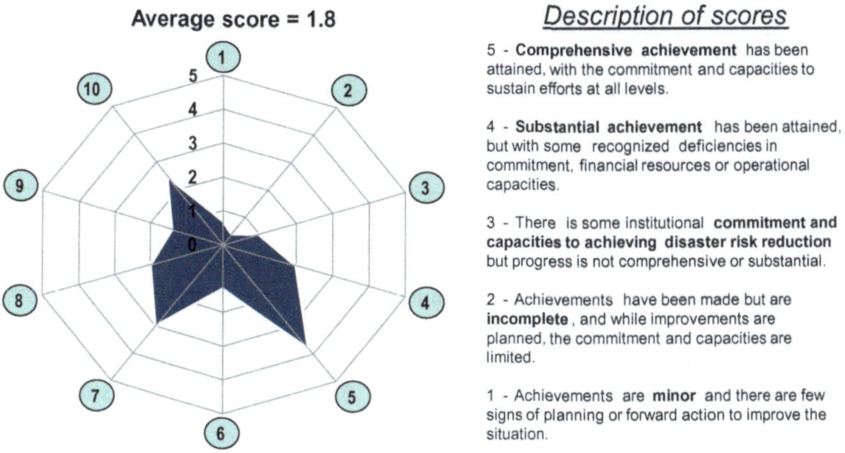

Fig. 12 The institutional capacity for Pemba city. *Source* CCAP (2015)

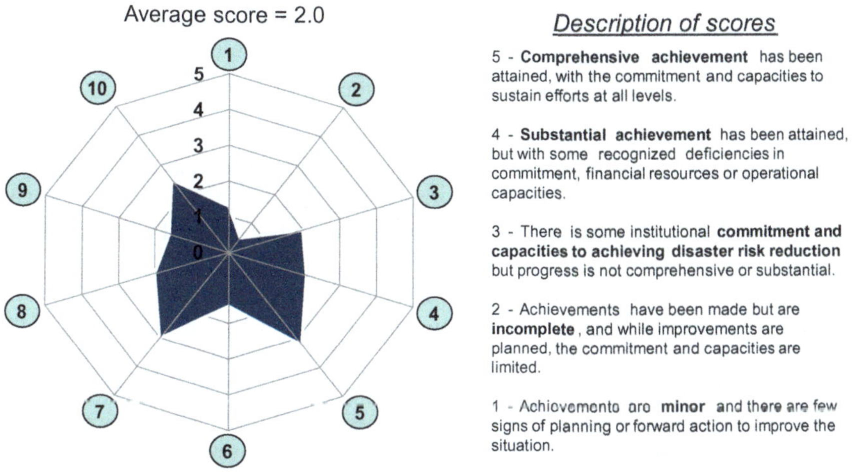

Fig. 13 The institutional capacity of Quelimane city. *Source* CCAP (2015)

local key government, academia, private sector and NGOs actors. This participatory process included:

1. Engaging the city mayors and the municipal councils in order to get the required political support and leadership.
2. Carrying out seminars with representatives from the municipalities, *the bairros,* the government, academia, NGOs and private sector to introduce the issue of climate change and impacts, cities' vulnerability, and need for adaptation

planning. This also included discussing the methodology and the road map for producing PLA and in setting urgent interventions.

3. Setting an advisory committee made of representatives of these key institutions to monitor and help municipalities and CCAP to carry out effective adaptation interventions.
4. Setting a municipal technical team responsible for the design of the PLA with technical support from CCAP staff and of the advisory committee. The technical teams were led by the municipal mayors.
5. Defining clear roles of the different actors involved (i.e. city major, municipal assembly, advisory committee, technical team, CCAP) in the process.
6. Conduct on-ground consultations, by using CVCA tools, with the city's citizens and local institutions on interventions to be carried for adaptation.
7. Submitting the preliminary draft of the PLA to the different actors and institutions for comments.
8. Carrying out seminars with the representatives to discuss and agree on the PLA interventions to be carried out, based on the assessment and the consultation process, with the municipalities either with or without CCAP support.
9. Submitting the PLA to the municipal assembly for approval and ratification so that it becomes institutionalized and interventions can be funded primarily by municipal budget.
10. Facilitating the annual planning and implementation of the PLA.

The process of consultations, designing and refining the resilient development agenda took about six months and the final product come out as Cities Local Adaptation Plans—(*Planos Locais de Adaptação—PLAs*). These were approved and endorsed by the respective assemblies by December 2015. The action plans included, for Pemba, a total of 30 interventions around five major objectives: (1) Promote resilient infrastructures; (2) Expand quality and environment friendly tourism; (3) Promote a resilient fishing industry; (4) Improve the resilience of the education and health sectors; and (5) Promote and protect city's environment. For Quelimane the PLA includes a total of 34 interventions around six major objectives: (1) Improve city's land use planning to foster resilience; (2) Improve the resilience of public infrastructures, especially related to education and health; (3) Protect the environment and expand the green infrastructure; (4) Prepare citizens for emergencies; (5) Improve governance systems to foster resilient development and (6) Promote sectoral climate proof development.

Implementing Interventions: Focusing on the Enabling Factors

Since the approval of the PLA, CCAP has been helping the cities in implementing the interventions and in mobilizing other actors to help cities implement their PLAs. The interventions are implemented by the municipalities with technical and financial support from CCAP and other partners with the relevant expertise for each of the approved intervention. To start with, the city councils, CCAP and other partners agreed in starting with interventions that will create the necessary conditions for the long term plan to take off. These interventions included:

- *Expanding the cities network/partnerships on climate resilience*—through CCAP city mayors and key municipal staff have visited a number of other cities such as Durban in South Africa and Bonn in German in order to share experience on climate adaptation and get support for the implementation of the PLA;
- *Improving urban land use management practices*—the municipalities have now managed to integrate risk and vulnerability maps into the cadaster system to inform site selection for different uses such as houses and infrastructure in the city. This responds to the objective of better land use planning. Key partner includes the National Urban Planning Directorate—a key government institution regarding urban planning and the National Institute for Disaster Management;
- *Integrated system for information management (SIGIU)—Through CCAP*—the municipalities have developed a tool where information about the city and city council activities are stored, updated and easily accessed. The tool is helping the municipalities to make informed decisions based on data and on creating a database which was not existent. This responds also to the objective of good governance for city resilience;
- *Integrated disaster information management system (SIGIC)*—is a SMS-based innovative feature for information sharing and data collection for emergency preparedness and response. This responds to the objective of preparing communities for emergency. Key partners in this intervention is the National Institute for Disaster Management (INGC);
- *Setting and equipping local committees for disaster management:* CCAP has provided training and equipment to 15 local committees for disaster response. This responds to the objective of preparing communities for emergency. Key partners in this intervention is the INGC;
- *3-2-1 on demand information system*—this is a mobile phone based apps that provides free and easy access to information about adaptation and resilience measures, disaster preparedness, and on a wide variety of topics ranging from health to agriculture throughout the year. This responds to the objectives of promoting sectoral climate proof development and, preparing communities for emergency. This system was fed by a number of experts on the areas of DRR, Climate change adaptation, smart agriculture, and health care, among others;
- *Socio-economic baseline study*—baseline study was conducted to provide essential data on demography; housing structure and conditions; household sources of income; expenditure and owned assets; environmental awareness; communication channels commonly used; current perceptions, attitudes, knowledge, and practices related to CCA and DRR; and water, sanitation and hygiene (WASH) practices. The results and analysis from the baseline are

helping the city managers and the project to better understand the socio-economic issues that should inform planning and intervention prioritization efforts. This was a jointly effort by the municipality, CCAP and academia (UEM Marine School, Catholic University and Pedagogical University). This responds to the objective of good governance for city resilience;

- *Training course on climate change adaptation and disaster risk reduction*—this course provided to the participants the knowledge of and access to tools that enable them to identify hazards affecting the city and maladaptive practices that are unsustainable for city resilient development or that contribute to increased risks and vulnerabilities. As a next step, this course will be transformed into an online e-learning course freely available for anyone with internet access and interest. This responds to the objective of good governance for city resilience. Key implementing partner was the academia (Eduardo Mondlane University—UEM);
- *Construction of a model of smart resilient housing*—the project in partnership with UN-Habitat is working with local stakeholders on the construction of models of resilient houses that include smart sanitation facility (compatible with areas with high water table) and rain water harvesting facility, all of them combined, will help the users deal with the effects of extreme climatic events. The construction techniques will be widely disseminated throughout the community. This responds to the objective of promoting resilient infrastructures. Key partners in this include UN-Habitat, Funding for Housing Expansion (*Fundo de Fomento de Habitação*), Provincial Directorate of Public Works and Water Resources (*Direcção Provincial de Obras Publicas e Recursos Hídricos*) and community members;
- *Green infrastructure (natural ecosystem) restoration*—the project is helping targeted cities restore green infrastructure, e.g., mangrove restoration and rehabilitation to ensure that cities and residents take advantage of ecosystem services such as city protection and in the long run livelihood for the citizens. This responds to the objective of protecting the environment and expands the green infrastructure. Key partners include academia, The Ministry of Sea, Inland Water and Fisheries (MIMAIP), the Ministry of Land, Environment and Rural Development (MITADER) and community members.
- *Behavior change campaigns*—the project is helping families through door-to-door campaigns, to raise awareness on climate change impacts and promote adaptation measures to deal with climate change impacts such as changes in water and energy use; building codes amongst others. This responds to the objectives of preparing citizens for emergencies and improving governance systems to foster resilient development.

Conclusions and Way Forward

Conclusions

This paper started by arguing that despite a growing discourse of climate adaptation, organizations promoting adaptation fall short in providing a coherent and long lasting resilient development. This process must start by a good understanding of the local context and in embedding adaptation into the local development agenda. Based in the CCA experience in Pemba and Quelimane, this paper presented the process of setting up a platform for developing a local climate resilient development agenda (PLA) and its intervention. The intervention focused on addressing the enabling factors that allows the plan to take off. The participatory approach that was followed by CCAP eased the uptake of the project by the city managers and other city actors to an extent that the local adaptation plans have been ratified by local city councils and institutionalized within the on-going annual planning and budgeting systems. Both cities are now in the process of implementing their adaptation plans and although is still early to assess, the local institutional capacity has expanded steadily. From the experience, we conclude that working through local governance systems and helping improve their institutional capacity for a long lasting development is an effective approach for fostering adaptive capacity. On the other, external actors such as CCAP that have a short long spam have to focus on addressing the enabling factors that will allow the process to continue after the lifetime of the project.

Way Forward

CCAP has settled the foundations required for a long lasting process. The way forward is to seek funding and partnership for a full implementation of the PLA. For this endeavor, there are a number of opportunities but also challenges that different actors will have to take into consideration:

Key Opportunities

- *The institutionalization of climate change planning through PLA*: Climate change and PLA are now institutionalized and this offers an opportunity for climate compatible development around these two cities. The interventions so far developed by the project helps also the process of institutionalization and of strengthening the understanding and implementing climate smart interventions.
- *The existing multi-stakeholders platforms*: the platforms created for the design and implementation of the PLA—which include city councils, private sector, academia, state and non-state actor—represent a wealth of resources to help the cities deal with climate change impacts.
- *New investments in pipeline*: Quelimane and Pemba are coastal cities with ports and considerable tourism industry. New planned investments related to gas and

oil for Pemba area and for charcoal shipment around Quelimane, may boost in mid-to long term, local economies and increase cities revenue that can be used to expand local resilience.

- *International networks*: The cities are progressively connecting to international networks and being furniture with opportunity to learn more about this topics, and tap into international available resources (including financial) to address climate change impacts.
- *Growing climate change and planning knowledge*: The trainings and the different tools and approach developed and being applied in the city councils are opportunities for climate change resilient planning and long term sustainable development.
- *Young staff*: Most of the municipal staff in the cities are young; this human resource pool, represents an opportunity for long term development.

Key Challenges

- *A fast changing socio-economic context*: Mozambique was, five years back, a fast growing and all blooming economy. Annual average growth rate was around 8%. Indeed, due to political turmoil and financial mismanagement, the economy sank and GDP growth since last 2 years has been below 5%. Public and private investments have decayed affecting livelihoods and lives across the country including in the 2 cities;
- *The novelty of climate change issue*: climate change is still a relatively new topic in the two cities. Available data indicate that nationwide most of the people are illiterate and have limited understanding of climate change. This influences their capacity to understand and take adaptation measure to deal with climate change impacts;
- *Prevailing poverty*: nearly half of the population in Mozambique is poor (about 46.1% lives below national poverty line). Therefore most of them face competing priorities, to deal with long term adaptation needs and immediate survival needs. This challenge can also be extrapolated to the city councils; in the 2 cities poverty is rampant and city councils face competing priorities to provide immediate wellbeing and security for the citizens and simultaneously a serious shortage of resources to respond to emerging needs to deal with climate change impacts;
- *Still limited institutional capacity*: The municipalities capacity to absorb new tools and approach is also limited for several reasons such as staff capacity, staff turnover, shortage of resources just to mention a some;
- *Power politics*: municipal councils are elected every 5 years based on the promises they have made but also through a very strong political/party backup. Mayors and councils have to make sure that they deliver what have promised at same time that responds to the party demands. This is relevant to ensure they are re-elected or promoted. This confluence of actors and political agendas plays a pivotal role on what happens in the ground and at what speed.

References

AEO. (2016). *African economic outlook 2016—Mozambique*. African Development Bank.

Brooks, N., Adger, W. N., & Kelly, P. M. (2005). The determinants of vulnerability and adaptive capacity at the national level and the implications for adaptation. *Global Environmental Change, 15,* 151–163.

CCAP. (2014). *CCAP annual report 2014*. Maputo.

CCAP. (2015). *CCAP annual report 2015*. Maputo.

Buys, P., Deichmann, U., Meisner, C., That, T., & Wheeler, D. (2007). *Country stakes in climate change negotiations: two dimensions of vulnerability*. Washington: The World Bank.

DFID. (1999). *Sustainable livelihood guidance sheets*. London, UK.

GIZ. (2012). *Let's Respond. A guide to integrating climate change risks and opportunities into municipal planning*. German.

Government of South Africa (GoSa). (2011). *Adapting South African Cities and Towns—A local Government Guide to Climate change adaptation planning*. South Africa.

GoM. (2014). *Plano Quinquenal do Governo 2014–2019*. Maputo.

IFRC. (2016). *World disaster report. Resilience, saving lives today, investing for tomorrow. Geneva.*

IFRC. (2004). *World disaster report*. Geneva: Focus on Community Resilience.

INGC. (2009). Synthesis report. In B. van Logchem & R. Brito (Eds.), *INGC Climate Change Report: Study on the Impacts of Climate Change on Disaster Risk in Mozambique*. INGC.

INE – Instituto Nacional de Estatística. (2007). CENSO, Maputo.

INGC, UEM, & Fewsnet. (2003). *Atlas for disaster preparedness and response in the Limpopo Basin*, Maputo, Mozambique.

IPCC. (2012). *Managing the risks of extreme events and disasters to advance climate change adaptation*. Cambridge: Cambridge University Press.

IPCC. (2007). *Fourth assessment on climate change*. Geneva.

Mayhew, P. J., Jenkins, G. B., & Benton, T. G. (2007). A long-term association between global temperature and biodiversity, origination and extinction in the fossil record. *Proceedings of the Royal Society B, 275*(1630), 47–53. https://doi.org/10.1098/rspb.2007.1302. pmc2562410. pmid 17956842.

Moss, R. H., Edmonds, J. A., Hibbard, K. A., Manning, M. R., Rose, S. K., van Vuuren, D. P., Carter, T. R., Emori, S., Kainuma, M., Kram, T., Meehl, G. A., Mitchell, J. F., Nakicenovic, N., Riahi, K., Smith, S. J., Stouffer, R. J., Thomson, A. M., Weyant, J. P., Wilbanks, T. J. (2010). The next generation of scenarios for climate change research and assessment. *Nature, 463*(7282), 747–56.

NEF and BCAS. (2002). *The end of development? Global warming, disasters and the great reverse of human progress*. UK: New Economics Foundation.

OECD. (2014). *Cities and climate change-policy perspectives: National governments enabling local action.* OECD Publishing.

Reason, P., & Bradbury, H. (Eds.). (2008). *The SAGE handbook of action research: Participative inquiry and practice*. London: SAGE Publication Ltd.

Stern, N. (2006). *The economic costs of climate change. The Stern review*. New York: Cambridge University Press.

UNDP. (2009). *Overcoming barriers. Human mobility and development.* Human Development 2009. New York.

UNU-EHS. (2011). *World risk report 2011*. Bonn, German.

Urban Climate Change Research Network (UCCRN). (2011). *Climate change and cities*. In: *First assessment report of the urban climate change research network*-center for Climate Systems Research, Earth Institute, Columbia University 2011.

Ypersele, J. (2008). *Climate change: What do we know, according to IPCC (WG II)?* Pre-COP IPCC workshop, Warsaw, 2 Oct 2008.

Webster, M., Ginnetti, J., Walker, P., Coppard, D., & Kent, R. (2009). *The humanitarian costs of climate change. Feinstein International Center*. US: Tufts University.
Wisner, B., Blaikie, P., Cannon, T., & Davis, I. (2004). *At risk: Natural hazards*. Routledge, London: People's Vulnerability and Disaster.
World Bank. (2011). *Guideline for Resilient Cities*. Washington DC.
World Bank. (2010a). *Cities and Climate Change*: An Urgent Agenda. Washington DC.
World Bank. (2010b). *Economics of Adaptation to climate change—Mozambique*. Washington DC.

Promoting Private Sector Engagement in Climate Change Adaptation and Flood Resilience—A Case Study of Innovative Approaches Applied by MSMEs in Mumbai, India

Caroline Schaer and Archana Pantakar

Abstract Recurring heavy precipitation and flooding cause extensive loss and damage in cities like Mumbai. Among the worst affected are Micro Small and Medium Enterprises (MSMEs), which suffer damage to physical structure and loss of business. These costs amount to millions of dollars and are borne by MSMEs in the absence of adequate insurance protection. With limited flood management services and inadequate infrastructure provided by the municipal authorities, MSMEs are implementing their own temporary measures for flood protection. These are often ineffective during heavy precipitation and create risks of mal-adaptation. As climate change is expected to worsen the risk of flash floods with changes in intensity, frequency and duration of rainfall, MSMEs need long-term solutions to build their adaptive capacity and resilience. This paper describes the business case for private sector engagement in flood risk reduction and climate adaptation from the perspective of MSMEs in Mumbai. Based on extensive field surveys of MSMEs located in industrial estates, the paper discusses the implications of floods for MSMEs. Moreover, the authors present a framework developed for MSMEs to make informed risk reduction and adaptation decisions and implement effective structural and non-structural measures to minimize the recurring adverse impacts of floods on their business operations.

Keywords Climate change adaptation · Private sector engagement
Flood resilience · Mumbai · India

The original version of this chapter was revised: Misplaced tables have been corrected. The erratum to this chapter is available at https://doi.org/10.1007/978-3-319-72874-2_33

C. Schaer (✉)
UNEP DTU Partnership, Marmorvej 51, 2100 Copenhagen Ø, Denmark
e-mail: cesc@dtu.dk

A. Pantakar
MP Ensystems Advisory Pvt. Ltd., Ground Floor, Pushpadhanwa Society,
Madan Mohan Malaviya Road, Mulund (West), 400080 Dwarka, Mumbai, India
e-mail: archana@mpensystems.com

F. Alves et al. (eds.), *Theory and Practice of Climate Adaptation*,
Climate Change Management, https://doi.org/10.1007/978-3-319-72874-2_10

Introduction

It is broadly acknowledged that floods, droughts, sea level rise and increased variability of weather patterns are increasingly and disproportionately affecting vulnerable regions, with the least capacity to tackle the manifestations of a changing climate (IPCC 2014). In 2014, 87% of disasters were climate-related (UNHABITAT 2016) and cities in developing countries are particularly and increasingly at risk. Due to a combination of physical exposure, increase in extreme event occurrence, rapid urban growth, poor urban planning, inadequate basic infrastructure and disaster preparedness, the implications in urban centers for households, businesses and the economy as whole are widespread (UNHABITAT 2016). By 2070, populations and assets in Asian cities, particularly in India, China and Thailand, are expected to be among the most exposed to coastal flooding, as a result of the rapid urbanization and economic growth projected in these countries (Nicholls et al. 2008).

This poses important challenges for the planning, financing and implementation of climate change adaptation measures, which has traditionally been considered as the sole responsibility of the public sector. Yet, global estimates show that the costs of adaptation in developing countries—which are estimated to rise to US$ 500 billion by 2050 (UNEP 2016)—will considerably surpass the public sector's financial resources. As a result, both public and private stakeholders are key to contribute to build climate-resilient societies (Dougherty-Choux et al. 2015). The private sector is thus increasingly recognized as critical to progress in adaptation, given its unique expertise, its capacity to innovate and produce new technologies, its financial leverage and its role in scaling up adaptation of communities, as well as in managing risk information (IPCC 2014). Harnessing the private sector's role in developing countries will thus be essential to help communities adapt to climate change.

Previous literature has mainly concentrated on companies' mitigation actions, while failing to give sufficient attention to their adaptation strategies (Okereke et al. 2012). Although studies on private sector adaptation have started to emerge the last 10 years (Linnenluecke et al. 2012; Agrawala et al. 2011; United Nations Global Compact 2015; Wedawatta et al. 2010; Haigh and Griffiths 2012; Okereke et al. 2012; Berkhout et al. 2006), these are largely limited to developed countries. In addition, while it is increasingly acknowledged that the private sector has an essential role to play in supporting societal resilience to climate change impacts and addressing the risks and opportunities posed by a changing climate, to date research has mainly focused on large multinational corporations (MNCs), private investors, and how they may contribute to close the gap in adaptation finance. Hence, current literature on private sector engagement only covers a fraction of the global private sector.

Thus far, little is therefore known on the role that Micro, Small and Medium sized Enterprises (MSMEs)[1] can play in relation to climate change adaptation in developing countries. Not only do MSMEs constitute more than 90% of businesses in developing countries (Hussain et al. 2012) and are thus contributing significantly to employment generation and supporting livelihoods, they also have a lower technical and financial capacity to respond to climate and disaster risks, and are therefore disproportionately affected by the negative impacts of climate change. MSMEs do for example not operate within the same planning timeframe as bigger companies, and do not necessarily have the needed climate information available nor the same incentives to implement risks minimization measures (United Nations Global Compact 2015). And even in cases and cities where climate information is accessible and risk awareness is high, there are few tools available for identifying, analyzing and selecting location and time-specific adaptation measures (Dougherty-Choux et al. 2015).

Acknowledging the significance and potential of MSME engagement in adaptation is thus essential and is receiving increasing attention in the international arena. However, the literature in the field is relatively under-developed and empirical evidence sparse. Beyond sector specific or firm-specific studies, the literature mainly presents broad conceptual insights and general recommendations, and only goes so far as to emphasize the gap between the need to improve MSME resilience and the existing knowledge on the issue.

This paper seeks to contribute to address these gaps by exploring the business case for adaptation engagement in the context of MSMEs in Mumbai, which are among the worst affected by heavy precipitation and flooding every year. This paper presents key results from a UNEP DTU Partnership supported initiative—the Admire project[2]—which aims at introducing innovative approaches to enhance MSMEs' adaptation capacity and support their flood resilience efforts.

The paper starts by presenting the causes and effects of recurrent flooding in Mumbai and the rationale for a focus on MSMEs. Subsequently, the methodology applied in the study is described followed by a depiction of MSMEs profile in Mumbai and owners' views on floods and risk management. The subsequent section presents an innovative and context specific framework developed to assess risks and facilitate MSMEs' flood resilience efforts and help actors make informed risk reduction and adaptation decisions.

---

[1]MSMEs are defined here as, 'micro enterprises: 1–9 employees; small: 10–49 employees; and medium: 50–249 employees' (Kushnir et al. 2010). However, it is important to note that often times this definition is inconsistent with the realities on the ground, and local definitions should take precedence.

[2]The ADMIRE Project is implemented in 14 countries worldwide and aims at developing commercially viable operational and financial frameworks engaging the private sector in climate mitigation and/or adaptation actions. http://www.admireproject.org

Flood risk (mis)management in Mumbai—causes, effects and prospects for MSMEs

The city of Mumbai is highly vulnerable to climate change impacts such as flooding, cyclones, sea-level rise and coastal erosion, due to its low-lying location on a peninsula, which is poorly drained and built on reclaimed land, mainly composed of landfill (Patankar and Patwardhan 2016; Schiller et al. 2007). Flooding is thus a chronic and recurrent problem in the city due to tidal variations, flat gradients and mud flats causing excessive siltation (MCGM 2014). Mumbai experiences severe events on a yearly basis, with the unprecedented 2005 floods as the most severe example, which caused damages estimated at USD 1.7 billion and 500 casualties (Ranger et al. 2011). Historically, during the monsoon season between June and October, the city experiences average rainfall surpassing 2200 mm (MCGM 2015). Since 2001, this average has increased to 2400 mm, with variations from the lowest of 1274 mm to highest of 3378 mm (MCGM 2015). The frequency, duration and intensity of rainfall and flooding have also changed over the years. Combined with other drivers of risk such as uncontrolled urban growth, encroachment of natural water ways, poor disaster risk management governance and inadequate drainage systems, this causes significant adverse consequences for public and private stakeholders. These include loss of property and livelihoods, disruption of economic and social activities, poor health, damage to public and private infrastructure etc.

In Mumbai, MSMEs constitute an important part of the local and regional economy and provide substantial employment opportunities to semi-skilled and unskilled workforce. These businesses have faced increasing risks and multiple vulnerabilities and future climate change is expected to worsen these risks further by changing the intensity, frequency and duration of rainfall (Field et al. 2012). Although they suffer from the impacts of recurrent flooding, in the form of damages to physical structure and loss of business, this sector has traditionally been neglected in disaster risk management planning and formal post flood loss assessments. Neither the local government nor industry associations representing MSMEs have attempted to assess the flood impacts on MSMEs in Mumbai or to suggest long-term measures for building flood resilience. The costs incurred are typically borne by MSMEs alone in the absence of adequate insurance protection. While many businesses in India—including MSMEs—are increasingly initiating mitigation measures, for example to improve energy efficiency, MSMEs do not yet consider climate risks as business as usual in order to deal with the direct and indirect impacts of climate change, by 'climate proofing' their businesses (GIZ and SIDBI 2013).

Methodology and Study Area

The study was carried out in three stages. First, a baseline mapping was undertaken to understand flood impacts, current responses undertaken by MSMEs and the attitudes of MSME owners and local actors. Secondly, suitable adaptation measures were designed based on relevant parameters for industrial estates; and lastly a framework was developed to provide guidance for MSMEs on designing, financing and implementing the suggested measures. The methodology used at each stage is described in Table 1.

Table 1 Research methodology

Objective	Research questions	Methodology	Data sources
Baseline mapping	• Nature, frequency and depth of flooding in the selected study area • Exposure to recurrent floods, impacts on MSMEs and monetary value of damages • Current coping mechanisms for flood protection • Relationship with local government and other stakeholders	• Land use mapping of study area to identify location of industrial estates vis-à-vis flooding spots • Primary surveys in randomly selected MSMEs and industrial estates in study area • Mapping of current structures and associations of MSMEs • Consultations with relevant stakeholders	• Base maps of study area available with Municipal Corporation of Greater Mumbai (MCGM) • GIS-based maps created using base maps to indicate location of MSMEs and flooding spots • Available databases on MSMEs with contact details and nature of business
Designing suitable adaptation measures	• Relevant parameters to design adaptation measures aimed at reducing recurrent floods • Challenges and barriers in implementing measures given institutional framework of industrial estates • Required investment and stream of benefits over lifetime	• Consultations with industrial estate representatives and site visits to identify relevant parameters • Identification and design of suitable hard (structural) and soft (non-structural) adaptation measures based on parameters • Evaluation of investment requirements for identified measures vis-à-vis benefits over their lifetime	• Consultations with industrial estate representatives • Literature review on private sector adaptation initiatives • Site visits to identify technical parameters • Cost-benefit analysis tool prepared by DTU experts

(continued)

Table 1 (continued)

Objective	Research questions	Methodology	Data sources
Sustainable operational and financial framework	• Step-by-step guidelines required for industrial estates to design and implement adaptation measures • Available options for financing of selected measures	• Preparation of robust operational framework with clearly defined roles and responsibilities for MSMEs and their industrial estate cooperative societies • Identification of suitable financing options	• Designs prepared by experts based on relevant parameters • Excel toolkit to input parameter values and • Investment outlays estimated for different measures • Consultations with banking and insurance professionals

In order to create a baseline for flood impacts on industrial estates and their current responses, the 'L ward'[3] representing a typical mixed land use prone to chronic flooding has been selected as the study area. L ward is prone to chronic flooding since it is located in the floodplain of the Mithi River, which experiences severe floods during rainy season each year. Five chronic flooding spots in the study area have been selected for the questionnaire-based primary surveys, among randomly selected MSMEs and industrial estates. The map of L ward with existing land use and location of flooding spots is given as Fig. 1. Circled areas are chronic flooding spots identified by the local government (MCGM) and purple areas denote industrial estates. The total area of the ward is 15.56 sq. km., out of which 38.69% is used for residential, 3.74% for commercial and 12.76% for industrial establishments (MCGM 2014).

100 MSME units were administered detailed questionnaires focusing on exposure to recurrent floods, associated damage costs and current response measures. These included questions related to MSMEs' business profiles in industrial estates, structural characteristics and location and maintenance of the premises. Moreover, specific questions addressed the frequency, depth of flooding and types of damage to business premises, infrastructure, equipment and products. The survey further captured information on the indirect and direct costs of flood events, the current measures applied to combat floods and their effectiveness and the respondents' willingness to pay for long-term flood resilience services.

In addition to MSME unit surveys, interviews were conducted with 15 industrial estate cooperative societies in the study area. Open-ended questions were used to elicit information on flood impacts and management, attitudes towards response measures and potential barriers were also addressed. Moreover, information on the

[3]For administrative purposes, Mumbai city is divided into 24 zones known as 'wards'. L ward is located in the Eastern suburbs of the city

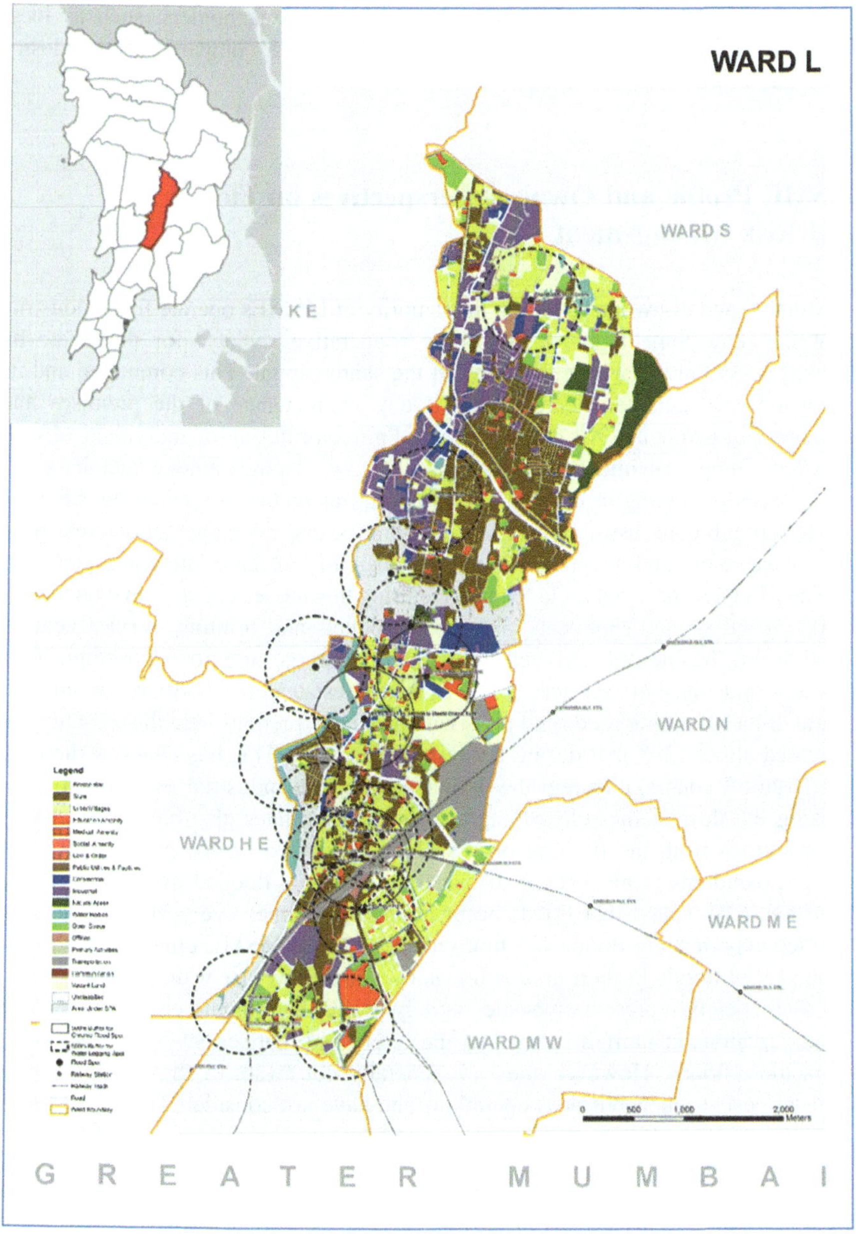

Fig. 1 Map of study area depicting land use and flooding spots.
Source GIS-based maps created for the study

inter-relationships between MSME actors and other stakeholders, such as local government officials, elected representatives, insurance companies and industry associations was also collected.

MSME Profile and Owners' Perspectives on Floods and Risk Management

In Mumbai and elsewhere in India, the majority of MSMEs operate from industrial estates.[4] They come together to form a cooperative society[5] for the estate by forming a committee and contributing to the share capital. This committee and its appointed managers take care of day-to-day maintenance of the premises and common infrastructure. All decisions regarding contribution of individual MSME members, annual maintenance, structural changes, disputes among members and other issues pertaining to common areas and infrastructure are taken by the committee through consensus. Hence, all collaborative decisions and actions related to risk management and adaptation is the responsibility of the estate committee.

The MSMEs surveyed include manufacturing businesses engaged in engineering works, metal works, chemicals and paper products and printing; service centers, warehouses, businesses engaged in garment exports, interior designing, food products and transport services. Most industrial estates are 30–40 years old. The height of the adjoining roads and plots with new constructions near these estates has increased almost 3–5 feet during the last few decades. This has changed the rain-water run-off course, affecting the drainage pattern of industrial estates and channelizing the flow to these low-lying estates. Many estates also have to deal with water springs from the flooring of MSME units located on the ground floor. All study respondents (100%) reported that their area gets flooded every year during monsoon. 54% stated that floods occur at least 3–4 times every year, while 22% reported experiencing floods 4–5 times a year. 12% of MSME units stated that the frequency of floods in their area is 6 instances and more per year. MSME owners and their society representatives are vaguely aware that climate change may bring about variations in rainfall patterns in the future and that sea level rise may bring about more floods. However, they are generally not aware of the implications in terms of costs to their business operations and have not considered potential future

[4]Industrial estate is a common area or site that houses a number of industrial units classified as SMEs that operate in similar or diverse businesses and co-exist for economies of scale in infrastructure facilities such as electricity, water, transport, banking, post office, canteen, etc.

[5]Cooperative society is a registered body under the Maharashtra Cooperative Societies Act 1960, where SMEs located in an industrial estate form the society and become members of it. Here, individual SMEs own the premises they operate from and the ownership of industrial estate (comprising common area and infrastructure) is vested with their cooperative society. Member SMEs make annual contribution to the cooperative society for performing various functions.

impacts of such changes on their businesses and possible risk protection measures for the future.

The study found that although flood risk management is one of the key functions of the local authority (Municipal Corporation of Greater Mumbai (MCGM)), MSMEs have a strong distrust towards the public sector to deliver essential services, including flood management. In low-lying areas, enhanced carrying capacity of storm water drainage network would help minimize the risks of floods. However, MSMEs do not have faith in the upgradation work, which is being undertaken by the local authorities. With limited flood management services and inadequate infrastructure provided by the municipal authorities, some MSMEs and industrial estates implement temporary measures for flood protection, such as the building of temporary barriers (small walls) outside business premises, shifting equipment and materials to higher ground, constructing platforms to elevate machinery, using dewatering pumps to drain floodwater etc. In many cases, the water pumped out of the industrial estates is simply released into adjoining drainage channels or next-door plot leading to serious maladaptation. They are not guided or supported by the local government or other agency to implement long-term solutions. As the measures applied are often temporary and ineffective during heavy precipitation, MSMEs suffer from significant impacts from floods and financial burden of losses. Moreover, as climate change is expected to worsen the risk of flash floods with changes in intensity, frequency and duration of rainfall, it is essential that MSMEs implement long-term solutions to build their adaptive capacity and resilience

For the businesses in Mumbai, it was found that—like everywhere else—size and geography constitute important determinants of risk exposure. Flood preparedness was thus found to differ across industrial estates depending on the type and size of businesses, flood experience, knowledge of flood management measures and the level of proactiveness of the estate managing committees. The study shows that the business case for proactive risk reduction and adaptation is obscured by a low capacity to respond and adapt. Interestingly, in contrast to common knowledge that access to finance is a main barrier for MSMEs, it was found that most industrial estate cooperative societies are in possession of an important corpus of funds that has been collected over last 30–40 years. Therefore, for these estate cooperatives access to finance does not constitute an important barrier for investing in disaster risk reduction measures and thus for building adaptive capacity. Once their members are convinced of the benefits of implementing adaptation measures, they would be able to make the initial investment through this corpus. However, while this may hold true for old industrial estates in Mumbai, it is very likely that other estates, which are newly developed and located elsewhere, will not have the same access to funding. Based on the findings from the baseline survey, it was found that owners—when given appropriate information—were also willing to share the costs of investing in flood reduction measures and collectively design and implement measures through their cooperative societies. The low capacity to respond and adapt to floods is thus mainly due to a lack of awareness of the business rationale for engaging in risk reduction and of the possible measures available to them. Moreover, uncertainty about the extent and timing of climate change impacts is also found to constitute a key barrier.

Consequently, for MSME owners and cooperative managers to consider the potential of involvement in flood risk management, they need to be aware of the costs and benefits of investing in adaptation measures and what specific options are available to them. The baseline study thus shows that there is a clear case for raising MSMEs' and cooperative managers' awareness of the business rationale for engagement in flood risk reduction. Against this background, the next phase of the project was to develop a step-by-step framework to guide the implementation of effective long-term adaptation measures to allow for MSME actors to plan, design and implement risk reduction measures on their own.

An Innovative Framework to Manage Flood Risks

While a number of useful tools and approaches have been developed to guide private sector actors' risk reduction actions and adaptation decision (see for example 'Climate Expert', GIZ; 'Aware for investment', Acclimatise; 'ND-Gain Country Index', University of Notre Dame), the baseline study showed that there is a clear need for specific guidance in the form of a tool that is adapted to the realities of MSMEs affected by flooding in an urban context like Mumbai. Some of the existing tools are very generic in their approach and often more suitable for bigger companies, that have the necessary skills and resources needed for applying the tools. To address this gap, a context and disaster specific hands-on tool to support MSME actors in planning and implementing risk reduction and adaptation measures was developed, in close collaboration with MSME managers and storm water- and structural experts. The step-by-step framework identifies data requirements, technical parameters and indicative structural and non-structural measures needed. A web-based toolkit was also developed to accompany the framework, and support the process of quantifying the costs and benefits of different measures for MSME representatives to achieve long-term flood resilience.

The key steps in the framework are shown in Fig. 2.

For the purpose of this article, the focus is on the first three steps of the framework, which are presented in the following.

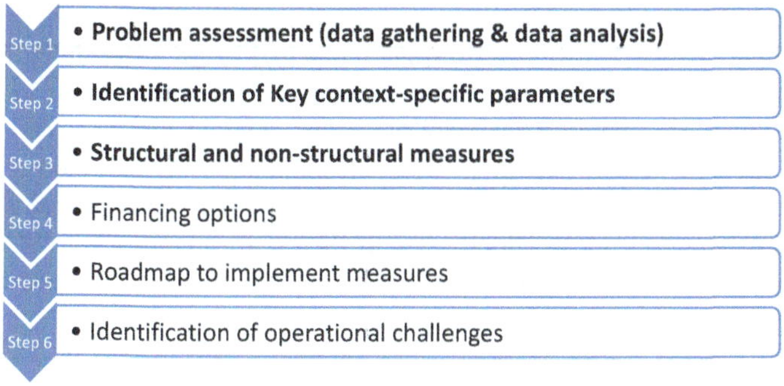

Fig. 2 Step-by-step operational framework for private sector funded resilience measures

Problem Assessment

The first step in designing the suitable flood resilience measures is to assess the nature of recurrent flooding and associated impacts on the industrial estate. Detailed analysis of impacts also helps in assessing how costs of the chosen measures would weigh against the potential benefits to the estates, in terms of reduced loss and damage from flooding. Problem assessment includes collecting and analyzing information pertaining to the frequency, duration and depth of flooding in industrial estate; the possible causes of flooding; the types of losses and damages and quantification of these; the current response measures and their effectiveness. Data Analysis focuses on calculating the different type of losses, such as damage to physical structure (flooring, walls, windows, etc.), infrastructure (air conditioning, electrical systems and communication systems), losses due to suspension of production and losses of productivity due to absence of employees from work during flooding events. Typically MSMEs do not consider intangible losses from workers' absence, interrupted power supply and health impacts in their overall risk perception and assessment. However, when both intangible impacts are taken into account, flood impacts on MSMEs are found to be considerable.

Identification of Key Context-Specific Parameters

During the second *the second step,* characteristic key parameters for the estate are estimated to inform the identification and design suitable adaptation measures. These include rainfall intensity per hour, height of the estate vis-à-vis surrounding plots[6], total plot area and external plot area, connection with the municipal storm water drainage network, distance from the municipal storm water drain, nature and types of MSME businesses which are located on the ground floor and directly impacted by floods. Rainfall intensity per hour, external plot area and height of surrounding areas determine total water accumulation in the industrial estate. Connection and distance from municipal storm water drainage network determines where the estates can release floodwater. If the estate premises are not connected with municipal drains, different set of measures have to be implemented.

Structural and Non-structural Measures

Based on the above parameters and suitable assumptions about costs etc., for *the third step* the study team designed structural and non-structural measures and cost estimates in consultation with cooperative societies, as outlined in Table 2. Each measure is described in detail along with the technical requirements. All structural measures have a lifetime of 25 years. Fixed and operating costs for each measure are approximate values estimated for assumed average values for different parameters. The innovative way in which these measures have been designed allows for context and location-specific changes. By changing input values of the parameters, recommended structural measures, alone or in combination, can be customized by industrial estates to build flood resilience.

Investing in structural and non-structural adaptation measures was found to be beneficial for the case estates. In one of the Industrial Estate, investments were for example projected to generate net economic benefits after a seven year payback period, with increasing benefits over time.

In addition to structural measures to help industrial estates protect their premises from current and future flood risks, non-structural measures, as depicted in Table 3, such as awareness raising, development of early warning system, emergency planning and improved solid waste management are proposed. These all have direct and/or indirect bearing on how effectively businesses can deal with flood risks.

The above mentioned structural and non-structural measures may in certain cases be implemented separately; in other cases they will only be effective when applied in combination, depending on the specific context. Key to the framework is the need for business owners' risk perception to be based on actual losses and damages from floods. Only with the demonstration of a clear business case for engagement in flood risk management engagement, will long-term risk reduction and adaptation measures be implemented.

---

[6]Most industrial estates in Mumbai are 25–35 years old and have seen heights of surrounding plots and adjoining roads increasing compared to their plots. This makes their plots low-lying compared to surrounding areas and more vulnerable to floods.

Table 2 Recommended structural measures

Measure	Description	Technical requirements	Approx. fixed and operating costs
Cut off channel at threshold	• Type of floor drain containing a channel shaped body used for the evacuation of surface water • Acts as a barrier for the flowing water to enter the lo-lying	• Channel size with specific width and depth • Precast RCC cover of specific thickness with drain holes	• INR 400,000 (USD 6061) fixed cost • Maintenance not required for next 10 years
Planned surface drainage and SWD drains	• Constructing surface drainage and SWD drains in the premises to remove the excess water in the area	• Channel size with specific width and depth • Precast RCC cover of specific thickness with drain holes • Adjusting surface slopes where required	• INR 8,500,000 (USD 128,788) fixed cost • Maintenance not required for next 10 years
Sump tank	• Capacity storage of approx. 100,000 litres concrete water tank beneath the ground	• Specific diameter and height	• INR 1,600,000 (USD 24,242) fixed cost • No maintenance as it is a concrete tank
Dewatering ring wells and dewatering pumps	• Dewatering well using cement concrete rings as per the technical requirements • Specific number of rings to be placed one on top of the other after excavation • Gap between two rings not to be sealed to facilitate sub-surface ground water to enter the dewatering well even during dry season • Pumping of water in the well using dewatering pump would help to keep ground water at a lower level so that no spring effect occurs in industrial sites on the ground floors	• Rings of specific diameter and depth • Total number of rings • Dewatering pump with specific capacity measured in litres per minute	• INR 1,000,000 (USD 15,152) of fixed cost • NR 5000 (USD 76) for overhauling of pump annually
Ground water cut off walls	• Contaminated ground water can cause damage to building foundations, load bearing columns etc.	• High Density Polyethylene (HDPE) liner and gravels inside to fill the hollow part	• INR 100,000 (USD 1515) fixed cost

(continued)

Table 2 (continued)

Measure	Description	Technical requirements	Approx. fixed and operating costs
	• Infiltration of sub-surface water from surrounding areas e.g. sea front, beach areas, or nullahs (urban streams formed due to rainwater flow, or overflow of lakes during monsoon, carrying wastes—both solid and liquid) into industrial areas • Construct a retaining wall between the source of seepage or infiltration and the source of ultimate damage		• No maintenance required
Recharging pit	• Recharging pit comprising honeycomb brick walls, interfaced with existing borewell through micro tunneling • When water fills in this pit and starts rising, it will flow through the micro tunnel into the borewell • Using submersible pumps, water from borewell can be drawn to use for industrial purpose • The water holding capacity of this filtration pit would be about 10,000 litres	• Specific diameter and depth	• INR 50,000 for one recharging pit as fixed cost • Maintenance INR 2500 annually

Table 3 Recommended non-structural measures

Non structural measures	Description	Cost	Lifetime
Early warning system	A tipping bucket rain gauge with sensor connected to computer will collect data every fifteen minutes	• INR 40,000 (USD 600) fixed cost • INR 2,000 (USD 30) for annual maintenance	7–8 years
Emergency planning & SOP	SOP guides to every SME unit & flyers with immediate steps	• INR 25,000 (USD 375) fixed cost	5 years (updated after 5 years)
Solid Waste Management	House Keeping will take care of waste on a daily basis, only dust bins, gloves & other essentials need to be bought	• INR 40,000 (USD 600) fixed cost for equipment • INR 75,000 (USD 1150) operating cost annually for housekeeping staff	2–3 years for housekeeping equipment
Awareness building	Brochures & workshop expenses in which a disaster management professional will carry out the session	• INR 15,000 (USD 225) fixed cost	Awareness campaign every 2–3 years

Conclusion

Little is yet known about the bottom-line repercussions of risk reduction and adaptation strategies implemented by firms (Pauw and Pegels 2013) and the kind of measures that are available and suitable for different contexts. To address this gap, studies have drawn attention to the need to further develop methodologies and tools for identifying and assessing adaptation strategies in a context of high uncertainty (Lempert and Groves 2010). To contribute to develop this area and unpack the rationale for—and the avenues through which—MSMEs may implement risk reduction measures, the present study has provided empirically-based insights from MSMEs impacted by recurrent flooding in Mumbai.

The study has shown that over the last ten years, frequent flooding from extreme precipitation has recurring negative impacts for industrial estates, where MSMEs are suffering from large annual or sometimes multi-annual damages. The study found that although there are limited direct short term financial incentives for MSMEs to invest in adaptation measures, the cost of inaction is substantial in the long term, when indirect impacts and costs are also taken into account. There is thus a clear case for applying a longer-term perspective to MSMEs' flood risk assessments. The benefits estimated through the proposed framework can provide a convincing argument for industrial estate committees and MSME owners to prioritize the implementation of flood resilience measures. The approach and framework developed through this study is expected to provide a structured entry-point to create greater awareness of the business rationale for engagement in flood resilience and an understanding of the processes needed to initiate adaptation initiatives from a business perspective.

In conclusion, this study demonstrates the need to encourage private sector participation in adaptation efforts, also in the case of smaller actors such as MSMEs. Irrespective of their size, there is a strong case for private sector actors to design, finance and implement suitable adaptation measures to build their long-term flood resilience. It was found that it makes business sense for MSMEs to participate collaboratively in this effort, given the benefits that would accrue from the short- to the long-term. The findings may provide a case example for other industrial estates and MSMEs to follow and there is thus great potential to adopt this approach and replicate it for different end-users and in different contexts. This approach provides a unique opportunity to identify locally designed solutions for local problems, rather than relying on a top-down approach where government and other entities prescribe solutions that may not be appropriate to the context and setting of the local users. The results from this study thus provide an opportunity for further work on developing and adapting frameworks to assess adaptation costs, benefits and suitable measures, which support risk reduction actions by small firms in diverse geographical contexts and for different types of disasters.

To make the approach relevant and replicable for MSMEs in different sectors, as well as upscalable in other geographical contexts, it is essential to capture relevant context specificities in the development of such approaches. This is a resource intensive exercise, which would therefore require a strong institutional mechanism or policy push to promote flood resilience among MSMEs and industrial estates. Industry associations may for example play an important role in creating awareness about the application of potential structural and non-structural measures by industrial estates, to build long-term flood resilience. Moreover, a policy push from the government in the form of incentives to new industrial estates or the inclusion of flood resilience in building codes also constitutes a key element in support of the adoption of resilience building measures by private sector players.

Bibliography

Agrawala, S., Carraro, M., Kingsmill, N., Lanzi, E., & Prudent-richard, G. (2011). Private Sector Engagement in Adaptation to Climate Change. *OECD Environment Working Paper No 39*.

Berkhout, F., Hertin, J., & Gann, D. M. (2006). Learning to adapt: Organisational adaptation to climate change impacts. *Climatic Change, 78*(1), 135–156. https://doi.org/10.1007/s10584-006-9089-3.

Dougherty-Choux, L., Terpstra, P., Kurukulasuriya, P., & Kammila, S. (2015). *Adapting from the ground up*. UNDP: WRI.

Field, C. B., Barros, V., Stocker, T. F., Qin, D., Dokken, D. J., Ebi, K. L., et al. (2012). *Managing the risks of extreme events and disasters to advance climate change adaption*. A Special Report of Working Groups I and II of the Intergovernmental Panel on Climate Change. Cambridge University Press, New York.

GIZ, & SIDBI. (2013). *Facing the impacts of climate change: Indian MSMEs and adaptation*. New Delhi.

Haigh, N., & Griffiths, A. (2012). Surprise as a catalyst for including climatic change in the strategic environment. *Business & Society, 51*(1), 89–120. http://doi.org/10.1177/0007650311427425.

Hussain, I., Farooq, Z., Akhtar, W. (2012). SMEs development and failure avoidance in developing countries through public private partnership. *African Journal of Business Management, 6*(4), 1581–1589.

IPCC. (2014). *Climate change 2014: Impacts, adaptation, and vulnerability. Part B: Regional aspects*. Contribution of Working Group II to the Fifth Assessment Report of the Intergovernmental Panel on Climate Change. In V. R. Barros, C. B. Field, D. J. Dokken, M. D. Mastrandrea, K. J. Mach, T. E. Bilir (Eds.), New York, NY, USA: Cambridge University Press, Cambridge, United Kingdom and 688 pp.

Kushnir, K., Mirmulstein, M. L., & Ramalho, R. (2010). Micro, small, and medium enterprises around the world: How many are there, and what affects the count? World Bank/IFC

Lempert, R. J., & Groves, D. G. (2010). Technological forecasting & social change identifying and evaluating robust adaptive policy responses to climate change for water management agencies in the American west. *Technological Forecasting & Social Change, 77*(6), 960–974.

Linnenluecke, M. K., Griffiths, A., & Winn, M. (2012). Extreme weather events and the critical importance of anticipatory adaptation and organizational resilience in responding to impacts. *Business Strategy and the Environment, 21*(1), 17–32. https://doi.org/10.1002/bse.708.

MCGM. (2014). Greater Mumbai City Development Plan 2005 to 2025. Available at http://www.mcgm.gov.in/irj/portal/anonymous?NavigationTarget=navurl://095e1c7b9486b1423b881dce8b106978.

MCGM. (2015). *Flood preparedness guidelines 2014*. Disaster Management Unit, Municipal Corporation of Greater Mumbai.

Nicholls, R. J., Hanson, S., Herweijer, C., Patmore, N., Hallegatte, S., Corfee-Morlot, J., et al. (2008). *Ranking port cities with high exposure and vulnerability to climate extremes: Exposure estimates* (Environment Working Papers No. 1). Paris.

Okereke, C., Wittneben, B., & Bowen, F. (2012). Climate change: Challenging business, transforming politics. *Business and Society, 51*(1), 7–30.

Patankar, A., & Patwardhan, A. (2016). Estimating the uninsured losses due to extreme weather events and implications for informal sector vulnerability: A case study of Mumbai, India. *Natural Hazards: Journal of the International Society for the Prevention and Mitigation of Natural Hazards, 80*(1), 285–310.

Pauw, P., & Pegels, A. (2013). Private sector engagement in climate change adaptation in least developed countries: An exploration. *Climate and Development, 5*(4), 257–267.

Ranger, N., Hallegatte, S., Bhattacharya, S., et al. (2011). *Climatic Change, 104,* 139.

Schiller, A., de Sherbinin, A., Hsieh, W.-H., Pulsipher, A. (2007). *The vulnerability of global cities to climate hazards*. George Perkins Marsh Institute, Clark University, Center for International Earth Science Information Network (CIESIN), Columbia University.

UNEP. (2016). *The adaptation finance gap report 2016*. Nairobi, Kenya: United Nations Environment Programme (UNEP).

UN-HABITAT. (2016). World Cities Report. UN-Habitat HS Number: HS/038/16E No of Pages: 262 ISBN Series Number: 978-92-1-133395-4.

United Nations Global Compact (UNGC). (2015). *The business case for responsible corporate adaptation: Strengthening private sector and community resilience a caring for climate report.*

Wedawatta, G., Ingirige, B., & Amaratunga, D. (2010). Building up resilience of construction sector SMEs and their supply chains to extreme weather events. *International Journal of Strategic Property Management, 14*(4), 362–375.

Can Young Olive Plants Overcome Heat Shock?

Márcia Araújo, Conceição Santos and Maria Celeste Dias

Abstract Climate change is bringing more frequent and intense heat waves over the last years. Under this circumstance, it is important to understand whether species can tolerate stress and which mechanisms are involved in this adaptation process. Olive tree (*Olea europaea* L.) have been known for centuries to be drought tolerant, but less is known about the impact on the physiological response of this species to heat. To understand how young olive plants deal with heat shock, one-year-old plants (cv. 'Arbequina'), grown at 23 ± 2 °C, were exposed to heat, 40 °C, for 2 h. Relative water content, gas exchange, carbohydrates content, cell membrane permeability and lipid peroxidation were assessed immediately after heat exposure. The heat shock treatment compromised plant water status, photosynthesis and induced stomatal closure. However, neither membrane damage nor carbohydrates contents (total soluble sugars and starch) were affected. The results indicate that young olive plants can overcome short heat shock episodes.

Keywords Climate change · *Olea europaea* L. · Photosynthesis
Abiotic stress

M. Araújo (✉) · M. C. Dias
Centre for Functional Ecology (CFE), Department of Life Science,
University of Coimbra, Calçada Martim de Freitas, 3000-456 Coimbra, Portugal
e-mail: marciaaraujo@fc.up.pt

M. Araújo · C. Santos
Center for the Research and Technology of Agro-Environmental
and Biological Sciences (CITAB), Universidade de Trás-os-Montes e Alto Douro,
Quinta de Prados, Edifício dos Blocos Laboratoriais, Sala C1.10, 1o Piso,
5001-801 Vila Real, Portugal

M. Araújo · C. Santos
Integrative Biology and Biotechnology Laboratory, Department of Biology,
Faculty of Sciences, University of Porto, Rua Campo Alegre, 4169-007 Porto, Portugal

M. C. Dias
QOPNA and Department of Chemistry, University of Aveiro, 3810-193 Aveiro, Portugal

C. Santos
LAQV & REQUIMTE, Faculty of Sciences, University of Porto, Porto, Portugal

F. Alves et al. (eds.), *Theory and Practice of Climate Adaptation*,
Climate Change Management, https://doi.org/10.1007/978-3-319-72874-2_11

Introduction

By the end of this century, the global temperature is expected to increase by 4 °C compared to the average 20th century values (IPCC 2014). This phenomenon will increase the frequency and intensity of high temperature episodes, which will impact on crops' yield and compromise agricultural production (Bita and Gerats 2013). Heat is considered one of the most deleterious abiotic stresses and severely affects plants (Bita and Gerats 2013; Zhao et al. 2014). Heat deleterious effects are observed at morphological, physiological, biochemical and molecular levels, which ultimately compromise plant growth, productivity and yield (Galán et al. 2001; Koubouris et al. 2009; Zhao et al. 2014; Bita and Gerats 2013). Whilst fluctuations in temperature occur naturally throughout the plant life cycle, during hot summers the elevated temperatures can induce severe stress in non-adapted plants.

Olive trees (*Olea europaea* L.) are known for centuries in the Mediterranean region for their economic value. About 98% of the world olive and 78% of world olive oil are produced in this region (Rigueiro-Rodríguez et al. 2009). This species is also known to tolerate a broad range of environmental stresses, including heat, mostly due to a variety of morphological and physiological adaptations (e.g. Bacelar et al. 2007b; Carr 2013; Tripepi et al. 2011), but less is known about the mechanisms regulating their adaptation to heat episodes. Due to climate change, the increase in temperatures is expected in the Mediterranean region (IPCC 2014). The quality of olive products (olive fruit and oil) is influenced by cultivar, soil and climatic conditions (Criado et al. 2007). Climate change may induce changes in the quality and production/yield of olive plants. Therefore, it is important for producers to select the fittest cultivars coping heat conditions and also to explore the way highly productive cultivars like 'Arbequina' interact with heat.

Heat is perceived by plants as a whole process, meaning that there is no specific molecule playing a thermosensor role (Brestic et al. 2013). In fact, diverse intracellular signalling molecules are expressed in response to heat exposure. These molecular signals are integrated with multiple signal transduction pathways involved in different cell compartments (Brestic et al. 2013). The changes in the regulation of gene expression followed by changes in the transcriptome, proteome and metabolome enable plants to acclimate and survive to heat stress (Brestic et al. 2013). A sensitive metabolic process in plant cells to heat stress is photosynthesis. The photosynthetic apparatus can tolerate moderate temperatures involving reversible changes or may be severely damaged as result of severe heat (Brestic et al. 2013).

In this study, our aim is to understand how young olive plants deal with heat shock. To achieve this objective, we exposed *O. europaea* cv 'Arbequina' to heat shock. Water status, gas exchange, carbohydrates content, cell membrane permeability and lipid peroxidation were measured. By investigating these physiological parameters, we expect to decipher which mechanisms are involved in the adaptation process of young olive plants to heat stress.

Materials and Methods

Plant Material and Experimental Conditions

O. europaea L. cv 'Arbequina' plants (1-year-old) were obtained at *Viveiros Vitor Santos* (Mealhada, Portugal). Plants were maintained for one month under controlled conditions in a phytotron: 23 ± 2 °C, with a photoperiod of 16 h/8 h (day/night), 30 ± 6% of relative humidity, a light intensity of ≈600 μmol m^{-2} s^{1} provided by lamps (Osram cool white fluorescent). All plants were watered to field capacity before the experiment. Plants were randomly divided into two groups as follows: (i) control, where plants were maintained in the same conditions mentioned above, and (ii) heat shock treatment, where plants were exposed for 2 h at 40 ± 2 °C. After the heat shock treatment, gas exchange, plant water status and cell membrane permeability were evaluated. Also, leaf samples were collected, frozen in liquid nitrogen and stored at −80 °C for further analysis of carbohydrates and lipid peroxidation.

Plant Water Status

As described by Araújo et al. (2016), leaf fresh weight (FW), turgid weight (TW) and dry weight (DW) were determined to calculate the relative water content (RWC). RWC is calculated as 100 × (FW-DW)/(TW-DW) and expressed as a percentage (%).

Gas Exchange

A portable infrared gas analyser (Infra-Red Gas Analyzer: IRGA, LCpro+, ADC, Hoddesdon, UK) was used to measure net CO_2 assimilation rate (P_N) and transpiration rate (*E*) in leaves. Also, stomatal conductance (g_s) was calculated according to (Caemmerer and Farquhar 1981).

Evaluation of Carbohydrates Content

To evaluate the total soluble sugars (TSS) (Irigoyen et al. 1992), 80% (v/v) ethanol was added to frozen leaves powder and incubated for one hour at 80 °C. After centrifuged (5000 × *g*, 10 min, 4 °C), the supernatant was incubated with an anthrone solution (40 mg anthrone: 1 mL dH_2O: 20 mL H_2SO_4) for 10 min at 100 °C. When cooled, samples were centrifuged (5000 × *g*, 10 min, 4 °C) and the absorbance of the supernatant was read at 625 nm using a Thermo Fisher Scientific spectrophotometer (Genesys 10-uv S). The concentration of TSS was obtained against a D-glucose standard curve.

Starch content was quantified according to Osaki et al. (1991). To the resultant pellet of TSS extraction was added 30% (v/v) perchloric acid and incubated for one hour at 60 °C. After centrifuged (10,000 × *g*, 10 min, 4 °C), the supernatant was incubated with an anthrone solution for 10 min at 100 °C. Then, the supernatant was cooled, centrifuged (5000 × *g*, 10 min, 4 °C), and the absorbance was read at 625 nm using a Thermo Fisher Scientific spectrophotometer (Genesys 10-uv S). The starch concentration was obtained against a D-glucose standard curve.

Determination of Cell Membrane Permeability and Lipid Peroxidation

Electrolyte leakage was used to evaluate cell membrane permeability as described by Dias et al. (2014a, b). Leaves were enclosed in vials with de-ionized water and incubated overnight at room temperature on a rotary shaker. Electrical conductivity was measured (L_t) after 24 h. Samples were autoclaved at 120 °C for 20 min and after reached room temperature, electrical conductivity (L_0) was read. Electrical leakage was calculated as L_t/L_0 and expressed as a percentage (%).

The production of malondialdehyde (MDA) was used to evaluate lipid peroxidation (Hodges et al. 1999). Samples were homogenised with 0.1% (w/v) trichloroacetic acid (TCA) and centrifuged (10,000 × *g*, 10 min, 4°C). An aliquot of supernatant was homogenized with 20% TCA (w/v) (−TBA) and another with 20% TCA (w/v) and 0.5% thiobarbituric acid (TBA) (w/v) (+TBA). Both samples (−TBA and +TBA) were incubated for 30 min at 95°C, cooled on ice and centrifuged as previously. Absorbance was read at 440, 532 and 600 nm using a Thermo Fisher Scientific spectrophotometer (Genesys 10-uv S). MDA were calculated as $(A - B/157\ 000) \times 10^6$, where A is $(\text{Abs } 532_{+TBA}) - (\text{Abs } 600_{+TBA}) - [(\text{Abs } 532_{-TBA}) - (\text{Abs } 600_{-TBA})]$ and B is $[(\text{Abs } 440_{+TBA}) - (\text{Abs } 600_{+TBA})] \times 0.0571$.

Data Analysis

The experiments were conducted with 3 plants per treatment and were analysed using a SigmaPlot (version 3.1) program for performing statistical analysis. Data were analysed by one-way analysis of variance (ANOVA) follow by the Holm-Sidak comparison test.

Results and Discussion

The emerging global climate change is already having a considerable impact on agricultural production, particularly in the Mediterranean region where *O. europaea* plays a crucial economic role. Several studies have shown that olive plants are tolerant to drought (Bacelar et al. 2006, 2007a, 2007b), nonetheless the physiological response of juvenile olive plants to heat episodes remains less studied. In this experiment all juvenile plants survived to the heat treatment, and showed no significant morphological changes (necrosis, chlorosis, leaf decay, etc.), supporting that even at an early stage they present tolerance mechanisms.

Similarly to other stresses, such as water deficit (e.g. Dias et al. 2014a) or high UV-B radiation (e.g. Araújo et al., 2016), the photosynthetic apparatus is one of the main targets of heat stress (Brestic et al. 2013). In this experiment, whilst no chlorosis was observed, our data support that the juvenile plants exposed to 2 h' heat shock (HS) episode suffered an impairment of the net CO_2 assimilation rate (decrease of 70.6%, $p \leq 0.05$, Fig. 1a), which supports some damage in the photosynthetic apparatus. Also, transpiration rate was reduced ($p \leq 0.05$) by around 71.6% and stomatal conductance by 79.6%, relative to control plants (Fig. 1b, c). These changes indicate that olive plants respond to HS by closing the stomata. Stomatal closure is a common defence mechanism in heat-stressed plants and can prevent the plant from water loss. At the same time, it also restricts CO_2 uptake, which conditions the availability of this gas in the intercellular spaces of mesophyll cells, which is the main cause of net CO_2 assimilation rate declination under stress (Araújo et al. 2016). In these young olive plants, the HS episode decreased ($p \leq 0.05$) only 7.5% leaf RWC (Table 1), indicating that stomatal closure prevented from severe water loss and putatively from cell dehydration. Our data are in line with Koubouris et al. (2015), who also reported a strong reduction in olive net CO_2 assimilation rate, stomatal closure and a decrease in the intercellular CO_2 concentration (C_i) after a longer (28 h at 38 ± 4 °C) HS treatment.

Plant growth and carbon metabolism are closely associated since carbohydrate (particularly sucrose), the final product of photosynthesis, is the main source of the production and maintenance of biomass. Abiotic stress leads, in some species, to significant reductions in photosynthesis in source tissues, leading to a reduction of the supply of soluble sugars to sink tissues (Wahid et al. 2007). In the present work, we analysed total soluble sugars and starch, a reserve carbohydrate (Table 1). Despite the strong decrease of net CO_2 assimilation rate observed in olive plants after the HS, the pools of starch and total soluble sugars were not significantly affected indicating that carbon partitioning/translocation was not affected. Also, Bacelar et al. (2007b) verified a similar amount of total soluble sugars after different water regimes applied despite a low P_N, indicating that olive plants accumulate osmotic compounds to maintain the maximum amount of water as possible. Similar strategies of maintaining the carbohydrates high are used by other species such as sugarcane (Satbhai and Naik 2014; Wahid et al. 2007) under heat stress.

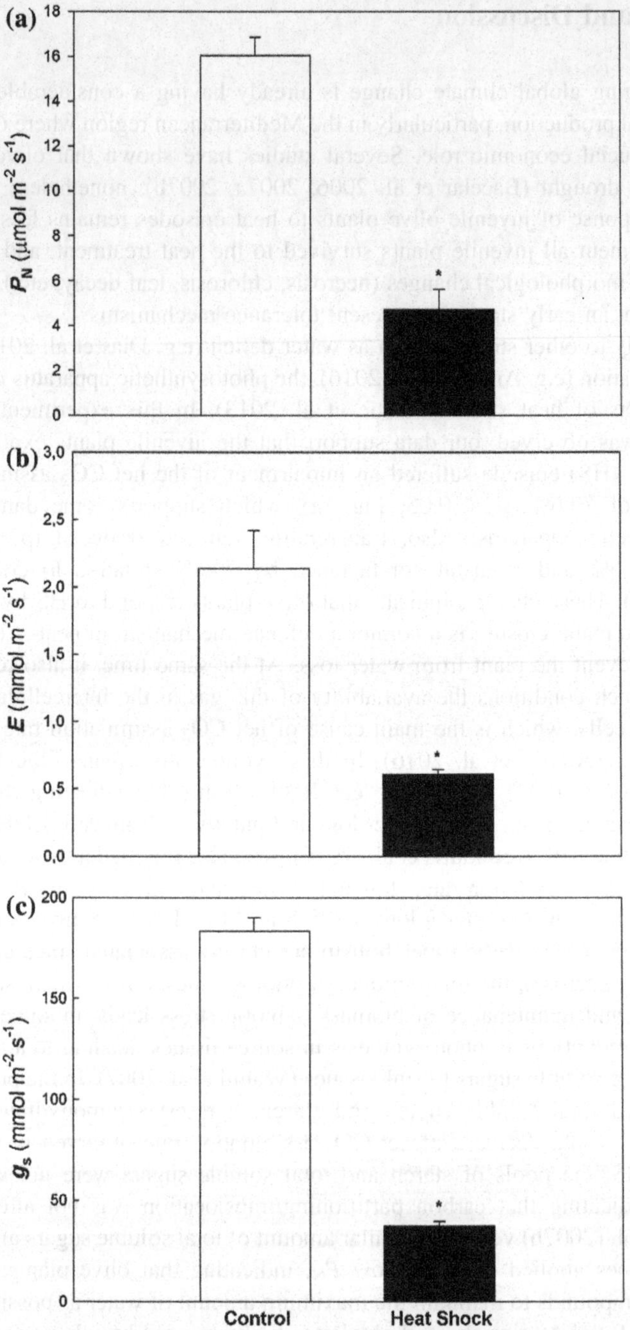

Fig. 1 Net CO_2 assimilation rate, P_N (**a**), transpiration rate, E (**b**) and stomatal conductance, g_s (**c**) in leaves of young *O. europaea* plants under control conditions and heat shock treatment. Data shown are mean ± SEM (n = 3). *, Significant differences between control and heat shock treatment ($p \leq 0.05$)

Table 1 Relative Water Content (RWC), total soluble sugars (TSS) and starch in leaves of young *O. europaea* plants under control conditions and heat shock treatment

Parameter	Control			Heat shock		
RWC (%)	85.3	±	0.14	78.9	±	0.49*
TSS (μg mg^{-1} DW)	105.6	±	2.80	98.1	±	2.63
Starch (μg mg^{-1} DW)	62.1	±	4.75	70.9	±	7.01

Data shown are mean ± SEM (n = 3). *, Significant differences between control and heat shock treatment ($p \leq 0.05$)

A common response of plants to heat stress is the increased production of ROS. ROS attack essential cellular components and structural elements, and accumulation of ROS are associated with lipid peroxidation, making cellular membranes particularly susceptible to oxidative damage (Feki et al. 2015; Mittler et al. 2011). Thylakoid membranes are the site where the light-dependent reaction of photosynthesis occurs and are particularly sensitive to ROS. Therefore, damages on these membranes can result in reduced photosynthetic activity. We analysed one parameter widely used as ROS induced oxidation in macromolecules, namely lipids (the MDA). The leaves of olive plants exposed to the heat episodes tested here showed to cope with the stress with no increase of MDA. Increases in cellular MDA levels may, indirectly, be correlated with parameters of cell membrane integrity, as is the CMP (Fig. 2a, b). Therefore, the absence of significant changes in the CMP ($p > 0.05$) of plants exposed to heat shock, is in line with the MDA results, again suggesting that these juvenile plants tolerate the imposed heat stress. Mechanisms underlying this tolerance may involve the increase of heat shock proteins, which are documented to have a crucial role in the heat shock response (McLoughlin et al. 2016). Heat shock proteins family act as first line of defence against heat stress (Assab et al. 2011; McLoughlin et al. 2016), and some preliminary data also support an increase of these proteins in these olive plants (data not shown), and may justify at least partially the absence of membrane damages and lipid peroxidation.

Conclusion

In conclusion, our data showed that after a short period of HS, young olive plants survived, but showed functional adjustments at the photosynthetic level. Plants closed their stomata to decrease the transpiration levels, but at the same time, this may reduce the amount of CO_2 availability, resulting in an impairment of net CO_2 assimilation rate. Despite the observed stomatal closure, plant water status decreased. However, changes seem not be deleterious to these plants as no

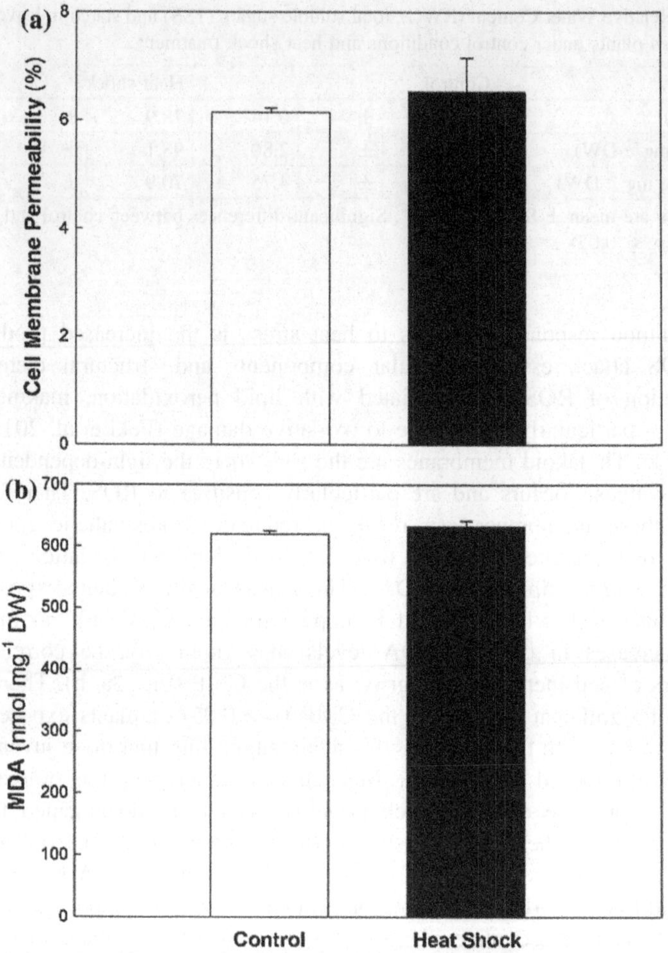

Fig. 2 Cell membrane permeability (**a**) and MDA content (**b**) in leaves of young *O. europaea* plants under control conditions and heat shock treatment. Data shown are mean ± SEM (n = 3). *, Significant differences between control and heat shock treatment ($p \leq 0.05$)

increased lipid peroxidation nor cell membrane damage were observed. These mechanisms performed by young olive plants shows that this species is capable to overcoming a short heat shock episode.

Acknowledgements This work was supported by Portuguese Foundation for Science and Technology (FCT) through a doctoral fellowship of Márcia Araújo (SFRH/BD/116801/2016) and a post-doctoral fellowship of Maria Celeste Dias (SFRH/BPD/100865/2014). This work was financed by FCT/MEC through national funds and the co-funding by the FEDER, within the PT2020 Partnership Agreement, and COMPETE 2010, within the projects UID/BIA/04004/2013, UID/QUI/00062/2013 and UID/AGR/04033/2013.

References

Araújo, M., et al. (2016). Plasticity of young *Moringa oleifera* L. plants to face water deficit and UVB radiation challenges. *Journal of Photochemistry and Photobiology B: Biology, 162*, 278–285. Available at: http://linkinghub.elsevier.com/retrieve/pii/S1011134416302251.

Assab, E., et al. (2011). Heat shock response in olive (*Olea europaea* L.) twigs: Identification and analysis of a cDNA coding a class I small heat shock protein. *Plant Biosystems, 145* (March 2016), 419–425 ST–Heat shock response in olive (*Olea e*).

Bacelar, E. A., Moutinho-Pereira, J. M., et al. (2007a). Changes in growth, gas exchange, xylem hydraulic properties and water use efficiency of three olive cultivars under contrasting water availability regimes. *Environmental and Experimental Botany, 60*(2), 183–192. Available at: http://linkinghub.elsevier.com/retrieve/pii/S0098847206001250.

Bacelar, E. A., Santos, D. L., et al. (2007b). Physiological behaviour, oxidative damage and antioxidative protection of olive trees grown under different irrigation regimes. *Plant and Soil, 292*(1–2), 1–12. Available at: http://link.springer.com/10.1007/s11104-006-9088-1.

Bacelar, E. A., et al. (2006). Immediate responses and adaptative strategies of three olive cultivars under contrasting water availability regimes: Changes on structure and chemical composition of foliage and oxidative damage. *Plant Science, 170*(3), 596–605.

Bita, C. E., & Gerats, T. (2013). Plant tolerance to high temperature in a changing environment: Scientific fundamentals and production of heat stress-tolerant crops. *Frontiers in Plant Science, 4*(July), 1–18. Available at: https://www.frontiersin.org/articles/10.3389/fpls.2013.00273/full.

Brestic, M., et al. (2013). Heat signaling and stress responses in photosynthesis. In K. R. Hakeem, R. Rehman, & I. Tahir, eds. *Plant signaling: Understanding the molecular crosstalk* (pp. 241–256). Springer.

Carr, M. K. V. (2013). The water relations and irrigation requirements of olive (*Olea europaea* L.): A review. *Experimental Agriculture, 49*(4), 597–639. Available at: http://dx.doi.org/10.1017/S0014479713000276.

Criado, M. N., et al. (2007). Comparative study of the effect of the maturation process of the olive fruit on the chlorophyll and carotenoid fractions of drupes and virgin oils from Arbequina and Farga cultivars. *Food Chemistry, 100*(2), 748–755.

Dias, M. C., Azevedo, C., et al. (2014a). *Melia azedarach* plants show tolerance properties to water shortage treatment: An ecophysiological study. *Plant physiology and biochemistry, 75*, 123–127. Available at: http://www.ncbi.nlm.nih.gov/pubmed/24440555.

Dias, M. C., Oliveira, H., et al. (2014b). Improving elms performance under drought stress: The pretreatment with abscisic acid. *Environmental and Experimental Botany, 100*, 64–73. Available at: http://linkinghub.elsevier.com/retrieve/pii/S0098847213002244.

Feki, K., Saibi, W., & Brini, F. (2015). Understanding Plant Stress Response and Tolerance to Salinity from Gene to Whole Plant. *Managing Salt Tolerance in Plants*, (May 2016), 1–18. Available at: http://www.crcnetbase.com/doi/10.1201/b19246-2.

Galán, C., et al. (2001). The role of temperature in the onset of the *Olea europaea* L. pollen season in southwestern Spain. *International Journal of Biometeorology, 45*(1), 8–12.

Hodges, D. M. et al., (1999). Improving the thiobarbituric acid-reactive-substances assay for estimating lipid peroxidation in plant tissues containing anthocyanin and other interfering compounds. *Planta, 207*(4), 604–611. Available at: http://link.springer.com/10.1007/s004250050524.

IPCC. (2014). *Climate change 2014 impacts, adaptation, and vulnerability. Part A: Global and Sectoral Aspects.* In C. B. Field et al., (Eds.), Contribution of Working Group II to the Fifth Assessment Report of the Intergovernmental Panel on Climate Change. Cambridge: Cambridge University Press. Available at: http://ebooks.cambridge.org/ref/id/CBO9781107415379.

Irigoyen, J. J., Emerich, D. W., & Sanchezdiaz, M. (1992). Water-stress induced changes in concentrations of proline and total soluble sugars in Nodulated Alfalfa (*Medicago sativa*) Plants. *Physiologia Plantarum, 84*(1), 55–60.

Koubouris, G. C., Metzidakis, I. T., & Vasilakakis, M. D. (2009). Impact of temperature on olive (*Olea europaea* L.) pollen performance in relation to relative humidity and genotype. *Environmental and Experimental Botany, 67,* 209–214.

Koubouris, G. C., et al. (2015). Ultraviolet-B radiation or heat cause changes in photosynthesis, antioxidant enzyme activities and pollen performance in olive tree. *Photosynthetica, 53*(2), 279–287.

McLoughlin, F., et al. (2016). Class I and II small heat-shock proteins protect protein translation factors during heat stress. *Plant Physiology, 172*(October), 00536.2016. Available at: http://www.plantphysiol.org/lookup/doi/10.1104/pp.16.00536.

Mittler, R. et al. (2011). ROS signaling: The new wave? *Trends in Plant Science, 16*(6), 300–309. Available at: http://dx.doi.org/10.1016/j.tplants.2011.03.007.

Osaki, M., Shinano, T., & Tadano, T. (1991). Redistribution of carbon and nitrogen compounds from the shoot to the harvesting organs during maturation in field crops. *Soil Science and Plant Nutrition, 37*(1), 117–128.

Rigueiro-Rodríguez, A., McAdam, J. & Mosquera-Losada, M. R. (Eds.). (2009). *Agroforestry in Europe: Current status and future prospects*, Springer Science + Business Media B. V.

Satbhai, R. D., & Naik, R. M. (2014). Osmolytes accumulation, cell membrane integrity, and antioxidant enzymes in sugarcane varieties differing in salinity tolerance. *Sugar Tech, 16*(1), 30–35.

Tripepi, M., Pöhlschroder, M., & Bitonti, M. B. (2011). Diversity of dehydrins in *Olea europaea* plants exposed to stress. *The Open Plant Science Journal, 5,* 9–13.

von Caemmerer, S., & Farquhar, G. D. D. (1981). Some relationships between the biochemistry of photosynthesis and the gas exchange of leaves. *Planta, 153,* 376–387.

Wahid, A., et al. (2007). Heat tolerance in plants: An overview. *Environmental and Experimental Botany, 61*(3), 199–223.

Zhao, X. X., et al. (2014). Effects of heat acclimation on photosynthesis, antioxidant enzyme activities, and gene expression in orchardgrass under heat stress. *Molecules, 19*(9), 13564–13576.

Author Biographies

Márcia Araújo completed her M.Sc. in Molecular and Cellular Biology at the University of Aveiro in 2015. Her master dissertation was on climate change and produced an ISI article entitled "Plasticity of young *Moringa oleifera* L. plants to face water deficit and UVB radiation challenges" published in Journal of Photochemistry and Photobiology B: Biology. Márcia obtained her bachelor degree in Biology at the University of Aveiro in 2013. She is currently a Ph.D. student at Faculty of Science of University of Porto, Portugal. In 2016, she awarded a Ph.D. fellowship by the Portuguese Foundation for Science and Technology.

Dr. Conceição Santos currently is a full professor of Biology at Faculty of Sciences, University of Porto. She graduated in Biology at the Faculty of Sciences, University Porto in 1986. Concluded her Ph.D. at University Aveiro in 1998 and the Habilitation for Full Professor in 2007. Her main scientific research: cell physiology under emerging risks (e.g. climate changes and toxicology), combining biotechnology tools and functional and genomic approaches. Prof. C Santos is a member of International/European Evaluation Commissions and scientific adviser for several meetings.

Dr. Maria Celeste Dias is graduated in Biotechnology Engineering at University of Algarve, Portugal in 2001. Concluded her Ph.D. in Biology at Johann Wolfgang Goethe Universität, Frankfurt am Main, Germany in 2005. She is an expert in olive physiology and photosynthesis. She is currently a Post-Doctoral researcher at Centre for Functional Ecology, University of Coimbra and at Department of Chemistry, University of Aveiro.

Microclimatic Adaptations That Occurred in Urban Area in the Brazilian Cerrado Between the Years 2011–2012 and 2016

Diana Carolina Jesus de Paula, Natallia Sanches e Souza, Marta Cristina de Jesus Albuquerque Nogueira and Flávia Maria de Moura Santos

Abstract The frequency and geographic distribution of extreme weather events began to change as a result of global climatic changes and anthropogenic changes in urban space that influence the urban microclimate, affecting the mesoclimate and the macroclimate. This research aimed to relate the microclimatic and anthropogenic adaptations that occurred in five portions in an urban area in the Brazilian Savanna named Cerrado between the years 2011–2012 and 2016. As a methodology, the nocturnal moving transect technique was used to measure micrometeorological variables such as air temperature and relative air humidity, and the maximum likelihood classification technique to classify and quantify the types of soil cover found in the two periods studied. In this way, it was possible to observe an increase of 1.45 °C in the intensity of the urban heat island and a decrease in relative air humidity of 3.32%, associated to areas with up to 17.173% of impervious materials such as concrete and asphalt, and decrease of up to 22.20% of vegetal cover. Evidenciating the importance of several studies on the use of materials with low thermal diffusivity and application of mitigating measures as an increase of vegetated areas, thus making the public awareness of the impacts of unplanned urbanization.

Keywords Urban climate · Vegetation · Urbanization

D. C. J. de Paula (✉) · N. S. e Souza · M. C. de J. A. Nogueira
F. M de M. Santos
Federal University of Mato Grosso, Fernando Corrêa Avenue, Cuiabá 78060-900, Brazil
e-mail: arqdiana.paula@gmail.com

N. S. e Souza
e-mail: natalliass@hotmail.com

M. C. de J. A. Nogueira
e-mail: mcjanp@gmail.com

F. M. de M. Santos
e-mail: flavia_mms@hotmail.com

F. Alves et al. (eds.), *Theory and Practice of Climate Adaptation*,
Climate Change Management, https://doi.org/10.1007/978-3-319-72874-2_12

Introduction

According to Wong and Yu (2005), with rapid urbanization, there has been a huge growth in population and buildings in cities. This increase leads to changes in the environment produced by man. The replacement of natural environments by urban areas ends up generating the urban climate, which occurs due to urban development, which usually occurs in a disordered way (Lima and Amorim 2010).

The frequency and geographic distribution of extreme weather events began to change as a result of global climatic changes and anthropogenic changes in urban space that influence the urban microclimate, affecting the mesoclimate and the macroclimate. The increase of large paved areas, suppression of green areas and intensification of the soil cover with impermeable materials ends up raising the temperature and reducing the relative humidity of the air.

The city ends up generating its own climate, the urban climate resulting from the interaction of urban factors with the regional climate and the preexisting physical environment (Monteiro 1976). The urban climate is a complex environmental, which is attached in such a way to the city, like the circulation, turbulence, and dispersion of air, in the albedo and in the storage of heat due to evaporation and energy balance at surface (Arnfield 2003; Kanda 2006).

The elevation of air temperature is a consequence of the heating of urban surfaces, due to the type of materials used in the buildings and to the land cover (Gartland 2010). Land cover refers to components that cover a land surface, encompassing soils and rocks, vegetation and forests, water or snow, reflecting the biophysical state of the terrestrial system (Prakasam 2010).

In this context, the energy balance undergoes intense changes, with the increase in the absorption of solar radiation, low albedo of the building materials and soil cover, which are associated with heat phenomena, especially in cities with a tropical climate.

This study encourages the inclusion of cities such as Cuiabá-MT-Brazil in the studies of urban climatology, offering an opportunity to disseminate information applied to regions of tropical climate, relevant in studies of urban climatology. Besides that, due to the biodiversity of the biomes near Cuiabá, such as wetland named Pantanal, Amazonian rainforest and Brazilian savanna “Cerrado”, it is necessary to study the climatic and environmental vulnerability in relation to urban and anthropogenic changes.

Cuiabá, one of the hottest cities in Brazil, with an average maximum temperature of 42 °C, had a significant population growth in the last 30 years, accompanied by an urban expansion without previous urban planning. Furthermore, in 2014 Cuiabá hosted the FIFA World Cup and to receive this event, several modifications were made to the road structure, and for this, there was a decrease of vegetation in the main avenues.

So, this research aimed to report the microclimatic and anthropogenic adaptations that occurred in five portions, in an urban area in the Brazilian Savanna "Cerrado", between the years 2011–2012 and 2016, through nocturnal mobile transect.

This article contributes to the analysis of urban climatology. First, assuming the relationship between changes in the urban environment, density, materials and land cover, and microclimatic variations. It also shows that the mobile transect technique is a good solution for the analysis of the spatialization of microclimatic variables.

Study Area

Cuiabá is on the Midwest region of Brazil, at latitude 15° 35′ 46″S and longitude 56° 05′ 48″W and average altitude below 200 m. It has an area of 3538.17 km^2, corresponding to 254.57 km^2 to the urban macrozone, with a population of 551,098 inhabitants, according to the *Brazilian Institute of Geography and Statistics*—IBGE (2010) (Fig. 1).

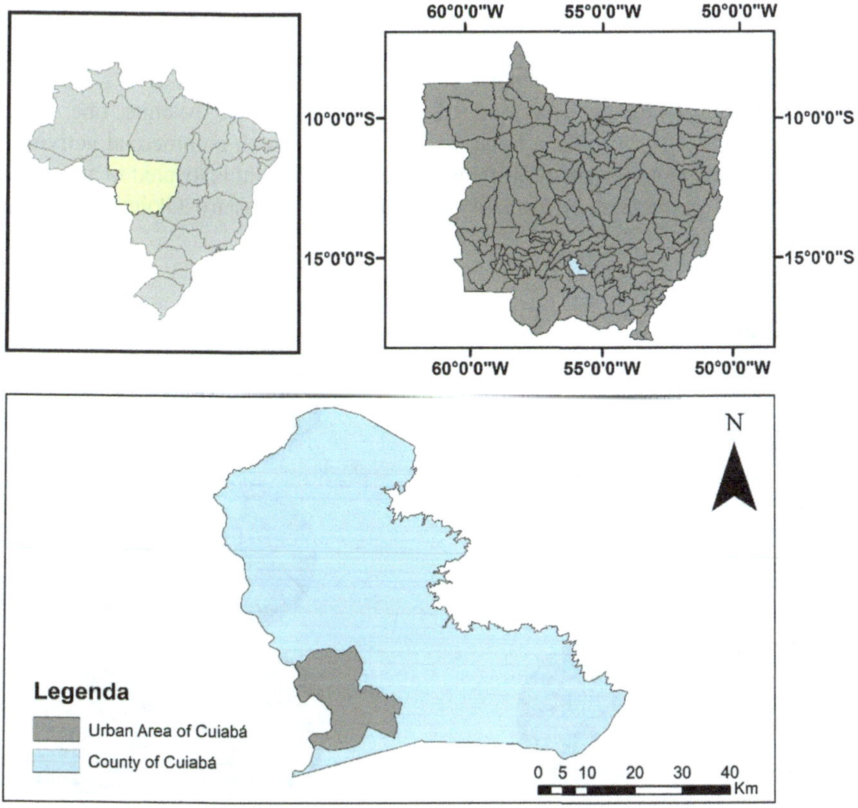

Fig. 1 Location of Mato Grosso in Brazil, Cuiabá in Mato Grosso and urban area of Cuiabá

The climate of Cuiabá region is in accordance with the climatic classification of Köppen (1948), "Aw type", tropical, humid, and typical of tropical savannas. It is characterized by presenting two well-defined periods: the dry period (May–Setember) and the humid period (October–April). Cuiabá has a small thermal amplitude, the annual average temperature is 26.8 °C, the maximum average temperature is 42 °C, the minimum average temperature is 15 °C, the average relative humidity is 78%, and the average total insulation of 2179 hours per year, *National Institute of Meteorology*—INMET (2010).

The study area consists of five urban portions in Cuiabá, pre-selected due to the difference in land cover. These urban portions and their land coverings were studied in the years 2011–2012 and 2016 (Fig. 2).

The P1 point, is located in an area characterized by commercial activities, services, residential, intense flow of vehicles and people. It suffered urban intervention in the implementation of the viaduct of access to the Federal University of Mato Grosso. The P2 point is located in the central area of the city, with activities of commercial predominance and services, with high flow of vehicles and people. In its vicinity, there are squares with tree cover.

The P3 point is located in an area that underwent urban intervention for urban crossing works with the construction of the Santa Rosa trench. It has high flow of vehicles and people and Presence of commercial activities, services and residential of the multifamily type in its majority.

The P4 point located at Historiador Rubens de Mendonça Avenue, one of the main avenues of the city, has a large flow of vehicles, with commercial activities, services and residential of the multifamily type. The P5 point is located in an area of environmental protection, in front of the Urban Park Massairo Okamura with an area of 53.77 ha.

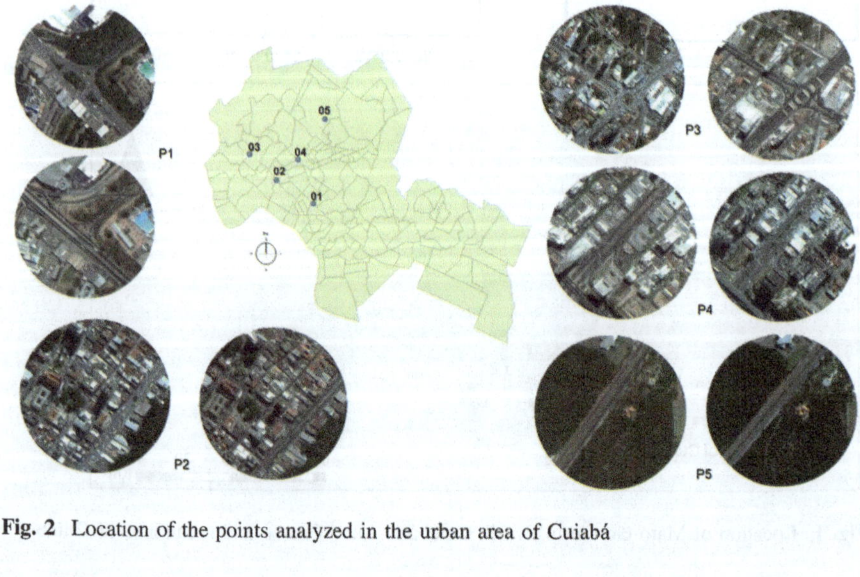

Fig. 2 Location of the points analyzed in the urban area of Cuiabá

Methods

The methodology for the acquisition of air temperature (T) and relative humidity (RH) data consisted of the use of a mobile night transect using a motor vehicle, passing through all the analyzed study points.

A GPS Datalogger, model GK_V02 was used to obtain the geographic coordinates. The protected sensors were used to obtain data on air temperature (RTL-10709) and relative humidity (AM2302). Installed on the side of the vehicle with an approximate height of 2 m. In vehicle data collection it is important that the time spent during the collection does not exceed one hour and the speed of the vehicle should not exceed 40 km/h (Araujo et al. 2010).

The measurements were carried out in the four seasons of the year in 2011–2012 and 2016, at night, after 8 pm, when temperatures are more constant due to the absence of solar radiation.

In order to classify the urban coverage, high resolution satellite images (Google Earth) of the study years were used, considering a radius of influence of 200 m around each point, according to the methodology of Oke (2004). Subsequently, the maximum likelihood classification technique was used to classify and quantify land cover, since according to Augusto-Silva et al. (2013) is the method that obtained the best performance aiming at the mapping of land cover.

For the comparison, the difference between the means of the data of air temperature and relative humidity of the air was used between the years in all the seasons of the year. Thermal maps were made using Surfer software v.13 (*Golden Software*) to demonstrate the distribution of differences between the four seasons and between the points studied.

Results

The maps developed by the maximum likelihood for the years 2011–2012 and 2016 used in surveying the type of land cover in order to measure urban interventions occurred in the period. For this, two classification classes were adopted: vegetation and impermeable cover (built and paved area). It is possible to observe in Fig. 3 that there was a considerable decrease in the vegetation, represented by the green coloration.

There was a reduction of 22.20% in vegetation cover at P1 and an increase of 13.85% in waterproof cover compared to 2012, as well as the increase of bare soil, represented by the orange color. At point P2, there was an increase in the amount of vegetation cover from 2012 to 2016, but there was also an increase in asphalt paving, as it is possible to see in Fig. 2.

At point P2 while there was an increase of 16.40% of paved area in relation to 2012, this was due to some expropriations carried out in the period (Fig. 4). However, the vegetation cover had an increase of 6.79%. It is possible to observe in

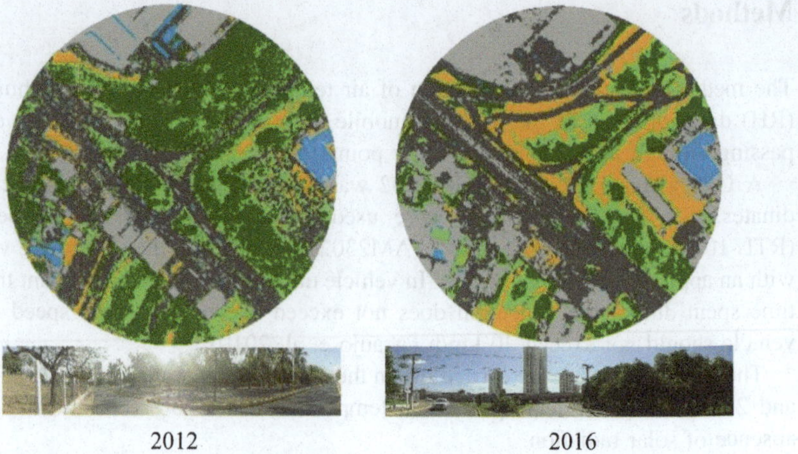

Fig. 3 Thematic maps of soil cover urban portion (Point 1) (P1)

Fig. 5 that there was a considerable increase in the paved area, and a decrease mainly in the cover of arboreal vegetation, represented by the darker green color.

In point 3 (Fig. 5) were observed a decrease of approximately 12.46% in permeable cover, with an increase of 8.46% in impermeable cover compared to 2012. In Fig. 6, representing the point 4, it was observed the increase in the map of the gray coloration, which represents buildings and the diminution of the green coloration, which represents vegetal cover.

A reduction of 14.97% in permeable cover and a 7.10% increase in impermeable cover was observed at point 4. In point 5 (Fig. 7) there is not much modification in

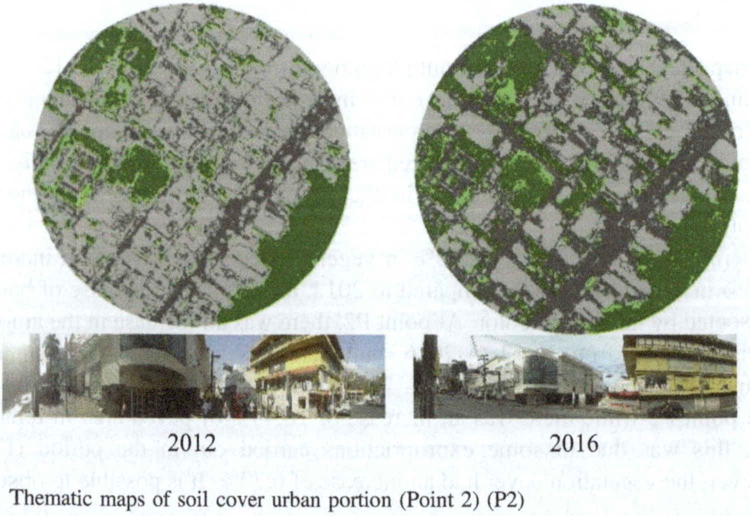

Fig. 4 Thematic maps of soil cover urban portion (Point 2) (P2)

Fig. 5 Thematic maps of soil cover urban portion (Point 3) (P3)

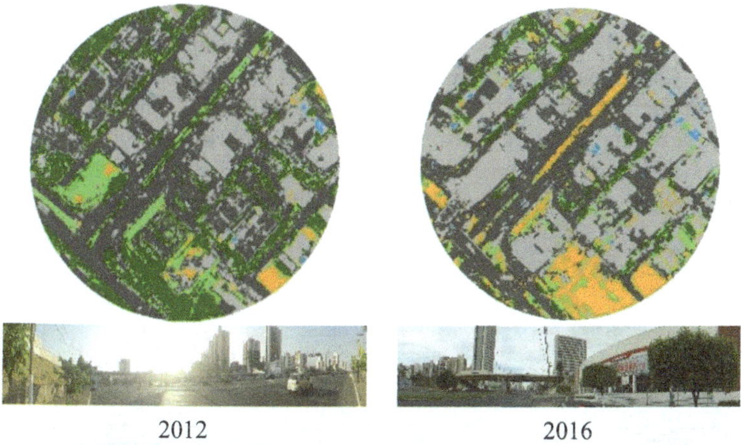

Fig. 6 Thematic maps of soil cover urban portion (Point 4) (P4)

the ground cover, since it is an area of environmental protection, that is, it is not allowed to build in the place. It is observed that the vegetal mass continues occupying almost all the urban portion.

Due to the fact that it is not possible to build in the vicinity of the urban park of point 5, despite being one of the main avenues of the city, there was a decrease of 1.71% of paved area and increase of 1.50% of tree cover.

It is noteworthy that in 2011–2012 and 2016, the highest air temperature averages were recorded at points P2 and P4. However, the lowest mean air temperature was recorded at P5 in both years of study.

Fig. 7 Thematic maps of soil cover urban portion (Point 5) (P5)

Though, it was noticed that there was an increase in the mean of the thermal amplitude in the points when compared between the years, the highest recording was 29.24 °C and the lowest was 27.20 °C in 2011–2012, observing amplitude of 1.54 °C. While in 2016 the larger register was 28.35 °C and the smaller was 25.36 °C, with amplitude of approximately 3 °C in 2016.

In order to show the microclimatic adaptation associated to direct urban interventions, thermal maps were used to show the differences in air temperature (Fig. 8)

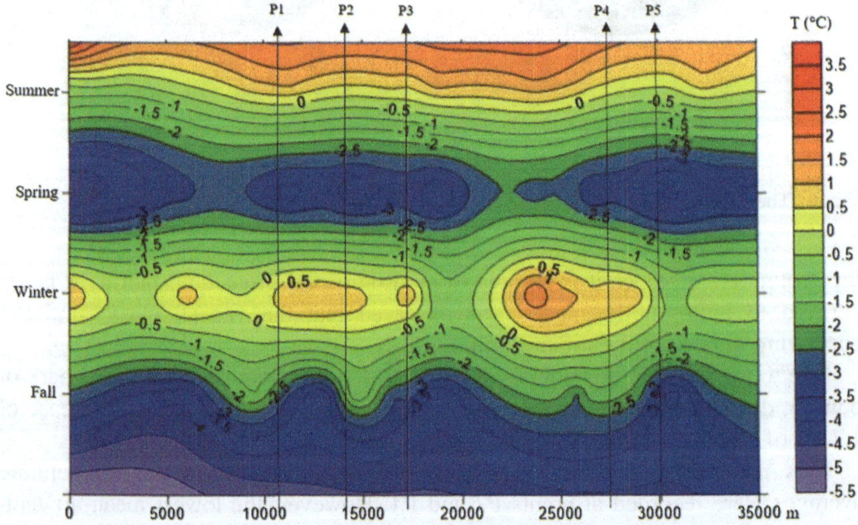

Fig. 8 Distribution of the air temperature difference in each season between the years

and relative humidity (Fig. 9) between the years 2011–2012 and the seasons of the years, as well as among the five points studied.

The largest differences (Fig. 8) between the years occurred in the summer season, where the differences were around 1.5 and 2 °C in all points. In winter, the differences are around 0.5 and 1 °C. The seasons of autumn and spring showed a decrease of air temperature in 2016 compared to 2012, being in both seasons around −2.5 and −3 °C.

Analyzing the thermo-hygrometric behavior, point 2 is configured as an urban heat island and point 5 is a urban fresh island. It is also noted that there was an increase in the heat island intensity of 1.45 °C when comparing the records between the years.

Regarding relative humidity, the greatest differences between the years occurred in winter and spring (Fig. 9).

In the summer between 2011–2012 and 2016 the differences were small, but in point 4 there was a decrease of 4% and in the other points there was a decrease of approximately 2% of relative humidity of the air.

In spring at all points there was an increase in relative air humidity of approximately 4–6%. In winter there was a decrease of relative humidity of approximately −12%. While in the fall there was a decrease in relative air humidity between the years, but with lower values than in the invention, the decrease was around −4% at all points.

For the analysis of averages of air temperature and relative humidity of the air between the years at the points, a student t test was obtained, obtaining a sig. value <0.05, demonstrating that the thermal amplitudes found are considerable at the local scale, and changes in soil cover may be associated.

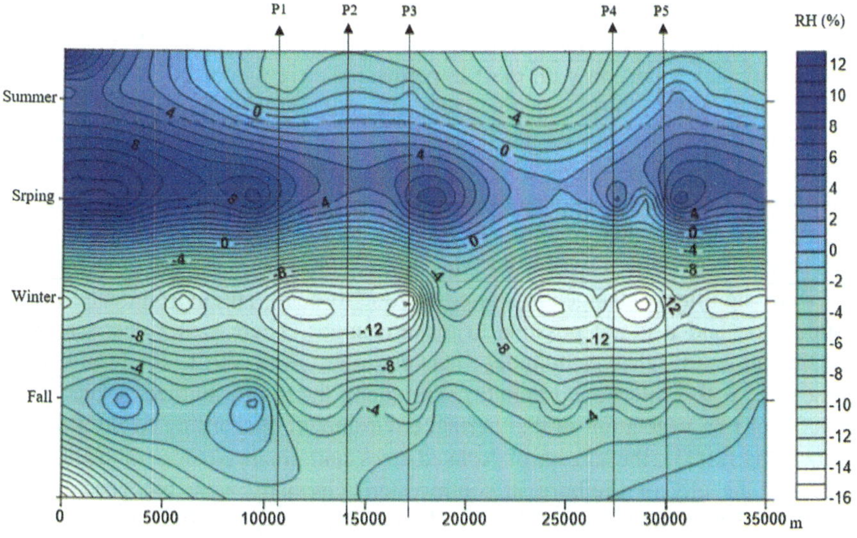

Fig. 9 Distribution of the relative humidity difference in each season between the years

Discussion

This study identified that urban interventions that lead to the removal of vegetation cover and the increase of the impermeable area, also increase the air temperature, corroborating with several studies that identified relationship between soil cover and microclimate realized in Cuiabá-MT (Franco 2012; Maciel et al. 2014). The points that presented higher increases of paved area and higher decrease of the vegetative area presented higher temperature differences between the years, like P1 and P2.

The urban portion P2, despite having a decrease in vegetated cover between the years studied, still presents a significant vegetative presence, however, before it was arboreal cover (dark green), now it is found as grass. It still had a significant difference and increased air temperature between the years. According to Stone and Norman (2006), the grass as a form of urban heat island mitigation is less effective than arboreal coverage, where in their research in Atlanta, replacing grass areas with 25% per arboreal coverage areas could reduce the heat of urban heat island by 13%.

The P3 portion was the most heated and dry portion of all points and seasons studied, but it was the P1 portion that obtained the higher difference in air temperature (increase) and relative humidity (decrease) between the years. The P1 is where the higher devastation of vegetation occurred, turning mostly to exposed soil, which increases the region's warming. According to Chen et al. (2006) on study in southern China's Guangdong province on the urban heat island and its influence on the regional climate found that exposed soil regions were warmer than other types of cover.

The P4 portion did not present a large difference between the years, considering the average of the seasons in air temperature (0.88 °C) and relative humidity (4.18%), but it was the point with the greatest change of coverage, despite the fact that it is already a portion with most impermeable coverage in 2011–2012, for the reforms of the FIFA World Cup, there was still an increase in the impermeable area. It presented itself as one of the hottest points. This is related to the fact that impermeable coverage exerts a strong influence on the thermal behavior of an environment, especially in areas covered with asphalt pavement, which absorb considerably more solar radiation (Borges and Pereira 2010).

The P5 urban portion, where the Massairo Okamura Urban Park is located, did not show significant changes in soil cover, due to being an environmental preservation area, and it is not possible to make changes. There was a decrease in air temperature and relative humidity between the years 2011–2012 and 2016. Presenting as the coldest and humid point among the points studied, related to the benefits of tree vegetation in the thermal environment. According to Streiling and Matzarakis (2003), the effects of isolated trees and small clusters of trees in the center of the city of Freiburg in southwestern Germany were analyzed and the positive effect of the trees in the thermal environment was verified.

The urban areas that the most natural systems undergo changes, caused by anthropic action, such as: removal of vegetation cover, waterproofing of the soil through thick layers of asphalt and cement, which prevent the infiltration of rainwater, modifying the regime of runoff And soil water evaporation, implying the reduction of relative humidity (Lombardo 1985).

This project provided the opportunity to analyze the relationship between urban interventions and microclimatic variables, as well as the interannual analysis to monitor climate change on a micro scale. The methodology proved to be successful in presenting urban heating and cooling due to anthropogenic modifications, especially in cities with a tropical climate.

Conclusion

It was concluded that there were large urban interventions in Cuiabá to host the FIFA World Cup, as well as increase of paved area and decrease of vegetated area. These interventions have a considerable influence on microclimatic variables, such as those studied here, air temperature and relative humidity.

The point with greatest difference between the years was also the point with the highest air temperature, presenting itself as an island of urban heat. While the point that underwent smaller urban changes, presented less difference between the years in both relative humidity and air temperature, presenting as an island of urban freshness.

It is hoped that with the methodology and results presented in this research, urban planning and studies of urban interventions and their positive and negative impacts in urban areas, especially in tropical cities, will become more important.

Acknowledgements The authors thank CAPES—Coordination for the Improvement of Higher Education Personnel for the financial support during this research.

References

Araujo, A., Aleixo, N., Menezes, B., Souza, C., Rivero, C., Montezani, E., et al. (2010). Methodological essay on the use of mobile transects in the daytime period in Presidente Prudente-SP. *Formação Magazine, 1*(17). Accessed January 28, 2016 (http://revista.fct.unesp.br/index.php/formacao/article/view/415).

Arnfield, A. (2003). Two decades of urban climate research: A review of turbulence, exchanges of energy and water, and the urban heat island. *International Journal of Climatology, 23*. Accessed January 28, 2016 (http://onlinelibrary.wiley.com/doi/10.1002/joc.859/epdf).

Augusto-Silva, P., Valerio, L., Santos, T., Alcântara, E., & Stech, J. (2013). *Analysis of classifiers for use mapping and soil coverage*. Paper for the XVI Brazilian Symposium on Remote Sensing, Vol. 1, pp. 2424–2430, National Institute of Space Research. Accessed January 17, 2016 (http://www.dsr.inpe.br/sbsr2013/files/p0115.pdf).

Borges, M., & Pereira, F. (2010). *Influence of the built environment in urban microclimate*. Paper for the XIII national meeting of Built Environment Technology, National Association of Built Environment Technology. Online, Available http://www.infohab.org.br/entac2014/2010/arquivos/409.pdf. Accessed January 28, 2016.

Brazilian Institute of Geography and Statistics—IBGE. (2010). *Demographic census*. Online, Available http://www.ibge.gov.br. Accessed January 28, 2016.

Chen, X., Zhao, H., Li, P., & Yin, Z. (2006). Remote sensing image-based analysis of the relationship between urban heat island and land use/cover changes. *Remote Sensing of Environment, 104*. Accessed January 20, 2016 (http://www.sciencedirect.com/science/article/pii/S0034425706001787).

Franco, F., & Nogueira, M. (2012). Microclimatic analysis as a function of soil use and occupation in Cuiabá-MT. *Mercator Magazine, 11*. Accessed January 28, 2016 (http://www.periodicos.ufc.br/index.php/mercator/article/view/674/653).

Gartland, L. (2010). *Heat Islands: How to mitigate heat zones in urban areas*. São Paulo, Brazil: Oficina de Textos Publisher.

Kanda, M. (2006). Progress in the scale modeling of urban climate: Review. *Theoretical and Applied Climatology, 84*. Accessed January 28, 2016 (http://link.springer.com/article/10.1007%2Fs00704-005-0141-4).

Köppen, W. (1948). *Climatologia*. México: Fondo de Cultura Económica Publisher.

Lima, G., & Amorim, M. (2010). Analysis of the nocturnal characteristics of the climatic elements in a summer episode in the municipality of Teodoro Sampaio-SP. *Geografia em Atos Magazine, 1*. Accessed January 28, 2016 (http://revista.fct.unesp.br/index.php/geografiaematos/article/view/217/gabrielan10v1).

Lombardo, M. (1985). *Island of heat in the metropolis*. Brazil: Hucitec Publisher.

Maciel, C., Luz, V., Santos, F., Nogueira, M., & Nogueira, J. (2014). Interaction of microclimatic variables and soil coverage in urban and bordering-urban regions in the city of Cuiabá/MT. *Caminhos de Geografia Magazine, 15*. Accessed January 15, 2016 (www.seer.ufu.br/index.php/caminhosdegeografia/article/download/25992/15470).

Monteiro, C. (1976). *Theory and urban climate* (Thesis). Institute of Geography, University of São Paulo, São Paulo, Brazil.

National Institute of Meteorology—INMET. (2010). *Climatologia*. Online, Available http://www.inmet.gov.br/climatologia. Accessed January 30, 2016.

Oke, T. (2004). *Initial guidance to obtain representative meteorological observations at urban sites* (Instruments and Observing Methods Report No. 81). Vancouver, Canada: World Meteorological Organization. Accessed January 28, 2016 (www.wmo.int/pages/prog/www/imop/publications/iom-81/iom-81-urbanmetobs.pdf).

Prakasam, C. (2010). Land use and land cover change detection trough remote sensing approach: A case study of Kodaikanal taluk, Tamil Nadu. *International Journal of Geometrics and Geosciences, 1*. Accessed January 28, 2016 (http://ipublishing.co.in/jggsvol1no12010/EIJGGS1015.pdf).

Stone, B., & Norman, J. (2006). Land use planning and surface heat Island formation: A parcel-based radiation flux approach. *Atmospheric Environment, 40*. Accessed June 12, 2016 (http://www.sciencedirect.com/science/article/pii/S1352231006001142).

Streiling, S., & Matzarakis, A. (2003). Influence of single and small clusters of trees on the bioclimate of a city: A case study. *Journal of Arboriculture, 29*. Accessed January 28, 2016 (ftp://149.222-62-69.ftth.swbr.surewest.net/TreePDF/INFLUENCE%20OF%20SINGLE%20AND%20SMALL%20CLUSTERS.pdf).

Wong, N., & Yu, C. (2005). Study of green areas and urban heat island in a tropical city. *Habitat International, 29*. Accessed February 15, 2016 (http://www.sciencedirect.com/science/article/pii/S0197397504000281).

Part III
Social Perceptions and Socio-cultural Dimensions of Climate Change Impacts and Adaptation

Outlining Community Perceptions of Climate Change in Local Adaptation Strategies Development: The Case of ClimAdaPT.Local

Luísa Schmidt, Adriana Ferreira Alves, Susana Valente and João Morais Mourato

Abstract ClimAdaPT.Local's goal was the development of 26 Municipal Adaptation Strategies for Climate Change (EMAAC). The project aimed to increase the capacity of these municipalities to incorporate adaptation to Climate Change (CC) in their planning instruments. To do so, several methodologies and tools were specifically developed. One of these, the local stakeholder engagement strategy, constitutes an innovation as far as the development of these strategies in Portugal is concerned. This paper focuses on this stakeholder engagement process and its achievements. ClimAdaPT.Local's stakeholder engagement was workshop based. These were put in place in order to understand perceptions of CC, its impacts, local risks, but also survey potential inputs into the ongoing EMAAC development. Each session comprised several discussion tables, intending to: (i) obtain a global reference framework on perceptions and sensitivity to CC at the local level; (ii) complement the vulnerabilities assessment made by the technical staff from each municipality; (iii) inform, adjust and optimise the local adaptation strategies. In addition, stakeholders were asked to share their visions of the future, namely how CC and local territorial identity will be articulated in the near, and not so near, future. The analytical findings encompassed in paper refer to point (i). All in all, in Portugal, there is a widespread perception that CC is happening, in particular among social

L. Schmidt (✉) · A. F. Alves · J. M. Mourato
Instituto de Ciências Sociais da Universidade de Lisboa,
Av. Professor Aníbal de Bettencourt, 9, 1600-189 Lisbon, Portugal
e-mail: mlschmidt@ics.ulisboa.pt

A. F. Alves
e-mail: adriana.alves@ics.ulisboa.pt

J. M. Mourato
e-mail: joao.mourato@ics.ulisboa.pt

S. Valente
Faculdade de Ciências da Universidade de Lisboa, Center for Ecology, Evolution & Environmental Changes (CE3C), C1, sala 1.4.05, Campo Grande 1749-016, Lisboa, Portugal
e-mail: smvalente@fc.ul.pt

F. Alves et al. (eds.), *Theory and Practice of Climate Adaptation*, Climate Change Management, https://doi.org/10.1007/978-3-319-72874-2_13

groups that engage in activities such as agriculture and fishing. And despite being considered to pose a threat, CC is, at the same time, broadly perceived as a window of opportunity to put in place measures that improve local living conditions.

Keywords Climate change · Local adaptation · Community perceptions Stakeholders · Adaptation strategies · Portugal · ClimAdaPT.Local

Introduction

Over the last two decades, Climate Change (CC) has become growingly visible, in the media, policy and politics. Accordingly, we witness a growing number of crosscutting, multidisciplinary debates and increasingly intense stakeholder engagement. Nevertheless, despite the apparent consensus on the phenomena of CC and the need to mitigate its causes and adapt to its impacts, little is known, in detail, about what people effectively think about CC, how they perceive it, deal with its risks, and foresee it evolution in the near and far future.

This paper builds on the assumption that if we are to design better mitigation and adaptation policies we must understand and factor in the dynamics of perception development and meaning-making of individuals and communities regarding CC. In detail, we explore the social construction of perceptions on CC as a dynamic process emergent from the individual and collective exposure to public policies, media and scientific communication against the backdrop of personal memories and experiences.

Structure wise, this paper consists of three parts. Initially, we will develop a conceptual framework to support the identification and inclusion of perceptions of CC in adaptation strategy development. Second, we will outline the methodological approach in terms of data collection strategy and analytical framework. This will allow us to identify how individuals perceive CC at a global and local scale, in the present, past and future time, the perceptions of threat that CC represents and also the opportunities that it encloses. The third and final part will systematise our input in terms of an overview of CC perceptions in Portugal.

Social Perceptions of CC; Conceptual Framework

CC is one of the main contemporary societal challenges. Perceptions of CC have been widely researched (Carvalho 2008, 2011; Capstick et al. 2015; Lorenzoni et al. 2007; Spence et al. 2012; Weber 2010, 2016), in view to determine what shapes them and how they influence and lead individuals and communities to concrete actions.

A better understanding of how perceptions evolve is fundamental. Schmidt and Delicado (2014) argue that CC mitigation and adaptation public policies depend of a change in people's practices, which in turn are deeply dependent of the political measures that enable those changes to happen. Thus, to discuss a shift in individual and collective practices, we must take into account that practices are, by nature, resilient, and based on routines and perceptions (Schmidt and Delicado 2014).

The underlying argument here is that if we are to design better mitigation and adaptation policies we must understand and factor in the dynamics of perception development and meaning-making of individuals and communities CC.

Let's begin by highlighting that the idea of climate "interacts with the human psyche and with cultural practice in more imaginative ways" (Hulme 2009, p. 12). Inseparable from feeling of national identity, climate is embraced in «social memory», with extreme events determining our own personal memory (often in a negative way). "The human experience of climate releases powerful emotions which can be both benign and threatening" (Hulme 2009, p. 12). As Hulme argues, climate, as well as describing a physical reality, can also be understood as "an imaginative idea—an idea constructed and endowed with meaning and value through cultural practice" (Hulme 2009, p. 14).

Like Hulme (2009), we agree that climate records can be measured not only by meteorological science, but also memory, texts, behaviour and identity, to which we add perceptions. The idea of climate exists both in the cultural practices and in the human mind (as perceptions) as in the physical world. "The multiple meanings of climate, and the ideological freightage we load onto interpretations of climate and our interactions with it, are an essential part of making sense of what is happening around us today in our CC discourses" (Hulme 2009, p. 28).

The media play a key role in this discourse-making process. In some countries, scepticism on CC has been a political weapon, manipulated by corporate interests and some political parties (Leiserowitz 2006; Dunlap 2008, cit. Schmidt and Delicado 2014), helping to spread messages of *denial* of CC. This seems to be the case of the United States of America, for instance. In fact, research shows that "Media coverage containing sceptical voices is much higher in the United States and United Kingdom than Brazil, China, France, or India; and right-leaning newspapers tend to feature more uncontested sceptical sources in their opinion pages" (Capstick et al. 2015, p. 54). In Portugal, the media seems to be much more aligned with scientists, and mostly consensual, undermining the rise of sceptical voices (Carvalho and Pereira 2008, cit. Schmidt and Delicado 2014).

However, and despite its overwhelming reach, media isn't the sole player in contemporary CC discourse-making processes. Which brings us to the role of scientific communication. Growingly, science is invited to contribute to public policies on CC, finding new ways to shape knowledge and making it accessible (Hulme 2009). Hulme perceives CC as a complex phenomenon, and the role adopted by scientists imply "adjustments to the way in which science is governed and how its knowledge is policed" (Hulme 2009, p. 92). In line with this argument, Schmidt (2008) states that citizens need scientists to help them understand the complexities and the urgencies of modernity, along with the risks they face. Schmidt then appeals to the scientific community, calling on their social responsibility in bringing together the scientific world with «the rest of the world».

In sync, Weber's (2010) research on what shapes perceptions of CC, argues that "better (environmental) science and statistics education can create the familiarity with the scientific presentation of information and the mental habits that will create citizens who give greater weight to the output of their analytic processing system, moving the risk perception of the general public and its officials closer to that of climate scientists" (Weber 2010, p. 339). Here, Weber (2010) makes a fundamental point, that a better understanding of the human contributions to CC can trigger policies to solve the problems caused by human action on the environment, and this with great relevance, given that "even though changes at a social or cultural level are harder to effect, they may have the most lasting consequences" (Weber 2010, p. 339). However, these change processes are far from linear.

In this sense, behaviourists alert to the difference between the learning of some climate phenomena from personal experience and the learning from a statistical description of possible climatic outcomes (Weber et al. 2004; Whitmarsh 2008). This distinction has "received much attention because ostensibly the same information about events and their likelihoods can lead to very different perceptions and actions" (Hertwig 2004, cit. Weber 2010). Learning from a repeated experience involves mental association processes that are fast and automatic, while learning from a description requires an analytical process involving some cognitive effort, therefore harder, as Weber argues. "When given the choice between attending to information provided in the form of statistical summaries or to information provided by personal experience, personal experience is far more likely to capture a person's attention, and its impact dominates the often far more reliable and diagnostic statistical information" (Erev and Barron, cit. Weber 2010, p. 333).

Hence, the impact of actual experiences on perceptions is bigger than the one obtained with any given demonstration, as Weber shows us. "Surveys conducted in Alaska and Florida, where residents in some regions have increasingly been exposed to physical evidence of CC show that such personal exposure greatly increases their concern and willingness to take action" (Weber 2010, p. 334). This author shows us that several international researches prove that the belief in CC increases if and when people have a direct contact trough personally experiences with CC demonstrations (Weber 2016, p. 126).

Strengthening this point, Reser et al (2014), state, in an article intituled «Encountering CC: 'seeing' is more than 'believing'», that risk perceptions and understanding of CC are changing at a global and national level, "reflecting changing social constructions and social representations of the 'environmental problem' and risk domain of CC, and both virtual and direct exposure to and experience of the unfolding biophysical impacts of CC, locally and globally" (Reser et al. 2014, p. 524).

Weber (2016) traces, through a revision of several surveys and research conducted throughout the world, individual, social and situational correlations on perceptions of CC. In this sense, she highlights (i) the role of ideology, whether on the denial of environmental problems or on the less willingness to assume pro-environmental actions; (ii) the role of gender and age, with women and young people being less sceptical; (iii) the author also underlines the importance of the regional dimension on different perceptions of the climatic risks; and (iv) the situational context and long term motivation leading to more engagement and concern with the future.

Climate Change Perceptions in Portugal

Contextual Framework

Portugal is one of the most vulnerable European countries to the impacts of CC (Santos et al. 2002; Miranda et al. 2006; Santos 2012). Its adverse impacts are widespread throughout the national territory. These include: sea level rise (which will affect the coastline areas); the desertification and drought (that will tend to spread, especially on the countryside); a temperature increase and frequent and intense extreme weather events (which are expected to take place all over the country), etc. In face of the short, medium and long term expected impacts, the urgency to create an effective adaptation strategy and the acknowledgement of a new climate reality is of paramount importance.

ClimAdaPT.Local was developed to address this need. The project's overall goal is the development of 26 Municipal Adaptation Strategies for CC (EMAAC[1]). These strategies imply the training of municipal officials and the engagement of local communities. The project's philosophy was to develop an approach that could be easily replicated throughout the whole of Portugal's territory. In order to do so, the 26 municipalities were strategically chosen, one per each inter-municipal community. The latter are a form of municipal association towards the promotion of a joint development vision. There are 26 inter-municipal communities in Portugal. ClimAdaPT.Local's ambition was to plant in each of these units an example of CC adaptation, in hope that fellow municipalities would engage in a copycat process. On top of this, the heterogeneity of the 26 selected municipalities allowed for a wide geographic, socio-economic and cultural diversity, enabling to consider the project's results as representative of Portugal's reality (Fig. 1).

Policy wise, building on its strong interdisciplinary dimension, the project's key objective was to increase the capacity for municipalities to incorporate adaptation in their planning instruments. Equally important was the project's goal to ensure the

[1]EMAAC Estratégia Municipal de Adaptação às Alterações Climáticas, stands for Municipal Adaptation Strategy to CC.

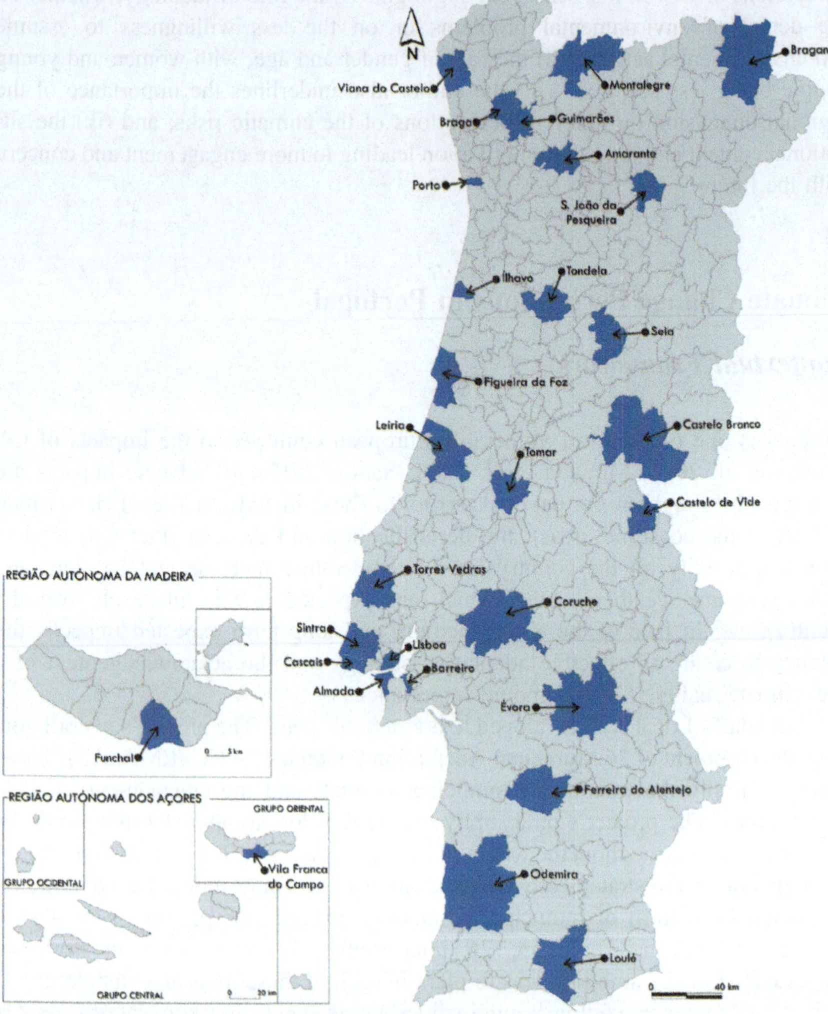

Fig. 1 Municipalities involved in the ClimAdaPT.Local project. *Source* www.ClimAdaPT-local.pt

political and institutional commitment of local decision makers, as well as to raise awareness and engage a wide and diverse group of local stakeholders in view of the elaboration and future implementation of the EMAAC (Schmidt et al. 2015b). Governance wise, the development of the EMAAC was underpinned by the principles of transparency and inclusivity. This implies, alongside stakeholder participation, openness to information/accountability and dialog, shared responsibility, efficiency and coherence (Lockwood 2010).

Methodological Framework

In our view, community wise, adaptation to CC is fundamentally a process of social-learning. It implies the construction of a wide social commitment that reinforces the legitimacy of the EMAAC, and gathers fundamental support for its implementation process. Stakeholder engagement is, therefore, crucial, in order to: (i) maximize synergies and ensure a good coordination and combination of answers and resources (UNDP 2010) and to (ii) promote the quality and the acceptance of the political choices taken, boosting the success of its implementation (Grothmann et al. 2014). ClimAdaPT.Local's stakeholder engagement strategy is twofold encompassing workshops and surveys.

In order to set the ground for ClimAdaPT.Local's social-learning process, 26 local workshops were undertaken between October 2015 and February 2016, totalling the participation of more than 1400 stakeholders. Apart from the overall goal of involving local stakeholders from diverse activity sectors in the EMAAC production, the workshops also aimed at: (i) increasing knowledge levels; (ii) helping clarify priorities; (iii) promoting transparency in the decision-making processes; (iv) promoting the basis of common understanding and encouraging sharing responsibilities; (v) creating synergies and ensuring a good coordination of answers and resources.

The workshops targeted a very diversified group of stakeholders in every one of the 26 municipalities they took place on. Participants came from different activity sectors (namely economy, agriculture, tourism, energy, forest), several members of the administration (central, regional, and local) NGO members, Church and Charity members, always in accordance with the issues identified in each municipality.

Lasting between 6 and 7 h, the workshops had two separate parts. The first, and more descriptive part, had a general scientific presentation on the impact of CC at a global and national level, focusing on the region of each specific workshop. Next, followed an exposition by the municipal technical staff of each municipality, on the local vulnerabilities, both present and future (projected), identifying a set of strategic measures of adaptation. On the second part of the workshop, the stakeholders were heard, and a debate was promoted. The latter was organized in different thematic tables where a moderator steered the discussion with a focus-group style script, and a rapporteur that would systematize and write down the key debate facts.

The script was based on three fundamental levels: (i) perceptions of the CC impacts already felt, or not, in the municipality; (ii) assessment of the viability of the proposals included in the strategy designed by municipal officials, as well as obstacles, responsibilities, suggestions and recommendations for the strategy; (iii) visions of the future: how CC and local identity will be articulated in the near future.

The retrieved data on the first level informs the reflexions and analysis we present in this paper. The information gathered in the workshops allows us to obtain a global reference framework on the sensitivity and the perceptions of CC at a local level, according to the geographic, professional and social diversity of the stakeholders engaged in this process.

Alongside the qualitative data retrieved in the workshops, there is an additional set of quantitative data resulting of the implementation of two surveys to the stakeholders. The first one was applied at the end of the workshop session. It was dedicated to the seriousness and the importance of CC, as well as to the participation in local measures related to CC. 1016 individuals responded to it. The second one was applied online, between August and September 2016. Composed by some identical questions, this survey enabled a comparison in terms of evolution over time, as well as a group of questions more focused on the necessary conditions and eventual blockages to the implementation of the EMAAC. This second round obtained 620 respondents, which, even if it doesn't match the total number of respondents of the first round, in statistical terms corresponds to a very representative sample (Guerra et al. 2017).

Analysis and Discussion of Results

CC is a phenomenon with a high level of technical-scientific complexity. Our analysis' underlying goal was to improve communication processes and enhance the knowledge appropriation capacity of local communities. This complexity often blocks the pro-activity and social engagement in the adaptation solutions to CC (Schmidt et al. 2013). To tackle such difficulty we must first look into CC from the local community's perspective. We therefore focused on three sequential questions: (a) **Is CC being felt?** (b) **How is it being felt?** (c) **What does it represent to the social actors?**

Understanding 'Glocal' Perceptions

Stakeholders generally recognise, in an almost unanimous way that CCs are occurring at a global level. This acknowledgement is more obvious at a global level than at a national and/or local level. Several studies refer to a perception at a global-distant level as being more serious than the one that is the closest to us (Dunlap 1998; Almeida 2000, 2004). According to research made in the United Kingdom the greater the distance to CC, the bigger their concern levels would be. "The evaluation of risky options under the repeated sampling in decisions from experience follows classical reinforcement learning that gives recent events more weight than distant events" (Weber 2010, p. 333). Weber (2016) goes on to identify that "A major obstacle to motivating action on CC is the fact that for many people the phenomenon appears not just abstract, but also personally distant in space and in time" (Weber 2016, p. 127). However ClimAdaPT.Local's stakeholder engagement seems to contradict such statement.

In fact, almost all workshops' stakeholders state the existence of CC effects in their everyday experience, whether they are unexpected extreme phenomenon (i.e. storms) or an impact brought upon economical activities (i.e. fishing and

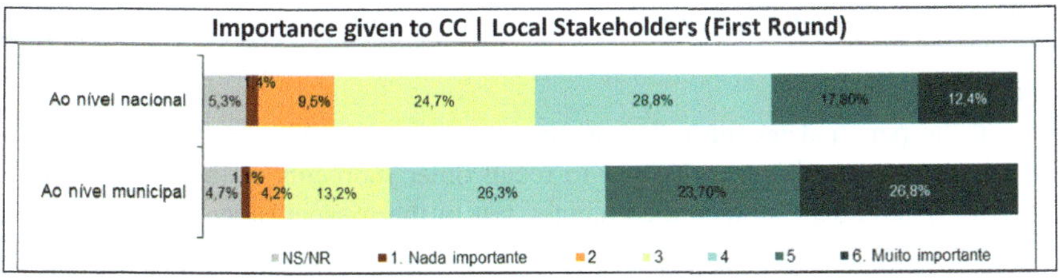

Fig. 2 The importance given to CC [red colour = not important; dark green = very important]. *Source* Guerra et al. 2017

agriculture). In particular, the first survey implemented to the stakeholders shows a high importance to CC at a municipal level (bottom line of Fig. 2).

Moreover, it emerges that, after some time had passed upon the workshop (second round of the survey), in the case of local stakeholders, the seriousness attributed to the CC problem at a local scale almost doubled (Guerra et al. 2017). The workshop has itself carried out an "awakening" effect on the issue, causing the stakeholders (on all regions of the country) to be more alert to the close signs of CC.

Capstick et al. (2015), in a revision paper on the research made on perceptions of CC, identifies several studies that reveal a direct influence on the local weather and the perceptions of CC. Reyes-Garcia et al. (2016) show us that research suggests that "local people evaluate global CC mostly in terms of extreme events, and that local people's perceptions of CC can be shaped by personal experience of recent increases in local temperatures, actual outdoor temperature at the moment of elicitation, or other short-term weather fluctuations, rather than long-term trends of global CC" (Reyes-Garcia et al. 2016, p. 110). The same applies to Portugal, were the empirical acknowledgement of CC was identified by the local stakeholders in almost all the regions where the workshops took place:

> We are in January and we haven't felt actual cold weather yet (Bragança, north).

> In January, I had planned to have a party at my house and I had to re-schedule it, because trees fell down on the main avenue. It is cyclonic winds and mini-tornados showing up (Ílhavo, centre).

> This is like Africa, on the same day we have rainstorm and then sun (Castelo de Vide, south).

In our research, we also verified that certain professional activities have a crucial influence on perceptions of CC. Thus, the farmers stated changes in the (early) maturing of fruits and vegetables, as did the fishermen on the fishery species:

> This winter the plants didn't go into dormancy, there wasn't enough cold weather (Montalegre, north).

> I noticed, this year, that my orange tree flowered twice, she must have been confused (Ílhavo, centre).

> The olive harvest started earlier, outside the usual period, at a time in which the mills weren't opened yet to receive them (Castelo de Vide, south).

We had tuna again in the Algarve, and the sardines are fatter and smaller (Loulé, south).

Climate Memory

Even if the personal recent and short term experiences are an important factor on the perceptions of CC, people tend to recall other moments of their biography in order to systematically place the changes felt in the present time:

I remember many more blizzards 17 years ago; in the winter we had around 2 straight weeks without (school) classes (Montalegre, north).

My age allows me to say climate as changed a lot. I make weather records everyday and I verify that we are currently at 2 degrees above the average temperature on previous years (Ílhavo, centre).

The clothes I used to wear as a child had nothing to do with the closes people wear nowadays; we used to wear thick clothing and now I feel that the temperature is much more mild (Ílhavo, centre).

The high tides and the heavy sea destroy the cliffs and it didn't used to be like that. We see a lot of changes in the ocean's behaviour (Odemira, south).

A relevant dimension on the social perceptions about CC framework is linked to the idea of the "fading out of the seasons" and the "disappearance of the intermediate seasons", caused by the experience felt of the increased average temperature and climate unpredictability/instability:

Before, the seasons were more bounded and reliable in terms of temperature (Montalegre, north).

I am sensitive to the changes in the seasons, they are a bit out of place (Ílhavo, centre).

These facts are nevertheless peculiar in a country where the four seasons of the year were traditionally well defined, being a part of the national imaginary.

Climate Change: Threat or Opportunity?

Despite a growing global awareness that CC is happening, whether at a global scale or at a local scale, it is important to understand if people perceive this phenomenon as a threat and/or as an opportunity, as well as the more or less pro-active way they stand themselves in light of the solutions.

Some researchers have considered the perception of CC as a threat, referring to a sense of "helplessness" facing such a wide and complex problem (e.g. Lorenzoni et al. 2007). In Portugal, a previous study noted that CC is mostly regarded as a threat, assuming local stakeholders as victims (Carvalho 2011).

Concurrently, in our research, the perception of CC as a threat also emerges, but in a limited way, confined to certain social groups, such as farmers and citizens living in coastal areas—both having already suffered damages due to CC—and also the elderly, more vulnerable to heat and cold weaves.

Throughout the workshops, the perception of threat and risk expressed by the participants about CC was mostly mentioned regarding four issues; the losses in agricultural output, forest fires, coastal erosion and sea level rise, and health.

As agriculture is concerned, CC's potential risk included:

> The trees bloom earlier and the frost in April or May burns everything; there will be no fruit after that, there will be nothing (Bragança, north).

> The chestnuts are dying at the cost of the warming in the winter (Bragança, north).

> Last spring, in May, the dams serving to hydrate the animals out on the fields didn't have water anymore (Castelo Branco, centre).

> Now, in the Algarve, we have three months of moist and nine of drought. That means that species like the carob tree and the almond tree will not make it (Loulé, south).

Forest fires were also one of the issues widely raised by the stakeholders all over the country, with a dramatic emphasis regarding this calamity that is no longer seasonal, given the frequency on which it occurs. This is shown rather eloquently by the fireman and the forest rangers, etc. attending the workshop.

> Because of the less rain, the trees present symptoms of hydric stress, in virtue of which the fires take on larger proportions (Bragança, north).

> With the temperature rise, the fire's critical period will spread year after year, and may start in February or March (Loulé, south).

Furthermore, CC exponentially increases coastal erosion. Portugal, with its extensive coastal line, presents a number of extremely vulnerable areas (Schmidt et al. 2013). Some of the areas that were most affected by violent sea storms sat on municipalities participating in the ClimAdaPT.Local project. This issue was a core discussion topic at the workshops, where there were even present some of the people whose assets and establishments had been destroyed in the extreme events of January 2014:

> I see the diminishing of the sand on the beaches. I remember the destruction of the beach clubs and the gangways (…) With the storms that occurred in the winter of 2013/2014, we witnessed a total transformation of the Barra beach (Ílhavo, centre).

> The quota increase on the Aveiro estuary is frightening. On one side, we have the the ocean, on the other, the estuary. All the people who own a house there realised already that they will have very serious flooding problems (Ílhavo, centre).

Finally, the threat to public health was stressed. Even if the increase of allergies and respiratory problems CC related, due to pollen and temperatures, were frequently mentioned, the elderly population was the most referred to, as being in risk, due to the extreme temperatures, in association with poor housing conditions, thermal comfort wise.

> The elderly suffer a lot of impacts, because there is not enough information and awareness-raising for the people to know how to answer this kind of situations (Bragança, north).

> The population is aged, as well as the buildings, which makes it difficult to achieve good acclimatization conditions (Ferreira do Alentejo, south).

Still, the perception of CC as a threat is not transversal or dominant. In the survey applied to the stakeholders at the end of the workshops, the answers actually point to a majority perception of CC as an opportunity for local development,

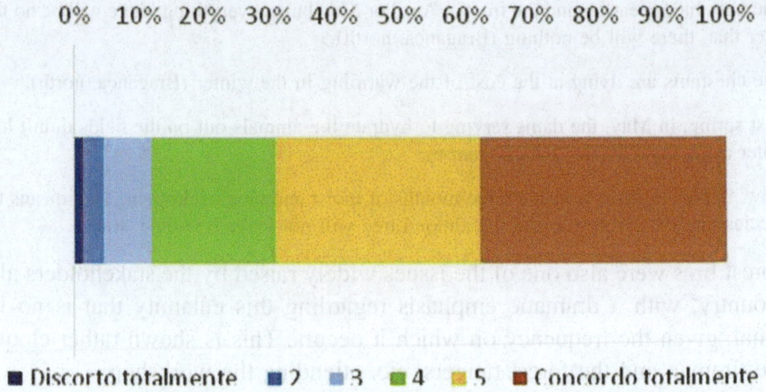

Fig. 3 CC as an opportunity for the local development—first round of the survey [red colour = totally agree; blue colour = totally disagree]. *Source* Guerra et al. 2017

therefore stating a willingness to act on a problem that needs participated solutions (Fig. 3).

Similar results emerged from the *Consulta Mundial sobre Clima e Energia*, the global consultation on Climate and Energy also held in Portugal, on the eve of the Paris 2015 COP. Here, 81% of the Portuguese participants considered that the measures that must be undertaken in order to fight CC are an opportunity to the improvement of living standards, whether by a restraining on consumption and the reuse of materials, or by the innovation possibilities and the technological development in a low carbon society paradigm (Schmidt et al. 2015a).

In line with this perception of CC as an opportunity, we find similar willingness in our research. A growing awareness on the need to adapt to CC in a context of local dynamic participation with information transfer and knowledge exchanges seems to promote an empowerment effect on the collective of stakeholders participating on the workshops. In this context, some groups of stakeholders showed greater willingness for action, namely some farmers, large companies, transition movements/organisations, teachers, and a set of public service entities, like fire-fighters, Civil Defence corps and the National Republican Guard (GNR).

These opportunities are glimpsed in several domains. For instance, in the agriculture case, there are, on one hand, farmers feeling threatened by CC, and on the other hand, there are farmers who see the opportunity for new crops, such as: wine production in places where it would be impossible before; the greater *vinho verde* wine quality, due to temperature increase; the "*rise in subtropical crops like avocatos*" (Loulé, south) or even "*the weather as an opportunity, for instance, for the winter grasslands (...) the grasslands are green this winter like in spring*" (Bragança, north).

The chance to invest in new crops and new more profitable varieties, offers an opportunity not only to attract investment and raising business, as well as to offer more training to the farmers, young people in particular, therefore raising a new generation of farmers, that may be themselves a way of alluring more population for the countryside, currently depopulated.

But this is only an example of a positive and creative approach to CC implications. Other cases could be made related to renewable energies or circular economy.

Final Remarks

In retrospect, this paper's input is twofold:

1. Conceptually, this paper contributes to a global reference framework on CC awareness, perceptions and meaning-making. To this effect, we can answer three core questions:

Firstly, is CC being felt? Yes. In Portugal, CC is felt and acknowledged at the local level. In other words, CC no longer stands as a faraway global problem visible only through media frameworks. Nowadays, individuals and communities witness a direct, often daily, contact with CC phenomena.

Secondly, how is it being felt? Present acknowledgement of CC is, on the one hand, visible through what can be named as a "memory effect"—the direct comparison of what is witnessed nowadays with the consolidated memories of past years or decades. On the other hand, daily behaviours and practices growingly embody a change in terms of clothing choices, etc. reflecting visible shifts in nature cycles including the dilution or even absence of yearly seasons.

Thirdly, what does it represent for social actors? CC is perceived as extremely serious. There is a clear notion of the need to swiftly mitigate its negative impacts and causes. In this sense, despite the consensual perception of a "threat", which renders several actors powerless and victimising themselves, CC is widely acknowledged as an "opportunity" for action and transformation of both policy making practices and local decision-making processes. CC perception, as a threat, is most commonly felt by those social groups at the forefront of its impacts (i.e. farmers, coastal areas' inhabitants and the elderly). Interestingly, farmers are also the social group that envisions CC as a window of opportunity for economic revival and local resilience.

2. Methodologically, the participatory approach underpinning the ClimAdaPT. Local project functioned as a social-learning platform and an exercise in policy co-production. The information-exchange dynamic established among policy makers, academics and local communities and the mutual trust that developed from it, set the ground for a proactive local stakeholder engagement. The physical presence at the workshops of top scientists and media public figures, experts on CC, only added to this trust development process. And trust is the essential element on CC communication, because people pay much more attention and are more willing to act, when they receive the information on CC

from a source they trust. Trust doesn't however guarantee a knowledge transfer. For the latter to take place, stakeholders need to perceive, debate and jointly decide to increase the efficiency/effectiveness of the options for adaptation to CC debated in the workshops.

In sum, ClimAdaPT.Local undoubtedly reinforced a local participatory culture and enhanced communication capacity amidst different audiences; It helped decode complexities of knowledge format, language and themes; It strengthened administrative and institutional confidence; and last but not least it fostered the advance in the co-creation of knowledge in a more consistent way.

Acknowledgements The authors wish to thank the Consortium of ClimAdaPT.Local led by professor Filipe Duarte Santos and Gil Penha-Lopes (CCIAM–FCUL) and EEA Grants, the Portuguese Carbon Fund and also the Agência Portuguesa do Ambiente (APA).

References

Almeida, J. F. (Ed.). (2000). *Os Portugueses e o Ambiente: I Inquérito Nacional às Representações e Práticas dos Portugueses sobre o Ambiente*. Oeiras: Celta Editora.

Almeida, J. F. (Ed.). (2004). *Os Portugueses e o Ambiente: II Inquérito Nacional às Representações e Práticas dos Portugueses sobre o Ambiente*. Oeiras: Celta Editora.

Capstick, S., Whitmarsh, L., Poortinga, W., Pidgeon, N., & Upham, P. (2015). International trends in public perceptions of CC over the past quarter century. *WIREs CC, 6*, 35–61. https://doi.org/10.1002/wcc.321.

Carvalho, A. (2011). Conclusões. In A. Carvalho (Ed.), *As Alterações Climáticas, os Media e os Cidadãos* (pp. 223–240). Coimbra: Grácio Editor.

Carvalho and Pereira. (2008). See Schmidt and Delicado 2014 (p. 123).

Dunlap, R. (1998). Lay perceptions of global risk—Public views of global warming in cross-national context. *International Sociology, 13*(4).

Dunlap. (2008). See Schmidt and Delicado, 2014 (p. 123).

Erev and Barron, see Weber 2010 (p. 333).

Grothmann, T., Prutsch, A., Schauser, I., McCallum, S., & Rob, S. (2014). Identify and cooperate with relevant stakeholders. In A. Prutsch, T. Grothmann, S. McCallum, I. Schauser & S. Rob (Eds.), *CC adaptation manual: Lessons learned from European and other industrialised countries* (pp. 119–152). London: Routledge.

Guerra, J., Travassos, D., Schmidt, L., & Ferrão, J. (2017). *Monitorização ClimAdaPT.Local: Relatório Final*. Projecto ClimAdaPT.Local. Instituto de Ciências Sociais da Universidade de Lisboa.

Hertwig. (2004). See Weber 2010 (p. 333).

Hulme, M. (2009). *Why we disagree about CC: Understanding controversy, inaction and opportunity*. Cambridge: Cambridge University Press.

Leiserowitz. (2006). See Schmidt and Delicado 2014 (p. 123).

Lockwood, M. (2010). Good governance for terrestrial protected areas: A framework, principles and performance outcomes. *Journal of Environmental Management, 91*(3), 754–766.

Lorenzoni, I., Nicholson-Cole, S., & Whitmarsh, L. (2007). Barriers perceived to engaging with CC among the UK public and their policy implications. *Global Environmental Change, 17*, 445–459.

Miranda, P. M. A., Valente, M. A., Tomé, A. R., Trigo, R., Coelho, M. F., Aguiar, A., et al. (2006) O clima de Portugal nos séculos XX e XXI. In F. D. Santos & P. Miranda (Eds.), *Alterações Climáticas em Portugal: Cenários, Impactos e Medidas d Adaptação – Projecto SIAM II* (p. 506). Gradiva, Lisboa.

Reser, J. P., Bradley, G. L., & Ellul, M. C. (2014). Encountering CC: "seeing" is more than "believing". *WIREs CC, 5*:521–537. https://doi.org/10.1002/wcc.286.

Reyes-García, V., Fernández-Llamazares, A., Guèze, M., Garcés, A., Mallo, M., Vila-Gómez, M., et al. (2016). Local indicators of CC: the potential contribution of local knowledge to climate research. *WIREs CC, 7,* 109–124. https://doi.org/10.1002/wcc.374.

Santos, F. D. (2012). *Alterações Globais: Os desafios e os riscos presentes e futuros.* Lisboa: Fundação Francisco Manuel dos Santos.

Santos, F. D., Forbes, K., & Moita, R. (Eds.). (2002). *CC in Portugal: scenarios.* Gradiva, Lisboa: Impacts and Adaptation Measures—SIAM Project.

Schmidt, L. (2008) 'Comunicar a Ciência: o papel dos media e os problemas científico-ambientais'. *Ciência e Cidadania - Homenagem a Bento de Jesus Caraça* (pp. 85–112). Lisboa: Imprensa de Ciências Sociais.

Schmidt, L., & Delicado, A. (2014). Alterações climáticas na opinião pública. In L. Schmidt & A. Delicado (Eds.), *Ambiente, Alterações Climáticas, Alimentação e Energia: a Opinião dos Portugueses no Contexto Europeu* (pp. 113–143). Lisboa: Imprensa de Ciências Sociais.

Schmidt, L., Delicado, A., & Fonseca, S. (2015a). *Consulta mundial sobre clima e energia: a opinião dos portugueses.* Policy Brief – Observa, Edição ICS – Instituto de Ciências Sociais da Universidade de Lisboa, Lisboa.

Schmidt, L., Ferrão, J., Guerra, J., Mourato, J. M., Alves, A. F., Baixinho, A., et al. (2015b). Adaptação às alterações climáticas nos municípios: um processo de aprendizagem social. In J. Ferrão & A. Horta (Eds.), *Ambiente, Território e Sociedade: Novas Agendas de Investigação* (pp. 15–131). Lisboa: Imprensa de Ciências Sociais. ISBN 978-972-671-363-0.

Schmidt, L., Prista, P., Saraiva, T., O'Riordan, T., & Gomes, C. (2013). Adapting governance for coastal change in Portugal. *Land Use Policy, 31,* 314–325.

Spence, A., Poortinga, W., & Pidgeon, N. (2012). The psychological distance of CC. *Risk Analysis, 32*(6), 957–972.

UNDP. (2010). *Designing CC adaptation initiatives: A UNDP toolkit for practitioners.* NY: United Nations Development Programme (UNDP). Available at: https://sustainabledevelopment.un.org/content/documents/951013_Toolkit%20for%20Designing%20Climate%20Change%20Adaptation%20Initiatives.pdf.

Weber, E. U. (2010). What shapes perceptions of CC? *WIREs CC, 1,* 332–342.

Weber, E. U. (2016). What shapes perceptions of CC? New research since 2010. *WIREs CC, 7,* 125–134. https://doi.org/10.1002/wcc.377.

Weber, E. U., Shafir, S., & Blais, A.-R. (2004). Predicting risk sensitivity in humans and lower animals: Risk as variance or coefficient of variation. *Psychological Review, 111*(2), 430–445.

Whitmarsh, L. (2008). Are flood victims more concerned about CC than other people? The role of direct experience in risk perception and behavioural response. *Journal of Risk Research, 11*(3), 351–374.

Website consulted. www.ClimAdaPT-local.pt.

Attitudes, Ability and Willingness: Rethinking Split-Incentives of Non-domestic Building Tenure to Overcome Energy Inertia

Kay Emblen-Perry

Abstract Split-incentives of non-domestic building tenure that divide the benefits of implementing energy efficient technologies and behaviours between property owners and users are generally recognised to act as a financial barrier to the adoption of energy efficient interventions in UK non-domestic properties. Despite the extensive availability of cost effective energy efficient interventions that could overcome financial spilt-incentives for UK non-domestic building owners and users, widespread energy inertia prevails. This suggests the barriers presented by split-incentives of ownership exert an influence beyond financial decision making. Rethinking the impact of split-incentives may therefore assist in unlocking energy inertia and contribute to the mitigation of climate change. This paper reports the findings of a qualitative survey undertaken to investigate the impact of non-domestic building ownership on the owners' and users' ability and willingness to adopt energy efficient and conservation technologies and behaviours. It explores the impacts of ownership beyond the reach of financial disincentives to adopt energy efficiency improvement and identifies four types of constraint affecting non-domestic building owners' and users' energy behaviours; ownership constraints, financial constraints, knowledge constrains and regulatory constraints. The paper extends the understanding of the scale and scope of split-incentives of non-domestic building ownership for energy performance improvement within non-domestic buildings and presents a wider scope of the split of incentives of ownership than previously established. It also explores the opportunity this new understanding offers for reforming UK energy policies. Findings suggest the impacts of tenure are influential beyond monetary considerations for non-domestic building owners and users and include practical and attitudinal barriers from relationships, contractual constraints and ownership concerns which drive energy inertia through the Owner-User Stalemate.

K. Emblen-Perry (✉)
Worcester Business School, University of Worcester,
City Campus, Castle Street, Worcester, UK
e-mail: k.emblenperry@worc.ac.uk

F. Alves et al. (eds.), *Theory and Practice of Climate Adaptation*,
Climate Change Management, https://doi.org/10.1007/978-3-319-72874-2_14

Keywords Energy efficiency · Non-domestic buildings · Tenure
Owners · Users · Energy inertia · Owner–user stalemate

Introduction

The current UK energy system and policy approach reflects long-term historical forces that have driven continual improvements in the availability and efficiency of energy over the last 150 years (Unruh 2013). Whilst these have traditionally provided low cost energy technologies and services, market forces have provoked an energy culture that considers energy to be in limitless supply and excessive consumption acceptable (Emblen-Perry 2016). However, in spite of increases to energy efficiency technologies these interventions have resulted in a vastly increased per capita consumption of energy (Warde 2010), creating an energy system and usage patterns that are now widely recognised as key contributors to carbon emissions.

Research into building energy consumption and efficiency has to date largely focused on domestic properties and energy intensive commercial sectors and views building type (Schleich and Gruber 2008; Janda 2008; de Groot et al. 1999) or commercial sector (Janda 2014) as the driver of carbon emissions. This approach mirrors the structure of Government policies whereby organisations below the intensive energy user threshold of the Carbon Reduction Commitment energy efficiency scheme have largely been excluded. Non-domestic buildings account for 300 TW (13%) of final energy consumption and contribute 18% of UK carbon emissions (Carbon Trust 2013).

Government non-domestic energy policy, based on attempts to provoke a rational economic response from consumers, has been split between "carrots" and "sticks". Greater importance on taxes and levies (sticks) has been harnessed to provoke change by non-domestic energy consumers within energy intensive buildings in an attempt to follow the Polluter Pays Principle. This has ensured that at least some of the costs of carbon pollution are borne by those responsible and to obtain political momentum for higher taxation levels (Environment Audit Committee 2011). However, for smaller, less energy intensive non-domestic users, on which this research focuses, a non-interventionist approach through incentives (carrots) such as low-rate loans, grants and the provision of information for good practice have been the principal policy mechanisms. Although these initiatives have applied a balance of "command and control" and encouragement they have proved unable to sufficiently lower energy consumption or change energy attitudes for smaller users to benefit from reduced energy expenditure and contribute to the mitigation of climate change. Since this was recognised by Lyon and Maxwell (2002) over decade ago little has changed to drive energy consumption reduction. Consequently, a state of energy inertia now exists within small and medium enterprises (SMEs). New approaches to unlock energy inertia within SMEs are now therefore required. This research suggests that rethinking the impact of split-incentives could achieve this.

Split-incentives of building ownership resulting in a simple financial disincentive for both owners and users to invest in energy improvements have been proposed as a driver of energy inertia (McAlllister et al. 2009; Bright 2010; Axon et al. 2012; Kontokosta 2016). However, despite the growth of green leases, a majority of SMEs in the UK continue to work within the Landlord-Tenant Divide proposed by Bright (2010) and the historically developed economic infrastructures that have built up around the availability of low cost energy.

This research, offers an alternative perspective on non-domestic building energy consumption, efficiency and conservation and considers the scope of split-incentives created by building ownership extends beyond the financial environment previously proposed. It investigates practical and attitudinal perspectives of non-domestic building owners and users as well as financial impacts of tenure to explore the full impacts of tenancy agreements and relationships on the ability and willingness of non-domestic building owners and users to adopt energy efficiency and conservation interventions.

The Energy Challenge from Non-domestic Buildings

Energy Policy

The role energy plays in the emission of carbon is generally accepted by the UK Government, politicians and the public, with energy access, efficiency, cost and security concerns high on the national political and business agendas. However, energy consumption reductions and adoption of lower carbon energy generation, targeted through existing conventional approaches of encouraging voluntary energy conservation actions, taxes and financial and non-financial incentives have fallen short of climate change requirements.

In response challenging Government targets have been agreed to mitigate climate change: Zero Carbon new non-domestic buildings by 2019 and 80% reduction in carbon emissions over 1990 levels by 2050 with at least 35% by 2020 (Committee on Climate Change 2015a). However, despite extensive energy information campaigns and financial and motivational incentives most writers and energy analysts agree that energy improvements have not been adopted as expected (DeCanio 1993; de Groot et al. 1999; Janda 2009, Warde 2010) and only a small proportion of the UK's smaller non-domestic buildings have adopted energy efficiency and conservations opportunities. This has resulted in energy inertia within SMEs, with an "Energy Inconsistency"[1] developing between the technically viable, cost effective and retrofit feasible interventions widely available and what is adopted (Emblen-Perry and Duckers 2016).

[1]The 'Energy Inconsistency' is a gap between proven energy efficiency improvements and what is actually implemented.

Greater sustainability in building design, driven through Building Regulations legislation and good practice from architectural practitioners, has delivered improvements in new buildings and large scale retrofits. However, as new non-domestic buildings (less than 5 years old) represent less than 2% of the present building stock, energy savings must also come from existing non-domestic or commercial buildings if climate change targets are to be met (Janda 2008; Kelly 2010; Carbon Trust 2013; Committee on Climate Change 2015b). Retrofitting energy improvements within existing smaller non-domestic buildings remains a voluntary intervention. This is a missed opportunity to provoke reduced expenditure on energy and contribute to the mitigation of climate change.

The current rate of progress on carbon emissions reductions and future emissions projections cast doubt on energy policy's ability to deliver UK targets. Whilst the UK is on track to meet the second carbon budget (2013–17) there is an acknowledged gap in the emission reductions required to meet the fourth carbon budget (2023–2027) which will jeopardise meeting 2050 targets (Committee on Climate Change 2015b). Withdrawal of funding for the Green Deal and Home Improvement Cashback Scheme in response to low take-up rates will put further pressure on energy policy to create new solutions to provoke carbon emissions reductions. Significant expansion of energy efficiency and reduction of energy demand within the stock of existing non-domestic buildings therefore provides a significant opportunity for carbon abatement in the UK (Carbon Trust 2013).

Non-domestic Property Ownership

The evolution in UK non-domestic property ownership over the last 150 years has created a complex pattern of building ownership and occupation within the overall sector (Dixon 2009). This evolution has resulted in a legacy building stock which is being replaced at only 1% per year (Scottish Government 2015); 70% of the non-domestic buildings standing in 2050 will have been built before 2005 to less energy efficient standards (Kelly 2010). The current building stock also exhibits diverse tenancy styles that have generated non-cooperative relationships between owners and tenants which frequently lead to energy inertia (DeCanio 1993).

This evolution of tenancy structures has also created a complex picture of energy supply routes, including non-domestic building users purchasing energy directly from the utility company, purchasing from the building's owner or receiving supplies within full-service contracts. This legacy of complexity may further widen scope of split-incentives that is explored by this research.

The author's experience of organisational energy behaviour suggests that the rates of adoption of, and involvement in, energy efficiency and conservation differs between SMEs due to varying levels of interest in energy management, willingness to invest in energy improvements and ability to control the changes to buildings necessary to achieve energy improvements.

Building type (de Groot et al. 1999; Janda 2008; Peacock et al. 2008; Schleich and Gruber 2008) and levels of building awareness (Lorenzoni et al. 2007; Fawcett 2010; UK Green Building Council 2011) have been allocated a contributory role to SME's energy inertia. However, non-domestic building ownership as a driver of energy inefficiency and source of carbon emissions has received less attention within energy research and by policy planners. This research aims to extend this understanding and offer alternative observations of barriers that hinder opportunities to improve energy efficiency and reduce energy demand which are widely considered to be the most promising, fastest, cheapest and safest means to mitigate climate change (Sorrell 2015).

The Design of the Study

A qualitative survey of owners and users of smaller UK commercial premises was undertaken to extend understanding of the impact of non-domestic building ownership on the owners' and users' ability and willingness to adopt energy efficient and conservation technologies and behaviours. The survey questions explored building owners' and users' practical and attitudinal perspectives of energy actions and behaviours, along with the financial impacts of their building tenure on energy performance. Together these established the full influence of tenancy agreements and relationships on the ability and willingness of non-domestic building owners and users to adopt energy efficiency and conservation interventions.

The online survey was distributed to more than 300 SMEs and owners of smaller non-domestic properties in the UK. The SME sector was selected for this research as SMEs have traditionally been excluded from non-domestic energy policies and initiatives in UK despite accounting for 5.5 million business premises and employing 60% of the nation's workforce (Federation of Small Businesses 2016). Together these SMEs represent an untapped source of energy and carbon emissions reductions (de Jong 2013) and could benefit from potential avoided energy cost and carbon savings of c. £13bn/86MtCO_2 by 2050 (Low Carbon Innovation Co-ordination Group 2012).

Participants were recruited from across a range of tenure structures and commercial sectors (manufacturing, service, retail and agriculture) which is representative of the UK SME sector recognised by the Department for Business Innovation and Skills (2015) and the non-domestic tenure profile estimated by the Federation of Small Businesses (2016). Responses were received from 60 participants.

Four areas of constraint that have impacted the behaviours of non-domestic building owners and users emerged from these responses; ownership constraints, financial constraints, knowledge constrains and regulatory constraints. These four themes have been used to structure the data analysis which has illustrated the wider scope of the split of incentives of ownership presented in this paper.

The Impact of Non-domestic Building Ownership

Ownership in Context

Overall, almost 80% of the SMEs participating in this research lease or rent non-domestic buildings from a private building owner. Fewer rent or lease their business premises from Local Authorities (7%), operate as franchisees with premises provided free of change (7%) or are owner occupiers (8%). All the owners lease their properties to SMEs. The research findings indicate that 80% of building users purchase their energy directly from the utility company. However, only 43% of them are solely responsible for energy efficiency within their premises. A further 7% of participants report they have a collaborative approach to energy management through a sharing of responsibility with their building's owner. This suggests, however, that half of SMEs have no influence in how energy is managed within their business premises despite paying for it.

The research also finds that only a small number of participants (8%) have installed renewable energy generation systems. This suggests the majority of SMEs' non-domestic buildings continue to operate within the historical energy culture described above.

Energy Improvement Incentives and Barriers

In order to explore barriers to energy consumption reduction non-domestic building owners and users are asked to identify factors that had acted as incentives and disincentives when considering the adoption of energy efficient technologies and behaviours during the previous twenty-four months. The findings suggest 90% of participants had evaluated interventions to improve their building's and organisation's energy performance. However, although participants reported experiencing incentives and disincentives, overall barriers outweigh incentives and prevent interventions to improve energy efficiency. When viewed discretely, users experience fewer incentives up improve energy than owners; an average of 0.4 incentives per user against 0.7 per owner. They also encounter more disincentives; an average of 1.5 per user against owners' 1.2.

Despite the recognition of some, largely financial, incentives (Fig. 1) not all owners and users prove able and/or willing to intervene to reduce energy consumption. For example, over a third of participants indicate they have no incentive to intervene in their own, or their building's energy performance. However, where changes to technologies and behaviours are adopted, users appear more sensitive to opportunities to benefit from energy consumption reduction than owners; owners implement an average of 1.14 interventions and users an average of 1.33. For both parties the majority of these interventions involves retrofitting less invasive energy efficient technologies such as lighting, IT equipment and kitchen appliances rather

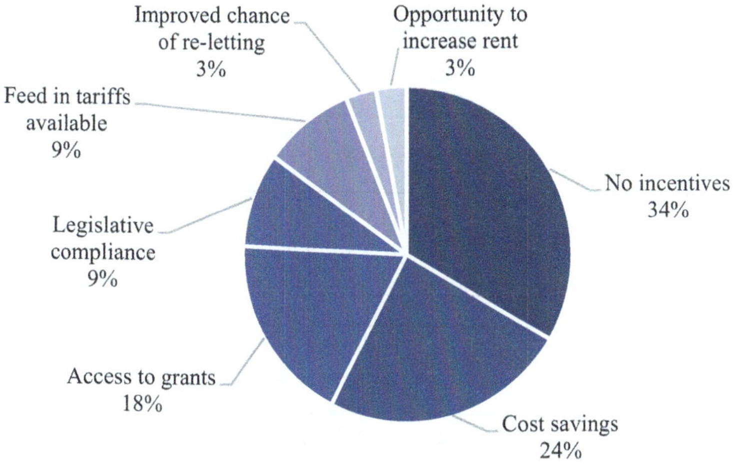

Fig. 1 Incentives to adopt energy efficient technologies and behaviours

than the more disruptive interventions such as insulation, glazing and heating systems which could achieve greater financial and carbon savings.

If the scope of split-incentives of building ownership are financial as previously proposed (McAlllister et al. 2009; Bright 2010; Axon et al. 2012), the majority of participants' disincentives and barriers to energy improvement would be expected to be financial. However, the research findings illustrate that this is not the case. Participants' responses suggest that whilst financial disincentives exist, non-financial disincentives present more significant practical and attitudinal barriers to the adoption of energy efficient technologies and behaviours within the non-domestic buildings surveyed (Fig. 2).

When explored in more detail the research findings suggest both owners and users face major barriers to the adoption of energy efficiency from the non-financial impacts of ownership; two-thirds of users and over half of owners appear constrained by the ownership structures of their buildings such as restrictive lease clauses including dilapidations requirements, other contractual issues and length and security of tenure. For non-domestic building users these are appreciably greater than the financial constraints (Fig. 3).

The research finds the majority of non-financial barriers to energy efficiency improvement align to the impacts of ownership stemming from the tenure of the properties. Over 50% of both owners and users identify barriers to improving their energy performance driven by contractual issues which prevent both parties from implementing changes that would be of financial benefit and contribute to carbon emissions reductions.

These ownership barriers appear more restrictive for users, with two thirds of participating users confirming they experience ownership barriers. For example:

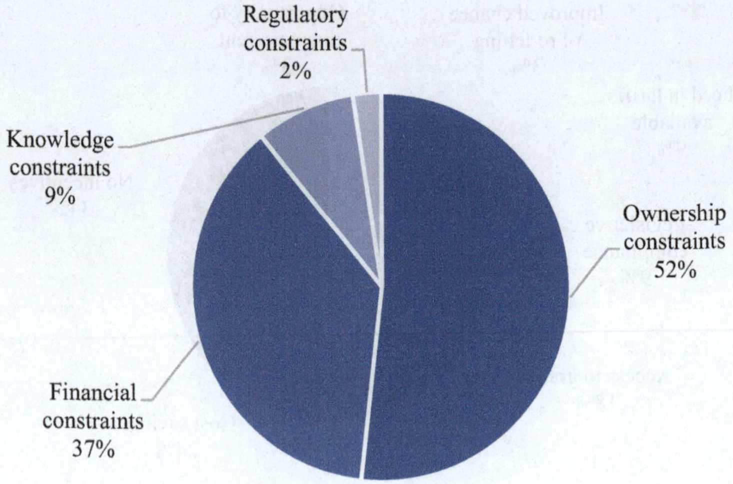

Fig. 2 Owners' and users' barriers to improving energy performance

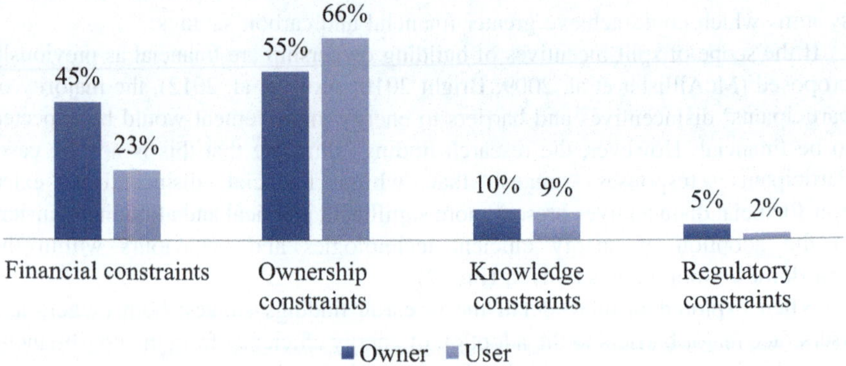

Fig. 3 Barriers to improving energy performance—owners versus users perspectives

"We wouldn't change things that we could not take with us if we were to leave the premises" (User)

"The only problem we see is that we are ground floor with flats above, the Local Authority own and rent out. If we want to put solar panels on we would have to go through tenants and local authority before [we would be allowed to do so]" (User)

"We wanted to install new local exhaust ventilation (LEV) with ventilation outside premises. The Landlord was very difficult over this. Even though it would save energy we decided it simply wasn't worth the hassle" (User)

These non-financial barriers to the adoption of energy efficiency and conservation interventions stemming from ownership suggest the scope of split-incentives is wider than previously proposed. Such constraints present practical and attitudinal

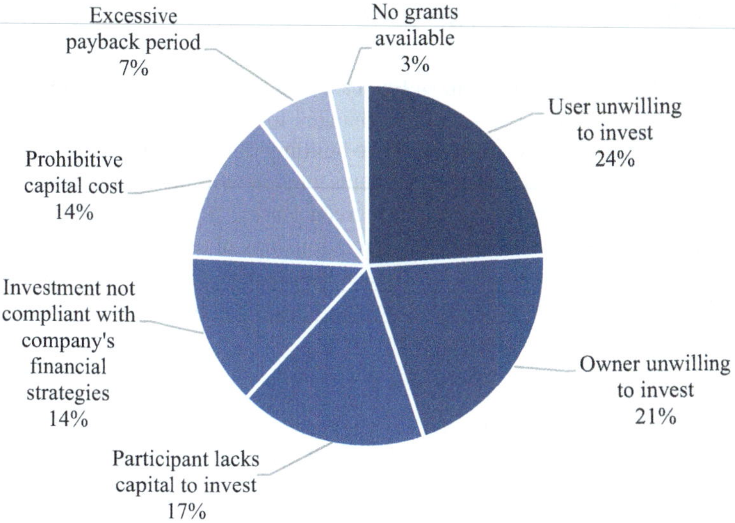

Fig. 4 Owners' and users' financial barriers to the adoption of energy efficient technologies and behaviours

influences that obstruct non-domestic building owners' and users' attempts to reduce energy consumption and obtain financial and carbon savings.

The research findings also illustrate the presence of attitudinal barriers embedded within previously recognised financial disincentives for the implementation of energy efficiency. Owners' and users' indirect barrier of an unwillingness to invest in energy efficiency improvements appear more influential than direct barriers such as the lack of internal and external financial resources and initiatives or compliance with organisational rules (Fig. 4). Almost 50% of participants recognise that unwillingness to invest is a distinct barrier to energy improvement. Monetary constraints to energy consumption reduction appear to act as attitudinal as well as financial barriers.

Participants have contributed examples of these direct and indirect financial barriers:

> "There are no incentives, just disincentives insofar as the costs far exceed any savings" (Owner)

> "Landlord is never interested in contributing to any improvements whatsoever" (User)

> "We've had a (funded) study done which said that the only things we could do all had a very long pay-back period" (User)

> "Would have done more in respect of renewable technology e.g. solar panels but no grants available" (Owner)

The Impact of Tenancy Agreements

The research finds that almost half of the lease and rental contracts held by research participants restrict users from making changes to their premises. A further 23% restrict owners from making changes to the building during the tenancy period. This suggests that traditional tenancy agreements prevail within this sample of SMEs surveyed, rather than the less restrictive green leases. It appears from the participants' responses that these leases and the relationships of owners and users continue to provoke the wider scope of split-incentives which incorporate practical and attitudinal barriers to energy improvement proposed by this research. Examples of these impacts of tenancy agreements include:

> "Lease clauses restrict tenant making changes so we are unable to get solar power" (User)

> "We lease, utility is included in the service charge and areas are not separately metered. It is particularly difficult to evidence actual use and to inspire occupants as the cost and use is invisible" (Owner)

It is encouraging that less than 10% of participants surveyed are tied by contracts subject to dilapidations clauses. Users, however, recognise the energy inefficient issues related to the actions required, for example:

> "Clauses are prohibitive and encourage waste…especially under dilapidations e.g. you have to put the place to a shell. When you leave you rip everything out even though it is in fine condition and replace with cheap items such as cheapest lighting and carpets which are not energy efficient" (User)

Owners and Users' Relationships Within Energy Management

The research finds that there is a fairly even split of cooperative,[2] collaborative[3] and preventative[4] energy relationships experienced by participants. It is encouraging that almost 30% of both owners and users feel that are able to collaborate to improve energy performance (Fig. 5).

However, when considered from the perspectives of owners and users, a larger proportion of users appear to have a preventative relationship with their building owner that creates a barrier to the adoption of energy efficiency and conservation improvements. Participants' responses provide examples of this:

---

[2]Owners agree to change without financial support provided.

[3]Owners and users work together to improve energy performance and share investment costs and benefits).

[4]Owners prevent users from making changes.

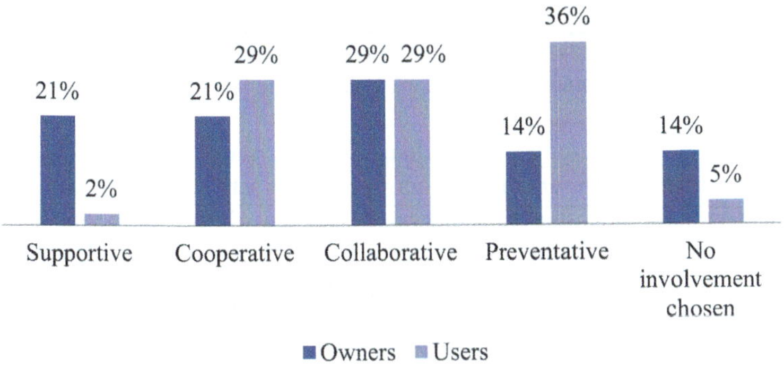

Fig. 5 Owners' and users' perceptions of owner-user energy relationships

> "So much depends on the relationship with the landlord" (User)
>
> "A great deal of tenants that would like to have a more energy saving building unfortunately their hands are tied" (User)
>
> "I choose to have no involvement in energy management" (Owner)
>
> "In the majority of properties we deal with, energy costs aren't a significant factor" (Owner)

Although it is not possible in this research to report on owners and users of individual buildings these responses suggest that at least one third of UK SMEs are prevented from making changes that could offer a financial benefit from reduced energy consumption and contribute to carbon emission reduction. A further third of participating users recognise that they are able to make changes at their own expense which may also benefit the building's owner despite no contributory investment.

These perceived relationships intensify the barriers to energy efficiency and conservation described above. The research findings illustrate the way in which the owner-user relationships widen the scope of split-incentives to restrict users and owners from improving energy performance, thus provoking energy inertia.

The Impact and Implications of Rethinking Split-Incentives

As shown above, energy policies based on financial and non-financial incentives and disincentives have not delivered the economically rational responses to energy price control expected or changes in energy behaviours planned by Energy Policy makers. This research suggests that split-incentives of non-domestic building ownership have introduced a number of barriers to energy efficiency and conservation beyond the previously recognised financial constraints. These practical and attitudinal barriers appear to result from constraints to energy consumption improvement related to ownership and behaviours rather than energy regulations and knowledge and have emerged as more influential that the financial constraints recognised by this and previous research (McAlllister et al. 2009; Bright 2010; Axon et al. 2012; Kontokosta 2016). The research findings suggest that for many SMEs these attitudinal and practical barriers combine to further split the incentives for change.

This research has presented an alternative context within which SMEs operate; a context in which behaviours and attitudes present practical and attitudinal barriers that are more influential than financial ones. The new view of the scope and extent of split of incentives for energy efficiency may help explain why UK SMEs are trapped in their on-going state of energy inertia that has been recognised over the last 25 years by DeCanio (1993), de Groot et al. (1999), Janda (2009) and Warde (2010). The findings of this research therefore offer an alternative view as to why energy policy has not achieved the energy consumption reduction required to meet UK carbon emissions reduction targets designed to support the mitigation of climate change.

This research has added to the knowledge of the impact of tenancy agreements on the ability and willingness of non-domestic building owners and users to adopt energy efficiency and conservation technologies and behaviours. Findings extend the scope of the previously recognised financial split of incentives to incorporate practical and attitudinal barriers to energy change; relationships between owners and users, disincentives presented by tenancy structures, disinterest in energy improvement and owners' and users' inability and unwillingness to make changes to building structure, fixtures and fittings and energy behaviours. Research findings suggest these wide-reaching barriers stemming from current tenancy structures have created an "Owner-User Stalemate" in which neither party is able or willing to adopt major improvements to substantially reduce energy consumption to benefit from financial savings and contribute to climate change mitigation through the reduction of carbon emissions. This results in energy inertia.

This presence of the "Owner-User Stalemate" appears to prevent change taking place or dilutes the incentives for change by splitting the rewards from energy consumption reduction between the building owners or users. This has dis-incentivised both parties from undertaking financially and environmentally beneficial interventions. Encouraging smaller non-domestic building owners and users to adopt energy efficiency technologies and behaviours through voluntary good energy practice may therefore not be the most effective route to provoke behaviour change.

A permanent change in energy attitudes, ability and willingness to overcome the Owner-User Stalemate, provoked by the wider scope of non-domestic split incentives illustrated here, is also required if energy inertia is not to be further perpetuated. An energy policy approach that unlocks the practical and attitudinal constraints presented by tenancy and owner-user relationships is required so that both SMEs and commercial building owners are encouraged to collaborate to improve energy performance, with the benefits of energy consumption reduction accessible to both parties.

As alternative drivers of change will be needed to overcome the ownership, relationship and behavioural impacts of tenancy which contribute to the practical and attitudinal barriers illustrated by this research, the author recommends that further research should be undertaken on policies and initiatives that more effectively target the practical and attitudinal changes that are key to long-term energy consumption reduction.

Conclusion

Although there has been a lengthy history of energy efficiency and conservation initiatives and widespread recognition of the need for carbon savings through energy management, the UK has failed to overcome energy inertia within smaller non-domestic buildings and deliver the energy consumption reductions required to mitigate climate change. Successive governments have relied on politically attractive but voluntary energy efficiency actions which have been considered cost effective, rational investments. However, rational economic responses have not been made by non-domestic energy consumers suggesting barriers exist to provoke energy inertia. This research suggests that this energy inertia stems from the impact of ownership which extends beyond the previously recognised financial influence of split-incentives.

The research findings illustrate that the scope of split-incentives of non-domestic building ownership extend beyond than previously proposed. More extensive impacts have been highlighted that provoke energy inertia through attitudinal and practical barriers which discourage or prevent non-domestic building owners and users from reducing energy consumption by adopting energy efficient technologies and behaviours. Research participants indicted that disincentives to energy efficiency are more prevalent the incentives; over half of the owners and users indicated they have faced disincentives to energy change based on their tenancy structures and agreements. The split of incentives from non-domestic building ownership appears to provoke the Owner-User Stalemate such that attitudes, ability and willingness of owners and users have a much greater contribution to energy inertia than previously envisaged.

References

Axon, C. J., Bright, S. J., Dixon, T. J., Janda, K. B., & Kolokotroni, M. (2012). Building communities: Reducing energy use in tenanted commercial property. *Building Research and Information, 40*(4), 461–472. http://www.tandfonline.com. Accessed October 14, 2016.

Bright, S, (2010). Carbon reduction and commercial leases in the UK. *International Journal of Law in the Built Environment, 2*(3), 218–223. http://www.emeraldinsight.com/. Accessed October 14, 2016.

Carbon Trust. (2013). *Energy efficiency in non-domestic buildings*. www.carbontrust.com/news/2013/09/energy-efficiency-in-non-domestic-buildings/. Accessed September 15, 2016.

Committee on Climate Change. (2015a). *2010 to 2015 government policy: Greenhouse gas emissions*. Report for Department of Energy and Climate Change. https://www.gov.uk/government/publications/2010-to-2015-government-policy-greenhouse-gas-emissions. Accessed November 3, 2016.

Committee on Climate Change. (2015b). *Reducing emissions and preparing for climate change: 2015 Progress Report to Parliament*. https://www.theccc.org.uk/publication/reducing-emissions-and-preparing-for-climate-change-2015-progress-report-to-parliament. Accessed December 12, 2016.

de Groot, H. L., Verhoef, E. T., & Nijkamp, P. (1999). Energy saving by firms: Decision-making, barriers and policies. *Energy Economics, 23*(6), 717–740. http://www.sciencedirect.com. Accessed November 22, 2016.

de Jong, R. (2013). Green premium or grey discount: What's the value of green workspaces for commercial building occupiers? www.2degrees.com. Accessed June 14, 2015.

DeCanio, S. J. (1993). Barriers within firms to energy-efficient investments. *Energy Policy, 21*(9), 906–914. http://www.sciencedirect.com. Accessed November 22, 2015.

Department of Business, Innovation and Skills. (2015). *Small business survey 2014: SME employers*. https://www.gov.uk/government/publications/small-business-survey-2014-businesses-with-employees. Accessed October 12, 2016.

Dixon, T. (2009). Urban land and property ownership patterns in the UK: trends and forces for change. *Land Use Policy, 26*, 43–53. http://www.sciencedirect.com. Accessed November 22, 2015.

Emblen-Perry, K. (2016). *Building owner vs. energy user: A split incentive for energy reduction*. Saarbrücken: Lambert Academic Publishing.

Emblen-Perry, K., & Duckers, L. (2016). Energy management in non-domestic buildings—a new perspective to provoke reduced carbon emissions. In *Sustainability Development Research at Universities in the United Kingdom World Sustainability Series*. Dordrecht: Springer.

Environmental Audit Committee. (2011). *Budget 2011 and environmental taxes*. http://www.publications.parliament.uk/. Accessed 23 October, 2016.

Fawcett, J. (2010). Department of energy and climate change: Non constrained sector research. http://www.decc.gov.uk. Accessed 12 October, 2016.

Federation of Small Businesses. (2016). *Small business statistics*. http://www.fsb.org.uk/stats. Accessed December 19, 2016.

Janda, K. B. (2008). Implications of the ownership of buildings in the U.S. and U.K.: An exploration of ENERGY STAR buildings and the energy efficiency accreditation scheme. http://www.eceee.org/library/conference_proceedings/ACEEE_buildings/2008/.../7_358. Accessed 14 June, 2015.

Janda, K. B. (2009). Buildings don't use energy: People do. *Architectural Science Review, 54*, 15–22. http://www.tandfonline.com/. Accessed 14 June, 2015.

Janda, K. B. (2014). Building communities and social potential; between and beyond organisations and individuals in commercial properties. *Energy Policy, 67*, 48–55. http://www.sciencedirect.com. Accessed November 22, 2015.

Kelly, M. J. (2010). Energy efficiency, resilience to future climates and long term sustainability: The role of the built environment. *Philosophical Transactions of the Royal Society, 368*, 1083–1089. http://rsta.royalsocietypublishing.org/content/368/1914/1083. Accessed November 29, 2015.

Kontokosta, C. E. (2016). Modeling the energy retrofit decision in commercial office buildings. *Energy and Buildings, 131*, 1–20. http://ac.els-cdn.com/. Accessed April 15, 2017.

Low Carbon Innovation Co-ordination Group. (2012). *Technology innovation needs assessment non-domestic building summary report*. https://www.carbontrust.com/media/218006/tina-non-domestic-buildings-energy-efficiency-summary-report.pdf. Accessed June 21, 2015.

Lorenzoni, I., Nicholson-Cole, S., & Whitmarsh, L. (2007). Barriers perceived to engaging with climate change among the UK public and their policy implications. *Global Climate Change, 17*, 445–459. http://www.sciencedirect.com. Accessed June 21, 2015.

Lyon, T. P., & Maxwell, J. W. (2002). Voluntary approaches to environmental regulation. In M. Franzini & A. Nicita (Eds.), *Economics institutions and environmental policy* (pp. 75–120). Vermont: Ashgate Publishing.
McAllister, I., Quartermaine, R., & McWilliams, F. (2009). *Costing energy efficiency improvements in existing commercial buildings*. http://www.ipf.org.uk/asset/5BF503F8-7798-4842-876C1A7F3605E2AE. Accessed June 23, 2016.
Peacock, A., Banfill, P. F., Turan, S., Jenkins, D., Ahadzi, G., Bowles. G., et al. (2008). *Reducing CO_2 emissions through refurbishment of non-domestic UK buildings*. https://researchportal.hw.ac.uk/. Accessed November 18, 2016.
Schleich, J., & Gruber, E. (2008). Beyond case studies: Barriers to energy efficiency in commerce and the services sector. *Energy Economics, 30*, 449. http://www.sciencedirect.com. Accessed November 12, 2016.
Scottish Government. (2015). *Improving energy performance and emissions in existing non-domestic buildings*. http://www.gov.scot/Topics/Built-Environment/Building/Building-standards/S63. Accessed October 28, 2016.
Sorrell, S. (2015). Reducing energy demand: A review of issues, challenges and approaches. *Renewable and Sustainable Energy Reviews, 47*, 74–82. http://www.sciencedirect.com. Accessed June 22, 2015.
UK Green Building Council. (2011). *Carbon reductions in existing non-domestic buildings*. http://www.ukgbc.org/. Accessed October 12, 2016.
Unruh, G. (2013). The sweet spot of sustainability strategy. *MIT Sloan Management Review, 55*(1), 16–19. http://sloanreview.mit.edu/article/the-sweet-spot-of-sustainability-strategy. Accessed October 19, 2016.
Warde, P. (2010). *Low carbon futures and high carbon pasts: Policy challenges in historical perspective*. http://www.historyandpolicy.org/. Accessed September 18, 2015.

Environmental Justice and Climate Change Adaptation in the Context of Risk Society

Maria Eduarda Medeiros da Silveira, Solange Buchele de S. Thiago, Larissa Pereira Cipoli Ribeiro, Ana Clara Medeiros da Silveira, José Baltazar Salgueirinho Osório de Andrade Guerra, Jéssica Garcia and Sthefanie Aguiar da Silva

Abstract The latest surveys show the importance of Environmental Justice in order to guarantee the rights of marginalized parts of society to live in a healthier environment and to provide equal treatment for people regardless of race, nationality, gender or income. Following the context of risk society, this article provides a literature review that aims to identify the importance of environmental justice in achieving climate change adaptation. Through a qualitative and interdisciplinary approach, the research illustrates how interconnected climate change adaptation, risk societies and environmental justice are, in the interest of drawing attention of stakeholders and decision-makers to the distinct levels of impact that climate change has on the various sectors of society. Overall, the study emphasize that climate change adaptation strategies can often be discriminatory. By having parts of the society vulnerable to climate change impacts, the pathway towards adaptation will never be fully reached. The course to prevent those issues from being even

M. E. M. da Silveira · S. B. de S. Thiago · L. P. C. Ribeiro · A. C. M. da Silveira
J. B. S. O. de Andrade Guerra (✉) · J. Garcia · S. A. da Silva
Universidade do Sul de Santa Catarina, Rua Trajano, 219, Centro,
Florianópolis, Santa Catarina 88010-010, Brazil
e-mail: Baltazar.guerra@unisul.br

M. E. M. da Silveira
e-mail: dudamaria.ms@gmail.com

S. B. de S. Thiago
e-mail: solangethiago@gmail.com

L. P. C. Ribeiro
e-mail: larissapcr@gmail.com

A. C. M. da Silveira
e-mail: anaclaramed3iros@gmail.com

J. Garcia
e-mail: jessica.sgarcia@outlook.com

S. A. da Silva
e-mail: sthefanie.sads@hotmail.com

F. Alves et al. (eds.), *Theory and Practice of Climate Adaptation*,
Climate Change Management, https://doi.org/10.1007/978-3-319-72874-2_15

more aggravated is to address climate change adaptation in a way that enables to benefit all groups of society by providing the information and resources for the adaptation of those living in the margins.

Keywords Environmental justice · Climate change adaptation · Risk society

Introduction

While human intervention in the environment goes back to the History of mankind, the accelerated process of industrialization led to an increasingly environmental deterioration which, along with rapid population and urban growth, marks the turning point in this process and invites to a debate about the limits of our current environmental exploration (Davies 2006; Kurtz 2003).

Along with the great potential for creativity and innovation, brought by scientific progress and economic growth, came global warming, destruction of the ozone layer, acidification of the oceans, loss of biological diversity, depletion of non-renewable resources, risk of nuclear accidents, among other phenomena, demonstrating the biosphere's resilience has been affected, which may lead to the disruption of the entire system that supports life on Earth (Lau et al. 2012; Truelove and Parks 2012).

As one of the greatest concerns of this century, climate change can be defined as "a change of climate which is attributed directly or indirectly to human activity that alters the composition of the global atmosphere and which is in addition to natural climate variability observed over comparable time periods" (United Nations 1992, p. 7).

The United Nations Environment Program (UNEP 2016c, p. 2) alerts for climate change as an effect that:

> […] adds considerable stress to our societies and to the environment. From shifting weather patterns that threaten food production, to rising sea levels that increase the risk of catastrophic flooding, the impacts of climate change are global in scope and unprecedented in scale. Without drastic action today, adapting to these impacts in the future will be more difficult and costly.

The rising global temperature will cause droughts, water stresses, storms and flooding distributed by the World (IPCC 2007). There is also evidence of the influence human activity, associated with climate change, has on the environmental degradation of the world water and land dynamics (Deng and Chen 2017; Li et al. 2016). For the United Nations Secretary General, climate changes is the greatest challenge currently regarding environmental regulations (UNEP 2016a).

Thus, the study of environmental justice, climate change adaptation and risk society and the link between them is also a discussion about survival and how to ensure a decent life for present and future generations. This debate has to deal with aspects related to regulations, which should assure the enjoyment of a balanced environment and are essential to a healthy quality of life, but also imply the discussion of a conception of justice seeking to harmonize the relationship between human and nature. At the same time, it deals with the consolidation and equalization of rights and allocation of responsibilities, with the basic premise of who is included and who is excluded from the circle of those who hold a fair distribution and reciprocal recognition (Fraser 2007).

This suggest that those living on the margins of society are clearly the most affected by the current exploration model of natural resources and the consequent environmental damage it causes (Maantay and Maroko 2009). Meanwhile, those belonging to the most disadvantaged social and economic classes integrate the portion of population which least gives rise to the occurrence of such damage and, concurrently, does not equally benefit from the results of such exploration and hardly participates in the process of making political and legal regulations regarding the environment (Teelucksingh et al. 2016; Larsen et al. 2011).

With this in mind, this article aims to draw attention to the climate change adaptation strategies being designed by governments and international organizations in order to raise awareness of the importance of a plan that seeks to reach these previously mentioned groups. It is also our expectation to foster the debates surrounding the subjects of this paper, since it is our belief that research is an important tool to understand how climate change goes beyond environmental damage, causing even more inequalities, which can be mitigated through long-term planning.

In order to explore the state of the art, scientific productions for Environmental Justice under the climate change adaptation spectrum and the connection of these issues in the context of a risk society, this study follows a literature review. It is through this literature review that we create a context on these subjects, and then discuss the issue of how climate change has diverse impacts on different groups of society, while at the same time, trying to understand the role of the State to mitigate these impacts.

The selected articles for this research were extracted mainly from three databases: Scopus, Web of Science (WoS) and Science Direct, given their relevance to the Academia. The main criteria for selecting the documents were: a time frame of the last eleven years of registration, in records from 2004 to 2015, to reflect the main ideas regarding environmental justice, climate change adaptation and risk society. The short-listed literature consisted of 51 peer-reviewed journal articles, after a verification of duplications and triplicates in the databases. Intending to complete the collection of articles selected through the literature review and to endorse the theoretical support of the research, an additional bibliographic research was carried out, also using books, reports, academic papers and journals cited in the selected publications of the databases.

Environmental Justice

Modern environmental movements began to take shape in the US along with concerns about environmental issues, such as hydrogen bomb testing, oil spills and the use of insecticides (Gordon 2012). However, American environmentalism was quite exclusive in terms of race, gender and income (Heinz 2005). Environmental justice (EJ) then emerged as a contrast to those movements, originating in a context marked by struggles for recognition of civil rights, held between the 50s and 80s (Legarda and Buendia 2011). It is possible to observe core principles of EJ in Title VI of the Civil Rights Act of 1964, which "prohibits discrimination on the basis of race, color or national origin in all programs or activities receiving federal financial assistance" (EPA 1964).

Emerging as an extension of the civil rights movement, EJ is built on the same goals: a quest for social justice, struggling against racism and poverty (Roberts 1998). At the same time, within an international level, the movement advocates that environmental injustices are unequally distributed, given the endurance of a colonial legacy stemming from neoliberalism (Gauna 2012). As a strong political and ideological project, neoliberalism allowed a great deal of environmental consequences, as natural resources are mostly evaluated on their economic value, which led to ecologically unsustainable exploitation (McCarthy and Prudham 2004; Fuentes-George 2013).

It was from multiple collective protests that the concept of EJ started taking shape: movements related to the recognition of civil rights, anti-pollution movements, student and academic movements, the struggle of American Indians, the workers and environmentalists movement, among others (Legarda and Buendia 2011; Cleere 2016; Ranco et al. 2011). Accordingly, it can be qualified as a building process or a "social-historical process", considering that since civilizations first appeared, the production of waste and pollution followed them (Legarda and Buendia 2011). The events in Warren County, North Carolina, constitute one of the cases that boosted the foundation of the movement for environmental justice. The protests raised by Warren County residents against a landfill of toxic materials led to a study, which proved that hazardous waste landfills are predominantly located in African-American communities (Bullard 2001).

Given this background, the concept of EJ refers to marginalized groups that are more exposed to environmental problems and do not benefit equally from positive environmental regulations and policy (Legarda and Buendia 2011). According to Maroko (2012, p. 533) "EJ describes a scenario where specific populations (e.g. racial and ethnic minorities, less educated populations, lower income populations) bear the burden of increased exposure to environmental pollutants [...] when compared to other populations."

Adding a different angle, Pearce and Kingham (2008), Todd and Zografos (2005) and Mitchell and Norman (2012) believe that the study of environmental justice is aimed at addressing the concerns and the promotion of even distribution of environmental risks of those aforementioned groups. Here, it is worth mentioning that some authors prefer the term "Climate Justice" when it comes to the harmful effects of climate change on vulnerable groups. However, the use of "Environmental Justice", as a broad term, was the one chosen for the purpose of this research.

Coupled with the context of struggles for recognition and representation, Environmental Justice may also be seen as a movement that advocates against "environmental racism". Studies conducted within the United States have shown that racial disparities in health are often linked with exposure to environmental pollution and poor quality built environments (Bravo et al. 2016). According to Jones et al. (2014), the residential location of ethnical minorities, associated with poor enforcement of environmental regulations and inadequate response to community complaints, are among the aggravated risks contributing to higher exposure to environmental hazards. Maantay and Maroko complement this idea stating that "Racial and ethnic minority households also tend to have less access to information about relief assistance and opportunities" (Maantay and Maroko 2009, p. 116).

Therefore, environmental justice investigates ways to include the communities most susceptible to environmental threats in the planning process and decision making, adjusting the disproportionality to which they are exposed, giving them voice and increasing their authority in those management processes (Shilling et al. 2009).

Ikeme (2003) brings an important discussion to the environmental justice field, showing the disparities of visions between the North and the South when it comes to addressing global emissions. The South suggests that "[...] the historical emissions legacy must be revisited for corrective or rectificatory purposes" (Ikeme 2003, p. 201), citing Brazil as one of the proposers for this kind of emission control. On the other hand, the northern view has been more consequentialist, aiming primarily at economic efficiency and disregarding differences in historical emissions (Ikeme 2003; Neumayer 2000).

For environmental justice, fairness is achieved through the protection of disadvantaged groups—in economic, social and ethnic-racial terms—who suffer the effects of environmental hazards without, however, even slightly giving rise to them, and compensate the groups that do not receive the profits from the environment exploitation (Shrader-Frechette 2005).

Thus, the concept of equity and distribution (or redistribution) are intertwined with the concept of environmental justice, as well as the principle of human dignity which, among other things, seeks to ensure access to a balanced environment and equal protection of environmental laws (Schlosberg 2004; Helfand and Peyton 1999).

However, EJ advocates beyond distributive justice. In fact, what can be observed is an Environmental Justice Tripod, relating three of its constituent principles and serving as an analysis axis of Environmental Justice: (I) the principle of distribution (equitable allocation of rights and responsibilities); (II) the principle of participation or procedural justice (role of affected individuals in decision-making); (III) the principle of recognition (acknowledgment of the diversity of each individual in affected communities) (Reese and Jacob 2015).

In addition, environmental justice is accomplished through regulations and government policies and actions taking into account the fair distribution of negative environmental burdens, as well as the environmental benefits (Bullard and Johnson 2000). Accordingly, it takes into account not only the environment and promoting the reduction and prevention of pollution and other degrading factors, but also aspects of social justice, throughout a process which increases participation and inclusion, regardless of ethnicity, class, gender, economic and social power, etc.

Environmental justice can also be seen as an urban discourse, given its specific engagement in local disfavored communities that are often forgotten by mainstream environmental organizations (Larsen et al. 2011). For this reason, Schlosberg (2004) puts activism as a major role in EJ, given that it encourages active community participation and their right to be included in decision-making.

Risk Society

The term "risk society" can denote content somewhat abstract. Many inquiries spring to the minds of the scholars focused on the theme, such as which areas of knowledge are concerned with it, or who or what exactly is at risk. For that reason, it is necessary to know from the first conceptualizations of the theory, up to their development, and consideration by the social sciences.

According to Beck (2006, p. 332), "Modern society has become a risk society in the sense that it is increasingly occupied with debating, preventing and managing risks that it itself has produced." Complementing that idea, Navickė (2015, p. 97) debates that

> as the modern 'class society' transforms into a 'risk society', the main concern of the contemporary welfare state is not redistribution of wealth, but rather redistribution and management of risks. In other words, the primary aim of the social protection system has shifted from the passive collection of taxes and administration of social payments to an active management of risks.

Ultimately, "[…] the paradigm of the Risk Society has emerged in european social theory as a device for analysing the way individuals, in post-industrial nations, are encouraged to become their own "risk managers", i.e. calculators of how their actions and choices will affect their own lives" (Aspinall 2016, p. 136).

Ulrich Beck (2010) assures that we live in the so called "tardy modernity", but we are not yet in a risk society. Making use of the author's words to clarify this central point theme, one can understand that modernization means key changes in social characteristics, reconfiguring the entire social fabric of certainty which nourishes life (Beck 2010).

In this modernization process, the problems are not restricted to making better use of natural resources available, nor to resolving conflicts of social distribution of wealth in a society also marked by shortage. What prevail at this stage are the problems that reflect this very model of economic and technical development arising from this production. Society then moves from the logic of distribution and social production of wealth to the "logic of the distribution of risks in tardy modernity", keeping in mind that "the social production of wealth is systematically accompanied by the social production of risks" (Beck 2010, p. 23).

> Modern societies are therefore confronted with the principles and limits of their own model, precisely to the extent that they do *not* change themselves, do not reflect on the consequences and pursue an industrial policy of more-of-the-same (Beck 1996, p. 29).

According to Beck, the result of using the largest ever seen productive technology is the emergence of risks related to radioactivity, pollution of the air, toxins in food and water and the irreversible effects of all these factors on the earth inhabitants (Beck 2010). When the characteristics of a globalized world where one currently lives are discussed, it is conceivable to assume that society is bound up with risk as industrial development has altered the World's climate, which brings uncertainty over what other changes and/or dangers it will bring (Giddens 2000).

On the other hand, the risks associated to climate change are frequently treated with skepticism, given that they are invisible to those economically privileged (Mayer 2016). Beck also alerts about the risks related to situations of social threats, as they go hand in hand with the differences of social and class inequalities (Beck 2010). However, in terms of distribution, they end up affecting indiscriminately all of society. For the author, the risks of modernity end up taking a "boomerang effect", as they reach those who gave rise or profited from the production of these risks, putting in check the social class differences (Beck 2010).

Alario and Freudenburg (2010) concentrated their efforts arguing that technological risks, described in the social theories developed by Anthony Giddens and Ulrich Beck as "transcendental"—for extrapolating the socioeconomic and geopolitical borders of a nation-state—are better understood as "Titanic Risk" (Alario and Freudenburg 2010). Referencing the victims of the tragedy of colossal impact in the Titanic, the authors state that "the entire ship did indeed sink, but the death rate was far higher for those who could not afford first-class tickets" (Alario and Freudenburg 2010, p. 501).

Alario and Freudenburg also draw a parallel between the events of the Titanic and the risks associated with technology, not only by the further damage that it caused to low income individuals, but also because both technological progress and the Titanic were considered to be reliable and short of infallibility (Alario and Freudenburg 2010). The debate here is to question the limit of the classification of transcendental risks, especially those that can be classified as environmental, given that "the risks associated with global climate disruption […] tend to be distributed in socially structured ways, reinforced by global political and economic inequalities that perpetuate the disproportionate use and abuse of the natural environment and its natural resources" (Alario and Freudenburg 2010, p. 508).

Neither the risks nor the impacts are distributed transcendentally and seeing them as being "without borders" only reinforces the structures of unequal exposure and patterns of social-geographic divisions (Alario and Freudenburg 2003). The social group that uses natural goods is not the same who receives profits from its exploitation or the one who suffers the effects of environmental disasters (Okereke 2006; Pearce and Kingham 2008). What reality shows, and what the Titanic tragedy reflects as an example, is that there are parts of the society that represent the most vulnerable, from a social and economic point of view, which really represent the risk society.

Nevertheless, there is a consensus among the literature on the existence of global risks, including the consequent environmental risks of climate change and technological progress (Giddens 2000; Beck 2010; Matten 2004; Mayer 2016; Bergkamp 2016). What is up for discussion is the extent of the burdens that marginalized groups carry opposed to those with better income generation. Short Jr. (1984) cites "the distributional boomerang", who designates the associated effect mainly to the environmental risks related to technology and ecological scarcity. Such risks may present characteristics of ubiquitous (omnipresence), affecting indiscriminately people of the globe, including the rich and powerful.

However, according to Short's understanding, what remains clear is that, despite the fact that "the distributional-boomerang" make sense, still lacks scientific evidence grounded by quantitative analysis for the measurement of the distribution of risks, costs and benefits. Possibly, for some categories of risks—such as those related to nuclear disasters—the effect, or boomerang distribution, may be better visualized. Nevertheless, the author himself confesses that "the distribution of risk has never been fair (it Normally falls disproportionately on the powerless and the poor)" (Short 1984, p. 718).

Other numerous analyzes of the risk were found. Many refer to the social perception of risk (risk perception), as exemplified by the above mentioned study by Short Jr. (1984), with a focus on "social transformation" of risk. In the same sense of social perception of risk, Kaperson et al. (1988) dedicated his thesis to the study of social and individual amplifications of risk, through research of behavioral responses that emerge from it, and the side effects generated by these perceptions, noting that the investigation of risk comprises a scientific activity and, at the same time, an expression of culture. Others are dedicated specifically to environmental risks and its governance, proposing ways to manage it through inclusive approaches and the participation of key stakeholders at all levels of public, private, political, social, individual and collective sectors (Bergkamp 2016; Matten 2004; Mythen 2015).

Nonetheless, even considering the relevance of those different approaches regarding risk, the analysis of risk, risk society and risk Titanic are the ones that matter most for the present study, because, as explained in the previous lines, they are interconnected with the "who" of the current concerns of environmental justice. It is imperative to identify this aspect of "who" (who gave rise to the damage? who increased the odds of risk? who was affected? which risks are we talking about?), so that responsibilities and risks can be distributed and the effects of damage can be diluted so, then, there is actually a fair equalization.

Discussion

The cause and the effects of climate change are unevenly distributed between poor and rich countries. Better developed in economy and infrastructure, the richest countries tend to emit more harmful gases to the atmosphere, while the developing countries—usually minor emitters—are more vulnerable to the impacts caused by climate changes (Mendelsohn et al. 2006; Padilla and Serrano 2006). An example of this unequal distribution of vulnerabilities can be noted in the fact that 94% of climate disaster related deaths occur in developing countries (Mathur and van Aalst 2004).

Bohle et al. (1994, p. 37) state that "global climate change presents a challenge to future livelihood strategies, especially for those social groups which are currently vulnerable", and that "any strategy envisaged as feasible for coping with future climate change must be rooted in a full understanding of the complex structure and causes of present-day vulnerability".

According to Proag (2014, p. 370), "the concept of vulnerability implies some risk combined with the level of social and economic liability, and the ability to cope with the resulting event. Vulnerability has been defined as the degree to which a system, or part of a system, may react adversely during the occurrence of a hazardous event".

Moss et al. (2001) defines vulnerability as the potential to be harmed, including harms caused by the depletion of environmental resources. When focusing on vulnerability to climate change, it could be defined as the degree in which the systems, or parts of the systems, are susceptible to the impacts of climate change, as climate variability and extreme events (Bizikova et al. 2009).

Confirming the idea that vulnerability to climate changes is unequally distributed between the poor and the rich, Olmos (2001) points out that the poorest populations and countries are the most vulnerable to be hardest hit by climate change impacts, as the poor have less access to resources able to respond to climatic effects.

Studies have proved that in developing countries—taking Brazil as an example—small scale food producers, who are usually poor in the country, will be the most affected by climate change impacts on food production and other resources (Burney et al. 2014; Obermaier et al. 2009; Ribeiro 2010).

The Intergovernmental Panel on Climate Change (IPCC) reports focus on adaptation strategies to help the most vulnerable populations dealing with the climate change impacts (IPCC 2007, 2014a, b, c). In Part A of the IPCC (2014a) document, it is declared that adaptation in human systems intends to moderate or avoid harm, also exploiting the beneficial opportunities, so that people can adjust to the climate change effects.

The 2007 report affirms that "a wide array of adaptation options is available, but more extensive adaptation than is currently occurring is required to reduce vulnerability to future climate change. There are barriers, limits and costs, but these are not fully understood" (IPCC 2007, p. 19).

According to Füssel and Klein (2006, p. 303), "adaptation primarily aims at moderating the adverse effects of unavoided climate change through a wide range of actions that are targeted at the vulnerable system".

Adaptation to climate change can also be defined as a process in which individuals, groups of individuals or countries seek to get along with climate change impacts (Lim et al. 2005). This process is not new—throughout History people have been adapting to climate variability in the short and long term—the innovative idea is to put strategies for adaptation in the center of policy-making (Lim et al. 2005).

The European Commission (2016) exemplifies some adaptation strategies as making efficient use of resources: invest in resilient buildings; build flood defenses and raise the levels of the dykes; develop crops that can resist to droughts; invest in forestry practices less vulnerable to storms and fires, and create land corridors to help species migrating.

The United Nations Environment Program (UNEP 2016b) also emphasizes the importance of knowledge and technical support and policy integration to help developing countries in adapting to climate changes. In this regard, adaptation for climate change is an essential tool for developed and developing countries to cope with the actual and future climate variations that will take place on their territories.

In this context, climate change adaptation is intensely connected with the fundamental human right to a balanced environment and quality of life, much like the main purposes of EJ, which is addressing environmental risks that unequally affect certain human communities. In fact, there are already initiatives that combine climate change adaptation strategies and EJ principles. As illustrated by Boone et al. (2014), the Baltimore City Sustainability Plan addresses pollution, resource conservation, transportation, education and awareness, among other themes, while paying attention to lower income communities regarding the production of waste, focusing on public engagement and assuring that all communities are being heard, proving to be an inclusive plan (Boone et al. 2014).

Unfortunately, this example is not the most common regarding climate change adaptation initiatives. According to Adger et al. (2006, p.12)

> Private adaptation measures are undertaken by those who have the knowledge and resources to do so and who typically are not particularly vulnerable to climate change impact. The aggregate outcome of private adaptation is that a society's resources are allocated to adaptation by the less vulnerable, for purposes they find beneficial for themselves. This is likely to accentuate inequality.

This clearly shows the importance of government involvement on assuring that climate change adaptation initiatives reach out the marginalized groups of society. It also shows that climate change adaptation can also be excluding if not coordinated to reach out to marginalized groups.

Back to the context of risk society, it is also accepted among academics in the field that low income individuals are the most vulnerable to those risks brought by technological modernization and the environmental threats related to it. In order to address climate change adaptation goals within a risk society scenario, a political economy restructuring is needed, coupled with the environmental prevention and more environmentally friendly and efficient production methods (Briska 2013).

Nonetheless, this is a complex matter and it brings out a lot of questionings about the way we relate to the scientific and technological advancement and how the productive and economic development will take place in the context of increasing scarcity of natural resources and the way we respond to risks, which are more and more unpredictable (Mayer 2016). In this scenario, environmental justice can function as a kind of moderator, as a privileged locus for the discussion of conflicts, debates and policy proposals.

McBride (2016) discusses how environmental disasters can lead to even more exacerbating inequalities by exposing what happened after hurricane Katrina in New Orleans within the risk society context. According to the author, poverty rate increased exponentially and other environmental problems linked with Katrina continued to grow (McBride 2016). In this case, the community development and their recovery policies were based on "normalized abusive and discriminatory policing, as well as historical interconnections of the private sector, on one hand, and the government and non-profit sectors on the other" (McBride 2016, p. 139). This is an accurate example of the menaces that can emerge along with environmental disasters, leaving heavy hand marks of inequality and disrespect for the community that could have been avoided if the regulator had paid attention to the characteristics of the region and consideration for the population surrounding it.

However, to come up with climate change adaptation strategies is not a simple task, and it cannot be done only in a national, broad context. Policymakers need to be willing to use local knowledge in order to build an effective strategy. For Jørstad and Webersik (2016, p. 987), "it is crucial that adaptation strategies are identified together with the communities and further adapted to fit into the local context." Thus, this goes beyond government action. An inclusive climate change adaptation strategy demands cooperation from the civil society, local or regional organizations, and universities. At the same time, the first step towards this cooperation should come from policymakers.

The International Fund for Agricultural Development suggests compensation programs as a way to incentive smallholders and poor communities to have an active role in climate change mitigation and to "help them overcome barriers such as very high transaction costs, insecure property rights, inability to afford investments, lack of information and risk aversion" (IFAD, n.d, p. 7). Because they lack information, they do not have an understanding of how their actions can help to aggravate climate change and what effect this will have on their own future. The idea of a compensation program does not necessarily needs to be monetary as, once again, information is key to promote their active role in climate change mitigation.

The biggest challenge here is to promote social change and long-term planning that is not only destined at benefiting the state or the economy market. For Blowers (1997, p. 867), environmental protection and the sustenance of resources will probably "[…] be achieved by a combination of market-based economic incentives and state regulation and standard setting" Unfortunately, the author claims that those actions are more likely to happen in developed countries (Blowers 1997). Based on this reality, the academic community owns the task to inform and push governments and stakeholders to address climate change adaptation into the least developed countries and to reach those vulnerable groups.

Even if the environmental damage is more perceptible and its impact is more severe in low-income communities, in the long term, the impacts of climate change will affect the entire society, for example, the increasing in temperature and flooding (Massey 2004; Power and Elster 2005). Scott et al. (2001), argues about human settlements, claiming that the impacts caused by the climate can also affect other sectors causing changes in productive capacity, health status and migration. Consequently, regulations and policies will be affected by the current scenario of uncertainty that Beck's risk society implies. In this case, climate change adaptation and EJ constitute fields that can evolve together in order to shape future decisions into a more equitable society allowing to face together the challenges imposed by climate change.

Final Remarks

Over time, the lens of justice has widened to include in its focus of analysis the protection of those social, economic and geographically disadvantaged and who are representative of the face and address of risk society. At the same time, environmental justice was born in the context of struggles for recognition of rights and larger representation of historically repressed and neglected groups, and has been seeking to identify, add, repair, protect and give voice to people who suffer the consequences of exacerbated consumerism, severe exploratory model of finite environmental resources, profit at all costs and individualism of a few who benefit from political, administrative and even justice molds still in force.

Environmental justice emerges to constitute an area of "humanization" of justice itself, to bring people affected by damage—in this case climate damage—closer to the process of making political decisions that directly affect them, and also to the process of dealing with conflicts originated from those damages, establishing itself as an inclusive justice.

Nevertheless, a change of culture is needed enabling environmental justice to be a discourse and effectively take shape. A culture willing to discuss, listen and opened to public democratic debate, which involves the participation of various sectors of the community. To fulfill that objective, access to information is a key component in the construction of a "citizen" Justice.

The interaction between Environmental Justice and Climate change adaptation is legitimate and justified by the idea that current and future generations should enjoy a healthy environment. Climate change adaptation strategies must follow legal guidelines related to the conservation and sustainable use of the environment. The principles of environmental justice are fully linked to these issues: the right to enjoy the benefits generated by the environment, its exploitation should be distributed, as well as environmental conservation duties should be equally distributed.

Similarly, it refers to liability with payment or compensation for environmental and climate damage caused, which not only ensures the survival of the current generation, but also the survival of an ecosystem subject to usufruct by future generations.

Within the risk society scope, modernization and consumerism brought us to a state of uncertainty when it comes to our natural resources, and the environmental and climate damage to which society might be exposed. The effects of climate change are more dangerous for certain marginal groups of society and the role of environmental justice is to specifically protect those in need.

Thus, the most important when planning for climate change adaptation strategies is to consider the environmental justice principles. Otherwise, climate change adaptation will end up creating more inequalities. This is where the article finds its greatest limitation. Although it was possible to determine that state regulation is one of the feasible ways to address climate change in an inclusive manner, there is also the need to involve the local community, universities and other organizations that can better assess the conditions and limitations of marginalized groups. This requires a complex and demanding action.

Climate change adaptation alone is already a complex subject, which the academic community is still at the heart of the debates. Thus, building strategies that can reach the most people faces even more obstacles. This paper was unable to identify in the literature an unrivalled guideline that would resolve how governments should act from now on. However, it was possible to identify the importance of cooperation and research as indispensable to get there.

Governments and world leaders must have in mind how important it is to inform and provide the resources to include vulnerable individuals on the realm of climate change adaptation. By caring and paying special attention to those marginalized groups, society is one step further in guaranteeing a healthier environment for future generations, and an adaptation serving all parts of the society.

Acknowledgements This study was conducted by the Research Centre on Energy Efficiency and Sustainability (Greens), from the University of Southern Santa Catarina (Unisul), in the context of the projects: Building Resilience in a Dynamic Global Economy: Complexity across scales in the Brazilian Food-Water-Energy Nexus (BRIDGE)), funded by the Newton Fund, Fundação de Amparo à Pesquisa e Inovação do Estado de Santa Catarina (FAPESC) and the Research Councils United Kingdom (RCUK).

References

Adger, W. N., Paavola, J., & Huq, S. (2006). Toward justice in adaptation to climate change. In N. Adger (Ed.), *Fairness in adaptation to climate change* (pp. 1–19). Cambridge: The MIT Press.

Alario, M. V., & Freudenburg, W. R. (2003). The paradoxes of modernity: Scientific advances, environmental problems, and risks to the social fabric? *Sociological Forum, 18*. https://doi.org/10.1023/a:1024043714247.

Alario, M. V., & Freudenburg, W. R. (2010). Environmental risks and environmental justice, or how titanic risks are not so Titanic after all. *Sociological Inquiry, 80*, 500–512. https://doi.org/10.1111/j.1475-682x.2010.00344.x.

Aspinall, R. W. (2016). Children's rights in a risk society: The case of schooling in Japan. *Japan Forum, 28*, 135–154. https://doi.org/10.1080/09555803.2015.1076871.

Beck, U. (1996). Risk society and the provident state. In: S. Lash, B. Szerszynski, & B. Wynne (Eds.), *Risk, environment & modernity: Towards a new ecology*. London: SAGE Publications Ltd. ISBN 080397937 1. ISBN 0 8039 7938X (pbk).

Beck, U. (2006). Living in the world risk society. *Economy and Society, 35*, 329–345. https://doi.org/10.1080/03085140600844902.

Beck, U. (2010). Sociedade de Risco: rumo a uma outra modernidade. São Paulo: Editora 34 Ltda.

Bergkamp, L. (2016). The concept of risk society as a model for risk regulation—Its hidden and not so hidden ambitions, side effects, and risks. *Journal of Risk Research*. https://doi.org/10.1080/13669877.2016.1153500.

Bizikova, L., Habtezion, Z., Bellali, J., Diakhite, M. M., & Pintér, L. (2009). *Vulnerability and climate change impact assessments for adaptation*. Available at http://www.unep.org/ieacp/files/pdf/ClimateChange_Manual_Final.pdf. Accessed on December 15, 2016.

Blowers, A. (1997). Environmental policy: Ecological modernisation or the risk society? *Urban Studies, 34*, 845–871. https://doi.org/10.1080/0042098975853.

Bohle, H. G., Downing, T. E., & Watts, M. J. (1994). Climate change and social vulnerability: Toward a sociology and geography of food insecurity. *Global Environmental Change, 4*(1), 37–48.

Boone, C. G., Fragkias, M., Buckley, G. L., & Grove, J. M. (2014). A long view of polluting industry and environmental justice in Baltimore. *Cities, 36*, 41–49. https://doi.org/10.1016/j.cities.2013.09.004.

Bravo, M. A., Anthopolos, R., Bell, M. L., & Miranda, M. L. (2016). Racial isolation and exposure to airborne particulate matter and ozone in understudied US populations: Environmental justice applications of downscaled numerical model output. *Environment International, 92–93*, 247–255. https://doi.org/10.1016/j.envint.2016.04.008.

Briska, I. (2013). Ecological Risks in the context of sustainable development: Statistical figres in Latvia. *European Integration Studies, 7*, 18–27. https://doi.org/10.5755/j01.eis.0.7.4328.

Bullard, R. D. (2001). Environmental justice in the 21st century: Race still matters. *Phylon, 49*, 151–171. https://doi.org/10.2307/3132626.

Bullard, R. D., & Johnson, G. S. (2000). Environmental justice: Grassroots activism and its impact on public policy decision making. *Journal of Social Issues, 56*, 555–578. https://doi.org/10.1111/0022-4537.00184.

Burney, J., Cesano, D., Russell, J., La Rovere, E. L., Corral, T., Coelho, N. S., et al. (2014). Climate change adaptation strategies for smallholder farmers in the Brazilian Sertão. *Climatic Change, 126*(1–2), 45–59.

Cleere, R. (2016). *Environmental racism and the movement for black lives: Grassroots power in the 21st century* (p. 140). Paper: Pomona Senior Theses.

Davies, A. R. (2006). Environmental justice as subtext or omission: Examining discourses of anti-incineration campaigning in Ireland. *Geoforum, 37*, 708–724. https://doi.org/10.1016/j.geoforum.2005.06.005.

Deng, H., & Chen, Y. (2017). Influences of recent climate change and human activities on water storage variations in Central Asia. *Journal of Hydrology, 544*, 46–57. http://dx.doi.org/10.1016/j.jhydrol.2016.11.006.

EPA—United States Environmental Protection Agency. (1964). Title VI of the Civil Rights Act of 1964.

European Comission. (2016). Adaptation to climate change. European Comission Website.

Fraser, N. (2007). Re-framing justice in a globalizing world. In T. Lovel (Ed.), *(Mis)recognition, social inequality and social justice* (pp. 17–35). London: Routledge, Taylor & Francis Group.

Fuentes-George, K. (2013). Neoliberalism, environmental justice, and the convention on biological diversity: How problematizing the commodification of nature affects regime effectiveness. *Global Environmental Politics, 13,* 144–163. https://doi.org/10.1162/GLEP_a_00202.

Füssel, H. M., & Klein, R. J. (2006). Climate change vulnerability assessments: An evolution of conceptual thinking. *Climatic Change, 75*(3), 301–329.

Gauna, E. (2012). Environmental law, civil rights and sustainability: Three frameworks for environmental justice. *Journal of Environmental and Sustainability Law, 19,* 34–59.

Giddens, A. (2000). *Runaway world: How globalization is reshaping our lives* (124p). London: Routledge. ISBN-13: 978-0415944878.

Gordon, E. L. (2012). *History of the modern environmental movement in America*. American Center—Environment.

Heinz, T. L. (2005). From civil rights to environmental rights: Constructions of race, community, and identity in three African American newspapers' coverage of the environmental justice movement. *Journal of Communication Inquiry, 29,* 47–65. https://doi.org/10.1177/0196859904269996.

Helfand, G. E., & Peyton, L. J. (1999). A conceptual model of environmental justice. *Social Science Quarterly, 80,* 68–83.

IFAD. (n.d.). *Climate change: Building the resilience of poor rural communities*. IFAD documents.

Ikeme, J. (2003). Equity, environmental justice and sustainability: Incomplete approaches in climate change politics. *Global Environmental Change, 13,* 195–206. https://doi.org/10.1016/S0959-3780(03)00047-5.

IPCC. (2007). Climate change 2007: Impacts, adaptation and vulnerability. In M. L. Parry, O. F. Canziani, J. P. Palutikof, P. J. van der Linden, & C. E. Hanson (Eds.), *Contribution of Working Group II to the Fifth Assessment Report of the Intergovernmental Panel on Climate Change* (976 pp.). Cambridge, UK: Cambridge University Press.

IPCC. (2014a). Climate change 2014: Impacts, adaptation, and vulnerability. Part A: Global and sectoral aspects. In *Contribution of Working Group II to the Fifth Assessment Report of the Intergovernmental Panel on Climate Change* (1132 pp.). Cambridge, UK: Cambridge University Press.

IPCC. (2014b). Climate change 2014: Impacts, adaptation, and vulnerability. Part B: Regional aspects. In *Contribution of Working Group II to the Fifth Assessment Report of the Intergovernmental Panel on Climate Change* (p. 688). Cambridge, UK: Cambridge University Press.

IPCC. (2014c). Climate change 2014: Mitigation of climate change. In *Contribution of Working Group III to the Fifth Assessment Report of the Intergovernmental Panel on Climate Change*. Cambridge, UK: Cambridge University Press.

Jones, M. R., Diez-Roux, A. V., Hajat, A., Kershaw, K. N., O'Neill, M. S., Gualiar, E., et al. (2014). Race/ethnicity, residential segregation, and exposure to ambient air pollution: The multi-ethnic study of atherosclerosis (MESA). *American Journal of Public Health, 104,* 2130–2137. https://doi.org/10.2105/AJPH.2014.302135.

Jørstad, H., & Webersik, C. (2016). Vulnerability to climate change and adaptation strategies of local communities in Malawi: Experiences of women fish-processing groups in the Lake Chilwa Basin. *Earth System Dynamics, 7,* 977–989. https://doi.org/10.5194/esd-7-977-2016.

Kaperson, R. E., Renn, O., & Slovic, P. (1988). The social amplification of risk: A conceptual framework. *Society for Risk Analysis, 8*, 177–187. https://doi.org/10.0272-4332/88/0600-0177 $06.00/1.

Kurtz, H. E. (2003). Scale frames and counter-scale frames: Constructing the problem of environmental injustice. *Political Geography, 22*, 887–916. https://doi.org/10.1016/j.polgeo.2003.09.001.

Larsen, K., Gunnarsson-Östling, U., & Westholm, E. (2011). Environmental scenarios and local-global level of community engagement: Environmental justice, jams, institutions and innovation. *Futures, 43*, 413–423. https://doi.org/10.1016/j.futures.2011.01.007.

Lau, L. C., Lee, K. T., & Mohamed, A. R. (2012). Global warming mitigation and renewable energy policy development from the Kyoto Protocol to the Copenhagen Accord—A comment. *Renewable and Sustainable Energy Reviews, 16*, 5280–5284. https://doi.org/10.1016/j.rser.2012.04.006.

Legarda, A. A., & Buendia, M. P. (2011). Environmental justice. *The State of the Art. Revista Internacional de Sociologia, 69*, 627–648. https://doi.org/10.3989/ris.2009.12.210.

Li, Q., Zhang, C., Shen, Y., Jia, W., & Li, J. (2016). Quantitative assessment of the relative roles of climate change and human activities in desertification processes on the Qinghai-Tibet Plateau based on net primary productivity. *CATENA, 147*, 789–796. https://doi.org/10.1016/j.catena.2016.09.005.

Lim, B., Spanger-Siegfried, E., Burton, I., Malone, E. L., & Huq, S. (2005). *Adaptation policy frameworks for climate change: Developing strategies*. United Nations Development Programme: Policies and Measures.

Maantay, J., & Maroko, A. (2009). Mapping urban risk: Flood hazards, race, & environmental justice in New York. *Applied Geography, 29*, 111–124. https://doi.org/10.1016/j.apgeog.2008.08.002.

Maroko, A. R. (2012). Using air dispersion modeling and proximity analysis to assess chronic exposure to fine particulate matter and environmental justice in New York City. *Applied Geography, 34*, 533–547. https://doi.org/10.1016/j.apgeog.2012.02.005.

Massey, R. (2004). Environmental justice: income, race, and health. Global Development and Environment Institute, Tufts University.

Mathur, A., & van Aalst, M. K. (2004). *An adaptation mosaic: A sample of the emerging World Bank work in climate change adaptation*. World Bank Global Climate Change Team.

Matten, D. (2004). The impact of the risk society thesis on environmental politics and management in a globalizing economy—Principles, proficiency, perspectives. *Journal of Risk Research, 7*, 377–398. https://doi.org/10.1080/1366987042000208338.

Mayer, S. (2016). Science in the World Risk Society: Risk, the novel, and global climate change. *Zeitschrift für Anglistik und Amerikanistik, 64*, 207–221. https://doi.org/10.1515/zaa-2016-0019.

McBride, D. (2016). Urban Power and community development in the 'World Risk Society': Post-Hurricane Katrina New Orleans (USA). *Perspectives on Global Development and Technology, 15*, 128–142. https://doi.org/10.1163/15691497-12341379.

McCarthy, J., & Prudham, S. (2004). Neoliberal nature and the nature of neoliberalism. *Geoforum, 35*, 275–283. https://doi.org/10.1016/j.geoforum.2003.07.003.

Mendelsohn, R., Dinar, A., & Williams, L. (2006). The distributional impact of climate change on rich and poor countries. *Environment and Development Economics, 11*(02), 159–178.

Mitchell, G., & Norman, P. (2012). Longitudinal environmental justice analysis: Co-evolution of environmental quality and deprivation in England, 1960–2007. *Geoforum, 43*, 44–57. https://doi.org/10.1016/j.geoforum.2011.08.005.

Moss, R. H., Brenkert, A. L, & Malone, E. L. (2001). *Vulnerability to climate change: A quantitative approach.* Prepared for the U.S. Department of Energy Under Contract DE-AC06-76RLO 1830.

Mythen, G. (2015). The problem of governance in the Risk Society: Envisaging strategies, managing not-knowing. In: *Risk Governance*, 43–57. https://doi.org/10.1007/978-94-017-9328-5_3.

Navickė, J. (2015). Between a risk society and a welfare state: Social risk resilience and vulnerability to poverty in Lithuania. *Socialinė politika, 10,* 95–109. https://doi.org/10.15388/STEPP.2015.10.4861.

Neumayer, E. (2000). In defence of historical accountability for greenhouse gas emissions. *Ecological Economics, 33,* 185–192. https://doi.org/10.1016/S0921-8009(00)00135-X.

Obermaier, M., Maroun, M. R., Kligerman, D. C., La Rovere, E. L., Cesano D., Corral, T., et al. (2009). Adaptation to climate change in Brazil: The Pintadas pilot project and multiplication of best practice examples through dissemination and communication networks [56kB]. *Proceedings of RIO, 9.*

Okereke, C. (2006). Global environmental sustainability: Intragenerational equity and conceptions of justice in multilateral environmental regimes. *Geoforum, 37,* 725–738. https://doi.org/10.1016/j.geoforum.2005.10.005.

Olmos, S. (2001). *Vulnerability and adaptation to climate change: Concepts, issues, assessment methods.* Climate Change Knowledge Network—Foundation Paper.

Padilla, E., & Serrano, A. (2006). Inequality in CO_2 emissions across countries and its relationship with income inequality: A distributive approach. *Energy policy, 34*(14), 1762–1772.

Pearce, J., & Kingham, S. (2008). Environmental inequalities in New Zealand: A national study of air pollution and environmental justice. *Geoforum, 39,* 980–993. https://doi.org/10.1016/j.geoforum.2007.10.007.

Power, A., & Elster, J. (2005). *Environmental issues and human behaviour in low-income areas in the UK: A report to the Economic and Social Research Council* (37p). Centre for Analysis of Social Exclusion. Available at http://sticerd.lse.ac.uk/dps/case/cr/CASEreport31.pdf. Accessed on July 08, 2016.

Proag, V. (2014). The concept of vulnerability and resilience. *Procedia Economics and Finance, 8,* 369–376.

Ranco, D. J., O'Neill, C. A., Donatuto, J., & Harper, B. L. (2011). Environmental justice, American Indians and the cultural dilemma: Developing environmental management for tribal health and well-being. *Environmental Justice, 4,* 221–230. https://doi.org/10.1089/env.2010.0036.

Reese, G., & Jacob, L. (2015). Principles of environmental justice and pro environmental action: A two-step process model of moral anger and responsibility to act. *Environmental Science & Policy, 51,* 88–94. https://doi.org/10.1016/j.envsci.2015.03.011.

Ribeiro, W. C. (2010). Impactos das mudanças climáticas em cidades no Brasil. *Parceriasestratégicas, 13*(27), 297–322.

Roberts, R. G. (1998). Environmental justice and community empowerment: Learning from the civil rights movement. *American University Law Review, 48,* 229–269.

Schlosberg, D. (2004). Reconceiving environmental justice: Global movements and political theories. *Environmental Politics, 13,* 517–540. https://doi.org/10.1080/0964401042000229025.

Scott, M., Gupta, S., Jauregui, E., Nwafor, J., Satterthwaite, D., Wannasinghe, Y. A. D. S., et al. (2001). Human settlements, energy and industry. *Climate change 2001: Impacts, adaptation and vulnerability* (pp. 381–416). Cambridge, England: Cambridge University Press.

Shilling, F. M., London, J. K., & Liévanos, R. S. (2009). Marginalization by collaboration: Environmental justice as a third party in a beyond CALFED. *Environmental Science & Policy, 12,* 694–709. https://doi.org/10.1016/j.envsci.2009.03.003.

Short, J. F., Jr. (1984). The social fabric at risk: Toward the social transformation of risk. *American Sociological Review, 49,* 711–725. https://doi.org/10.2307/2095526.

Shrader-Frechette, K. (2005). *Environmental justice: Creating equality, reclaiming democracy.* Environmental Ethics and Science Policy Series (288p). ISBN: 9780195183573.

Teelucksingh, C., Poland, B., Buse, C., & Hasdell, R. (2016). Environmental justice in the environmental non-governmental organization landscape of Toronto (Canada). *Canadian Geographer,* 1–13. https://doi.org/10.1111/cag.12278.

Todd, H., & Zografos, C. (2005). Justice for the environment: developing a set of indicators of environmental justice for Scotland. *Environmental Values, 14*, 483–501. https://doi.org/10.1016/j.apgeog.2008.07.006.
Truelove, H. B., & Parks, C. (2012). Perceptions of behaviors that cause and mitigate global warming and intentions to perform these behaviors. *Journal of Environmental Psychology, 32*, 246–259. https://doi.org/10.1016/j.jenvp.2012.04.002.
UNEP. (2016a). Climate Change—Introduction. UNEP Website.
UNEP. (2016b). Knowledge Support and Policy Integration. UNEP Website.
UNEP. (2016c). Climate change Factsheet. UN Environment. UNEP Website.
United Nations. (1992). *United Nations framework convention on climate change*. New York: UNFCCC.

Fish Farmers' Perceptions, Impacts and Adaptation on/of/to Climate Change in Africa (The Case of Egypt and Nigeria)

M. L. Adeleke, D. Al-Kenawy, A. M. Nasr-Allah, S. Murphy, G. O. El-Naggar and M. Dickson

Abstract Perception is the bed rock to really apprehend the assertiveness and interpretations of the farmers which are the grass root receptors or benefactors of the effects of climate change. Individual perception and knowledge on climate change varies according to geographical location, occupation, political and socio-economics, ecological, cultural background of the entity. Empirical observations and climate models both indicate that global climate and ocean conditions have been changing over the last 100 years and will likely change more rapidly in the future. Fish production and supply in Africa could not meet up with the demand of the consumers as a result of financial constraints, low capital investment, high cost of fish stocking and feeding, dry season, pollution and climate change. This research therefore, assesses the fish farmers' perceptions, impacts and adaptation on/of/to climate change in Africa, using Egypt and Nigeria as archetypal examples. It also annotates the precautionary measures taken by the fish farmers to ameliorate the negative impacts of climate change in the continent. KoboCollect method was used to design relevant questions and analyzed the fish farmers' responses. The results revealed that climate change has both positive and negative impacts on African aquaculture and it is believe that aquaculture is a way of adapting to the adverse effect of climate change on fisheries. 93% of the fish farmers in Africa have the ideal of climate change, 64% believe that the change will linger and persevere in the next 10–20 years. In Nigeria, 61% of the respondents relied on stream and river while in Egypt, 99% of the fish farmers cultured their fish on earthen ponds and depend on the use of agricultural drainage water. Fish production could not meet up with the demand of the consumers in Nigeria as a result of pending constraints unlike Egypt which has achieved the scale of aquaculture expansion compare to

M. L. Adeleke (✉)
Department of Fisheries and Aquaculture Technology,
The Federal University of Technology, Akure, Nigeria
e-mail: mosunmolalydia@yahoo.com

D. Al-Kenawy · A. M. Nasr-Allah · S. Murphy · G. O. El-Naggar · M. Dickson
WorldFish, Abbassa, Abou-Hammad, Sharkia 44662, Egypt

F. Alves et al. (eds.), *Theory and Practice of Climate Adaptation*,
Climate Change Management, https://doi.org/10.1007/978-3-319-72874-2_16

other African countries. It is therefore, expedient that efforts should be geared towards regional and continental integration in order to encourage aquaculture practices in other part of Africa and climate change investment should be encouraged.

Keywords Perceptions · Fish farmers · Aquaculture · Climate change KoboCollect · Egypt · Nigeria and Africa

Introduction

Climate change to different people means different things and individual perception and knowledge on climate change varies according to geographical location, occupation, political and socio-economics, ecological, cultural background of individual in question. Climate change is a threat and loss, other view it has a positive and gain while it has no effect to some depending on their socio-cultural belief. Africa as one of the most vulnerable regions to climate change, they are mostly located in regions that are more susceptible to its impact. Most of north and southern Africa will be subject to water stress while east, central and West Africa is projected to receive heavy rain resulting in increased flooding (IPCC 2007a, b). Africans generate low carbodioxide emission but it effect is very high, this result in the depletion of the ozone layers which result in Global Warming GW. The continent is now experiencing extreme drought which leads to crop failure and shortage of water supply in the North, excessive rain fall which lead to floods that has rendered both man and animals homeless. Though, there has been more research on climate change since it has become a global issue. Globally, the issue of climate change is address depending on it effect and phase. Mitigation, Adaptation and Resilience are what countries are clamoring for to proffer solutions to climate change positive or/and negative effects whereby some losing and some gaining.

The Intergovernmental Panel on Climate Change (IPCC), (established in 1988 by the World Meteorological Organization, the United Nations Environment Programme to provide information on periodic basis (every five years) about the risk of climate change and its potential environmental and socio-economic consequences) predicted that during the next decades, billions of people, especially those in developing countries will face changes in rainfall patterns that will contribute to severe water shortages or flooding, rising temperatures that will cause shifts in crop growing seasons and aquatic organism distribution (FAO 2008a, b, c; Adeleke and Matthias 2016). Despite various studies on climate change on Agriculture and other sector, there has not been much research that schoolwork climate change in/and Aquaculture in Africa.

Perception is the bed rock to really apprehend the assertiveness and interpretations of the farmers which are the grass root receptors or benefactors of the effects of climate change. The attitude of the farmers are often neglected, therefore, emphasis should be laid on their perceptions to know the right strategy to implore in solving climate change issues (Cherif et al. 2017). Is climate really changing in African Aquaculture? To answer the research question, the study focused on the fundamental knowledge and perceptions of the fish farmers in Africa on the past, present and future impacts of climate change in aquaculture.

Broadly, the research assesses the fish farmers' perceptions on climate change in Africa. The case of irrigated and rain fed aquaculture using Egypt and Nigeria as archetypal examples respectively. Looking at the increase and decrease effects of some climatic variables such as temperature (surface water temperature), rainfall, sunlight intensity, humidity, wave and tides on aquaculture, some perception statements were designed and administered on the fish farmers. Specifically, current and future awareness of climate change were, adaptation strategies to support prevailing evidences and critical issues/institutional supports were dogged. The results of this study will further be compared with satellite and meteorological records and further inference can be made. The findings from this study will add to the list of growing literature and scientific discovery on climate variability and change.

Climate Change Impacts in Africa

The earth's climate is predicted to change over time as a result of human activities which alter the chemical composition of the atmosphere through the buildup of greenhouse gases, primarily carbon dioxide, methane, and nitrous oxide. Climate change may also result from natural factors, such as, changes in the sun's intensity or slow changes in the earth's orbit around the sun. Africa's riverine systems are considered highly vulnerable to climate change effects involving extreme droughts, floods, coastal storms and rising sea levels. Although wetter and drier trends will vary widely between regions, impacts are likely to be severe across these systems where there are high levels of agricultural dependence, food insecurity, rapid population growth and a limited capacity to adapt. Africa has been estimated "the continent most vulnerable to the impacts of projected climate change" by the UN Intergovernmental Panel on Climate Change. The continent has one of the most volatile water systems on the planet, and its rivers routinely experience wild fluctuates in flow. This situation will only worsen with climate change (IPCC 2007a, b).

Egypt is one of the potential vulnerable countries to the effect of Global Warming. Egypt is located in an arid—to semi-arid zone. Its only source of water is the River Nile; which provides more than 95% of all water available to the country. The source of this water lies far to the south, from rainfall on Ethiopian hills (86%) and equatorial lakes (14%). Global warming is expected to affect Egypt in many ways. In particular, water resources, agricultural resources and coastal zones are expected to be adversely affected. Both water supply and demand are expected to be affected by climate change. Impacts on the supply side are likely to arise from possible changes of precipitation patterns over the Ethiopian hills (which accounts for around 85% of water flow into the River Nile), and equatorial lakes such as Lake Victoria (15%). The demand for water in Egypt is dominated by three major user groups: agricultural irrigation, domestic use and industry.

Global warming results in sea-level rise due to the melting of glaciers and arctic ice. Consequently, the world's fresh water resources decline while salt water intrudes into underground reservoirs. Egypt is particularly susceptible due to its low-altitude Nile Delta. Livestock and fisheries are also vulnerable to the impacts of climate change, though changes in climatic conditions and sea level rise are expected to affect populations and various species differently. Climate change on fisheries productivity would be negatively affected in African countries. Various attempts to quantify the possible effects of climate change in Egypt divulged the fact that, the country faces serious risks as a result of its geographical situation, its population distribution, and its sole reliance on Nile River water. Consequently, agricultural production could decrease by 8–47% by 2060, with employment losses of up to 39%, as a result of losses of agricultural land and temperature rises.

Regional scenario forecasting across West Africa predicts a 50% decline in fisheries related jobs. Nigeria's coastline is undergoing significant morphological changes as a result of sea surges. Global warming-induced accelerated sea level rise (ASLR) of 0.5 m is likely to exacerbate these coastal economies, where about 35% of the highly-productive delta could be lost (French et al. 1995). Climate change impacts, resulting in increasing floods and droughts, soil erosion, and rises in sea level, pose a significant threat to agricultural economy and national food security which might lead to agricultural productivity decline of between 10 and 25% by 2080.

Research Methodology and Approach

The Study Area

The study was carried out in Egypt and Nigeria, the field survey was between the period of January and March, 2017.

Egypt

Egypt is the most populous country in North Africa and ranked the third position in Africa after Nigeria and Ethiopia. Based on the United State estimates the current population is 94,660,721 which is equivalent to 1.27% of the total world population (Worldometers 2017). By Latitude and Longitude of 30°06N and 31°25E (Fig. 1), the total land area is 995,560 Km^2 (384,388 sq. miles) which is divided into two unequal regions by the northward-flowing of the River Nile. The climate is better describe has arid/desert where most of the rains fall in the winter months with an average of 2–5 mm (0.1–0.2 in.) per year with the exception of the Northern coast which sometimes can be as high as 410 mm (16.1 in.) between the period of October and March. The temperatures range from 9.5 to 23 °C (49.1–73.4 °F) in low wind and sometimes average high temperatures vary from (62.6 °F) in wintertime to 32 °C (89.6 °F) in summertime were recorded(Wikipedia, World Meteorological Organization WMO).

Egypt is indeed blessed with the Nile River along with some natural and artificial lakes. The Nile River is the father of African rivers and the longest river in the world. The Nile is the primary water source of Egypt and provides more than 95% of all water available to the country. Over 60 million people are connected with the agricultural sector which constitutes 20% of Gross National Products (GDP) and consumes about 80% of the water budget. Aquaculture in

Fig. 1 Egypt map showing Latitude and Longitude. *Source* www.mapsofworld.com accessed 15/3/2017, Author's creation, 2017

Egypt is supported by water from irrigated agriculture and lakes (Burullus, Manzala and Edku). The water system has been so regularized and according to law in Egypt, Aquaculture cannot make direct use of the Nile water. Waters from the lakes and agricultural waste canals are channeled to feed the fish farms where about 99% of the fish farmers cultured their fish on earthen ponds using lands that are not seemly for agricultural cultivation and depend on the use of agricultural drainage water.

Nigeria

Nigeria lies exclusively within the tropical zone in West Africa, it shares land borders with the Republic of Benin in the west, Chad and Cameroon in the east, and Niger in the north. Its coast lies on the Gulf of Guinea in the south and it borders Lake Chad to the northeast. The current population is 190,423,532 which is equivalent to 2.55% of the total world population and ranked number 7 in the list of countries by population (Worldometers 2017). The total land area is 910,802 km^2 (351,662 sq. miles) with 9.0820°N, 8.6753°E coordinates (Fig. 2). The northeast and southwest winds are the two principal wind currents that make Nigeria to be distinguished with two distinct seasons: a wet season from April to October, with generally lower temperatures, cloudy and rainy weather upshot of the southwest wind current. And a dry season from November to March instigated by the northeast wind current; it is hot and dry, it carries a reddish dust from the desert; it causes high temperatures during the day and cool nights this is specially refers to as the harmattan. There are wide climatic variations in the country depending on the season and region.Temperatures vary according to the seasons of the year, in some part of the country temperature rarely exceed 44, 38 °C (100 °F), 32 °C (90 °F) at midday and drops as low as 12 °C (54 °F) at nights (Wikipedia). Average rainfall along the coast varies from about 180 cm (70 in.) in the west to about 430 cm (170 in.) in certain parts of the east. Inland, it decreases to around 130 cm (20 in.) in the extreme north. The Niger River is the principal river of western Africa, extending about 4180 km (2600 miles). Its drainage basin is 2,117,700 km^2 (817,600 sq. miles) in area. Nigeria is the final downstream country through which the Niger River flows, and discharges through a massive delta, known as the Niger Delta or the Oil Rivers, into the Gulf of Guinea in the Atlantic Ocean. Nigeria got her name from the river. The Niger Basin extends across 20 of the 36 states of Nigeria and comprises two main rivers, the Niger and the Benue, and 20 tributaries. Almost 60% of Nigeria's population, or about 67.6 million inhabitants, live in the Basin. Apart from the Niger River, Nigeria is indeed blessed with marine, brackish and freshwater in addition to lakes, rivers, deltas, reservoirs, dams, rain, floodplains, inland and coastal waters to support fisheries and aquaculture in the country.

Fig. 2 Nigeria Map showing Latitude and Longitude. *Source* www.mapsofworld.com accessed 20/3/2017, Author's creation, 2017

Aquaculture has been the fastest growing animal protein production sector for the last 20 years with an all-time high record of 59.9 million metric tons of fish for human consumption in 2010, which is an estimated value of US$119 billion (FAO 2012). Aquaculture's contribution to world food fish production steadily increased from 9% in 1980 to 47% in 2010 and is projected to surpass that of capture fisheries by 2015 (OECD-FAO 2011). Aquaculture was introduced in Nigeria and Cote d'Ivoire in the 1950s based on indigenous species of tilapias and catfishes. Aquaculture production in Nigeria at present is fairly insignificant at 5000 tons per year; FAO estimates that this could be increased to about 650,000 tons (FAO 2000). Though, aquaculture has potential to contribute to both food security and economic development of the country but is still in a promising state.

The Design

The study was so designed that the area and respondents were selected based on the objectives of the research. In Egypt, the governorates selected were the one that best represent the area where fish farming are practice in the country and good representation of the green water/irrigated aquaculture. Similarly, four states were purposively selected from southwest Nigeria representing the blue water/rain fed type of aquaculture and the states fall within the same geopolitical zone in the country. The study exhibit and data collection is a full gender representation because in the data collection of six paired (Egypt 4, Nigeria 2) groups each pair comprise a male and a female trainees who were trained to administer the questionnaire on the respondents.

The design of the questionnaire was in two languages: English and Arabic. The English version was administered in Nigeria while the Arabic version was administered on the respondents in Egypt. The questionnaire was sectionalized into three broad sections: perception on climate change, production and cost activities and socio economic parameters of the respondents. The questions were pre-test and tested before the proper administration on the respondents. Data collection was between the month of January and March, 2017.

Sample and Sample Techniques

Multistage sample techniques were used in selecting the study area and sample size in the study. At the first stage, the countries; Egypt and Nigeria were purposively selected because they respectively represent the water type (green/irrigated and blue/rain fed) of aquaculture practice needed in the study from the continent (Africa). At the second stage, the Governorates in Egypt and States in Nigeria were purposively selected. Four governorates (KFS, Bahera, Shakia and Port Said) were purposively selected in Egypt because they represent the area were aquaculture is practice in the country. Likewise in Nigeria, four states (Ekiti, Ondo, Osun and Oyo) were purposively selected from the southwest representing the Geopolitical zone with the rain forest area of the country. At the third stage, the villages/towns/settlements and respondents/fish farmers were randomly selected. In Bahera, Shakia and Port Said 15 respondents/fish farmers were randomly selected in each of the governorates while in KFS 35 respondents/fish farmers were randomly selected because 50% of fish farms and fish produced in Egypt is from KFS. On the other, 20 respondents/fish farmers were randomly selected in each of the states in Nigeria. 80 respondents/fish farmers were randomly selected in each of the countries, making a total of 160 respondents. This sample size represents a great deal to get adequate and necessary information on the perceptions of the fish farmers to climate change, hence the sufficiency of the survey instruments/parameters/questions.

Data Collection

Data collections were done by a 6 paired (male and female) trained enumerators who were trained on the use of kobocollect and questionnaire administration. Data collected were done using Kobocollect application on the tablets/I phones, in some cases, structured questionnaires (Appendix) the completed questionnaire were later keyed into the KoboCollect. Questions were asked on the perception of the fish farmers on climate change, production and cost activities and socio economic parameters. Interview schedule and Focus Group Discussions (FGD) were also employed to get responses from the fish farms as regards the subject matter.

Kobocollect was used to analysed the data, the data were sieve and the results of the survey were interpreted accordingly. KoboToolBar supported the participants to collect data online and off line especially during a logistical constraint. It gives room for backed up into secure servers.

Results and Discussion

The study revealed that 76 and 95% of the respondents/fish farmers in Egypt and Nigeria have the idea of what climate change means respectively. The percentage of those that does not have idea of what climate change means were 24 and 5% in Egypt and Nigeria respectively. The level of awareness might have been to the fact that the respondent exposure and education gave them edge on the changes that are occurring in their environment either positively or otherwise. Nhemachena et al. (2014) ascertained that rural communities had noticed and aware of climate change and variability in Africa (Table 1).

Table 2 represent the period/time that the respondents started to experience/observe climate change in their fish farms. 58.2 and 47.5% represent the percentage of the fish farmers in Egypt and Nigeria respectively who observed that climate started changing between the periods of 0–5 years. 12.7 and 13.8% started observation between 6–10 years in Egypt and Nigeria respectively. 3.8 and 3.8% observed changes between 11–20 years respectively, only 3.8% of the fish farmers from Nigeria observed change in climate between 21 and 30 years while observation between 31 and above years were 25.4 and 31.3% in Egypt and Nigeria

Table 1 Respondents idea on what climate change means

Parameters	Egypt	Nigeria
Do you have idea of what climate change means	Frequency (%)	Frequency (%)
Yes	60 (76)	76 (95)
No	19 (24)	4 (05)
Total	79 (100.0)	80 (100.0)

Source Computed Survey Data, 2017

Table 2 Respondents period/time to first/start observe climate change

Parameters	Egypt	Nigeria
Value (years)	Frequency (%)	Frequency (%)
0–5	46 (58. 3)	38 (47.5)
6–10	10 (12.7)	11 (13.8)
11–20	3 (3.8)	3 (3.8)
21–30	–	3 (3.8)
31 and above	20 (25.4)	25 (31.3)
Total	79 (100.0)	80 (100.0)

Source Computed Survey Data by Authors, 2017

respectively. The study revealed that a greater percentage (75 and 65%) of fish farmers in Egypt and Nigeria respectively started to experience climate change between the periods of 0–30 years while 25 and 35% represent the percentages of the respondents who started to experience climate change between the period of 31 years and above in both countries respectively. The result of the study implies that the fish farmers in both countries might have been observing change in climate variables in recent years. This implies that climate change in aquaculture is a recent incident and most of the fish farmers might observed change in climate not only from their fish farming experiences but other dictates of the environment. Climate is considered to be changing after a long period of time precisely above 30 years (IPCC 2007a, b).

The result revealed that the fish farmers in both countries got to know about climate change via various means. Their sources of awareness vary from media, history, personal observations and fish farming experiences. In Egypt, 36.7 and 81% were aware of climate change through media and history respectively. While in Nigeria, 82.5 and 66.3% represent the percentage of the fish farmers that got the awareness from media and history respectively. In addition, 12.5% were through personal observation and fishing experience and 6.3% were aware through workshop and studies. The multiple responses and level of awareness might be to the fact, that climate change is now a global concern which has brought much anxiety to all countries and has become an important agenda in both public and private discourse. Therefore, the level of awareness in all possible medium of communication will be extraordinary. Moreover, the level of education of the respondents will also contribute to their greater awareness. Although, Maddison (2007), harangued that if farmers perceived climate change, some may still fail to adapt optimally due to information limitation (Table 3).

Boxes 1 and 2 spectacle the fish farmers observed evidences to show that climate is changing. Perception is the bed rock to really apprehend the assertiveness and interpretations of the fish farmers on the subject matter. For this, emphasis was laid on their perceptions to know the right strategy to implore in solving climate change issues as opined by Cherif et al. (2017). In Egypt the respondent's presented up to nine (16) evidences that climate is changing while in Nigeria, the respondents bared eight (9) evidences of climate change (Boxes 1 and 2). Though, the study allowed

Table 3 Source of climate change awareness

Source of awareness	Egypt	Nigeria
Value (source)	Frequency (%)	Frequency (%)
Media	29 (36.7)[a]	66 (82.5)[a]
History	64 (81.0)[a]	53 (66.3)[a]
Personal observation and fish farming experience	–	10 (12.5)[a]
Workshop/studies	–	5 (6.3)[a]

[a]Multiple responses *Source* Computed Survey Data by Authors, 2017

the fish farmers to state the evidences in their own point of view to really ascertain the fact. It was observed from the presented evidences that there were similarities and disparities in climate change indications in both countries. Extreme temperature change as a result of world weather change which has negative impact on the fish is common to both countries. Climate change in Europe is also a resultant experience and evidence of the fish farmers in Egypt that climate is changing because of the extreme cold; (less than 13 °C) during winter this result in fish mortality and negative change in fish feeding habit. Change in raining pattern which result in flood at times reduction in water level is also an evidence that climate change is affecting fish production in Nigeria. The result of the study further revealed that 98.7 and 96.25% of the fish farmers in Egypt and Nigeria respectively perceived that temperature is one of the climate variables that is changing in recent years. Other climate variables and perceived percentages of the fish farmers on the change in Egypt and Nigeria respectively were given as follows: rainfall 63.3 and 93.8%, wind speed 54.4 and 56.3%, solar radiation 60.8 and 81.3%, waves and tides 17.7 and 53.8%, humidity 92.4 and 81.3% (Table 4).

Climate variables have adverse effect on fish and fish farmers in both countries as revealed in Boxes 3 and 4 some of the effects include fish mortality which result in low fish production, reduction in stocking density, high cost of inputs, frozen fishes during winter and more stress on the fish farmers. The result also revealed that good management practice can restraint or reduce the effect of climate change on both the fish and the fish farmers but this require extra cost which most of the

Table 4 Do you perceived any change in any change in the following climatic variables

Parameters	Egypt	Nigeria
Value (climate variables)	Frequency (%)	Frequency (%)
Temperature	78 (98.7)	77 (96.3)
Rainfall	50 (63.3)	75 (93.8)
Wind speed	43 (54.4)	45 (56.3)
Solar radiation	48 (60.8)	65 (81.3)
Waves and tides	14 (17.7)	43 (53.8)
Humidity	73 (92.4)	65 (81.3)

Source Computed Survey Data by Authors, 2017

farmers could hardly afford as a result of not getting assistance from the government/extension agents and low savings (Table 7). The result is similar to other studies on agricultural farms in the semi-arid environments and other parts of Africa (Maddison 2007; Slegers 2008; Rao et al. 2011; Nyanga et al. 2011; Adeleke and Balogun 2013). This implies that enough evidences had supported the occurrence of climate change in Africa. Depending on the form of farms productivity; agriculture or aquaculture the effect or impacts might be positive or negative.

Table 5 Do you think these changes will continue in the next 10–20 years?

Parameters	Egypt	Nigeria
Think these changes will continue in the next 10–20 years	Frequency (%)	Frequency (%)
Yes	48 (60.8)	51 (63.8)
No	31 (39.2)	29 (36.3)
Total	79 (100.0)	80 (100.0)

Source Computed Survey Data by Authors, 2017

Table 6 Any future threat of climate change on your fish farm

Parameters	Egypt	Nigeria
Any future threat of climate change on your farm	Frequency (%)	Frequency (%)
Yes	71 (89.9)	58 (72.5)
No	8 (10.1)	22 (27.5)
Total	79 (100.0)	80 (100.0)

Source Computed Survey Data by Authors, 2017

Table 7 Critical issues/limitation and institutional framework supports

Parameters	Egypt	Nigeria
Has there been any change in fish production in recent years	Frequency (%)	Frequency (%)
Yes	60 (95.9)	23 (28.8)
No	19 (24.1)	57 (71.3)
Total	79 (100.0)	80 (100.0)
Does your fish production meet the demand of the consumers		
Yes	35 (44.3)	44 (55.0)
No	44 (65.7)	36 (45.0)
Total	79 (100.0)	80 (100.0)
Major challenges faced in fish farming		
Finance	–	43 (53.8)
Labour/man power	–	14 (17.5)
Fish feed cost and other inputs	27 (34.2)	6 (7.5)
Water shortage/water quality	32 (40.5)	11 (13.8)

(continued)

Table 7 (continued)

Parameters	Egypt	Nigeria
Has there been any change in fish production in recent years	Frequency (%)	Frequency (%)
Fish disease	20 (25.3)	–
Vices (thief and flood)	–	6 (7.5)
Total	79 (100.0)	80 (100.0)
Do you belong to any cooperative society		
Yes	6 (7.6)	17 (21.3)
No	75 (92.4)	63 (78.8)
Total	79 (100.0)	80 (100.0)
Do you have access to extension agents		
Yes	17 (21.5)	8 (10.0)
No	62 (78.5)	72 (90.0)
Total	79 (100.0)	80 (100.0)
Purpose for rearing fish		
Small scale (Household)	–	50 (62.5)
Medium scale (Household and sales)	1 (1.3)	24 (30.0)
Large scale (commercial/industrial/exports)	78 (98.7)	6 (7.5)
Total	79 (100)	80 (100)
Types of pond		
Earthen ponds	78 (98.7)	70 (87.5)
Others (concrete, plastic, collapsible tanks, vats)	1 (1.3)	10 (12.5)
Total	79 (100)	80 (100)
What type of fish do you produce		
Catfish	8 (10.1)	64 (80.0)
Tilapia	77 (97.5)	15 (18.8)
Mullet	57 (72.2)	–
Flathead mullet	56 (70.9)	–
Others (carps, etc.)	7 (8.9)	1 (1.3)
Total	79 (100)	80 (100)
Source of water		
Stream		29 (36.3)
River	5 (6.3)	9 (11.3)
Ground water/well/boreholes		28 (35.0)
Rain catchment/reservoirs		16 (20.0)
Lake water	10 (12.7)	–
Agricultural drainage water	61 (77.7)	–
Mixed drainage and Nile water	3 (3.8)	
Total	79 (100)	80 (100)

Source Computed Survey Data by Authors, 2017

Boxes 1 and 2 Evidences observed to support that climate is changing

Egypt

1. Reduction in rain and increase in level of temperature particularly during summer period (July)
2. Movement of the fish in the pond
3. Sudden suffocation/floating of fish and death.
4. Low oxygen
5. Extreme cold during winter.
6. Increase rate of mortality
7. The world and weather is changing
8. Negative change in fish feeding habit
9. Observable differences in weather temperature
10. Breeding time changes as the climate change
11. Outbreak of diseases that affect the fish resulting into fish death.
12. Sudden death of fish without reasons
13. Change in climate in Europe
14. Appearance of disease
15. Increase humidity
16. Extreme temperature

Source Survey Data by Authors, 2017

Nigeria

- Change in raining pattern and hot temperature
- Pollution of air
- Observation of weather
- Reduction in water level
- The world and weather is changing
- Negative change in fish feeding habit
- Flood
- Present of climate is affecting hatchery
- Fish habit

Source Survey Data by Authors, 2017

Boxes 3 and 4: How changes in climatic variables affect fish production

Egypt

1. Death of fish
2. Reduction in production
3. Frozen of fish in winter
4. Change in production season
5. Rise in cost
6. Weakness/low appetite
7. Decrease in temperature
8. Mortality
9. No bad effect because of the good management
10. Growth rate reduction

Source Survey Data by Authors, 2017

Nigeria

1. Low production
2. Low feeding in fish
3. More stress for farmers
4. Reduction in stocking density
5. Colour of water changes
6. Reduce growth rate
7. Flooding
8. Death of fish

Source Survey Data by Authors, 2017

Fish Farmers Perception on Climate Change in the Study Areas

Figure 3 indicates the descriptive representation of the perception of the fish farmers in Egypt on climate change. The results revealed that fish farmers have different opinions to both increasing and decreasing climate variables. 47% agreed (A) that temperature increase would lead to increase fish production, 28% strongly disagree (SD), 16% disagreed (D), 9% strongly agreed (SA) while none (0%) undecided (U). Also for increase in rainfall 53% (A), 18% (SD), 15% (SA), 14% (U), 0% (D). For wind 57% (A), 17% (U), 13% (SA), 12% (SD), 0% (D) on the fact that increase wind will lead to increase fish production. 39% (A), 25% (D), 19% (SD), 13% (U), 9% (SA) that increase in solar radiation will also result in increased fish production. For increase in waves and tides as a parameter for increased fish

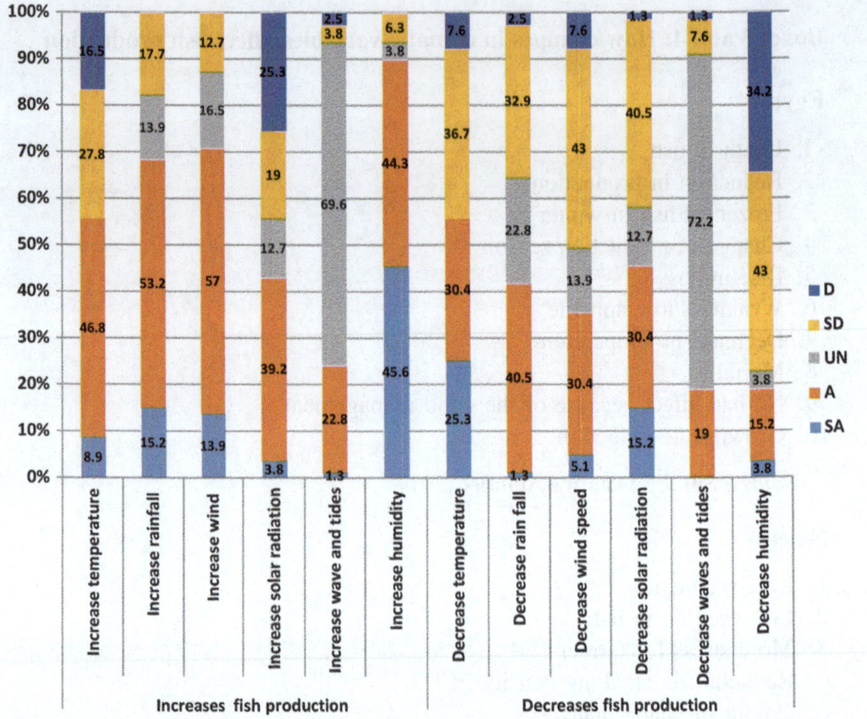

Fig. 3 Descriptive responses of the fish farmers in Egypt. *Source* Computed Survey Data by Authors, 2017

production in Egypt, 70% (U), 23% (A), 4% (SD), 3% (D) while 1% (SA). For humidity, 46% (SA), 44% (A), 6% (SD), 4% (U) while 0% of the respondents disagreed that increase humidity will lead to increase fish production.

Conversely, the same perception questions were also modelled to understand the opinions of the fish farmers as regards decrease in the climate variables. Fish farmers responses are presented as follows: Temperature; 37% (SD), 30% (A), 25% (SA), 8% (D) while 0% (U). Rainfall; 40% (A), 33% (SD), 23% (U), 3% (D) while 1% (SA). Wind; 43% (SD), 30% (A), 14% (U), 8% (D), 5% (SA). Solar radiation; 41% (SD), 30% (A), 15% (SA), 13% (U), 1% (D). Waves and tides; 72% (U), 19% (A), 8% (SD), 1% (D), 0% (SA). Humidity; 43% (SD), 34% (D), 15% (A), while the percentages of respondents that strongly agreed (SA) and undecided (U) were 4% each.

Furthermore, Fig. 4 shows the descriptive responses of the fish farmers in Nigeria to some perception questions. The result on the perception of the fish farmers on 'increase temperature leads to high fish production' divulged that 39 and 30% of the fish farmers agreed and strongly agreed that increase in temperature would lead to increase in fish production. 18 and 6% disagreed and strongly disagreed that increase temperature would increase fish production. Only 8% of the fish farmers were

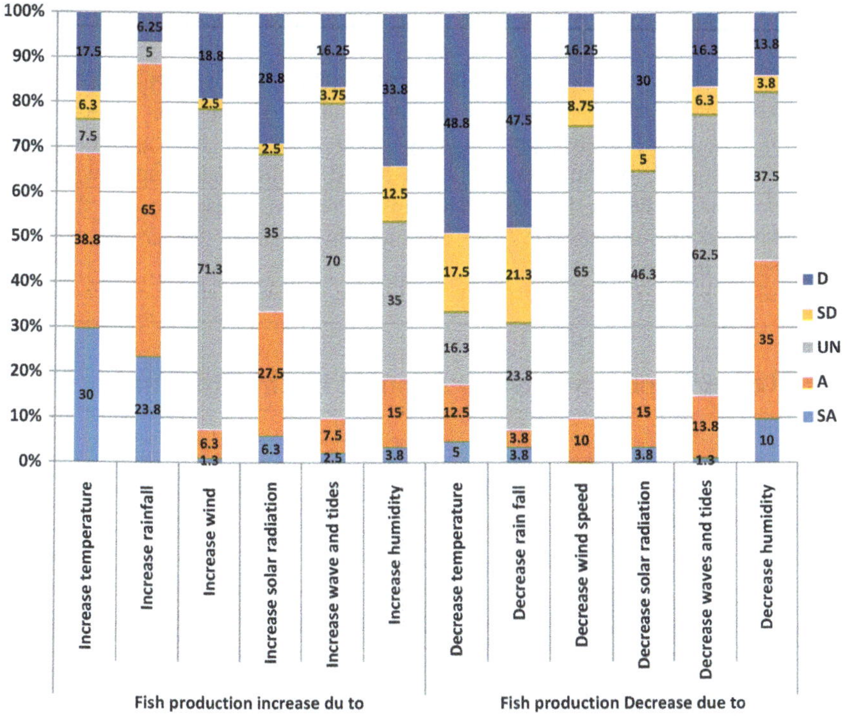

Fig. 4 Descriptive responses of the fish farmers in Nigeria. *Source* Computed Survey Data, 2017

undecided. For perception question on increase rainfall on fish production, 65% strongly agreed (SA), 24% agreed (A), 6% disagreed (D), only 5% were undecided. 71% Undecided (U), 19% (D), 6% (A), 3% (SD) while only 1% (SA) that increase in wind leads to high fish production. In the same vein, 35% (U), 29% (D), 28% (A), 6% (SA) while 3% (SD) on the fact that increase in solar radiation leads to high fish production. 70% (U), 16% (D), 7% (A), 4% (SD), 3% (SA) on the perception question that increase in wave and tides would lead to high fish production. The fish farmers also show their responses on the effect of increase in humidity on fish production; 35% (U), 34% (D), 15% (A), 12% (SD) while 4% (SA).

On the other hand, 49% (D), 18% (SD), 16% (U), 12% (A) while 5% (SA) on the acuity that decrease in temperature leads to high fish production. The fish farmers' responses on the perception statement on 'decrease in rain fall leads to high fish production' spectacles that 47% (D), 24% (U), 21% (SD), 4% (A) while 4% (SA). The fish farmers perception on the fact that decrease in wind speed leads to high fish production is as follows; 65% (U), 16% (D), 10% (A), 9% (SD), 0 (SA). Also, their perception on decrease in solar radiation leads to high fish production showed that 46% (U), 30 (D), 15% (A), 5% (SD) while 4% (SA). Likewise, 62% (U), 16% (D), 14% (A), 6% (SD) while only 2% (SA) that decrease in waves and tides leads to high fish production. The percentages 37% (U), 35% (A), 14% (D),

10% (SA) and 45 (SD) shows the respondents' perception on decrease in humidity leads to high fish production.

Previous studies and empirical evidences showed that African communities have been experiencing negative impacts of climate change as a result of some pronounced and multiple stressors (poverty, food insecurity and unemployment) on their livelihood (Nhemachena et al. 2014). It implies that despite what their perceptions might be on the climate variables there are some stressors that need individual and institutional attention to contest the negative effects/impacts of climate change in Africa.

Adaptation/Coping Strategies Adopted by the Fish Farmers

The respondents have taken time to find to arrive at some sustainable adaptation strategies to cope with the changing effect of climate variables. Boxes 5 and 6 explicitly highlight the coping strategies to ameliorate the effect of climate change in aquaculture/fish farming in Egypt and Nigeria respectively. Most of the respondents stated some intentional and inadvertent adaptive strategies which include; low stocking density, change feeding pattern and time, pumping or alternative water source, study/observing weather before feeding, minimizing feeding in the morning and more in the evening, good feeding, add additive/treatment on the fish feeds, add lime and salt for 40–45 days, continuous changing of water, increase the depth of the pond to increase water level, deal with specialist, harvest before time, the use of organic fertilizer, change water/aerations, fertilizer to keep water parameters, stop/prevent feeding, stocking with larger size of fish, green house to control the temperature, use superphosphate and lime, water recycling and adequate pond preparation. Although, as revealed by the study, some of the fish farmers do not have any method of adaptation. Most of these coping strategies were as a result of either personal effort on discovery or research result. Climate variables are changing, coping strategies remain the only option to stay balance with the changes. The significant difference in the responses of the fish farmers implies that there are various current inadvertent and intentional adaptation strategies adopted for maximum fish production as supported by Adeleke and Balogun (2013) and Nhemachena et al. (2014).

Meanwhile, 60.7 and 63.8% of the fish farmers in Egypt and Nigeria respectively believed that the change will continue and persist in the next 10–20 years. Basically 39.2 and 36.3% of the respondents in Egypt and Nigeria did not believe or agree with the fact that the changes will continue in the next 10–20 years. As perceived by the fish farmers some reasons for the persistent in climate change were given as presented in Boxes 7 and 8. It implies that the respondents were aware of climate change and that the change will persist adversely unless it is checkmates by sustaining adaptation measures in both countries. 89.9 and 72.5% of the respondents in Egypt and Nigeria respectively believed that climate change is a future threat phenomenal in aquaculture in Africa (Table 6). Several reasons were highlighted by the fish farmers on climate change future pose in the countries (Table 5).

Boxes 5 and 6: Fish farmers coping strategies on climate change adaptation

Egypt

1. No plan
2. Changing water
3. No solution
4. Reduction in nutrition/feeding
5. Add additive/treatment on the fish feeds
6. Add lime and salt for 40–45 days
7. Continuous changing of water
8. Increase the depth of the pond to increase water level
9. Deal with specialist
10. Raise water level
11. Harvest before time
12. The use of organic fertilizer
13. Change water/aerations
14. Fertilizer to keep water parameters
15. Stop/prevent feeding
16. Stocking with larger size of fish
17. Green house to control the temperature
18. Use superphosphate and lime
19. Water recycling
20. Adequate pond preparation

Source Survey Data by Authors, 2017

Nigeria

1. No method
2. Change in feeding time
3. Low stocking density
4. Change feeding pattern and time
5. Pumping or alternative water source
6. Changing water daily
7. Study/observing weather before feeding
8. Minimizing feeding in the morning and more in the evening
9. Stop stocking
10. Fresh water and good feeding
11. Removing fish to other ponds
12. Applying alternative means

Source Survey Data by Authors, 2017

Boxes 7 and 8: Perceived reasons for future climate change in the study areas

Egypt

1. Reduction in production
2. Fish death
3. Spread of diseases
4. Good management
5. No good environment to raise fish
6. Change in weather leads to mortality
7. Change in humidity
8. Appearance of diseases
9. Change in production season
10. Fish with high weight die
11. Suffocation of fish
12. Change in the environment requirement to raise fish
13. Experience in the last 4 years
14. Ammonia level

Source Survey Data Authors, 2017

Nigeria

1. Can't predict
2. Change is constant and paramount
3. Industrial/air pollution
4. It is natural
5. Think it happens every 5 years
6. It takes time before it changes
7. No adequate preparation to checkmate the change
8. Urban expansion
9. Ways of doing things is affecting the weather
10. Change will always occur but nobody knows when
11. Human activities
12. Everything is subject to change
13. Change is a must
14. Strategies and moderation
15. Changes due to ways we exploit our resources
16. Man can't control it
17. Climate variable always change after sometimes
18. Things are not stable
19. Less awareness of the change and how to combat them

Source Survey Data by Authors, 2017

Box 9: Reasons for climate change future threat in the study area

Egypt/Nigeria

1. Global warming
2. Change in ozone whole
3. Previous years' experience
4. Pollution of industries
5. Following changes in water
6. Depletion in the ozone layer
7. Death of fish in previous year and impact on fish
8. Increase in cold to extreme level
9. Farm near to Burullus lake resulting to bad impact
10. It is worsen every year
11. Interference of humans and contamination of water
12. According to international studies
13. Increase in weather temperature every year
14. We can't predict
15. God knows
16. I don't know
17. Don't know the future

Source Computed Survey Data by Authors, 2017

Critical Issues/Limitation and Institutional Framework Supports

Despite climate change issues in Africa, only 29% of the fish farmers in Nigeria attest to the change in fish production in recent years. The reason for the above fact might be as a result of the type of fish cultured in both countries. In Nigeria, Catfish is the most common fish species cultured in the country. Due to it prevailing features and characteristics which include: hardiness, resistance, freshwater habitat and small space among others the fish farmers were not noticing any adverse effect of climate change on their fish production and most of the fish farmers (50%) produces in large scale. Tilapia species is the fish commonly cultured in Egypt, the fish can thrive well in brackish water and enjoy larger spaces. Most of the production is in large scale (99%). As observed recently, Egypt is now experiencing pronounced winter (December-March). Tilapia cannot survive the temperature below 13 °C. and 96% of the respondents attested to recent changes in fish production.

Fish production in both countries cannot meet the demand of the consumers as presented on Table 7. The reason might be to the fact that the population explosion in both countries especially in Nigeria contributed to the inadequate supply. Also, the cost of acquiring the inputs especially fish feed is skyrocket. Other challenges as experienced by the fish farmers include; finance, cost of labour, water shortages and quality, fish diseases and vices/mugging. Most of the respondents do not belong to any cooperative society and no government assistance as regards subsidy and extension agents. Most of the fish farmer produces in small scale.

Conclusion

This study examined the perceptions and adaptation of fish farmers to climate change impacts in Africa; Egypt and Nigeria as a case study. The study revealed that climate change has both positive and negative impacts on African aquaculture. The fish farmers perception to climate change varies and the adaptation strategies employed were both inadvertent and intentional methods depending on the extent of the impacts. Unlike other agricultural sectors, climate change is a recent issue in aquaculture and the change will persist adversely if adequate management measures are taken. The result of the socio-economic characteristics of the respondents revealed that over 70% of the fish farmers were in the active and productive age bracket of 26–55 years, 92% were male, 83% were married, and 64% had their level of education above secondary school. The mean and standard deviation of the number of person in the household were 4 and 1.6 respectively. In Nigeria, 61% of the respondents rely on stream and river for their source of water and 54% had other occupation apart from taking aquaculture as their major occupation. Fish production and supply in the area could not meet up with the demand of the consumers as a result of financial constraints, low capital investment, high cost of fish stocking and feeding, dry season and pollution, only 32% were ready to take precautionary measures. Although, 93% of them have the ideal of climate change, 64% believe that the change will linger and persevere in the next 10–20 years. Egypt is the only African country to have achieved the scale of expansion observed elsewhere in the world. However, efforts should be geared toward regional integration in order to encourage aquaculture practices in other part of Africa and climate change investment should be encouraged.

Based on the results of this study, climate change will pose future threat on aquaculture, it is therefore expedient that efforts should be geared on improving the institutional supports in addressing critical issues facing the fish farmers. Efforts should be geared by the Governmental and Non-Governmental Organizations (NGOs) toward regional integration in order to encourage aquaculture practices in other part of Africa and climate change investment should be encouraged.

Acknowledgements CGIAR Research Program on Climate Change Agriculture and Food Security (CCAFS) and CGIAR Research Program on Fish Agri-Food Systems (FISH) for their assistance in supporting preparation of the paper and my participation in the Conference.

Appendix

See Tables 8 and 9.

Table 8 Fish farmers perception on climate change in Egypt, 2017

S/N	Perception questions	Strongly Agree		Agreed		Undecided		Strongly Disagree		Disagreed	
	PQ	Freq.	%	Freq.	%	Freq.	%	Freq.	%	Freq.	%
1.	Increase in temperature leads to high fish production	7	8.9	37	46.8	0	0.0	22	27.8	13	16.5
2.	Increase in rainfall leads to high fish production	12	15.2	42	53.2	11	13.9	14	17.7	0	0.0
3.	Increase in wind leads to high fish production	11	13.9	45	57.0	13	16.5	10	12.7	0	0.0
4.	Increase in solar radiation leads to high fish production	3	3.8	31	39.2	10	12.7	15	19.0	20	25.3
5.	Increase in wave and tides leads to high fish production	1	1.3	18	22.8	55	69.6	3	3.8	2	2.5
6.	Increase in humidity leads to high fish production	36	45.6	35	44.3	3	3.8	5	6.3	0	0.0
7.	Decrease in temperature leads to high fish production	20	25.3	24	30.4	0	0.0	29	36.7	6	7.6

(continued)

Table 8 (continued)

S/N	Perception questions	Strongly Agree		Agreed		Undecided		Strongly Disagree		Disagreed	
	PQ	Freq.	%	Freq.	%	Freq.	%	Freq.	%	Freq.	%
8.	Decrease in rain fall leads to high fish production	1	1.3	32	40.5	18	22.8	26	32.9	2	2.5
9.	Decrease in wind speed leads to high fish production	4	5.1	24	30.4	11	13.9	34	43.0	6	7.6
10.	Decrease in solar radiation leads to high fish production	12	15.2	24	30.4	10	12.7	32	40.5	1	1.3
11.	Decrease in waves and tides leads to high fish production	0	0.0	15	19.0	57	72.2	6	7.6	1	1.3
12.	Decrease in humidity leads to high fish production	3	3.8	12	15.2	3	3.8	34	43.0	27	34.2

Table 9 Fish farmers perception on climate change in Nigeria, 2017

S/N	Perception questions	Strongly agree		Agreed		Undecided		Strongly disagree		Disagreed	
	PQ	Freq.	%	Freq.	%	Freq.	%	Freq.	%	Freq.	%
1.	Increase in temperature leads to high fish production	24	30.0	31	38.75	6	7.5	5	6.25	14	17.5
2.	Increase in rainfall leads to high fish production	19	23.75	52	65.0	4	5.0	0	0	5	6.25
3.	Increase in wind leads to high fish production	1	1.25	5	6.25	57	71.25	2	2.5	15	18.75

(continued)

Table 9 (continued)

S/N	Perception questions	Strongly agree		Agreed		Undecided		Strongly disagree		Disagreed	
	PQ	Freq.	%	Freq.	%	Freq.	%	Freq.	%	Freq.	%
4.	Increase in solar radiation leads to high fish production	5	6.25	22	27.5	28	35.0	2	2.5	23	28.75
5.	Increase in wave and tides leads to high fish production	2	2.5	6	7.5	56	70.0	3	3.75	13	16.25
6.	Increase in humidity leads to high fish production	3	3.75	12	15.0	28	35.0	10	12.5	27	33.75
7.	Decrease in temperature leads to high fish production	4	5.0	10	12.5	13	16.25	14	17.5	39	48.75
8.	Decrease in rain fall leads to high fish production	3	3.75	3	3.75	19	23.75	17	21.25	38	47.5
9.	Decrease in wind speed leads to high fish production	0	0.0	8	10.0	52	65.0	7	8.75	13	16.25
10.	Decrease in solar radiation leads to high fish production	3	3.75	12	15.0	37	46.25	4	5.0	24	30.0
11.	Decrease in waves and tides leads to high fish production	1	1.25	11	13.75	50	62.5	5	6.25	13	16.25
12.	Decrease in humidity leads to high fish production	8	10.0	28	35.0	30	37.5	3	3.75	11	13.75

Source Computed Survey Data by Authors, 2017

References

Adeleke, M. L., & Matthias, W. (2016). Adaptation of the Artisanal Fisher folks to climate change in the coastal region of Ondo State, In W. Leal Filho (Wd.), *Innovation in climate change adaptation, climate change management*. Nigeria: Springer. https://doi.org/10.1007/978-3-319-25814-0_13.

Adeleke, M. L., & Balogun, A. A. (2013). Occurrence of climate change and its effects on the fishing activities in the coastal region of Ondo State, Nigeria. *European-American Journals, IJDES-101*; *International Journal of Development and Economic Sustainability, 1*(1).

Cherif, S., Walter, L. F., Azeiteiro, U. M. (2017). The role of farmers' perceptions in coping with climate change in Sub-Saharan Africa. *International Journal of Global Warming, 12*(3/4), 483–494. http://dx.doi.org/10.1504/IJGW.2017.10005907.

FAO, Fisheries and Aquaculture Information and Statistics Service—web site (http://www.fao.org/fishery/statistics/global-aquaculture-production/query/en).

FAO. (2000). Fishery country profile, Nigeria. Food and Agriculture Organization of the United Nations, Rome, Italy. FID/CP/NIR Rev. 4.

Food and Agricultural Organisation. (2008a). Climate change, energy and food high. In *Level Conference on Food Security: The Challenges of Climate Change and Bio-energy: Climate Change for Fisheries and Aquaculture*. Technical Background Document HLC/08/BAK/6.

Food and Agricultural Organisation. (2008b). Food and agriculture organization report of the FAO expert workshop on climate change implications for fisheries and aquaculture, April 7–9, 2008.

Food and Agriculture Organization (FAO). (2008c). Report of the FAO expert workshop on climate change implications for fisheries and aquaculture: Rome, Italy, April 7–9, 2008. FAO Fisheries Report no. 870 (Food and Agriculture Organization of the United Nations, 2009.2008);http://tiny.cc/fisheries7.

FAO. (2012). The state of World Fisheries and Aquaculture 2012. FAO: Rome, Italy.

French, G. T., Awosika, L. F., & Ibe, C. E. (1995). Sea level rise and Nigeria: Potential Impacts and consequences. *Journal of Coastal Research,* Special Issue 17. Level Conference on Food Security: The Challenges of Climate Change and Bio-energy: Climate Change for Fisheries and Aquaculture Technical Background Document HLC/08/BAK/6.

Intergovernmental Panel on Climate Change (IPCC). (2007a). Climate change 2007: Impacts, adaptations and vulnerability. In M. L. Parry, O. F. Canzian, J. P. Palutikof, P. J. van der Linden, J. Paul & C. Hanson (Eds.), *Contribution of working group II to the 4th assessment report of the IPCC*. Cambridge: Cambridge University Press.

Intergovernmental Panel on Climate Change (IPCC). (2007b). Climate change: impacts, adaptation and vulnerability. In Working group II contribution to the intergovernmental panel on climate change. Summary for policy makers. IPCC Secretariat, Geneva, Switzerland.

Maddison, D. (2007). The perception of and adaptation to climate change in Africa, Development Research Group, Sustainable Rural and Urban Development Team, the World Bank, Washington, DC. (Policy Research Working Paper 4308).

National Aquaculture Sector Overview. Egypt. National Aquaculture Sector Overview Fact Sheets. Text by Salem, A. M., Saleh, M. A. In: FAO Fisheries and Aquaculture Department [online]. Rome. Updated 16 November 2010. Cited 17 April 2015. http://www.fao.org/fishery/countrysector/naso_egypt/en.

Nhemachena, C., Mano, R., Mudombi, S., & Muwanigwa, V. (2014). Perceptions on climate change and its impact on livelihoods in Hwange district, Zimbabwe. *Jàmbá: Journal of Disaster Risk Studies, 6*(1), Art. #123, 6. http://dx.doi.org/10.4102/jamba.v6i1.123.

Nyanga, P. H., Johnsen, F. H., Aune, J. B., & Kalinda, T. H. (2011). Smallholder farmers' perceptions of climate change and conservation agriculture: Evidence from Zambia. *Journal of Sustainable Development, 4*(4), 73–85. https://doi.org/10.5539/jsd.v4n4p73.

OECD-FAO. (2011). Agricultural outlook 2011–2020. OECD: Paris, France. National Population Commission (NPC), Government Notice of No.21 on 2006 Census, (2007), Federal Government Printer, Lagos, Nigeria.

Rao, K. P. C., Ndegwa, W. G., Kizito, K., & Oyoo, A. (2011). Climate variability and change: Farmer perceptions and understanding of intra-seasonal variability in rainfall and associated risk in semi-arid Kenya. *Experimental Agriculture, 47,* 267–291. https://doi.org/10.1017/S0014479710000918.

Slegers, M. F. W. (2008). "If only it would rain": Farmers' perceptions of rainfall and drought in semi-arid central Tanzania. *Journal of Arid Environments, 72,* 2106–2123. https://doi.org/10.1016/j.jaridenv.2008.06.011.

UNFCCC. (2015). UNFCCC (United Nations Framework Convention on Climate Change) INDCs as Communicated by Parties (2015) http://www4.unfccc.inta.

Worldometers: (www.Worldometers.info). (2017). Elaboration of data by United Nations, Department of Economic and Social Affairs, Population Division. World Population Prospects.

Environmental Migrations Without Environmental Migrants? Perceptions and Policies on Environmental and Mobility Issues

Inês Vieira

Abstract Environment, migration and asylum are among the most socially constructed issues of the 21st century. The intersection of these issues has recently gained institutional recognition in diverse UN frameworks and conventions. Despite the reiterated importance to research and develop policies on environmental migrations, the identification of "environmental/climate refugees/migrants" has been discouraged (IPCC, Foresight), considering the difficulty of isolating the environmental factor within mobility drivers, and the non-identification with such labels by some of the people which could be paradigmatically affected by environmental and climate changes. This paper brings into discussion the results of qualitative Ph.D. research work (the author benefited from a Ph.D. grant by FCT from 2011 to 2015 (SFRH/BD/68730/2010). This article was made with the support of CICS. NOVA—Interdisciplinary Centre of Social Sciences of the Universidade Nova de Lisboa, UID/SOC/04647/2013, with the financial support of FCT/MCTES through National funds) developed with Ethiopians and Eritreans in Italy and Cape Verdeans in Portugal. The aim is to explore the reflexivity of migrants and refugees from countries with environmental risk, observing their representations of environmental factors in the countries of origin, interrelated with other drivers of mobility (mainly economic and political), and the framing within their experiences of mobility towards Southern Europe. Putting these issues into perspective can be useful for the reflection about policies addressed to environmental migrations, which will be reviewed in this paper, highlighting some gaps between global political aims and meanings of mobility and the environment for migrants and refugees.

Keywords Africa · Europe · Environmental migrations · Migrants and refugees Policy development · Qualitative research

I. Vieira (✉)
Human Ecology, CICS.NOVA, Interdisciplinary Centre of Social Sciences, FCSH-UNL, Av. de Berna, 26-C, 1069-061 Lisbon, Portugal
e-mail: ines.vieira@fcsh.unl.pt

F. Alves et al. (eds.), *Theory and Practice of Climate Adaptation*,
Climate Change Management, https://doi.org/10.1007/978-3-319-72874-2_17

Introduction

Environment, migration and asylum are among the most socially constructed issues of the 21st century. The growing importance of environmental issues occurs in a context of affirming the environment as a global social problem. Although the "global" label is only attributed to certain environmental issues (namely climate change), visualizing the world as a whole, a lone planet as seen from space in the 1970s, has been important for developing consciousness of a world that needs to be protected (Yearley 2009). The present context is also one of enlargement of the political scheme to regimes of global governance that unite States, non-governmental organizations and communities of knowledge around issues of environmental risk (Beck 2015).

Within the "globalized themes" of the 20th and 21st centuries, the environment occupies a remarkable place—let us recall, for example, the Paris Agreement of 2015, the first agreement to be signed by all 196 parties of the 21st Conference of the Parties (COP21) of the United Nations Framework Convention on Climate Change (UNFCCC 2015a), turning it into the first global agreement. This process of globalizing environmental issues must be read through the success of the concept of sustainability and of the project of ecological modernization, which enabled the highlighting of opportunities to develop environmental policies along with economic modernization and technological innovation, in a framework of moderate political action and institutional reform (Giddens 2015; Hajer 1995).

Regarding the globalization of mobilities it is possible to observe a double emergence. On the one hand, in modern society, the emergence of a paradigm of mobilities (Sheller and Urry 2006; Urry 2007) implied important social changes, attributing new meanings to space and time and generating the multiplication and integration of complex systems of mobilities. It is important, nonetheless, to observe that this globalization generalized the unequal possibility to move (Baptista 2012; Kaufmann et al. 2004). On the other hand, despite the remarkable growth of international migrations[1] (Castles et al. 2014; King et al. 2010), it is estimated that only around 3% of the world population are international migrants, corresponding to 214 million people in 2010 (for the year 2000, the percentage was 2.9% corresponding to 175 million people, and in 1965 it was 2.3%, corresponding to 75 million people) (King et al. 2010, pp. 40–41). Of these, only around 10% are refugees, or 0.3% of total population—21.1 million people[2] out of the total world population of 7.4 billion (de Haas 2016, paragraphs 4–5). So, the percentage of international migrations is generally stable in the last decades (despite the many differences between regions and countries), notwithstanding its social and territorial impacts and its growing construction as a social problem.

[1] Among other experiences of mobility, from which we can highlight the growing mass experience of tourism (Urry 1995).

[2] The author includes both refugees under UNHCR's mandate (around 16.1 million) and Palestinian refugees (around 5 million).

Recently, the consensual acceptance at the United Nations of the New York Declaration for Refugees and Migrants (UN 2016) can be considered an important moment for the globalization of mobilities, particularly at the political horizon of protecting the rights of migrants and refugees and sharing responsibilities for this matter at the global scale. Similar to environmental topics, issues of human mobilities have witnessed the internationalization of their political actors and structures. Yet, it is important to observe that a truly global regime for mobilities would conflict with the territorial and frontier management model of Nation-states in effect since the French revolution (Mazzella 2014).

The intersection of environmental and mobility issues has also gained international institutional recognition. On a more longitudinal level, we can consider the long-lasting development of policies related to issues of environmental migrations in the succession of international processes for migration, environmental and climate issues since the beginning of the 1970s (Ionesco et al. 2016). More recently, though, these issues have been directly identified in diverse UN frameworks and conventions, mainly regarding climate change (UNFCCC 2011, 2015a, b) and desertification (UNCCD 2016). Another important mention is the creation of a specific division[3] in the International Organization for Migration, the leading intergovernmental migration agency, that aims "to bring environmental migration to the heart of international, regional and national concerns, in collaboration with its Member States, observers and partners."

Despite the reiterated importance of researching and developing policies on environmental migrations, the identification of "environmental/climate refugees/ migrants" has been discouraged (IPCC 2014; Foresight 2011), considering the difficulty of isolating the environmental factor within mobility drivers, and the non-identification with such labels by some of the people who could be paradigmatically affected by environmental and climate changes (for example, inhabitants of Tuvalu—Farbotko and Lazrus 2012; Mortreux and Barnett 2009). On the one hand, for empirical research, this opens a breach in which to explore the perceptions of migrants and refugees on environmental factors along with other drivers of mobility. This research can be developed with people coming from countries with slow onset issues like those related to water stress, without the need to focus solely on those coming from contexts with immediate natural catastrophes, thus allowing the exploration of different types of environmental issues. On the other hand, regarding policy development, specific legal instruments might be needed for situations of forced migration and for cases of loss of sovereignty over territory (as in Small Island States), reinforcing the need to improve governance mechanisms for the adaptation to—and mitigation of—climate changes, as well as to build resilience for the protection and empowerment of climate-induced migrations (IPCC 2014, p. 771).

[3]IOM Migration, Environment and Climate Change Division, website consulted on 21/04/2017 (http://environmentalmigration.iom.int/iom-and-migration-environment-and-climate-change-mecc).

This leads us to three questions: Can we think about environmental migrations without identifying environmental migrants? Which are the gaps between global political aims and meanings of mobility and the environment for migrants and refugees? How can this discussion contribute to policies addressed to environmental migrations?

Methodological Approach

This paper brings into discussion the results of qualitative Ph.D. research work developed with Ethiopians and Eritreans in Italy and Cape Verdeans in Portugal. The aim of this work was to explore the reflexivity of migrants and refugees from countries with environmental risks, observing their representations of environmental issues in their countries of origin, the interrelations with other drivers of mobility (mainly economic and political), and the framing within their own experiences of mobility towards Southern Europe.

Considering these objectives, the chosen method for the main part of this research was the interview, which can capture different senses that social actors attribute to their practices and experiences, as well as retrace past processes and occurrences. Gathering information through semi-structured interviews allows for an interesting degree of depth and flexibility without loosing the horizon of the interview guide that stabilizes its scope (Quivy and Campenhoudt 2008).

The interview guide was built before holding the interviews, following a phase of theoretical reading and review of empirical studies (Vieira 2010). The guide included six sections: (1) identification of the interviewee; (2) former pathways of mobility; (3) social networking; (4) work on the land and strategies of adaptation in the context of origin; (5) factors in the context of origin that influenced migration; (6) memories and present life in Italy and Portugal. The guide was written in Portuguese and Italian and, when needed (particularly with recently arrived refugees in Italy), was translated into English.

Sampling was initiated through contact with migrant associations and through participant observation in significant spaces for Ethiopian and Eritrean groups in Bologna, Italy, and for Cape Verdean groups in Lisbon, Portugal. Interviews were conducted using a research process of contextualization and interaction followed by verification that suggested saturation of the information gathered (Ferrarotti 2011).

Following the aforementioned guide, 32 interviews were held in Bologna, Italy (19 refugees and 13 migrants interviewed in 2012) and 22 interviews were held in Lisbon, Portugal (22 migrants, of which 9 were interviewed in 2013 and 13 in 2015). Interviews had a medium length of one hour and were held in places chosen by each interviewee: associations, shops, cafes and bars, gardens and other public spaces, work and residential places.

All interviews were preceded by an introductory presentation of the research project and subject to a written consent declaration to hold the interview and to maintain the anonymity of interviewees. Whenever authorized, interviews were audio recorded and transcribed. Interviews were subject to content analysis with a thematic focus, proceeding to an analytical synthesis of the gathered texts through a regrouping of content into structural categories of analysis (Bardin 2004). The software MAXQDA was used to frame a system of codes and sub-codes.

Besides this primary source of data, for the specific aim of this article it was necessary to consult secondary sources to update the political framing of environmental migrations. Secondary sources include documentation on international agreements and declarations, and the wide-ranging *Atlas des Migrations Environnementales* by Ionesco et al. (2016).

Regarding the theoretical framework, this research derived from a Ph.D. program on Human Ecology in FCSH/Universidade Nova de Lisboa, Portugal. This approach to Human Ecology, on the one hand, is sustained in the urban ecological Chicago School (Park et al. 1992; Wirth 1938), interpreting the city as having an innovative character as human habitat, generating new social dynamics, and generating an environmental effect considered determinant on society. On the other hand, it also embeds an approach to socially focused environmental sciences (Catton and Dunlap 1978; Hawley 1950) that proposes to look not only at the adaptation of human beings to environmental factors, but also at the ways in which humans impact the environment.

Two concepts are considered central in the theme of this work: mobilities and environment, which are considered allusive to the social production of movement and nature, in a context of modern abstraction of space and time (Cresswell 2006) and predominantly on a constructivist reading of environmental problems. Such concepts are considered against a background of growing affirmation of the environment as a global social problem (Yearley 2009) and of rising assertions about a paradigm of mobilities, including but not limited to international migrations (Sheller and Urry 2006; Urry 2007). Environmental issues are taken as part of different potential drivers of multi-causal migrations (Black et al. 2011).

The Role of Environmental Issues for the Mobility of Migrants and Refugees: Results from the Interviews with Ethiopians and Eritreans in Italy and Cape Verdeans in Portugal

The interviews referred to in the previous section, held with Ethiopian and Eritrean refugees and migrants and with Cape Verdean migrants, revealed different perceptions on the role of environmental issues for mobilities. In this section, the results of this research will be briefly presented regarding general perceptions about environmental issues in the countries of origin, interrelated with other drivers of mobility, and considering interviewees' experiences of mobility towards Southern Europe.

Table 1 General perceptions about environmental issues in the countries of origin

Problems in the physical environment	Problems of soil quality and water quality and quantity/drought
	Slow environmental degradation
	Abrupt natural disasters
Environmental issues and "visible" social problems	Drought, food production and hunger
	Health issues related to environmental problems
Environmental problems associated to a human cause	Deforestation (for infrastructural development; as a consequence of war)
	Environmental problems related to urban development or resource overuse
Environmental/territorial management	Issues related to land-use planning
	Resource management and problems of access to natural resources
Environment and social change	Education and growing actions on environmental and climate issues
	Urbanization and less dependence on the environment
	Hope in ecological modernization

To provide a general view on the perceptions about environmental issues in the countries of origin (Table 1), five main dimensions have been identified.

Most of the interviewees directly identified problems in the physical environment in some regions of Ethiopia, Eritrea and Cabo Verde, particularly regarding drought and effects associated with slow environmental degradation. Soil and water quality and quantity were perceived as problems mainly by the interviewees that had concrete memories of agriculture (mainly for subsistence, as was the case for the majority of the interviewed Cape Verdeans). Occasional abrupt natural disasters were also identified, related to floods (in the aftermath of droughts), parasites and volcanic eruption (the latter only in Cabo Verde), which some interviewees associated with short-term displacement (rather than long-lasting migration).

There was a general association of environmental issues with "visible" social problems related to hunger and health issues (mainly regarding water-borne diseases). Yet, not very often have these problems been identified within the interviewees' direct experience, but rather from the media and past family memories. Despite this, the rapid and uncritical association between the interviewees' countries of origin and problems of hunger, namely by Italians regarding Ethiopians and Eritreans, has been criticized for its stigmatizing pressure and consequent impact on viewing these migrants and refugees as permanently in need of assistance, rather than as integrated citizens.

The interviewees associated certain environmental problems with a human intervention cause. Interviewed Ethiopians and Eritreans particularly referred to deforestation, which is put both for the aim of infrastructural development (changes in land-use mainly for housing, industry and mobility) and as a consequence of war (with direct effects derived from generalized deforestation to avoid ambushes, as a war strategy, as a consequence of bombing, or reflecting the lack of available workforce for agriculture and other environment-related jobs when the majority of the population is enrolled in the military service). Some Ethiopians and most Cape Verdeans also referred to the memory of deforestation in periods of intense urban development, which is linked to the overuse of certain resources (namely sand in Cabo Verde).

The question of environmental/territorial management refers to indirect causes of environmental degradation and/or unequal access to and distribution of natural resources, linking the environment to structural and political determinants rather than observing it directly in a "soft policy" *milieu*. In the periods of colonial and other types of dictatorship regimes in Cabo Verde, Ethiopia and mainly (up to the present) Eritrea, this question emerges in issues related to land-use planning, resource management and problems about access to natural resources. Regarding the subsequent breakdown of traditional resource management systems within the colonial regimes in these African countries, some of the interviewees discussed the agrarian structure and the regimes of land tenure in the context of inequitable systems of rights, namely citing memories of expropriation/expulsion and land grabbing.

In a second indirect reading of the environment, many of the collected perceptions can be linked to a context of cumulative effects of education and social change. Particularly in the case of the interviewees that held graduate degrees, education was referred to as a core factor for the rising number of public and private investments in environmental and climate issues in these countries of origin. In another reading of the effects of education, the increasing urbanization of mainly the Cape Verdean and Ethiopian societies has been linked to a lower dependency on environment/agrarian jobs and subsistence. Finally, many interviewees referenced hope in ecological modernization and technological development to solve existing problems in the environment.

An important finding is that the majority of the interviewees considered that environmental factors, despite their importance and relation to social problems in the countries of origin, are not directly related to international migrations. Only one Cape Verdean interviewee considered herself an environmental migrant, an attribute that she extended to every other Cape Verdean migrant, citing environmental issues as a root cause of the difficulty of living in her country of origin.

The majority of the interviewees noted that their main mobility drivers were more related to economic or political factors rather than environmental ones, despite their importance for their fellow citizens who remain in Ethiopia, Eritrea and Cabo Verde. Economic and political drivers are actually important poles of interpretation of environmental matters; in other words, rather than reflecting directly about environmental problems, many of the interviewees alluded to political and economic problems with reference to environmental issues.

Regarding the economic dimensions of environmental problems, the majority of respondents cited insufficient resources (namely land for agriculture and animal production) and the impact of environmental degradation on the quality of life, depending on the economic activities of each region or island. In regards to the political dimensions, the responses focused on conflicts over natural resources and issues of environmental/territorial management, including situations of land expropriation and forced displacement (mainly in Ethiopia and Eritrea).

Another important dimension of the interviewees' interpretation on the drivers of mobility and their relation to environmental issues regarded economic factors both in the countries of origin and in the countries of arrival. Despite the deluded expectations of higher income and quality of life in the countries of arrival, many of those interviewed mentioned the importance of the possibility to send remittances after migrating, thus alleviating their families from poverty and food and water insecurity. Remittances, like international aid, are seen as economic possibilities for coping with the effects of environmental issues. Additionally, some of the interviewees (mainly Ethiopians and Eritreans) mentioned that the travel cost of the long migration reduces the likelihood of mobility for those most impoverished by environmental issues. On the other hand, long-lasting mobilities in the same post-colonial pathways benefit from the existence of previously established networks, which have been identified by some Cape Verdean interviewees as arising from times of cyclical drought, hunger and lack of work, leading to many generations of international migrants.

An important point must be made about the framing of migrations for both migrants and refugees in their countries of origin, during transit and upon arrival. The experience of refugees is of particular concern: Ethiopians and Eritreans share memories of long and traumatizing pathways from their countries of origin (from where they had to escape, in the Eritrean case migration being broadly forbidden), then crossing Sudan, Libya and the Mediterranean Sea. Once arrived in Italy and considering that all interviewees had their asylum requests confirmed (as refugees or under subsidiary or humanitarian protection), many complained about the sense of immobility deriving from the Dublin regime, which requires refugees to remain in the country that accepted their asylum request despite various obstacles to their integration in that society.

Not only for refugees, but also for migrants, mostly those in the least favourable sectors of socio-economic life, there are important recurrent issues of integration in the countries of arrival, mainly regarding labour, housing, social security and language. In most interviews, participants stated that after arrival, these issues became the focus of their social problems, with significance for the present, and also for the growing distance regarding memories of their countries of origin (of environmental factors, among others)—memories that some interviewees mentioned as fading and being lost over time.

Policies Addressed to Environmental Migrations

Putting all these issues into perspective can be useful for reflecting about policies addressed to environmental migrations, highlighting some gaps between global political aims and meanings of mobility and the environment for migrants and refugees.

In what regards the existing governance and political answers, it is possible to observe that environmental migrations have been predominantly developed in environmental policy, despite the transversal challenge and the growth of diverse political domains. So far, migrations have been included more in the agenda of environmental and climate change, and to a lesser degree the environment has been included in the agenda of migrations. The *Atlas des Migrations Environnementales* by Ionesco et al. (2016) provides a wide view of the development of this policy field.

As a first thought, despite its main development during the 21st century, environmental migration may be partially framed in previous, non-binding international legal devices, particularly those sustained by the Universal Human Rights. It is possible to observe the increase in international political processes since the 1970s, particularly the consolidation of the political agenda around issues such as global governance of the environment, biodiversity, climate, desertification, humanitarian response, migration, development and reduction of the risks of catastrophe (Ionesco et al. 2016, pp. 122–123). Nonetheless, the Atlas authors suggest that existing juridical mechanisms need to be revised and updated, or even replaced by a specific binding framework on environmental migrations.

Ionesco et al. (2016) highlight the importance of regional regulatory frameworks, counselling forums and international organizations. The regional scale is considered of particular significance in what regards the governance of migrations, at the same time being a good scale at which to develop legal solutions that protect human rights, the right to a safe environment and special rights for those internally displaced (or migrating to neighbouring countries, as is the case for most African migrants). Besides, the multiplication of regional and international forums during the 21st century contributed to maintaining this issue in the international political agenda. Many international organizations play important roles in the conceptualization, dialogue and advocacy for environmental migrations, with a pioneer role attributed to IOM and UNHCR. This growing international dimension allowed for the construction of common policy grounds and clustered operational strategies coordinated between different agencies. The international turn is particularly observable since the Cancun Adaptation Framework (CAF, within COP16, the Cancun Agreements at the 2010 Climate Change Conference in Cancun, Mexico), when adaptation was given the same level of priority as mitigation.

Adaptation became one of the key areas in the development of the field of environmental migrations. Along with the previously mentioned policy advance, seeing migration as an adaptive approach to environmental issues highlights the agency of people on the move and their potential contribution to development initiatives in their countries of origin, namely through the application of remittances, which in Sub-Saharan African countries was estimated around 40 billion dollars in 2010 (compared with 23.5 billion dollars of development aid for that region in the same year—Ionesco et al. 2016, p. 114).

When estimating the global costs of environmental migrations, different calculations must be made. It would be very difficult, and ethically complex, to provide an account of the human costs associated with uprooting, morbidity and mortality. Evaluation becomes more feasible when used to question the cost of material damage and reparation due to natural catastrophes and of the needed adaptation to environmental and climate change. The global cost of catastrophes alone has been estimated as having increased from 18.4 billion dollars in the 1960s to 903.8 billion dollars in the 2010s. In 2010, around 75% of these costs were attributed to emergency response to natural disasters, around 10% to the reduction of the risks of catastrophe and around 10% to rebuilding or rehabilitation (Ionesco et al. 2016, p. 109). The distribution of costs has changed over the last years, with increasing amounts spent on long-term prevention and adaptations, even though the funds most easily mobilised in recent years are those used for responding to emergency situations.

A growing field of study and investment, mainly between governments and reinsurance companies, regards financing mechanisms and instruments used to cope with risks and damages related to natural disasters. Approaches and mechanisms for long-term migrations associated with desertification, coastal erosion and sea level rise (slower environmental changes, rather than abrupt natural disasters) are particularly insufficient. But even when these financing mechanisms are more directed to actions of adaptation and long-term development, they are still not attributing a key role to migrations (Ionesco et al. 2016).

Migrations, indeed, tend to be seen as a failure while part of the irreversible effects of environmental and climate change, and this can be considered an important challenge during the process of rethinking policy strategies regarding environmental migrations. For example, many African, Asian and South American countries have implemented National Adaptation Programmes of Action (NAPAs[4]) in order to identify priority areas for adaptation to climate change and foster urgent plans of action, thus avoiding further vulnerability and costs related to catastrophes. Yet, many of these plans highlight a negative reading on migrations, seeing them as contributing to the exponential growth of cities and to increase the vulnerability of the affected populations (Ionesco et al. 2016, pp. 128–129).

However, migrations could contribute positively to adaptation strategies if properly integrated into a global approach; a well-planned management of human mobility could reduce the risk of catastrophes and migrations could actually play an important role in the strategies of development and adaptation to climate change (Ionesco et al. 2016, pp. 130–133). Besides presenting good practices regarding

[4]NAPAs are framed within the United Nations Framework Convention on Climate Change (http://unfccc.int/adaptation/workstreams/national_adaptation_programmes_of_action/items/7567.php).

adaptation and development, the Atlas makes three suggestions for policy development to facilitate migrations: (1) to promote the creation of juridical, financial, administrative and economic environments that foster the positive effects of migration in the countries of origin; (2) to boost planned resettlement of populations in high risk zones, ensuring that their resettlement is accompanied by policies of social and economic integration; and (3) to open regular international channels for migrations, particularly considering temporary work, seasonal or circular migrations, and extending the reach of bilateral agreements (Ionesco et al. 2016, p. 133).

In the path of the challenge to deepen the debate with a positive view of migrations, we propose to return to our introductory question: can we think about environmental migrations without identifying environmental migrants? It is important to keep in mind that it is difficult or even inaccurate to differentiate/isolate environmental migrants from other migrants (Foresight 2011), and the assumption of multi-causal migrations could contribute to going beyond the focus on environmental triggering factors. A more integrated view in this regard could possibly mean more "space" for migrants undergoing environmental migrations. Notwithstanding the importance and growing development of environmentally focused policy forums, instruments and other devices, the development of migrant-oriented policies could have many positive outcomes. Thus, even if not directly identifying "environmental migrants" but simply migrants and refugees coming from environmentally pressured areas/high risk zones, hearing their perspectives and perceptions on critical issues related to the environment and to migrations could enlighten potential directions for policy development.

Challenges and Final Remarks

At this point we propose to return to our second question: which gaps can be identified between global political aims and meanings of mobility and the environment for migrants and refugees? We have seen that the policy field of environmental migrations has been developing mainly in regards to: environmental policies; international political processes concerning environmental/climate and migration issues; advances in regional and international regulation and counselling organizations; developing financial mechanisms and instruments; providing international aid and reflecting on the distribution of the costs of environmental migrations, thus pointing to a growing long-term investment in adaptation measures to environmental/climate threats, which can include and benefit from migrations. Considering the results presented in this paper (from a small sample, yet with a valid qualitative input), some points related to the meanings of mobility and the environment for migrants and refugees can invite further inquiry.

Looking to the meanings here conferred to environmental issues, we may go beyond the interpretation of problems directly identified in the physical environment. Both slow environmental degradation and abrupt natural disasters are focuses of worry for the interviewees, but they invite us to frame this worry particularly regarding social, political and economic problems with reference to environmental issues. Therefore, even more than worrying about the natural catastrophe per se, migrants and refugees have placed their attention on the observable social problems to which it is related, or to political and economic issues that turn environmental problems into hunger, health issues, insufficient resources, and conflicts over land and natural resources, among other issues. Even though this can be considered an expected result for such research, and not a result that helps to analyse separately the different drivers of migrations (environmental, social, political, economic, etc.), it could be useful to inspire the "policy imagination" on environmental issues.

This "policy imagination" could be important, in the first instance, for intervention within the contexts of origin. We ponder the possibility of: reinforcing measures to improve the access to food, water and healthcare; supporting actions focused on the environment and climate with a long-term view of economic sustainability; and problematizing economic and political structures and strategies to better cope with regular or long-term environmental problems. Additionally, supporting the recognition of local practices and forms of knowledge about environmental and territorial issues could help with the engagement of the affected populations in adaptation and mitigation strategies. None of these proposals is necessarily in contradiction with existing actions, namely engagement in international platforms such as V20.[5]

Thinking about the meanings conferred to mobility by the interviewees, we can highlight the importance attributed to networks, remittances and integration issues in the arrival context (or, in the case of most refugees, both in transit and arrival contexts). Mobility can thus be read as the main referent in the double question of environment and migration: for many migrants and refugees, factors related to the pathway or to the country of arrival are more important (as social problems and possibilities to cope with those problems) rather than the drivers of mobility in the countries of origin (environmental drivers, among others). This reading, despite providing the needed framing of a sample that does not reflect direct victims of natural catastrophes but rather reflects the general population of environmentally pressured countries, can also pose interesting challenges for the "policy imagination".

On a general level, a focus on mobility can contribute to the existing effort to increase the consideration of migrations in a field that has been predominantly framed as an environmental issue. This could benefit, for example, from a consistent policy direction so as to ensure that environmental issues are broadly taken into consideration in the follow-up of the New York Declaration; if 193 member States of the United Nations agreed with this Declaration's expression of political will regarding large movements of refugees and migrants, a legally binding

---

[5]The Vulnerable 20 (V20) is a Group of Finance Ministers of the Climate Vulnerable Forum (CVF), constituted after the Paris Agreement in 2015, contributing to the compromise of limiting temperature growth through financing solutions for adaptation and mitigation of climate change in the most vulnerable countries (http://www.v-20.org, 27/12/2016).

consequent pact focused on mobilities but integrating environmental issues (among others) could be crucial for the global agenda of environmental migrations.

Finally, it is important to acknowledge that both environmental and mobility issues are framed in a context of social change. They can be seen as problematic areas but also as areas of political investment (considering the present study, both in the African countries of origin and in the arrival EU countries). The present context of extended education and urban growth might simultaneously mean a lower dependence on agriculture and other types of work directly related to land and natural resources, an increase in the search for qualified jobs (also through previously established international networks), an increased number of unemployed, vulnerable new urban populations, and a growing investment in ecological modernization strategies.

The way forward for the policy agenda on environmental migrations is certainly complex. It benefits from a growing political attention, accompanied by an expanding body of research, both of a quantitative and a qualitative nature. The latter (where this research project might be considered) can positively inspire the "policy imagination" (1) on environmental issues, in this case by overcoming direct environmental readings and embracing important economic and political issues related to the environment; and (2) on mobility issues, namely by highlighting migrants' and refugees' perceptions and meanings attributed to the environment and mobility issues, as well as by reinforcing the need to increase the consideration of migrants and migrations in the broad field of environmental migrations.

References

Baptista, L. (2012). Le monde des mobilités. Globalisation, territoire et capital de mobilité. In Diogo, F., Gonçalves, R. & Tomás, L. (Dir.) *Les nouvelles configurations de la mobilité humaine* (pp. 15–23). Fribourg, Switzerland: Academic Press Fribourg.

Bardin, L. (2004). *Análise de conteúdo*, Edições 70, Lisbon, Portugal.

Beck, U. (2015). *Sociedade do risco mundial: em busca da segurança perdida*, Edições 70, Lisbon, Portugal.

Black, R., Adger, W., Arnell, N., Dercon, S., Geddes, A., & Thomas, D. (2011). The effect of environmental change on human migration. *Global Environmental Change, 21S*, S3–S11.

Castles, S., Haas, H., & Miller, M. (2014). *The age of migration International population movements in the modern world* (5th ed.). Hampshire, United Kingdom & New York, USA: Palgrave Macmillan.

Catton, W., & Dunlap, R. (1978). Environmental sociology: A new paradigm. *The American Sociologist, 13*, 41–49.

Cresswell, T. (2006). *On the move: Mobility in the modern western world*. London, United Kingdom: Routledge.

Farbotko, C., & Lazrus, H. (2012). The first climate refugees? Contesting global narratives of climate change in Tuvalu. *Global Environmental Change, 22*(2), 382–390.

Ferrarotti, F. (2011). *L'empatia creatrice. Potere, autorità e formazione umana*, Rome, Italy: Armando Editore.

Foresight. (2011). *Foresight: Migration and global environmental change*. Final Project Report, The Government Office for Science, London, United Kingdom.

Giddens, A. (2015). *The politics of climate change* (2nd ed.). Polity, Cambridge, United Kingdom & Malden MA, USA.

Haas, H. (2016). Refugees: A small and relatively stable proportion of world migration. Blog *Hein de Haas*. http://heindehaas.blogspot.pt/2016/08/refugees-small-and-relatively-stable.html. Accessed 21 April 2017

Hajer, M. (1995). *The politics of environmental discourse: Ecological modernization and the policy process*. New York, USA: Oxford University Press.

Hawley, A. (1950). *Human ecology: A theory of community structure*, New York, USA: Ronald Press Co.

IPCC. (2014). 12.4 migration and mobility dimensions of human security. In *IPCC Fifth Assessment Report 2014, WP2 II 'Impacts, Adaptation, and Vulnerability'*. https://www.ipcc.ch/report/ar5/. Accessed 6 Jan 2015.

Ionesco, D., Mokhnacheva, D. & Gemenne, F. (2016). *Atlas des Migrations Environnementales*, Paris, France: SciencesPo/IOM.

Kaufmann, V., Bergman, M., & Joye, D. (2004). Motility: Mobility as capital. *International Journal of Urban and Regional Research, 28*(4), 745–756.

King, R., Black, R., Collyer, M., Fielding, A., & Skeldon, R. (2010). *The atlas of human migration, global patterns of people on the move*. Brighton, United Kigdom: Earthscan.

Mazzella, S. (2014). *Sociologie des Migrations*, Col. 'Que sais-je?'. Paris, France: Presses Universitaires de France.

Mortreux, C., & Barnett, J. (2009). Climate change, migration and adaptation in Funafuti, Tuvalu. *Global Environmental Change, 16*(1), 105–112.

Park, R., Burgess, E., & McKenzie, R. (1992). *The city—Suggestions for investigation of human behavior in the urban environment*. USA: The University of Chicago Press Chicago.

Quivy, R., & Campenhoudt, L. (2008). *Manual de Investigação em Ciências Sociais*, Gradiva, Lisbon, Portugal.

Sheller, M., & Urry, J. (2006). The new mobilities paradigm. *Environment and Planning A, 38*, 207–226.

UN. (2016). *New York declaration for refugees and migrants*. Draft resolution referred to the high-level plenary meeting on addressing large movements of refugees and migrants by the General Assembly at its seventieth session. http://www.un.org/ga/search/view_doc.asp?symbol=A/71/L.1. Accessed 16 June 2017.

UNCCD. (2016). *Report of the Conference of the Parties on its twelfth session, held in Ankara from 12 to 23 October 2015*. http://www.unccd.int/Lists/OfficialDocuments/cop12/20add1eng.pdf. Accessed 27 Feb 2017.

UNFCCC. (2011). *Report of the Conference of the Parties on its sixteenth session, held in Cancun from 29 November to 10 December 2010*. https://unfccc.int/resource/docs/2010/cop16/eng/07a01.pdf. Accessed 27 Feb 2017.

UNFCCC. (2015a). *Paris Agreement*. Authentic text. http://unfccc.int/files/essential_background/convention/application/pdf/english_paris_agreement.pdf. Accessed 16 June 2017.

UNFCCC. (2015b). *Report of the executive committee of the Warsaw International Mechanism for loss and damage associated with climate change impacts*. http://unfccc.int/resource/docs/2015/sb/eng/03.pdf. Accessed 27 Feb 2017.

Urry, J. (1995). *Consuming places*. London, United Kingdom & New York, USA: Routledge.

Urry, J. (2007). *Mobilities, polity*. Cambridge, USA: United Kingdom & Malden MA.

Vieira, I. (2010). *Migrações e ambiente. Imigrantes ambientais no contexto europeu*, Master Thesis (Project Work) on Human Ecology and Contemporary Social Problems, FCSH/NOVA, Lisbon, Portugal.

Wirth, L. (1938). Urbanism as a way of life. In LeGates, R. & Stout, F. (Eds.), (1996) *The city reader*. London, United Kigdom & New York, USA: Routledge.

Yearley, S. (2009). *Cultures of environmentalism. Empirical studies in environmental sociology*. Hampshire, United Kingdom & New York, USA: Palgrave Macmillan.

Can 3D Visualizations Really Convince Small Island Coastal Communities About the True Risks of Sea Level Rise?

Attard Claudia, Galdies Charles and Conrad Elisabeth

Abstract Inundation due to sea level rise is among the most expensive and deadly coastal hazards that can gradually impact small island coastal communities. The intent of this research is to look at ways how coastal island communities can be convincingly informed about the impacts of sea level rise; an understanding of these potential impacts is critical for people to engage in related adaptation actions. Effective climate communication may require more than just using language; the use of virtual reality, 3D spatial technology, and digital elevation models can prove equally or even more powerful to enable the portrayal of risks and dangers posed by climate change. Using a case study in Malta, we mapped impacts of the effects of sea-level rise on valuable real estate, critical public infrastructure, and natural resources. Coastal communities were then asked for their views on sea level rise after looking at precise but different modes of visualization of the same impact at specific coastal locations. Results showed a general preference for 3D over 2D visualizations, for various reasons, including a perception that these better reflect reality; 3D visualisations were also shown to be more effective in convincing respondents about the significance of sea level rise impacts. The results of this study provide valuable insights for local authorities to understand what may be needed to communicate messages related to climate change in an effective manner, ultimately contributing to enhancement of coastal resilience and climate adaptation.

Keywords Climate change · Sea level rise · 2D & 3D visualisations
Public perception · Malta · Small island states

A. Claudia · G. Charles (✉) · C. Elisabeth
Institute of Earth Systems, University of Malta, Msida MSD 2050, Malta
e-mail: charles.galdies@um.edu.mt

F. Alves et al. (eds.), *Theory and Practice of Climate Adaptation*,
Climate Change Management, https://doi.org/10.1007/978-3-319-72874-2_18

Introduction

Climate change is an inevitable and pressing global challenge with irreversible and dangerous impacts. To mitigate these effects, significant and sustained reductions in greenhouse gas emissions (GHGs) are urgently needed (IPCC 2014). Tackling such a global problem requires unprecedented cooperation, innovative polices, novel technologies, difficult trade-offs, and new ways of thinking and behaving. Proper communication of such a serious and complex problem to wider society is pivotal to achieve success in global climate actions, since effective mitigation and adaptation measures require public support; in turn, the degree of public support for such measures is dependent on how the public processes information and perceives threats and possible impacts (Hagen et al. 2015).

Sea level rise (SLR) has been identified as one of the major impacts of climate change (Lambeck et al. 2004). A major factor contributing to sea level rise is the melting of glaciers, ice caps and ice sheets, as the water once contained inside ice will now find its way into the ocean (Paul 2011). Another factor is that of thermal expansion of the oceans, due to the relationship between temperature and density (Church et al. 2001). SLR is associated with increased flooding, coastal erosion, storm surge damage, and saltwater intrusion. In its latest Assessment report, the Intergovernmental Panel on Climate Change (IPCC 2014) estimates that globally, sea levels will rise by 0.26–0.82 m by the end of the 21st century. There is also the possibility that sea levels would increase further than those stated (Church et al. 2013). Hansen et al. (2016) argue that the 2100 SLR projections are too modest due to the fact that ice sheet models are not highly developed due to limited knowledge in this field. Similar conditions prevailed some 120,000 years ago, when huge swathes of polar ice melted, resulting in sea level rise of up to 9 m. This height does not take into account storm surges and extreme weather events such as tropical cyclones, during which sea levels may rise more and be extremely damaging to small island communities. Hansen et al. (2016) also state that SLR increase will not be linear but exponential and accelerating—a situation that may require as little as 50 years to unfold. The resultant costs incurred by small island coastal communities are yet unknown.

Communicating the Impacts of Sea Level Rise

Climate change communication requires a strong synergy between correct information, its mode of presentation, and its intended audience. A number of challenging traits make climate communication difficult, including the fact that understanding of climate change continues to evolve, the fact its implications are open to interpretation, the lack of a single solution, and the need for the involvement of multiple actors who often tend to approach the issue from different political, philosophical and technological perspectives. Moser (2010) eloquently explains these traits as being the lack of visibility and immediacy between cause and effect, peoples' insulation from the environment, delayed climate action, disbelief that man can alter the global climate, the complexity and uncertainty of climate change, and inadequate signs indicating the need for global action, among others. This makes climate change difficult to perceive and be understood by most lay audiences, highlighting the need for communicators to use clear, simple metaphors, imagery, and information technology.

The need for effective communication is especially evident when considering research findings that appear to indicate that the level of public concern about climate change may have already peaked (e.g., Haavisto 2010; Smith and Leiserowitz 2012), and that there is now less conviction about the issue than there was a decade ago (Department of Transport 2010). There is evidence that climate change communication has often underperformed (Hagen et al. 2015), with reasons include failure to consider risk perceptions and factors of trust (Wardekker 2004). Conversely, however, effective climate change communication can further public understanding of the issue, provide insight into possible solutions, mobilize individuals to make modifications to their own lifestyles, encourage support, and provide political capital for action (Ockwell et al. 2009). Recent communication programmes have started to emphasize more the link between the target population and climate change (Gifford et al. 2011), and to reflect more explicitly the audience's own worldviews (Hagen et al. 2015). Moreover, the communication of climate impacts at a more local level is often more successful in raising awareness (Herring et al. 2016).

Framing Impacts Through the Use of Spatial Visualisation Models

Effective climate change communication can be facilitated through an understanding of values, knowledge and attitudes, and preferred communication modes, sources and channels of our audience (Moser 2014). Visualisation and spatial mapping as a preferred, modern mode of communication can be one such element to look for. Retchless (2014) identifies visualisation and mapping as strong tools able to overcome barriers to climate change communication by activating 'experiential processing' and doing away with indifference and inaction. This approach has the power of encoding reality into graphical representations of the impacts of climate change at the level of locally familiar landscapes, thus provoking a strong affective response. Patel et al. (2013) state that the goal of visualisation is that of improving what the viewer understands when presented with information. This links very well with what Nicholson-Cole (2005) and Sheppard (2005) claim, that static, 2D representations of climate change impacts are capable of expressing the personal relevance to climate change more efficiently than would be possible by using text alone. If we consider elevating this approach one step higher by using the latest 3D visualisation technology, then we should be able to portray climate change impacts from different spatial viewpoints in a highly compelling way. Modern visual technology provides us with a general assumption that it would be easier to understand and interpret spatial problems when viewed in 3D over 2D ones. In fact, Balke (2009) tries to prove that 3D visualisations are able to alleviate some of the communication issues which exist between experts and citizens when it comes to spatial planning. The use of modern visualisation technologies to frame spatial problems is therefore very promising and could be used as one of the key elements of the overall communication process in favour of climate action. There is already evidence of how cinematic simulations in climate/environment disaster films can generate strong emotions and increase the public profile of climate science (Svoboda 2014; Sakellari 2014).

SLR is particularly amenable to visual representation, more so than other climate change-related phenomena such as ocean acidification. The communication of SLR via visual representation allows demonstration of current and projected future linked to place-based identities. The rationale of this research originates from the lack of local studies that focus on the communication of climate change impacts in Malta, with a particular emphasis on locally vulnerable communities that reside in low-lying coastal areas, and are therefore more vulnerable to coastal flooding. The aim of this paper is two-fold: (i) to project SLR inundation scenarios at the local level using spatial modelling and visualisation techniques, including virtual 3D (Stephens et al. 2014), and (ii) to analyse the effectiveness of the developed visualization products via a public survey. The research therefore seeks to explore visualization options that could be used for effective communication of sea level rise impacts to the public; although focused on the study area of Malta, the approach used has potential applicability elsewhere. In this investigation, current conditions were compared against incremental increases of SLR in order to highlight the various risks associated with SLR within a spatial context at high resolution. This study provides insights that will help ensure effective climate change communication, based on people's perception, interpretation and preferences. What we try to achieve here is to use the survey results in order to help local climate and social scientists interrogate their assumptions about user needs and understandings and modify their communication practices accordingly. This study provides a sufficient basis for local climate change communicators to develop a more tailored approach to communicating coastal flood risk information to the public, geared towards a proactive response approach.

Methods

Impact Visualisation and Mapping

The methodological approach used in this study is based on the work of Stanchev et al. (2009) and Sutherland and Seeram (2011); it has also been shaped by countless forum posts and videos, which helped in the development of the present visualisations. The most critical issue for agreeing on the best approach to communicate local coastal flood risk due to SLR is how to define an effective communication of the risk resulting from the impact. This idea stems out of a sense that many of the islanders' interest and understanding about climate change, and more specifically on SLR and related coastal flood risks, are not par with the risk. From the scanty literature available about the subject of public perception to climate change impacts in Malta, Akerlof et al. (2010) shows that 67% of the Maltese are uncertain whether the standard of living will decrease due to climate change (36% somewhat likely; 31% somewhat unlikely). As in other small island states, the Maltese public may lack an understanding on the likely changes in climate and human systems at the *local* scale. This led to the idea of researching the best approach to communicate this risk to lay audiences with visual models that can provoke the public's sensitivity on the issue. Based on the feedback, we can make logical inferences and identify consistency in preferences based on multiple factors, including for example, demographics.

In this study, we opted to look at the extent to which 2D and 3D mapping and visualisation can potentially enhance local experiential processing of projected SLR impacts in landscapes that are well-known by the target audience, to explore whether such visualisation techniques can communicate the personal relevance of SLR impacts more effectively than textual communication.

The Maltese archipelago lies approximately at the centre of the Mediterranean Sea between 36° 00′ 00″ and 35° 48′ 00″ north-south latitude and 14° 35′ 00″ and 14° 10′ 30″ east-west latitude (Fig. 1). The Islands have a population of 431,333 people in an area of just 316 km^2 (The World Bank 2017), together with a tourist population that generally exceeds 1 million per annum (Malta Tourism Authority 2016). The climate of the Maltese islands shows an annual mean temperature of 18.6 °C, a mean maximum of 22.3 °C and mean minimum of 14.9 °C (Galdies 2012). Such a temperature regime together with annual total precipitation of 553 mm make the islands relatively hot and dry.

For this study, four coastal localities were chosen on the basis of their importance to the Maltese economy, society, and/or environment (such as tourism, business and agriculture, nature conservation, urbanisation, etc.). The most common ecological risks communicated were potential flooding in ecologically sensitive areas such as parks and natural reserves, and visualising coastal areas at risk from erosion and possible locations of coastal saltmarsh migration. Examples of SLR risks to human communities or the built environment included socioeconomic vulnerability, intersection of inundation with key infrastructure connectivity

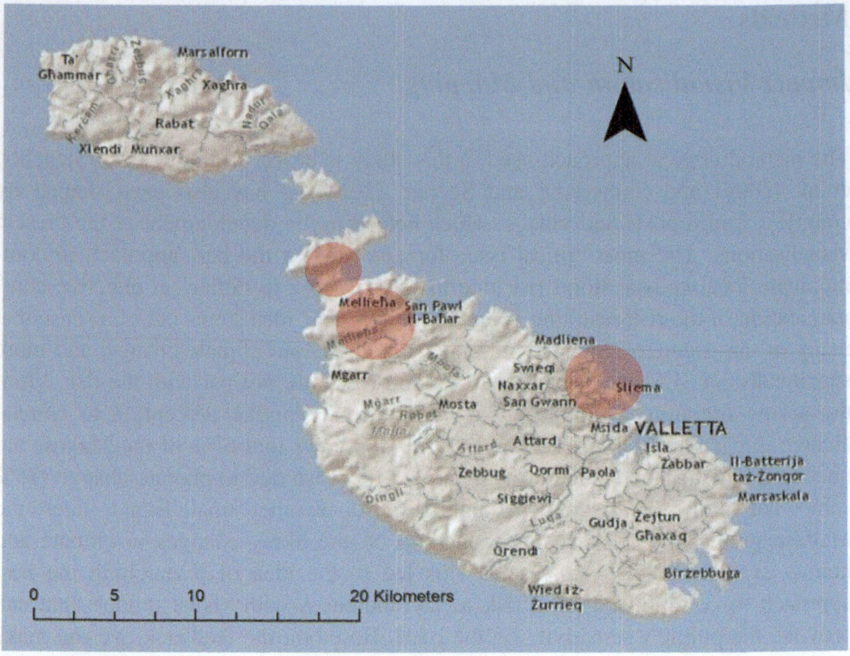

Fig. 1 Annotated map of the Maltese islands showing general terrain information and study areas in red (*Source* ESRI)

(such as roads), and type of residents affected, socio-economic value. Selected localities included parts of Sliema and St Julian's (both intensely urbanised, with a mix of residential, leisure and commercial uses), Mellieħa (touristic site with nature reserves and a residential area) and San Pawl il-Baħar (with recreational fishing, agriculture, and leisure facilities, as well as a residential area). This approach was intended to enable the target audience to visualise potential SLR impacts in a geographic area that is personally significant to them and with which they may have affective attachments. An island-wide contour dataset was used to provide the elevation of the Maltese islands at intervals of 5 m, from which the 5-, 10- and 15 m contour heights above sea level (EEA 2016) were selected to vectorise spatial polygons, representing the projected classes of SLR over the selected areas. To assist in the visual communication of the spatial impact, high resolution satellite imagery (using ESRI basemap) was draped and existing buildings, roads and prominent landmarks were digitised. In order to make visualisation more realistic, coastal features were transformed into 3D using ESRI ArcMap by making use of height information embedded in Google Maps in 3D. Using ESRI ArcScene, a TIN for each of the four localities was extrapolated from the respective polyline datasets, to assist in 3D rendering of SLR so as to give the land the required elevation. Additional datasets were finally draped on top of the TIN layer. For the 2D visualisations, a basemap was loaded in ESRI ArcMap and relevant shapefiles were added to form the basis of 2D visuals.

ESRI ArcGlobe was used to render the 3D visualisations which used the generated TIN as an elevation layer, so that the Globe's map could be draped over it. The relevant shapefiles were then added. Screenshots and, videos of the resulting landscapes were captured at very high detail using ArcGlobe's animation tool. Additional application software were used such as Adobe Photoshop, Adobe Premier Pro and Windows Movie Maker to enhance communication effectiveness by guiding the respondents of the survey to insights through the process of discovery using two main cognitive, visual layers (Stephens et al. 2015).

Surveying Preferred Communication Technologies and Modes of Projected Local Sea Level Rise

The questionnaire was hosted online and advertised by posting on social media and informing the local councils of the four localities. The design of the questionnaire was principally based on Weitekamp (2012). Videos and images generated were embedded into the questionnaire, which was made available in both Maltese and English, taking into account the fact that Malta is officially bilingual. The design catered for a mixed audience in terms of climate perception and understanding, beliefs and values, level of engagement and preferred visualisation mode to frame our narrative. The questionnaire was targeted at local residents aged from 18 upwards, giving a survey population of 353,880 (National Statistics Office 2016). A total of 373 respondents participated in this online survey, giving a margin of error of around 5 points at a 95% confidence level. Figure 2a–c show the main demographics and other related information of the complete sample.

Principal questions in the survey (Table 1) addressed the respondents' general thoughts about SLR, the impact of site-specific 2D and 3D visualisations on their overall perception of SLR, their visualisation preference and reasons for it, and their opinion about the effectiveness of the visualizations. Another important question was that related to the grading of suggested threats resulting from SLR in order to detect whether site-specific visualisations were effective in displaying the impact of SLR. Table 1 shows the questionnaire structure and thematic questions in order of appearance.

To address experiential processing and affectiveness, we embedded both static images and animated visual rendering of the local landscapes of the four chosen sites. According to cognitive science, static graphics can facilitate comprehension, learning, memorisation, problem solving, and communication (Fabrikant et al. 2010). To capture how respondents' subjective SLR predictions changed from

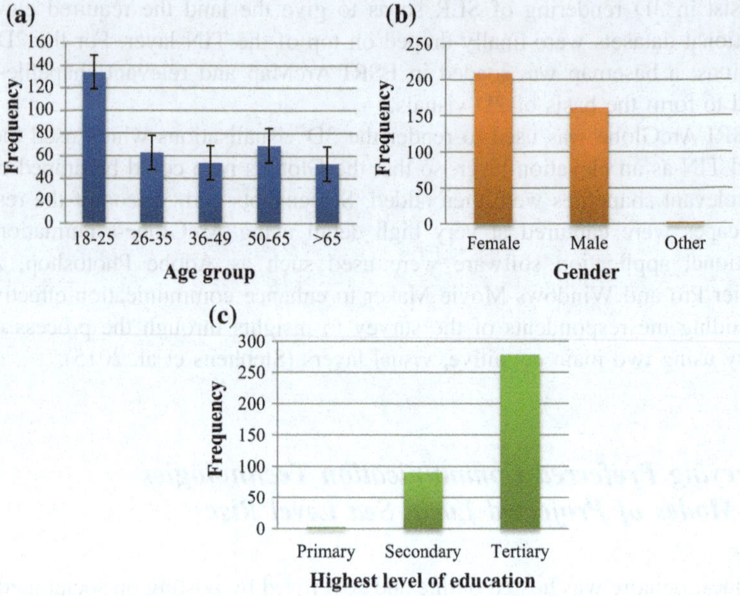

Fig. 2 **a–c** Demographics, gender and level of education of the respondents participating in the present survey

Table 1 Questionnaire structure and thematic questions in order of appearance

Purpose	Number and type of questions
Introduction	NA
Baseline for SLR awareness	1 Likert scale questions
Perception of SLR	1 Likert scale questions
Site preference with familiar local landmarks	1 Analog selection question
Affectiveness of 2D visualisation and mapping	1 Multiple choice question
Experiential processing of 2D visualisation and mapping	1 Likert scale questions
Affectiveness of 3D visualisation and mapping	1 Multiple choice question
Experiential processing of 3D visualisation and mapping	1 Likert scale questions
Visualisation preference and reason	2 Multiple choice question
SLR visualisation as key communication element and its local use in favour of climate action	2 Multiple choice question
Acknowledgment and choice of potential threats	2 Multiple choice question
Demographics	4 Multiple choice and numerical input questions

pre- to post-visualisation, a graded scale was used to note this change. Because of its ordinal nature, the change score records the direction of change (yes, no, not sure) but not its magnitude. Locally, it is not yet known whether a stronger rendering of SLR impact simulation can be more effective than use of static images and maps. This is one of the research unknowns of the present study.

Respondents were given the choice of which of the four localities they wished to see visualizations for. Respondents were distributed as follows: Mellieħa—n = 67, Sliema—n = 129, St. Julian's—n = 37, and San Pawl il-Baħar—n = 22).

All responses were deemed valid and retained for analysis. Results were imported to IBM SPSS software in order to statistically analyse the inter-dependence of survey responses. The chi-square test was used to assess the association between SLR perception and modes of visualisation, with the H_0 specifying no association at 0.05 level of significance.

Results and Discussion

Through an inductive content analysis approach, this study generated a framework to better understand SLR risk communication. This section presents and discusses the two main aspects of this study, namely (1) how can spatial modelling be used to depict potential flooding magnitude, and (2) how effective are the two main types of spatial simulations in contributing to the respondents' overall mental model of SLR. In this section, we offer recommendations for SLR impact communicators stemming from our study findings.

Visualisation and Mapping of Projected Local SLR

What follows are some examples of the mapping of projected SLR at the selected sites. Figure 3a–b show a representation of different levels of inundation superimposed over digitised features and land marks, including buildings and road network. This is an example of a 2D rendering of one of the sites (Mellieħa). It is important to note that with a rise of 10 m, the island's main road connection with the northern part of the island will be significantly jeopardised. Moreover, on the other side of Mellieħa bay, a 10 m SLR will isolate one of the national wastewater treatment plants from main transportation routes (Fig. 3a) and Malta's most important natural reserve for birds is expected to be fully inundated by a rise of 5 m (Fig. 3b).

A 2D visualisation of different SLR scenarios for Sliema (Figs. 4 and 5) and St Julian's (Fig. 6) shows the approximate impact of SLR on urbanisation and general infrastructure such as the road network.

The Sliema coast road hosts a good number of retail outlets and related services. It can be shown that a 5 m SLR can be detrimental to most of the tourist hotels, amenities and restaurants extending all the way to St Julian's. In Fig. 4 it can be seen that a rise of 15 m will cut off a considerable amount of the peninsula, while Fig. 5 shows that a 5 m SLR will result in the loss of substantial road network of

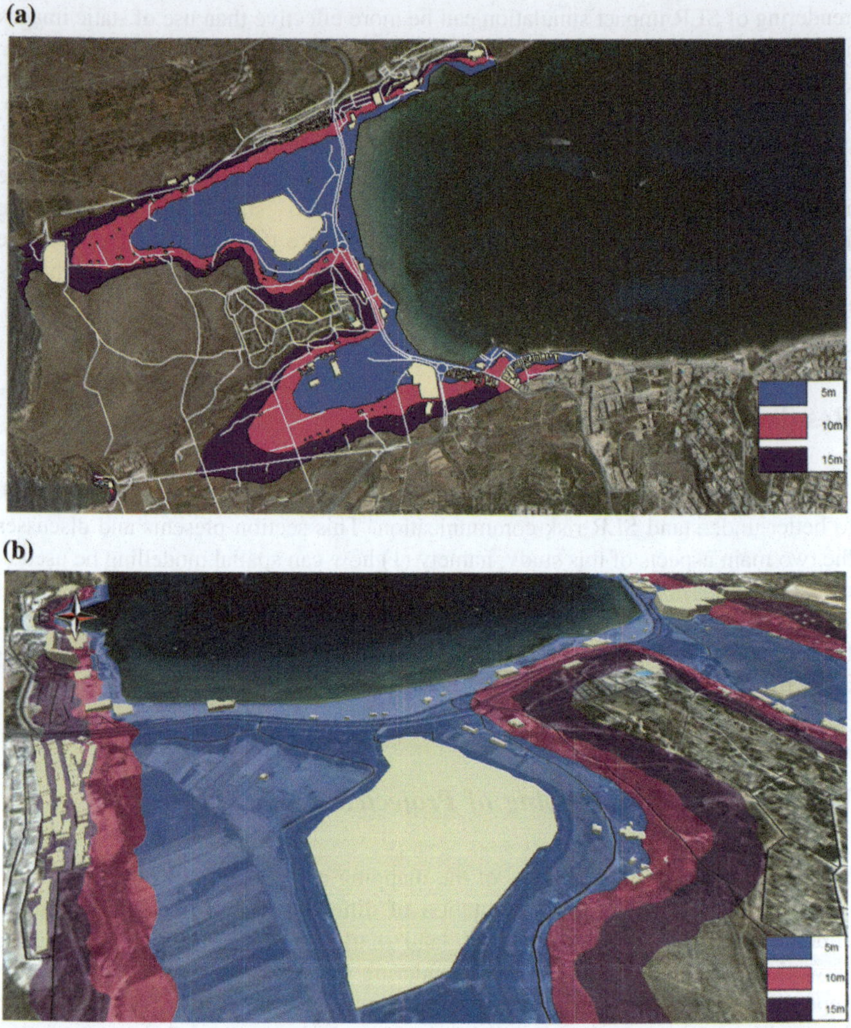

Fig. 3 **a** 2D visualisation of Mellieħa, showing different levels of inundation (*Source of basemap* ESRI) **b** Oblique view of parts of Mellieħa showing the Għadira Natural Reserve (central light brown polygon) and the surrounding environment, including road network, tourist facilities and local residences (*Source of basemap* ESRI)

the Sliema area. Under such a scenario, the important ferry link of the Sliema and Valletta waterway will have to be relocated. An extensive part of the urban grid network in Sliema (Fig. 5) is completely flooded.

Figure 6a–b show views of St George's Bay in St Julians and its neighbourhood. This area is an important upmarket site and a highly touristic destination. The historic Dragonara Casino situated on one of the peninsulas will be completely

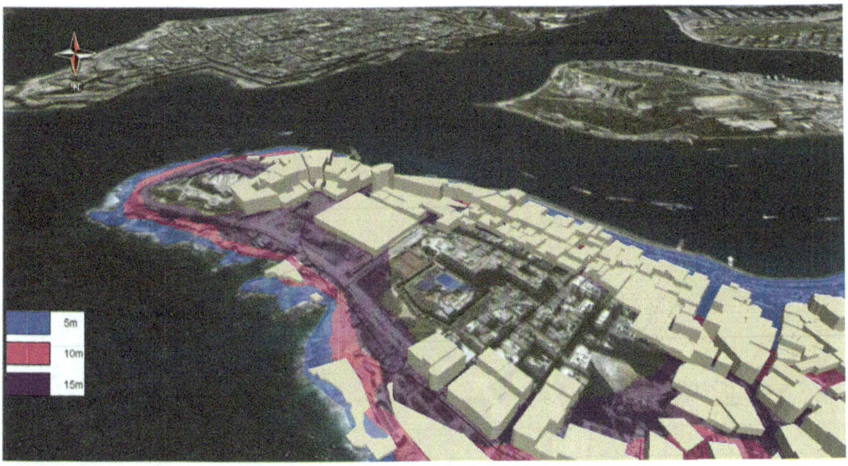

Fig. 4 Oblique view of Tigne' Point (Sliema) in 3D showing inundation of the peninsula due to different SLR scenarios (*Source of basemap* ESRI)

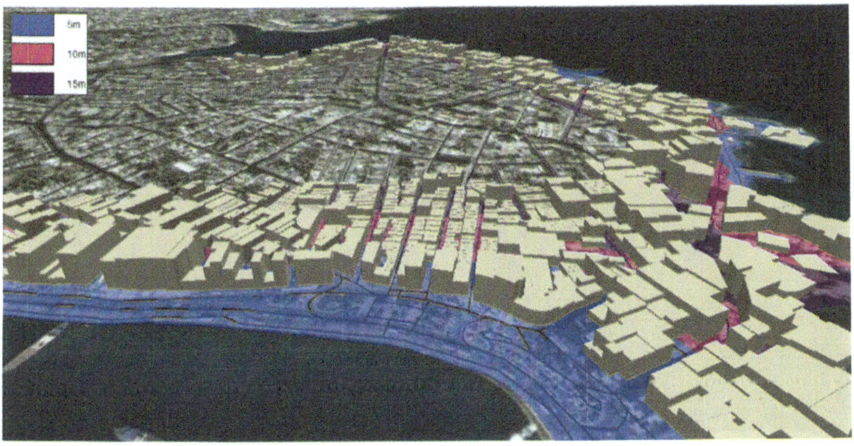

Fig. 5 Oblique view of the Sliema Strand in 3D, showing extensive inundation of main roads, side streets, and of the dense urban network. This location also hosts the Sliema-Valletta waterway ferry (*Source of basemap* ESRI)

inundated by a rise of less than 5 m. The sandy beach in St. George's bay will also be completely inundated.

Figure 7a–b show SLR impact at San Pawl il-Baħar, specifically in Xemxija-Pwales area. It can be seen that seawater will flow inland towards Malta's prime agricultural (irrigated) sites and a nature reserve site.

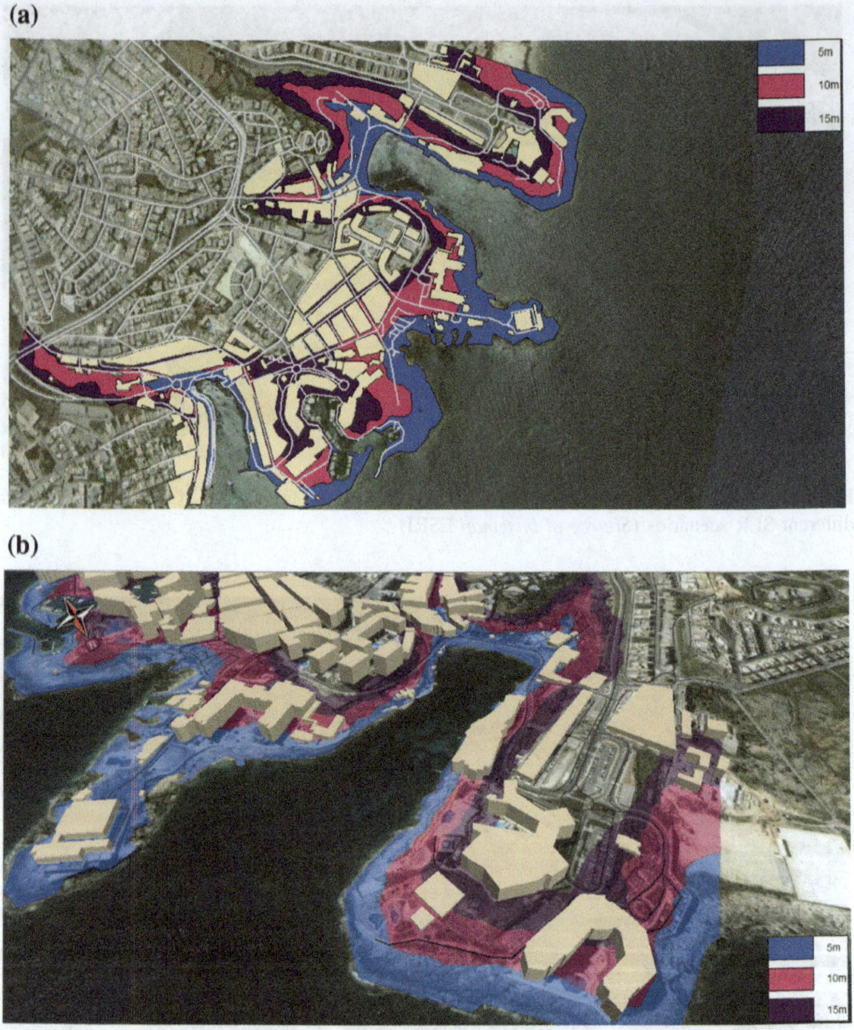

Fig. 6 **a** 2D visualisation of inundation of St Julian's based on different SLR scenarios (*Source of basemap* ESRI). **b** Oblique view of St George's Bay in 3D, showing different SLR scenarios for the bay and its surroundings in St. Julian's area (*Source of basemap* ESRI)

Public Perception of Local Threats Arising from SLR

The survey results show that the most significant threat of SLR is that of loss of valuable land (Fig. 8). The negative impacts on tourism were ranked highest for St. Julian's (49.1%) and Sliema (44.7%) as they are both prime tourist destinations and very popular residential areas. This is followed by Mellieħa in view of its leisure, touristic and aquatic facilities. On the other hand, the most significant SLR impact

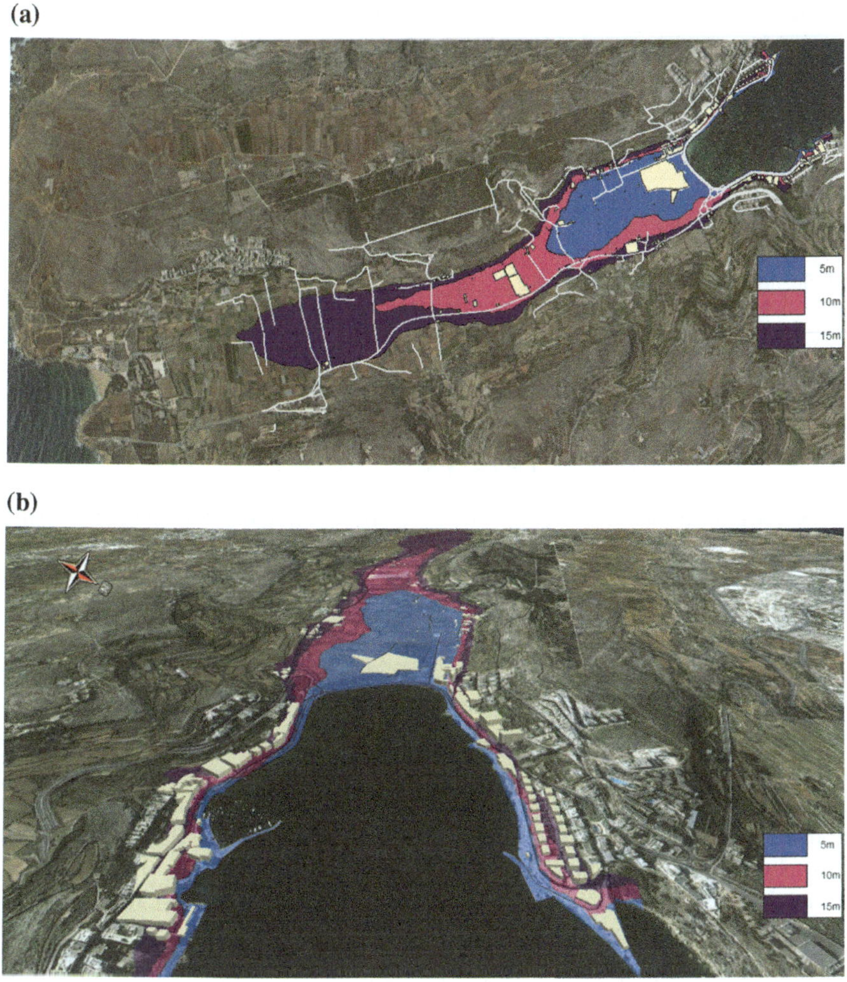

Fig. 7 **a** Overhead view of Xemxija-Pwales area at San Pawl il-Baħar, showing different levels of inundation (*Source of basemap* ESRI). **b** Oblique view of Xemxija Bay at San Pawl il-Baħar in 3D (*Source of basemap* ESRI)

on local agricultural land area was primarily assigned to San Pawl il-Baħar (81.5%) and Mellieħa (55.6%).

The threat to commerce and business was similar to the threat to the tourism sector, with St. Julian's (54.4%) ranking the highest. This is followed by Sliema (49.2%), Mellieħa (35.6%) and San Pawl il-Baħar (25.9%). This is due to the fact that St. Julian's and Sliema are two prime tourist areas.

Mellieħa (52.2%) and San Pawl il-Baħar (51.9%) were the top two localities in which loss of historical sites was considered to be of a significant threat. Mellieħa is home to historical fortified towers and an artisanal fisheries museum.

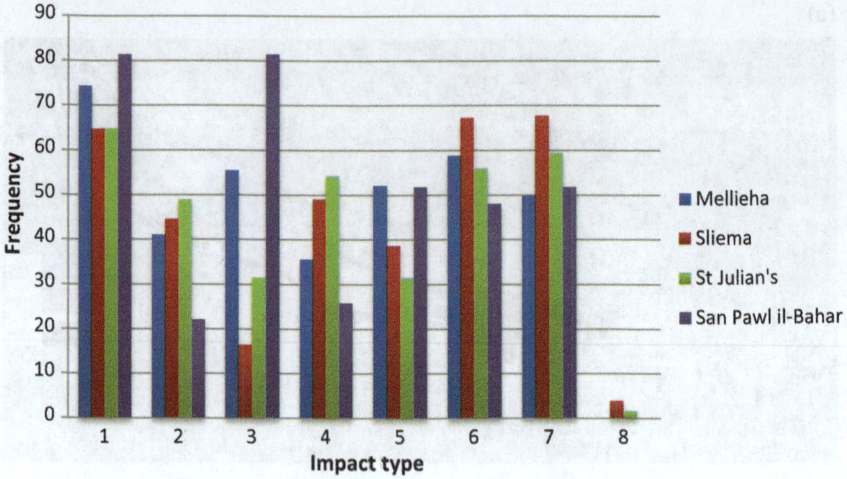

Fig. 8 Potential threats due to sea level rise per locality. Impact type: 1: Loss of valuable land; 2: Negative impacts on tourism; 3: negative impacts on agriculture; 4: harm to commerce and businesses; 5: loss of historical sites; 6: loss of main roads and connectivity; 7: Permanent displacement of local residents; 8: Other

Loss of road infrastructure is considered to be worst in Sliema (67.3%). Surprisingly, the public does not see the loss of connectivity in the Xemxija-Pwales area as highly significant (48.1%), even though this will practically disconnect the main island from its northern region. The inundation of this area will be extremely troublesome to those travelling to the sister island of Gozo with a population of around 30,000 inhabitants. The Sliema area was also considered the worst hit when it comes to permanent displacement of local residents (67.8%; most probably attributed to its population density compared to the other three sites), followed by St Julian's (59.6%), Xemxija-Pwales area in San Pawl il-Baħar (51.9%) and Mellieħa (50%).

Preferred SLR Visualisation Modes

Survey results show that the most preferred type of visualisations are the ones in 3D (57.9%). A total of 23.6% of the respondents answered that they had no preference, whilst 18.5% preferred the 2D visualisations (Table 2). Reasons given for the preference of 2D representations included simplicity (78.3%), the perception that these are quicker to understand (71%), and that virtual 3D visuals provided unnecessary detail (14.5%; Table 3). On the other hand, the virtual 3D

Table 2 Visualisation preferences

		Frequency	Percentage (%)
Which type of visualisation did you prefer?	2D visualisation	69	18.5
	3D visualisation	216	57.9
	No preference	88	23.6

Table 3 Reasons why 2D visualisations were preferred

		Frequency	Percentage (%)
Why did you prefer the 2D visualisations over the 3D ones?	They were more simple to interpret	54	78.3
	They were more visually pleasing	20	29.0
	All necessary information was provided	34	49.3
	It only takes one glance to 'absorb' its message	49	71.0
	The 3D visualisations were distracting	16	23.2
	The 3D visualisations provided unnecessary details	10	14.5

Table 4 Reasons why 3D visualisations were preferred

		Frequency	Percentage (%)
Why did you prefer the 3D visualisations over the 2D ones?	They were closer to reality	162	75.0
	They were more visually pleasing	55	25.5
	The 2D visualisations were rather dull	18	8.3
	The 2D visualisations were lacking detail	31	14.4
	Having the buildings in 3D helped in recognizing important/large structures	134	62.0

representation was the preferred choice on the basis of its local rendering of real landmarks (75%), and the ease by which respondents are able to recognise local features and distinctive landmarks (62%). Moreover, 8.3% of the respondents who opted for the virtual 3D approach found 2D visualisation to be dull (Table 4).

Perception of SLR and Affectiveness of Particular Visualisation Modes

This study used the chi-square test to examine the effectiveness of 2D and 3D visualisations. A significant relationship (*p*-value = 0.002) was found between perceived threat of SLR and change in perception induced by the 2D visualisation (Fig. 9). Conversely, the degree of association between local perceived threat of SLR and change in perception induced by 3D rendering was found not to be significant (*p*-value = 0.063). Despite the lack of relationship between local perceived threat of SLR and change in perception after viewing 3D visuals, 3D visualisations were more effective in changing the opinion of the respondents than 2D ones. This means that 3D visualisations were much more effective in changing the opinion of the respondents than 2D visualisations. Reference is made to Table 4, which shows that 75% of the total respondents (who preferred 3D visualisations over 2D ones) stated that one of the reasons for this preference was that the 3D rendering was closer to reality. If such visualisations are perceived to be more realistic, then this can be exploited in future studies and awareness-raising campaigns on other climate change impacts, and not just for SLR (Fig. 10).

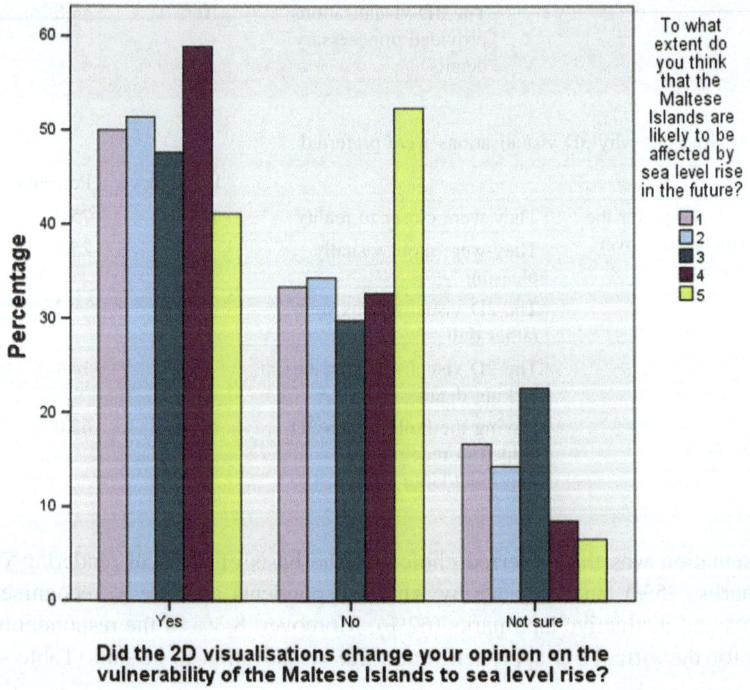

Fig. 9 Affectiveness of 2D visualisations of projected SLR at the chosen sites

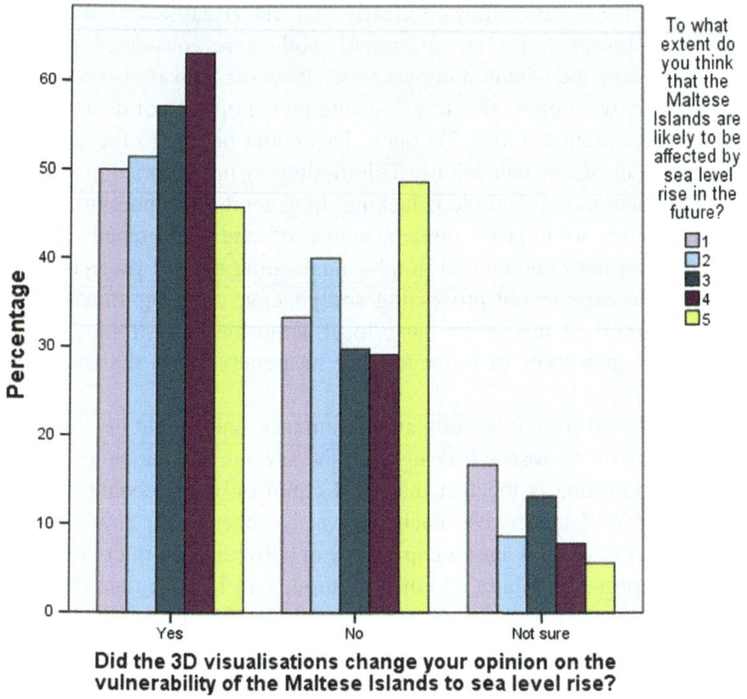

Fig. 10 Affectiveness of 3D visualisations of projected SLR at the chosen sites

Conclusion

This study aims at improving local climate change communication by understanding personal engagement with climate change impacts (namely SLR) *at the local level*. Communicating to increase public-level engagement requires an understanding of how best to contextualise scientific information in the form of personal experience in ways that help shape people's thoughts, their emotions, and ultimately their actions (Lorenzoni et al. 2007; Sheppard 2005; Marx et al. 2007; CRED 2009). Moreover, research shows the need for communicators to focus on the local scale, showing how climate change impacts can be relevant to the living space and lifestyle of local coastal communities (Herring et al. 2016) and how such impacts can be detrimental to insular countries such as small island states. It is highly striking how a rise of a couple of metres can lead to the loss of all beaches along the low-lying coastal parts of the Maltese islands. Equally striking is the potential fragmentation of territory, with substantial projected losses in connectivity that could lead to displacement of the local population, and disruptions related to transportation of goods and services, among others.

The results of this study showed clearly that 3D visualisations were preferred over 2D ones; however, for the most part, both were considered to be highly effective in showing the potential impact of sea level rise. It was also concluded that 3D visualisations were more effective in changing the opinion of respondents (after viewing the visualisations) than 2D ones. This could be due to the increased element of reality in 3D visualisations. This finding is not surprising, even though previous information on this topic is lacking. In general, 3D rendering supports the ability of graphics to improve understanding of and engagement with spatial problems. Studies have shown that graphical encoding of SLR projections is more likely to activate experiential processing and generate a strong affective response (Slovic et al. 2004). If this is the case, local authorities are strongly encouraged to make use of such tools in order to raise awareness about the risks posed by local SLR.

The lessons learnt from this study are preliminary, and should be viewed in light of the limitations of the research design. Of the several limitations to the research, the most important one is the fact that SLR could only be accounted for in 5 m intervals. A gap of 5 m between each interval is rather large, and it would have been much better to have at least a gap of 1–2 m between each interval. This can be significantly improved if Malta's high resolution, 1 m LIDAR data (Galdies et al. 2015) was used instead as the basis for graphical rendering. Another limitation is that respondents had no control over the visualisations, but instead were presented with pre-established videos and image compilations of the study areas. Moreover, the visualisations somewhat lacked a higher element of reality and more sophisticated and expensive graphical tools are required which were not available to the authors. Another limitation is that the current narrative does not distinguish the SLR risk attributed by IPCC's four main Representation Concentration Pathways (IPCC 2014) nor any particular time-framed projection for this century. This would obviously have constituted a fuller scientific presentation of the potential projections of local SLR.

Despite its limitations, the research is a valuable first attempt at exploring an important but scarcely researched area. Barriers to effective climate change communication threaten efforts to mitigate and adapt to serious local impacts. Combining the use of surveys with computer-aided visualisations, climate change communication has the potential to improve understanding of related science and increase public engagement to take climate actions at the local and national level. This research has shown that the use of 3D spatial visualisation and mapping may be one means of doing this; however, additional research is still needed to identify other effective means. One suggested avenue for future research is that of combining such visualisation techniques with communication of concrete mitigation actions, focusing on positive opportunities and potential solutions.

References

Akerlof, K., DeBono, R., Berry, P., Leiserowitz, A., Roser-Renouf, C., Clarke, K.-L., et al. (2010). Public perceptions of climate change as a human health risk: surveys of the United States, Canada and Malta. *International Journal of Environmental Research and Public Health, 7,* 2559–2606.

Balke, K. (2009). *Perception and role of 3D visualisations in planning: A case study of the Northwest passage scenic byways viewshed protection and visualisation project* (Theses, dissertations, Professional Papers. Paper 740). http://scholarworks.umt.edu/cgi/viewcontent.cgi?article=1759&context=etd. Accessed April 30, 2017.

Center for Research on Environmental Decisions (CRED). (2009). *The psychology of climate change communication: A guide for scientists, journalists, educators, political aides, and the interested public.* New York: Earth Institute, Columbia University.

Church, J. A., Clark, P. U., Cazenave, A., Gregory, J. M., Jevrejeva, S., Levermann, A., … Unnikrishnan, A. S. (2013). Sea level change. In T. F. Stocke, D. Qin, G. K. Plattner, M. Tignor, S. K. Allen, J. Boschung, … P. M. Midgley (Eds.), *Climate change 2013: The physical science basis. Contribution of working group I to the fifth assessment report of the intergovernmental panel on climate change.* Cambridge, United Kingdom, New York, NY, USA: Cambridge University Press.

Church, J. A., Gregory, J. M., Huybrechts, P., Kuhn, M., Lambeck, M. T., Nhuan, D, … Woodworth, P. L. (2001). Changes in sea level. In J. T. Houghton, Y. Ding, D. J. Griggs, M. Noguer, P. J. van der Linden, X. Dai, K. Maskell, & C. A. Johnson (Eds.), *Climate change 2001: The scientific basis. Contribution of working group I to the third assessment report of the intergovernmental panel on climate change.* Cambridge, United Kingdom, New York, NY, USA: Cambridge University Press.

Department of Transport. (2010). *Public attitudes towards climate change and the impact of transport: 2006, 2007, 2008, and 2009.* London: GOV.UK.

European Environment Agency. (2016). *Global and European sea-level.* From https://www.eea.europa.eu/data-and-maps/indicators/sea-level-rise-4/assessment-2. Accessed May 5, 2017.

Fabrikant, S. I., Hespanha, S. R., & Hegarty, M. (2010). Cognitively inspired and perceptually salient graphic displays for efficient spatial inference making. *Annals of the Association of American Geographers, 100*(1), 13–29.

Galdies, C. (2012). Temperature trends in Malta (central Mediterranean) from 1951 to 2010. *Meteorology and Atmospheric Physics.* https://doi.org/10.1007/s00703-012-0187-7.

Galdies, C., Azzopardi, D., & Sacco, A. (2015). *Large scale erosion mapping using LIDAR and USLE technique for landscape management in the island of Gozo, Malta. RSPSoc, NCEO and EOI-ST joint annual conference 2015.* The University of Southampton. September 8–11, 2015.

Gifford, R., Kormos, C., & McIntyre, A. (2011). Behavioral dimensions of climate change: Drivers, responses, barriers, and interventions. *WIREs Climate Change, 2*(6), 801–827.

Haavisto, I. (2010). *Työelämän kulttuurivallankumous. EVAn arvo-ja asennetutkimus 2010* [The cultural revolution of work life. EVA's study of values and attitudes 2010]. EVA & Taloustieto Oy. Available from http://www.eva.fi/wp-content/uploads/2010/06/tyoelaman_kulttuurivallankumous.pdf. Accessed April 30, 2017.

Hagen, B., Middel, A., & Pijawka, D. (2015). European climate change perceptions: Public support for mitigation and adaptation policies. *Environmental Policy and Governance, 26,* 170–183.

Hansen, J., Sato, M., Hearty, P., Ruedy, R., Kelley, M., Masson-Delmotte, V., … Velicogna, I. (2016) Ice melt, sea level rise and superstorms: Evidence from paleoclimate data, climate modeling, and modern observations that 2 °C global warming could be dangerous. *Atmospheric Chemistry and Physics, 16*(6), 3761–3812.

Herring, J., VanDyke, M. S., Cummins, R. G., & Melton, F. (2016). Communicating local climate risks online through an interactive data visualization. *Environmental Communication,* 1–16.

IPCC. (2014). In Pachauri, R. K., Allen, M. R., Barros, V. R., Broome, J., Cramer, W., Christ, R., … van Ypersele, J. P. (Eds.), *Climate change 2014: Synthesis report. Contribution of working groups I, II and III to the fifth assessment report of the intergovernmental panel on climate change* (p 151). Geneva, Switzerland: IPCC.

Lambeck, K., Anzidei, M., Antonioli, F., Benini, A., & Esposito, A. (2004). Sea level in roman time in the central mediterranean and implications for recent change. *Earth and Planetary Science Letters, 224*(3), 563–575.

Lorenzoni, I., Nicholson-Cole, S., & Whitmarsh, L. (2007). Barriers perceived to engaging with climate change among the UK public and their policy implications. *Global Environmental Change, 17,* 445–459.

Malta Tourism Authority. (2016). *Tourism in malta 2015*. Available from http://www.mta.com.mt/research Accessed April 20, 2017.

Marx, S. M., Weber, E. U., Orlove, B. S., Leiserowitz, A., Krantz, D. H., et al. (2007). Communication and mental processes: Experiential and analytic processing of uncertain climate information. *Global Environmental Change, 17,* 47–58.

Moser, S. C. (2010). Communicating climate change: History, challenges, process and future directions. *WIREs Climate Change, 1,* 31–53.

Moser, S. C. (2014). Communicating climate change adaptation: The art and science of public engagement when climate change comes home. *Wiley Interdisciplinary Reviews—Climate Change, 5,* 337–358.

National Statistics Office. (2016). *Demographic review 2014*. Available from https://nso.gov.mt/en/publicatons/Publications_by_Unit/Documents/C3_Population_and_Tourism_Statistics/Demographic_Review_2014.pdf. Accessed April 30, 2017.

Nicholson-Cole, S. A. (2005). Representing climate change futures: A critique on the use of images for visual communication. *Computers, Environment and Urban Systems, 29*(3), 255–273.

Ockwell, D., Whitmarsh, L., & O'Neill, S. (2009). Reorienting climate change communication for effective mitigation: Forcing people to be green or fostering grass-roots engagement? *Science Communication, 30*(3), 305–327.

Patel, V. M., Dholakia, M. B., & Singh, A. P. (2013). Tsunami risk 3D visualizations of Okha coast, Gujarat (India). *International Journal of Engineering Science and Innovative Technology, 2,* 130–138.

Paul, F. (2011). Melting glaciers and icecaps. *Nature Geoscience, 4*(2), 71–72.

Retchless, D. P. (2014). Communicating climate change: spatial analog versus color-banded isoline maps with and without accompanying text. *Cartography and Geographic Information Science, 41*(1), 55–74. https://doi.org/10.1080/15230406.2013.826479.

Sakellari, M. (2014). Cinematic climate change, a promising perspective on climate change communication. *Public Understanding of Science, 24*(7), 827–841.

Sheppard, S. R. J. (2005). Landscape visualisation and climate change: the potential for influencing perceptions and behaviour. *Environmental Science & Policy, 8,* 637–654.

Slovic, P., Finucane, M. L., Peters, E., & MacGregor, D. G. (2004). Risk as analysis and risk as feelings: Some thoughts about affect, reason, risk, and rationality. *Risk Analysis, 24*(2), 311–322.

Smith, N., & Leiserowitz, A. (2012). The rise of global warming skepticism: Exploring affective image associations in the United States over time. *Risk Analysis, 32*(6), 1021–1032.

Stanchev, H., Palazov, A., & Stancheva, M. (2009). 3D GIS model for flood risk assessment of Varna bay due to extreme sea level rise. *Journal of Coastal Research*, 1597–1601.

Stephens, S. H., DeLorme, D. E., & Hagen, S. C. (2014). An analysis of the narrative-building features of interactive sea level rise viewers. *Science Communication, 36,* 675–705. https://doi.org/10.1177/1075547014550371.

Stephens, S. H., DeLorme, D. E., & Hagen, S. C. (2015). Evaluating the utility and communicative effectiveness of an interactive sea-level rise viewer through stakeholder engagement. *Journal of Business and Technical Communication, 29,* 314–343. https://doi.org/10.1177/1050651915573963.

Sutherland, M., & Seeram, A. (2011). Sea level rise modelling in support of socioeconomic impact analysis: Grande riviere, trinidad and tobago. In *FIG working week 2011 bridging the gap between cultures*, Marrakech, Morocco, May 18–22, 2011.

Svoboda, M. (2014). *The long melt: The lingering influence of the day after tomorrow*. Yale Climate Connections. https://www.yaleclimateconnections.org/2014/11/the-long-melt-the-lingering-influence-of-the-day-after-tomorrow/. Accessed April, 30 2017.

The World Bank. (2017). http://data.worldbank.org/country/malta. Accessed April 30, 2017.

Wardekker, J. A. (2004). *Risk communication on climate change* (Doctoral dissertation). Utrecht University, Utrecht, NL.

Weitekamp, B. (2012). *Engagement through visualization: Sea level rise in the Matanzas River Basin* (Unpublished M.Sc. thesis). University of Florida. http://ufdc.ufl.edu/AA00013310/00001/pdf. Accessed April 30, 2017.

Overview of Contextual Factors When Designing and Implementing Climate Risk Communication with Rural Communities in the Global South

Elvis Modikela Nkoana

Abstract Climate change is the biggest threat of the 21st century. Perception of the climate change risk is an important step towards coping with and adaptation to it. The perception of risk is influenced by numerous factors among them-risk communication. Successful communication of climate change risk, like any other environmental risk, hinges on many factors. Careful consideration of these factors when planning risk communication initiatives is crucial particularly in developing countries steeped in cultures, customs, traditional ecological knowledge, and poverty. This theoretical paper uses a traditional literature review and backward snowballing techniques to highlight the important factors that need to be considered when planning and implementing climate change risk communication exercises in rural communities in a developing country context. This chapter is crucial in the context of scientific findings that climate change adaptation tools-a form of decision support tool-are poorly designed to engage rural communities in adaptation efforts. A well designed climate change risk communication initiative represents an entry point for scientists and practitioners to engage with rural people on climate change adaptation.

Keywords Climate change · Developing countries · Frames · Narratives
Participatory integrated assessment · Risk communication · Rural communities
Traditional ecological knowledge · Typology · Visualisation · Worldviews

E. M. Nkoana (✉)
University of South Africa (UNISA), P.O. Box 392, Pretoria 0003, South Africa
e-mail: nkoana.elvis6@gmail.com

E. M. Nkoana
University of Antwerp, Universiteitsplein 1, 2610 Antwerp, Belgium

F. Alves et al. (eds.), *Theory and Practice of Climate Adaptation*,
Climate Change Management, https://doi.org/10.1007/978-3-319-72874-2_19

Introduction

Human activities of excessive production and consumption are accelerating global climate change (IPCC 2012) that is negative affecting regions in the global South Africa (Niang et al. 2014). Adaptation and mitigation of anthropogenic climate change requires individual and collective efforts from actors and sectors that are degrading the environment. Communication of climate change risks is the first step of engaging the public in planned adaptation and mitigation. Bryan et al. (2009) agrees individuals and communities must first perceive the occurrence of climate change in order to not only cope with but adapt to it (Adger et al. 2009). However, many factors including uncertainty, framings, (Morton et al. 2011), and risk perception (Nkoana et al. 2016; Patt and Schroter 2008; Maponya et al. 2013) affects the climate risk communication process and its message. Generally, more uncertainty about climate change impacts might increase psychological defenses in a particular behavioural domain and while loss and gain-frames also affect the public's potential to engagement in climate mitigation and adaptation (Morton et al. 2011). Risk perception is also shaped by complex factors that include experience (personal and vicarious), psychological (heuristics and biases), social, moral, institutional, and cultural (Adger et al. 2009; Dessai et al. 2004; Grothmann and Patt 2005; Frank et al. 2010; Patt and Schroter 2008; Wiid and Ziervogel 2012; Marx et al. 2007; Perry and Lindell 2008; Paton et al. 2008; Gaillard 2008; Schmuck 2000). This complex milieu had inspired countless studies on climate risk communication in different contexts and such circumstances often have a bearing on the message been communicated. Most climate risk communication studies have focused on post-industrial societies of Western Europe and North America (Spence and Pidgeon 2010; Morton et al. 2011) ignoring the global South. This chapter addresses this shortfall by highlighting importance factors that have to be heeded when planning a climate change risk communication for a rural audience in the global South.

Climate Change Risk Communication as Part of Participatory Integrated Assessment (PIA)

Climate change risk communication is the first step in climate adaptation utilising a Participatory Integrated Assessment (PIA) approach that acknowledges the input of affected stakeholders. PIA consists of problem framing, risk assessment, and response options assessment (Salter et al. 2010; Siebenhüner and Barth 2005; Schlumpf et al. 1999). Nkoana et al. (2016) customised these stages to fit the climate adaptation context as: problem framing (climate change risk communication), risk assessment (impact and vulnerability assessments), response options assessment (adaptation assessment), and sustainability assessment. For the purpose of this chapter, we focus on climate change risk communication stage only but requests the reader to always bear in mind that this phase is part of a larger PIA process. Locating climate change risk communication within the broader PIA process does assist when framing positive messages of opportunities that can be derived from climate adaptation instead of inducing anxiety and fear with negative messages that trigger avoidance and risky behaviour detrimental to the environment.

Climate Change Risk Communication Typology

Defining a climate change risk communication typology is important to clarify what communication is and what it is not. The climate change risk communication typology (in Table 1) is made up of three steps namely: information-sharing, consultation (extraction), and communication. Information-sharing is when outsiders (policy makers and practitioners) enlighten insiders (affected stakeholders and/or community members) about the climate change phenomenon and its impacts, typically through awareness raising campaigns. Consultation is when outsiders solicit the local knowledge of insiders on climate change monitoring, impacts, coping, adaptation, and mitigation techniques. In information-sharing and consultation, climate information and knowledge flows in one direction from outsiders to insiders and vice versa. Communication entails a two-way flow of climate information and knowledge from both parties involved in a collaborative process. In reality, communication is often used to refer to information through climate awareness endeavours aimed at the public. Strictly speaking, this is a misuse of the term. Like participation, this is not surprising because the implementation of genuine communication (two-way flow of climate information and knowledge) requires enormous resources and can become complicated by many factors. These factors have to be considered when designing and implementing climate change risk communication in the global South.

Table 1 Climate change risk communication typology

Stages of participation (i.e. degree of participation)	Main purpose of participation (i.e. objectives of participation)	Characteristics of the stage (i.e. direction of communication flows)	Type of participation (i.e. theoretical basis/ pragmatic participation etc.)
Information-sharing	Informing (awareness raising)	Information flows in one direction from the outsiders to insiders	Passive participation
Consultation	Extraction (soliciting perceptions & local knowledge)	Information flows in one direction from insiders to outsiders	
Communication	Collaboration (integration of Western science & Local knowledge)	Flow of information is arranged in both directions	Interactive participation

Source Author's own synthesis inspired by Arnstein (1969)

Contextual Factors in Climate Change Risk Communication in the Global South

Livelihoods, Local Champions, and Traditional Ecological Knowledge

Most countries in the global South did not realise the expired Millennium Development Goals (MDGs) and their focus will be on achieving the new Sustainable Development Goals (SDGs) by the year 2030. Realising the SDG goals on education and livelihoods security has a bearing on the design and implementation of climate change risk communication messages for local communities (Morton et al. 2011; Spence and Pidgeon 2010). Low levels of literacy requires outsiders to design and implement messages that can be understood by local community members that subsist on rain-fed agriculture and livestock rearing-sectors which are highly vulnerable to negative climate change impacts. Also, the struggle to secure livelihoods is an everyday reality of these local communities in the global South (Artur and Hilhorst 2012). Climate change risk communication messages should target the livelihoods sector in order to gain a buy-in from these insiders. Outsiders can appoint and train local champions from within the community to work alongside them when designing and implementing climate change risk communication efforts (Speranza et al. 2010; Roberts 2008). Local champions speak the native languages and can negotiate access to other community members including elders, community leaders, and gate keepers. Over and above, local champions can highlight important cultural dos and don'ts within the local community. Soliciting access to the local knowledge of community members is an integral part of any genuine climate change risk communication exercise. Generally, local knowledge is institutionalised knowledge accumulated from interactions with the local social and physical environments over long periods of time and these experiences are passed on from one generation to another through word of mouth (adapted from Nyong et al. 2007). In this chapter we focus on traditional ecological knowledge as a particular form of local knowledge that derives from people's interaction with their ecological environment. Traditional ecological knowledge refers to common and specialist knowledge that is location and culture-specific, is acquired through long-term interaction and observation of the natural environment (i.e. changes in flora, fauna, sea and weather conditions, and astronomy constellations), and is passed down through oral traditions from one generation to another (Codjoe et al. 2014; Egeru 2012; Alexander et al. 2011; Mengistu 2011). Traditional ecological knowledge is of particular relevance in the context of climate change risk communication because it serves as frames through which people interpret and understand the world around them (Speranza et al. 2010; Nyong et al. 2007; Lemos et al. 2002; Roncoli et al. 2002; Luseno et al. 2003). Local knowledge shapes a worldview that is different from a Western worldview derived from scientific knowledge. Western science has dominated socio-political and economic development and knowledge generation since colonialisation of

Southern countries leading a cultural hegemony (Van Opstal and Hugé 2013). It is then not surprising that "development projects are known to have been created, funded and managed by outside resources and introduced into rural communities with the hopes and promises of impacting lives. These projects did not take into consideration the culture of people and resulted in low participation and success rates" (Nyong et al. 2007, p. 794). It can be argued that the type of development projects described above focuses only on the information part of climate change risk communication as they do not go further to solicit the local knowledge of insiders. If they did, they will understand that culture constitute local frameworks that governs decisions made on a daily basis-including to partake in development cooperation projects. Van Aalst et al. (2008) agree that outsiders are often surprised by risk and adaptation priorities of insiders which are often shaped by immediate socio-economic, political, and cultural contexts. The climate change risk communication typology we propose attempts to ameliorate this shortcoming by incorporating a two-way flow of information and knowledge between insiders and outsiders.

Visualisation, Narratives, Cultural Biases, and Frames

This two-ways interactive process espouses the normative, substantive, and instrumental principles of participation (Voinov and Bousquet 2010; Salter et al. 2010). Crucially, this collaborative process incorporates cultural concerns of insiders through the guidance of the local champion. Cultures in the global South value elders as custodians of community values and rely on them to pass down local knowledge to younger generations. These elders, in most cases illiterate, are the gate keepers and experts in these local communities. Their wisdom of the local environment acquired over long periods from interaction with their surroundings should serve as another entry point for engaging in climate change risk communication. The use of participatory platforms and methods such as narratives and visuals can enhance the climate risk communication message in this ways.

Visualisation

Visuals have a great potential of motivating and engaging people in climate action (Nicholson-Cole 2005) and the success of Greenpeace is a case in point (Doyle 2007). The potential for visuals such as images and photographs is inherent in their ability to simplify complex issues such as climate change into a universal and easy to understand visual language. The abstract nature, long temporal and large spatial scales, and the uncertainty of climate change makes it an issue of little personal concern to many people all over the world (Nicholson-Cole 2005). Unfortunately, this include the educated elite that know about climate change but do not take personal actions to halt it-leading to an attitude-behaviour divide (Morton et al. 2011). It is noteworthy that responding to climate change will require actions from all actors and sectors in any society. The reliance of rural communities in developing countries on climate sensitive natural resources make them vulnerable to the negative impacts of climate change (Niang et al. 2014). These actors have to perceive climate change in order to respond effectively to it (Adger et al. 2009; Bryan et al. 2009). Planned adaptation is crucial to protect rural livelihoods and preserve their cultures that are under threat from climate change. Genuine climate change communication should use images and photographs to encourage participation of these actors in planned adaptation efforts. The development, selection, and use of visuals is not without its own challenges as Nicholson-Cole (2005) warns that interpretation of images and photographs is influenced by a variety of factors including prior risk perceptions, heuristics and biases, and the demographic, judgemental, and social contexts and so forth (see Table 2). Attention has to be paid to ethical and methodological issues by acknowledging the inherent subjectivity, incorporating realism while maintaining simplicity, and using emotive imagery and so forth (Nicholson-Cole 2005).

Despite this, there are numerous benefits of using visuals in climate change risk communication that include conveying strong and easy to remember messages, condensing complex information, providing the basis for personal thoughts and conversations, and contributing to people's memory and issue-awareness (Nicholson-Cole 2005). However, some best practices for the development and use of visuals in climate change risk communication exists across the environmental visualisation literature. For example, Nicholson-Cole (2005) found that using locally relevant 'emotive' imagery of national, local, and personal significance can have more resonance with the locals and increases their self-efficacy. In light of this, we suggest using locally sourced visuals (images, photographs, and maps) in the context of rural communities in developing countries which might overcome some challenges inherent in the climate change risk communication domain highlighted in the previous paragraphs. It should always be borne in mind that the symbolism carried in visuals can trigger emotional reactions that attract people's attention and may motivate them to act but they can also arouse defensive psychological responses that leads to climate inaction. Hence, it is crucial to develop, select, and use visuals with the aid of the local champion(s) who are familiar with the cultural dos and don'ts.

Narratives

Traditional ecological knowledge is an integral part of the oral culture of rural people in the global South. This traditional ecological knowledge is passed on from one generation to another through narratives (Codjoe et al. 2014; Egeru 2012; Alexander et al. 2011; Mengistu 2011). Recognising this inter-generational transfer of knowledge through oral narratives presents an opportunity for collaboration of

Table 2 Factors influencing risk perceptions

Factors that influence risk perception			Risk perception references
Psychological (Dessai et al. 2004)	Socio-cognitive (Personal/ Vicarious experience & Behaviour)	Heuristics	Marx et al. (2007), Grothmann and Patt (2005), Patt and Schroter (2008), Frank et al. (2010), Wiid and Ziervogel (2012), Adger et al. (2009), Paton et al. (2008) and Paton et al. (2001)
		Biases	
	Knowledge	Local knowledge	Adger et al. (2009)
		Scientific knowledge	
Social (Dessai et al. 2004)	Socio-economic	Household structure	Frank et al. (2010) and Grothmann and Patt (2005)
	Economic	Private property ownership	Grothmann and Patt (2005) and Perry and Lindell (2008)
	Socio-physical	Hazard environment (hazard proximity, frequency, magnitude)	Perry and Lindell (2008), Paton et al. (2008), Gaillard (2008), Adger et al. (2009), Frank et al. (2010), Grothmann and Patt (2005) and Wiid and Ziervogel (2012)
	Demographic	Age, gender, household structure, education level	Frank et al. (2010)
Moral (Dessai et al. 2004)	Values	Culture (hierarchical, individualistic, egalitarian, fatalism, & hermitic)	Wiid and Ziervogel (2012), Leiserowitz (2006), Weber and Hsee (1998) and Weber (2006)
Institutional (Dessai et al. 2004)	Trust deficit	Trust of public opinion	Wiid and Ziervogel (2012) and Paton et al. (2008)
	Participation	Community participation in risk management	Paton et al. (2008), Grothmann and Patt (2005), Gaillard (2008) and Paton et al. (2008)
	Risk management	Risk communication (information)	Perry and Lindell (2008), Grothmann and Patt (2005: 209) and Paton et al. (2008)
Cultural (Dessai et al. 2004)	Religious beliefs	Natural disasters are acts of gods and ancestors	Schmuck (2000), Grothmann and Patt (2005) and Gaillard (2008)
	Local customs	Environmental cues or signs used for traditional predictions and climate forecasts	Marx et al. (2007)

Source Author's own synthesis

multiple actors in climate change risk communication. Additionally, traditional ecological knowledge is an important source of climate information in areas with limited instrumental records (Alexander et al. 2011). Crucially, this knowledge can fill in the gap left by climate information tools that focus on international, regional, and country-level spatial resolutions only. This climate information can be useful to outsiders contemplating climate adaptation initiatives in local communities. Oral narratives (including storylines) from insiders including local elders should be incorporated in the two-ways climate change risk communication exercises.

Frames

Narratives emanating from the local residents (insiders) might be less difficult for experts (outsiders) to decipher but the ability of insiders to interpret climate risk messages developed by outsiders might be complicated by many factors including illiteracy and prior perceptions (see Table 2). Morton et al. (2011, p. 103) acknowledge that "communicating climate change in ways that inspire individual action is complicated by a number of features inherent to the climate change message" such as the intangible nature of climate change, the publics' expectations of science and scientists, and the uncertainty in climate change science that invites public doubts. Empirical studies (Morton et al. 2011, p. 104; Spence and Pidgeon 2010) have shown that the way in which a climate risk communication message is framed has a bearing on human agency. Frames affects the feelings of self-efficacy-which is an ability to take effective action (i.e. through taking insurance). Prospect theorists Tversky and Kahneman (1985) found that framing a message positively had induced their study participants to take caution when confronted with difficult and life threating choices. They also discovered that framing risk communication messages in a negative manner encouraged their participants to take risky decisions in the face of life threating choices. Their findings are corroborated by Morton et al. (2011) who discovered that participants were willing to make lifestyle changes in order to curb the negative impacts of climate change when they heard positively framed messages. Negatively framed messages (i.e. it is 20% chance likely that global warming of 2 °C will make a quarter of all species extinct) had an opposite effect leading participants to cope rather than adapt to climate change. Typically, coping can include defensive psychological triggers such as avoidance, denial, and "drop in the ocean" sentiments leading to an unhinged risky behaviour (Morton et al. 2011).

In general, (climate) risk communication scholars agree that positive frames encourage feelings of efficacy and negative frames induces powerlessness to do anything about the wicked problem of climate change. In all of these, the message been communicated is the same but subtle variations in how it is framed has proved to have a significant effect on the audience (Morton et al. 2011; Spence and Pidgeon 2010; Tversky and Kahneman 1985). Encouraging a climate friendly behaviour within local communities can be realised by moving beyond the climate risk communication phase to highlight the opportunities that can accrue from taking climate action. Since climate adaptation does not only include dealing with the negative impacts but also taking advantage of the positive opportunities that it brings along (World Bank 2012; Morton et al. 2011). Also, highlighting local capacities through asset-based community development can create pathways towards realising such opportunities and increasing feelings of self-efficacy.

Concluding Remarks

The negative impacts of climate change are been felt by rural communities in the global South with drought and famine causing tribal conflicts in the horn of Africa and floods inducing catastrophic landslides in South Asia. Development cooperation organisations are assisting rural communities in the global South to adapt to these changes. But, climate change risk communication messages developed from a Western worldview might fall on deaf ears. We propose a climate change risk communication typology that creates a dialogue between outsiders and the locals (insiders). We went further by highlighting the contextual factors that should be heeded when designing a climate risk communication message for a rural audience in the global South. These include acknowledging their daily livelihoods struggles, the important roles of elders and community leaders that are experts in traditional ecological knowledge, the role of local champions in steering the adaptation endeavours, the relevance of locally-sourced and culturally-sensitive visuals and narratives, and positive frames manipulation. Confirmatory empirical studies are required to validate the climate change risk communication typology and these contextual factors.

References

Adger, W. N., Goulden, S., Dessai, M., Hulme, M., Lorenzoni, I., Nelson, D. R., et al. (2009). Are there social limits to adaptation to climate change? *Climatic Change, 93*, 335–354. https://doi.org/10.1007/s10584-008-9520-z.

Alexander, C., Bynum, N., Johnson, E., King, U., Mustonen, T., Neofotis, P., … Vicarelli, M. (2011). Linking indigenous and scientific knowledge of climate change. *BioScience, 61*(6), 477–484.

Arnstein, S. R. (1969). A ladder of citizen participation. *Journal of the American Institute of Planners, 35*(4), 216–224.

Artur, L., & Hilhorst, D. (2012). Everyday realities of climate change adaptation in Mozambique. *Global Environmental Change, 22*(2), 529–536.

Bryan, E., Deressa, T. T., Gbetibouo, G. A., & Ringler, C. (2009). Adaptation to climate change in Ethiopia and South Africa: Options and constraints. *Environmental Science & Policy, 12*, 413–426.

Codjoe, S. N. A., Owusu, G., & Burkett, V. (2014). Perception, experience, and indigenous knowledge of climate change and variability: The case of Accra, a sub-Saharan African city. *Regional Environmental Change, 14*(1), 369–383.

Dessai, S., Adger, W. N., Hulme, M., Turnpenny, J., Köhler, J., & Warren, R. (2004). Defining and experiencing dangerous climate change: An editorial essay. *Climatic Change, 64,* 11–25.

Doyle, J. (2007). Picturing the clima (c) tic: Greenpeace and the representational politics of climate change communication. *Science as Culture, 16*(2), 129–150.

Egeru, A. (2012). Role of indigenous knowledge in climate change adaptation: A case study of the Teso Sub-Region, Eastern Uganda. *Indian Journal of Traditional Knowledge, 11*(2), 217–224.

Frank, E., Eakin, H., & López-Carr, D. (2010). Risk perception and adaptation to climate risk in the coffee sector of Chiapas, Mexico. In: Conference on international research on food security, natural resource management and rural development. ETH Zurich, September 14–16, 2010.

Gaillard, J.-C. (2008). Alternative paradigms of volcanic risk perception: The case of Mt. Pinatubo in the Philippines. *Journal of Volcanology and Geothermal Research, 172,* 315–328.

Grothmann, T., & Patt, A. (2005). Adaptive capacity and human cognition: The process of individual adaptation to climate change. *Global Environmental Change, 15,* 199–213.

IPCC. (2012). Managing the risks of extreme events and disasters to advance climate change adaptation. In C. B. Field, V. Barros, T. F. Stocker, D. Qin, D. J. Dokken, K. L. Ebi, M. D. Mastrandrea, K. J. Mach, G.-K. Plattner, S. K. Allen, M. Tignor, & P. M. Midgley (Eds.), *A special report of working groups I and II of the intergovernmental panel on climate change* (p. 582). Cambridge, UK, New York, NY, USA: Cambridge University Press.

Leiserowitz, A. (2006). Climate change risk perception and policy preferences: The role of affect, imagery, and values. *Climatic Change, 77*(1–2), 45–72.

Lemos, M. C., Finan, T. J., Fox, R. W., Nelson, D. R., & Tucker, J. (2002). The use of seasonal climate forecasting in policymaking: Lessons from Northeast Brazil. *Climatic Change, 55*(4), 479–507.

Luseno, W. K., McPeak, J. G., Barrett, C. B., Little, P. D., & Gebru, G. (2003). Assessing the value of climate forecast information for pastoralists: Evidence from Southern Ethiopia and Northern Kenya. *World Development, 31*(9), 1477–1494.

Maponya, P., Mpandeli, S., & Oduniyi, S. (2013). Climate change awareness in Mpumalanga Province, South Africa. *Journal of Agricultural Science, 5*(10), 273.

Marx, S. M., Weber, E. U., Orlove, B. S., Leiserowitz, A., Krantz, D. H., Roncoli, C., et al. (2007). Communication and mental processes: Experiential and analytic processing of uncertain climate information. *Global Environmental Change, 17,* 47–58.

Mengistu, D. K. (2011). Farmers' perception and knowledge on climate change and their coping strategies to the related hazards: Case study from Adiha, central Tigray, Ethiopia. *Agricultural Sciences, 2*(02), 138.

Morton, T. A., Rabinovich, A., Marshall, D., & Bretschneider, P. (2011). The future that may (or may not) come: How framing changes responses to uncertainty in climate change communications. *Global Environmental Change, 21*(1), 103–109.

Niang, I., Ruppel, O. C., Abdrabo, M. A., Essel, A., Lennard, C., Padgham, J., et al. (2014). Africa. In V. R. Barros, C. B. Field, D. J. Dokken, M. D. Mastrandrea, K. J. Mach, T. E. Bilir, M. Chatterjee, K. L. Ebi, Y. O. Estrada, R. C. Genova, B. Girma, E. S. Kissel, A. N. Levy, S. MacCracken, P. R. Mastrandrea, & L. L. White (Eds.), *Climate change 2014: Impacts, adaptation, and vulnerability. Part B—Regional aspects. Contribution of working group II to the fifth assessment report of the intergovernmental panel on climate change* (pp. 1199–1265). Cambridge, UK, New York, NY, USA: Cambridge University Press.

Nicholson-Cole, S. A. (2005). Representing climate change futures: A critique on the use of images for visual communication. *Computers, Environment and Urban Systems, 29*(3), 255–273.

Nkoana, E. M., Waas, T., Verbruggen, A., Burman, C. J., & Hugé, J. (2016). Analytic framework for assessing participation processes and outcomes of climate change adaptation tools. *Environment, Development and Sustainability*, 1–30.

Nyong, A., Adesina, F., & Elasha, B. O. (2007). The value of indigenous knowledge in climate change mitigation and adaptation strategies in the African Sahel. *Mitigation and Adaptation Strategies for Global Change, 12*(5), 787–797.

Paton, D., Johnston, D. M., Beddington, M. S., Lai, C-D., & Houghton, B. F. (2001). Direct and vicarious experience of volcanic hazards: Implications for risk perception and adjustment adoption. *Australian Journal of Emergency Management*, 58–63.
Paton, D., Smith, L., Daly, M., & Johnston, D. (2008). Risk perception and volcanic hazard mitigation: Individual and social perspectives. *Journal of Volcanology and Geothermal Research, 172,* 179–188.
Patt, A. G., & Schroter, D. (2008). Perceptions of climate risk in Mozambique: Implications for the success of adaptation strategies. *Global Environmental Change, 18,* 458–467.
Perry, R. W., & Lindell, M. K. (2008). Volcanic risk perception and adjustment in a multi-hazard environment. *Journal of Volcanology and Geothermal Research, 172,* 170–178.
Roberts, D. (2008). Thinking globally, acting locally—Institutionalizing climate change at the local government level in Durban, South Africa. *Environment and Urbanization, 20*(2), 521–537.
Roncoli, C., Ingram, K., & Kirshen, P. (2002). Reading the rains: Local knowledge and rainfall forecasting in Burkina Faso. *Society & Natural Resources, 15*(5), 409–427.
Salter, J., Robinson, J., & Wiek, A. (2010). Participatory methods of integrated assessment—A review. *Wiley Interdisciplinary Reviews: Climate Change, 1*(5), 697–717.
Schlumpf, C., Behringer, J., Dürrenberger, G., & Pahl-Wostl, C. (1999). The personal CO_2 calculator: A modeling tool for participatory integrated assessment methods. *Modeling and Environmental Assessment, 4*(1), 1–12.
Schmuck, H. (2000). "An act of Allah": Religious explanations for floods in Bangladesh as survival strategy. *International Journal of Mass Emergencies and Disasters, 18*(1), 85–96.
Siebenhüner, B., & Barth, V. (2005). The role of computer modeling in participatory integrated assessments. *Environmental Impact Assessment Review, 25*(4), 367–389.
Spence, A., & Pidgeon, N. (2010). Framing and communicating climate change: The effects of distance and outcome frame manipulations. *Global Environmental Change, 20*(4), 656–667.
Speranza, C. I., Kiteme, B., Ambenje, P., Wiesmann, U., & Makali, S. (2010). Indigenous knowledge related to climate variability and change: Insights from droughts in semi-arid areas of former Makueni District, Kenya. *Climatic Change, 100*(2), 295–315.
Tversky, A., & Kahneman, D. (1985). The framing of decisions and the psychology of choice. *Environmental impact assessment, technology assessment, and risk analysis* (pp. 107–129). Berlin. Heidelberg: Springer.
Van Aalst, M. K., Cannon, T., & Burton, I. (2008). Community level adaptation to climate change: The potential role of participatory community risk assessment. *Global Environmental Change, 18*(1), 165–179.
Van Opstal, M., & Hugé, J. (2013). Knowledge for sustainable development: A worldviews perspective. *Environment, Development and Sustainability, 15*(3), 687–709.
Voinov, A., & Bousquet, F. (2010). Modelling with stakeholders. *Environmental Modelling and Software, 25*(11), 1268–1281.
Weber, E. U. (2006). Experience-based and description-based perceptions of long-term risk: Why global warming does not scare us (yet). *Climatic Change, 77,* 103–120. https://doi.org/10.1007/s10584-006-9060-3.
Weber, E. U., & Hsee, C. (1998). Cross-cultural differences in risk perception, but cross-cultural similarities in attitudes towards perceived risk. *Management Science, 44*(9), 1205–1217.
Wiid, N., & Ziervogel, G. (2012). Adapting to climate change in South Africa: Commercial farmers' perception of and response to changing climate. *South African Geographical Journal, 94*(2), 152–173.
World Bank. (2012). *Mainstreaming adaptation to climate change in agriculture and natural resources management projects-guidance notes* (pp. 1–24). New York. Retrieved from http://siteresources.worldbank.org/EXTTOOLKIT3/Resources/3646250-1250715327143/GN8.pd.

Part IV
Food Security and Climate Change: Adapting Agriculture to Climate Change

Improving Water Productivity in Irrigated Agriculture: Challenges from Climate Change and New Water Resources Paradigms

José Manuel Gonçalves and Isabel Pedroso de Lima

Abstract The increasing pressure on water resources is due to continuously growing consumption, which itself results from demographic and economic developments and the effect of expected climate change. Therefore, action to encourage the more efficient and productive use of water is urgent. Water management needs to focus on regulating the use of this limited resource by multiple users; the allocation of water to non-agricultural users, in particular, is increasing steadily. Moreover, increased variability in climate regimes is expected to modify the volume, temporal and spatial distribution of water storage and fluxes, and therefore to affect the distribution, availability and sustainability of regional water resources. Irrigated agriculture is one of the sectors also facing new challenges. Given that water scarcity is expected to worsen in large agricultural production areas, one such challenge is how to maintain or increase sustainable agricultural production under growing water use restrictions. Improvements in agricultural water management should seek to conserve not only water but energy and soil, too, while still satisfying society's relentless demand for high quality food and fiber crops, and livestock, aquatic and forest products. This work addresses the main problems related to this topic, illustrated with case studies from Europe and Asia.

Keywords Agricultural water management · Climate change · Irrigation water productivity · Water use indicators

J. M. Gonçalves (✉)
College of Agriculture, Polytechnic of Coimbra, 3045-601 Coimbra, Portugal
e-mail: jmmg@esac.pt

J. M. Gonçalves
Centro de Investigação em Agronomia, Alimentos, Ambiente e Paisagem (LEAF), Instituto Superior de Agronomia, Universidade de Lisboa, Tapada da Ajuda, 1349-017 Lisbon, Portugal

I. P. de Lima
Marine and Environmental Sciences Centre (MARE), Department of Civil Engineering, University of Coimbra, Campus 2, Rua Luis Reis Santos, 3030-788 Coimbra, Portugal
e-mail: iplima@uc.pt

F. Alves et al. (eds.), *Theory and Practice of Climate Adaptation*,
Climate Change Management, https://doi.org/10.1007/978-3-319-72874-2_20

Introduction

Water resources are under the pressure of continuously growing water consumption, due in equal measure to demographic and economic development and to expected changes in the climate. The sustainability of this resource demands a more efficient and productive use of water, which should appreciate that society must share water between domestic, industrial, agricultural, environmental and recreational purposes. While water management is adjusting the limited resource to its multiple users, it is likely that the allocation to non-agricultural users is progressively increasing. Along with this, increased variability emerging in climate systems and regimes and in the hydrological cycle is expected to modify the volume and the temporal and spatial distribution of water storage and fluxes. This is expected to affect the distribution, availability and sustainability of regional water resources, thus impacting several countries' water resources (e.g., UNDP 2006). Although it might also aggravate extreme climate events, this is usually of less concern to the water sector. Expected changes in the climate of anthropic origin are likely to affect surface air temperature and precipitation regimes, for example, which will affect society and the environment. The local impact of those changes on the water cycle is bound up with a large range of spatial and temporal scales and is often poorly understood, but it will include water availability and demand issues. Thus, there is an urgent need to establish measures to adapt to and mitigate the impact of climate change scenarios anticipated by different models.

Despite the uncertainty in model predictions, it is very likely that climate change will produce an increase in water demand for agricultural production, significantly influence agricultural practices and water management, and cause disturbances in the demand for water for natural ecosystems. These disturbances will produce imbalances, especially in regions facing water scarcity, which are also highly vulnerable to climate change impacts (Rosenzweig and Hillel 1998). Since water scarcity can have physical, economic or management-induced causes, impacts might include altered phenological cycles, changes in the growth of crops, crop water requirements, water availability for crop production, crop irrigation requirements, biotic stresses such as competition from weeds, pests and diseases, and economic and social farming issues (Pereira 2011). Rainfed crops are more vulnerable than irrigated ones due to changes in precipitation, infiltration, evapotranspiration and soil moisture regimes. Food security is therefore threatened in the more vulnerable regions and countries of the world. Changes in rainfall regimes will induce changes in streamflow regimes; in many regions, lower base flows are expected. Water availability is expected to decline, thus enhancing competition between users and making it more difficult to satisfy the increasing urban demand. Coping with water scarcity requires appropriate mitigation and adaptation measures.

Improvements in agricultural water management require water conservation actions to be taken. Of the various sectors that depend on water, irrigated agriculture is one of those facing new challenges, since water scarcity is expected to worsen in areas of large agricultural production (e.g., IPCC 2014). It is mandatory that these measures include a paradigm shift in water resources management strategies.

This paper addresses the main problems related to water use and productivity and adaptations in irrigated agriculture. The discussion is illustrated with case studies from Europe and Asia that show different types of solutions that could be applied to mitigate water scarcity problems.

Literature

The dynamics of water use in irrigation is presented in Fig. 1. The agricultural fields receive water from precipitation and diverted from rivers or groundwater. The water is transported by canals or pipes through conveyance and distribution systems. Seepage and runoff occur during this process. In open channels, non-crop evapotranspiration also occurs. The water effectively applied to the cropped field can also include reused drainage water. The beneficial use corresponds mainly to the crop evapotranspiration required for crop production. "Wasted" water is the non-consumed fraction of good quality water collected in the drainage network that is not reused. Water losses correspond to non-beneficial evaporated water and water that is non-reusable and non-consumptive because of its degraded poor quality. The relative importance of the several flow paths depends on the climate, hydrology,

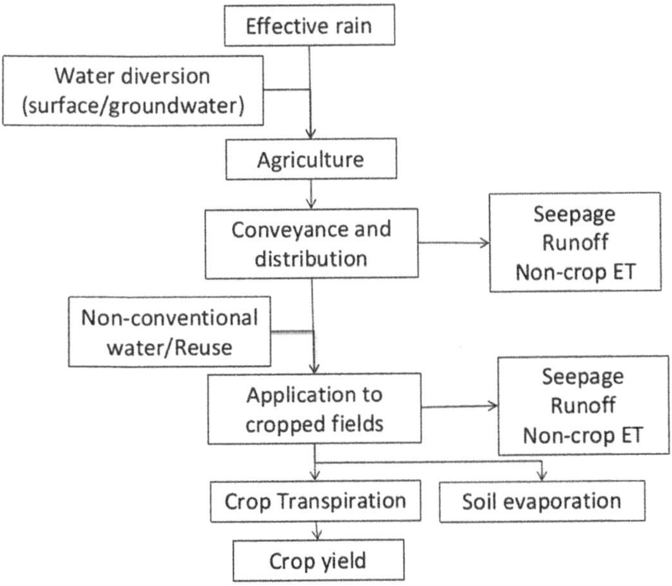

Fig. 1 Water flows in irrigated agriculture

agriculture and irrigation systems' characteristics. This analysis is very important to knowing which and how much water is lost or can be reused on the farm or by other downstream users (Pereira et al. 2009).

Society, mainly influenced by the expansion of urban systems but also by more awareness of the potential climate change impacts, is urging irrigated agriculture to save water. Examples of measures are the (i) decrease of water consumption, (ii) payment of the full cost of water spending in different uses, (iii) improvement of irrigation efficiency and water productivity. These measures are generally accepted, but their effective application needs a major adaptation to local and real conditions. Nevertheless, irrigation will play a leading role in feeding the ever-growing world population (Alexandratos and Bruinsma 2012). Thus, increasing water productivity, the concept of "more crop per drop", is accepted as a main objective of irrigated agriculture. The challenge is how to maintain or increase sustainable agricultural production when there is less available water. Additionally, water savings in agriculture must be compatible with the farmers' technical know-how and their farms' economic sustainability. A first question to answer is therefore understanding how the water is used, and its relationship with technological, cultural and social aspects (Lenton 2014).

The severe reduction of water supply to irrigated agriculture often implies serious social, economic and environmental impacts, particularly for poor peasant farmers. So, water use in agriculture requires an assessment of its value to understand the full effects of change. It is also necessary to find technical solutions related to agronomic and irrigation practices that can adapt these systems to cope with climate change. However, despite our current understanding of global warming and its anthropic origin, the scenarios of climate change foreseen by different models are still uncertain at regional or local scales. This is important for defining appropriate measures for agriculture, which are typically influenced by local scale considerations. One difficulty in studies of the atmosphere is the untangling of natural and anthropic variability in climate variables (e.g., temperature, precipitation); another difficulty is the forecasting of atmospheric processes at larger scales: fortnightly, monthly, seasonal scale (see, e.g., Lovejoy and de Lima 2015; Lovejoy et al. 2017). The benefits of such forecasting for agriculture and water management would be manifold.

Water productivity has to be considered in water use analyses, but it is a concept that depends on several factors and the scale of analysis. In broad terms, water productivity concerns the benefits resulting from using water; however, the issue is often tackled by specialists who do not use a common framework to discuss water productivity. It is generically defined as the ratio between the actual yields achieved and the water used, a concept with several specific definitions related to yield measurements (e.g., physical or economic) and the water use (e.g., transpiration, irrigation or the total use). Yield and water productivity are functions of the water available; however, they depend on the cropping, soil and climate characteristics, with the yield and water productivity maxima corresponding to different amounts of water applied (Molden et al. 2003). This implies that it is not possible to simultaneously optimize water and land (crop yield) productivity, which pursue different objectives related to socioeconomic farming conditions. Water and land productivity respond differently to water management that aims to optimize the value of water (sometimes not just derived from crops but from the multiple users of water), while farmers aim to optimize their income through improved land productivity.

Water productivity concept is applied to different scales and for different processes and perspectives, including the crop, field, farm, irrigation project or drainage basin (see, e.g., Molden et al. 2003). For example, at the field scale the process focuses on the cropping and irrigation systems. The physical production is highlighted and transpiration and evaporation are the dominant water fluxes. On the other hand, at the irrigation project scale the process refers to the distribution of water to farmers, the system's operation and maintenance, fees and drainage. The production terms are both physical and economic, and the water terms concern the irrigation deliveries, soil water depletion and available water. These are related to the many scales in water use and management in agriculture, which might even extend to national and international scales. At these large scales, for instance, trade and virtual water are relevant. Overall, when the scale of analysis increases, the complexity and competition for water also increase. However, processes between scales are interlinked. For example, an increase in water productivity at a smaller scale (e.g., field) may lead to a lower productivity at larger scales (Molden et al. 2003); on the other hand, the solution to many problems at the basin scale involves actions that are implemented at local scales.

Methods

The adaptation of irrigated agriculture to global changes and water scarcity involves improving: (i) agronomic practices relative to crop and soil management; (ii) irrigation scheduling methods; (iii) on-farm irrigation systems; (iv) water supply and (v) off-farm delivery systems.

(i) Agronomic practices relative to crop and soil management embrace a wide range of technologies for increasing land productivity, reducing crop water demand and conserving soil and water. Examples centered on crop characteristics are: increased photosynthesis through genetic improvements; new fertilization and crop cultivation practices; use of crops of higher economic value; reduction of the crop water demand by using low demand crop or shortening the crop growing period. The adoption of better farming practices creates compatibility between increased water productivity and soil and water conservation: timely sowing, weed control, fertilizer and nutrient management, best crop rotation, application of organic matter to increase soil water-holding capacity, tillage and soil management techniques, surface mulch evaporation and runoff control, weed management to control non-crop evapotranspiration and deep crop rooting, and supplementary irrigation for an effective use of water storage.

(ii) Improved irrigation scheduling is crucial to increasing water productivity. It leads to higher yields, prevents crop stress, particularly at critical periods, and by adopting rational deficit irrigation it cuts water and energy losses. The development of irrigation scheduling models has provided valuable computing tools to support farmers in optimizing their irrigation operation. These new technologies use remote sensing, automation and Web services and can offer new perspectives to improve irrigation water use (Car et al. 2012; Backeberg 2014).

(iii) On-farm irrigation systems have a high potential for improvement, first because current irrigation systems have significant design, maintenance and operation problems, and second because new technologies offer multiple solutions to increase efficiency, while saving water and energy. System field evaluation techniques can help to improve the farm management of these irrigation systems. Other techniques involve design modeling tools, precise irrigation, automation and remote control, fertigation, chemigation and energy saving procedures.

Surface irrigation systems generally require the implementation of a precise land levelling operation. This reduces the advance time and the volume of water needed to complete the advance, allowing greater infiltration uniformity and higher yields. Additional precise land levelling prevents ponding, improves drainage, and allows the adoption of deficit irrigation and better control of the leaching fraction. The operation of field systems should improve control of the inflow rate and cutoff time to minimize runoff and deep percolation. A wide range of techniques have been adapted for the main systems (furrows, borders and basin) to improve irrigation performance: (a) keeping shallow water depths in rice basins, for lower seepage and percolation losses, higher yields and better conditions to store any storm rainfall; (b) furrow irrigation by alternate furrows, or with surge-flow system; (c) graded border or furrows, combined with the reuse of tail water runoff.

Sprinkler irrigation systems comprise the solid set, center-pivot, and traveler-gun types, which might comprise many variants in terms of equipment. The main design and operating techniques of these systems aim to achieve the greatest uniformity of distribution with minimum water losses, optimize the overlapping of sprinkler jets, minimize the discharge variations within each operation system, minimize evaporation and wind drift losses, and adjust the precipitation rate with soil infiltration to avoid runoff losses.

Micro-irrigation systems comprise drip emitters and micro-sprinklers. The good quality of emitters and the maintenance procedures are essential to high irrigation performance. The best practices for irrigation performance optimization should fit the following guidelines: (a) use a single drip line for a double crop row to reduce water use and cost with soils whose lateral water transmission is good enough to wet both rows, and micro-sprayers when the soil has high infiltration to avoid deep percolation losses; (b) adjust the duration and timing of the water application according to the soil and crop characteristics, to control percolation losses and salt distribution and accumulation in the soil; (c) control the pressure and discharge variations within each operating set, by adopting pressure regulators in sloping areas and self-compensating emitters in long and sloping laterals; (d) use filters that are appropriate to the water quality and emitter characteristics and use chemicals periodically to clean the system and treat the clogged emitters.

(iv) The water supply management for water conservation in agriculture could consider, for example, the following approaches to increase the water available for use according to the specific hydrological and geographical conditions: (a) small reservoirs for surface runoff storage, and allowing groundwater recharge from excess runoff; (b) rainwater harvesting and vegetation management to control runoff; (c) micro-catchments, land forming, terracing and conservation tillage; (d) non-conventional resources use, such as urban treated water or drainage water of poor quality, which might involve coping with crop salinity tolerance, or mixing poor and good water quality alongside crop management to reduce salinity effects.

(v) Off-farm conveyance and distribution systems have a wide range of hydraulic structures and equipment (e.g., Masseroni et al. 2017). The systems' performance could be improved by: canal lining; improving water regulation and control devices, particularly through automation and remote control; adopting low pressure pipe distributors; changing from supply oriented to demand oriented delivery schedules; using intermediate storage in canals, reservoirs, or farm ponds; involving farmers in delivery schedule planning to cope with limited supply; adopting demand delivery scheduling in pressurized systems; pricing water as a function of the volumes of water diverted and irrigation scheduling; applying information technology to optimize delivery schedules.

The process of irrigation design to adapt to global changes and water scarcity could be developed through the support of mathematical computer models. It allows to take into consideration the main factors that influence systems' performance, namely the water available, field topography, soil type, and crop water requirements. The design process could be improved by adopting the decision support system methodology or the multicriteria analysis, to also take into consideration economic and environmental constraints and issues.

Following this approach, the methodology applied to simulate alternatives for several irrigation systems (surface, drip and sprinkler systems) considers the models SADREG (Gonçalves and Pereira 2009), MIRRIG (Pedras et al. 2009) and PROASPER (Rodrigues et al. 2010). Each alternative was then characterized by appropriate performance indicators relative to water saving and economic results. To compare and select the alternatives, the multicriteria analysis was chosen because it combines various criteria, often contradictory, of economic, environmental and technical nature, as well as relative to water saving. Furthermore, it admits various schemes for prioritizing and ranking the technical solutions being compared (Ishizaka and Nemery 2013; Rinaldi and He 2014).

Results and Discussion

Technology is available to improve existing irrigation practices, at both farm and system levels. They aim at increasing water productivity by improving biological and economic output and the social and economic conditions of farmers. However, knowledge and technology progress much faster than technology transfer, in particular with regard to peasants with a small family farm. Technology is easily available to farmers when manufacturers have an easy market, as is the case for a variety of irrigation scheduling equipment used by commercial farms and for sprinkler and micro-irrigation equipment. But these more efficient uses of irrigation methods need to be supplemented by the appropriate quality control of equipment and the implementation of management tools to support farmers modernizing their systems.

The farmers' constraints are well known and need to be taken into consideration. They can include inappropriate delivery conditions, high cost of production factors and irrigation equipment, low commodity prices, low incomes, lack of technical assistance and poor system design.

These issues, pertinent to irrigated agriculture facing water scarcity problems, particularly in relation to water use and adaptation measures and solutions, are illustrated with three case studies from Europe and Asia. These examples refer to recent research activity carried out at LEAF center; they comprise several agricultural and climatic conditions. The methodological approach considered the modeling of irrigation systems and the economic and environmental analysis through multicriteria, allowing a general comparison of the several case studies.

Case Study 1—Furrow Irrigation in Fergana Valley, Uzbekistan

The first case study concerns furrow irrigation in Fergana Valley, Uzbekistan, which is a large irrigated oasis in the Aral Sea Basin (Horst et al. 2007; Gonçalves et al. 2011). The area is characterized by an arid climate; the main crops are wheat and cotton, which are furrow irrigated. The excessive (irrigation) water use practised in the Aral Sea Basin led to enormous changes in the Aral Sea and the river ecosystem. Therefore, the sustainability of the system depends on reducing the irrigation demand. With this objective, alternative processes were evaluated together with improvements in irrigation scheduling, and some results are shown in Fig. 2. Since furrow irrigation is by far the dominant method, it was considered that every furrow continuous flow (EC) maximizes yields with minimal water use ratios. In contrast, alternate surge irrigation (AS) does not achieve maximum yields but it does give the best irrigation water productivity, the largest beneficial water use fraction and the largest water consumed fraction (Fig. 2). Alternative improvements were compared by multicriteria analysis using utilities; a dimensionless expression

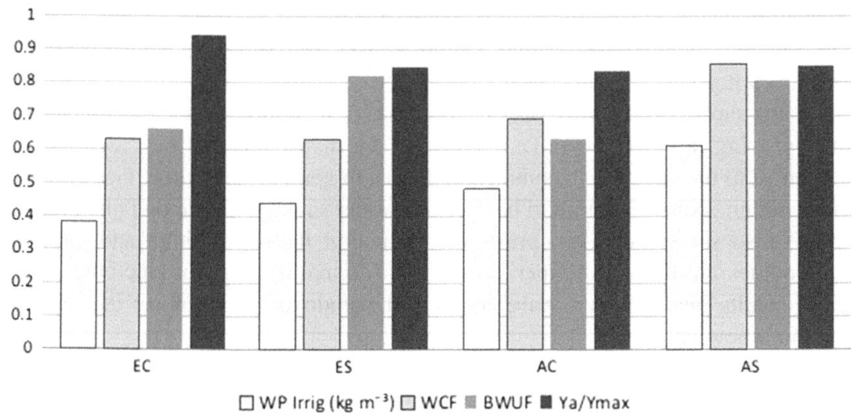

Fig. 2 Case study from Fergana Valley, Uzbekistan: Relative yield (Ya/Ymax), beneficial water use fraction (BWUF), water consumed fraction (WCF), and irrigation water productivity (WP) relative to four improved furrow irrigation conditions: EC and ES—every furrow irrigation with continuous and surge flow, respectively; AC and AS—irrigation of alternate furrows with continuous and surge flow (based on Horst et al. 2007)

of the value of a criterion or alternative according to the decision-maker's perspective was normalized into the [0–1] interval, with zero for the most adverse and 1 for the most advantageous result. Figure 3 presents the utilities, showing that for the actual farming economic conditions in Fergana, the best alternatives are the current one—no change—or adopting an improved irrigation schedule with alternate furrow irrigation (Sched + AC, in Fig. 3). If the multicriteria option were

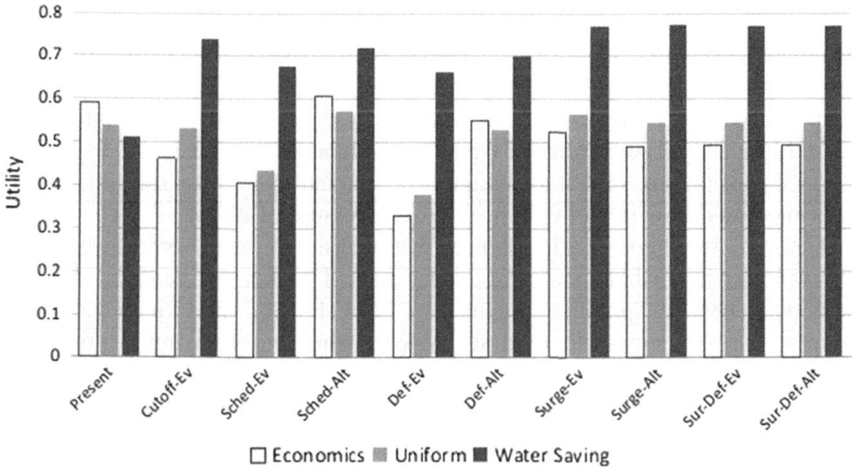

Fig. 3 Case study from Fergana Valley, Uzbekistan: Global utilities relative to several alternative furrow irrigation improvements when criteria relate to water saving priorities, farm economic results priorities or a uniform weighting is adopted (from Gonçalves et al. 2011)

dictated by water saving criteria, then the best solutions are those relating to surge flow applied to alternate furrows (Surge + Alt). Furthermore, deficit irrigation (Def) does not lead to better results.

Results show that the price of some products (e.g., cotton), does not compensate for improving farm irrigation, i.e. paying for the investment and increased labour. In contrast to the situation in commercial farms, in general surge flow does not offer good enough economic return to be used as a water saving practice on family farms. Results also show that when prices of production factors are high and those of commodities are low, the farmer's option is to maximize land productivity. Once again, results bear out the above mentioned paradoxes and support the need to develop a new paradigm.

Case Study 2—Supplemental Winter Wheat Irrigation in Ras-El-Ain, Northeast of Syria

The second case study concerns supplemental irrigation of winter wheat in the Ras-El-Ain area, in the semi-arid northeast of Syria, which is a great challenge for sustainable water use in agriculture (Darouich et al. 2017). The study compared set sprinkler and border irrigation methods, taking into account their performance in terms of water saving and economic returns when considering mild and moderate-deficit irrigation. The adoption of multicriteria analysis is justified because two central but contradictory issues have to be made compatible: water saving and farm economic returns. Alternative solutions for surface irrigation and for sprinkler systems were developed using design models and two deficit irrigation strategies. Graded borders with and without precise land levelling assumed border lengths of 50, 100 and 200 m and inflow rates from 0.8 to 3.6 $l\ s^{-1}\ m^{-1}$. Semi-permanent solid set sprinkler systems were considered with: (i) two layout types: L1, with the laterals in the field length direction, and L2, with them in the width direction; (ii) two sprinkler spacings: G2, of 12 × 12 m, and G3, of 12 × 15 m; and (iii) two sprinkler types: sp1, with nozzle diameter of 2.38 mm and a flow rate of 0.30 $m^3\ h^{-1}$ at a pressure of 25 m, and sp2, with nozzle diameter of 3.18 mm and a flow rate of 0.55 $m^3\ h^{-1}$ at a pressure of 27 m.

The study (see Figs. 4 and 5) showed that precision levelled borders were better for water saving, while non-precise ones ranked higher for economic returns. Semi-permanent set sprinkler systems have been shown to perform better in economic terms and similar to solid set systems when water saving is prioritized. Semi-permanent sprinkler systems rank first among all types of systems, regardless of the deficit irrigation strategy adopted. It is likely that border irrigation is appropriate when wheat is in rotation with cotton, if the latter is surface irrigated. Overall, it was concluded that appropriate economic incentives and training for farmers are required when implementing innovative approaches.

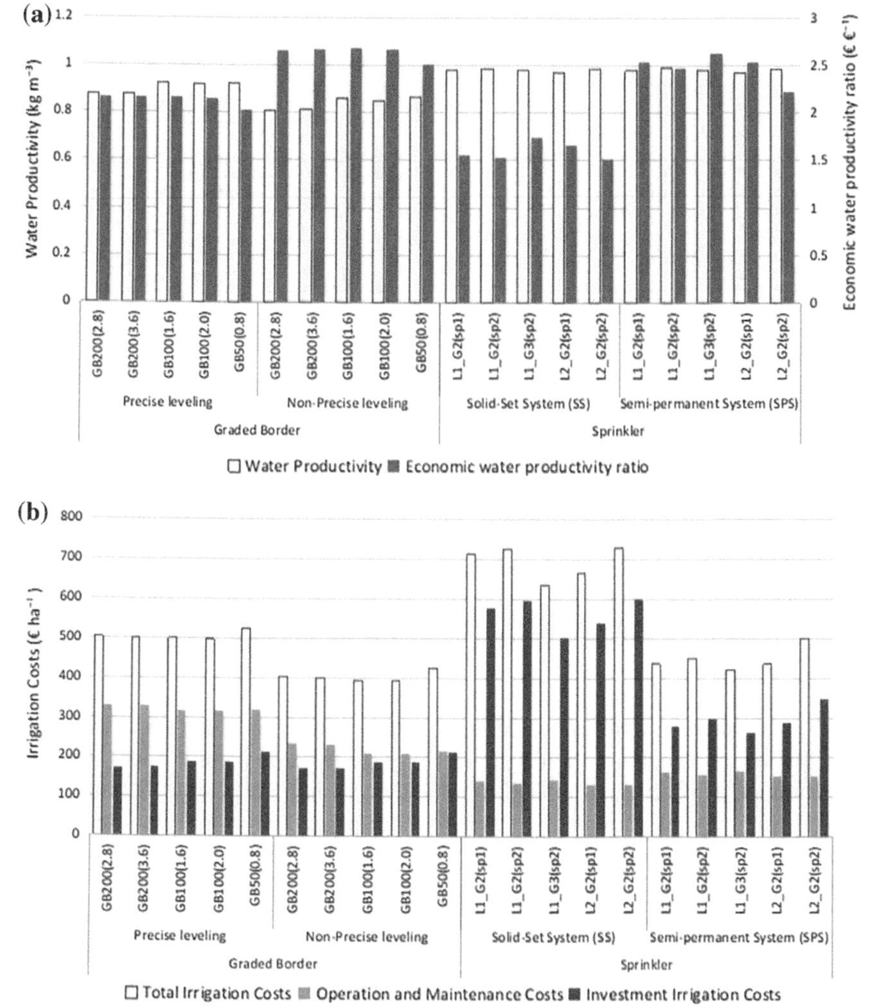

Fig. 4 Case study from Ras-El-Ain area, Syria: Comparing graded borders (GB) and sprinkler irrigation alternatives for **a** water productivity, WP (kg m^{-3}), and economic water productivity ratio, EWPR; and **b** irrigation costs (€ ha^{-1}). Border lengths are 50, 100 and 200 m and inflow rates (l s^{-1} m^{-1}) are in brackets. Sprinklers refer to layouts L1 and L2, spacings are G2 and G3 and sprinkler types are sp1 and sp2 (from Darouich et al. 2017)

Case Study 3—Full and Deficit Irrigated Maize in the South of Portugal

The third case study concerns maize crops in the south of Portugal (Rodrigues et al. 2013). This study aims to assess the economic feasibility of full and deficit irrigated maize using center pivot, set sprinkler systems and drip tap systems. Different

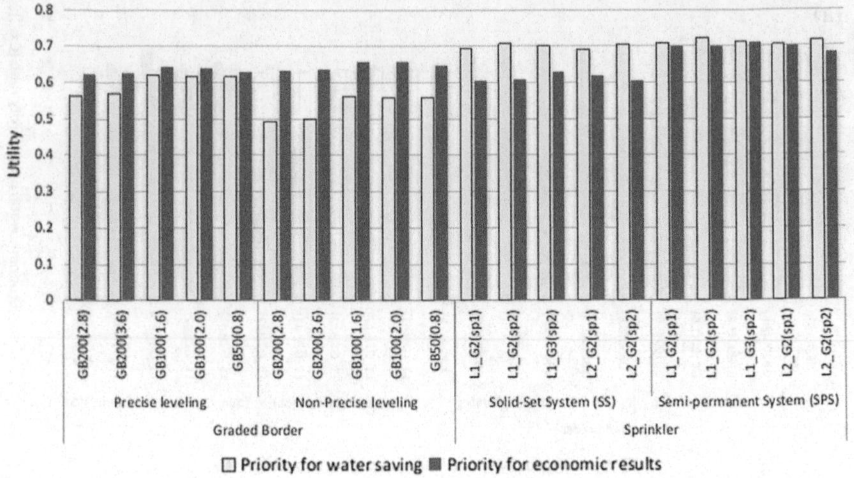

Fig. 5 Case study from Ras-El-Ain area, Syria: Comparing global utilities for solid-set and semi-permanent sprinkler systems with those for precision and non-precision levelled graded borders when priority is assigned to water saving or to economic returns. Sprinkler systems refer to layouts L1 and L2, spacings G2 and G3 and sprinklers sp1 and sp2. Graded borders have 50, 100 and 200 m length and inflow rates (in brackets) are from 0.8 to 3.6 $l\ s^{-1}\ m^{-1}$ (from Darouich et al. 2017)

irrigation treatments were evaluated and compared in terms of beneficial water use and physical and economic water productivity for two commodity prices and three irrigation system scenarios applied to a medium (5 ha) and a large (32 ha) farm.

The study results show that economic water productivity indicators may be an appropriate approach for assessing the impacts of deficit irrigation, mainly considering commodity prices. The economic water productivity ratio EWPR, relating yield values per unit of farming costs, was found to be suitable for assessing the feasibility of deficit irrigation as influenced by commodity prices and irrigation systems (see Fig. 6). Results show that the viability of deficit irrigation strategies is extremely dependent on commodity prices. If low commodity prices are considered, then all the treatments for all the irrigation systems and field sizes lead to a negative income. However, for higher commodity prices, most scenarios lead to a positive income. Drip irrigation systems were found to lead to better water use performance in terms of beneficial water use and water productivity than sprinkler systems did. However, the EWPR were lower for drip than for both set and center-pivot sprinkler systems due to their respective investment costs. Results were also different for the 5 and 32 ha fields; the best results for all treatments were from the set sprinkler for the smaller field and from the center-pivot for the large one, thus evidencing the influence of the higher cost of center-pivot systems when a small field is considered.

This study demonstrated that the adoption of well-designed and managed irrigation systems may lead to contradictory results when the achieved water saving

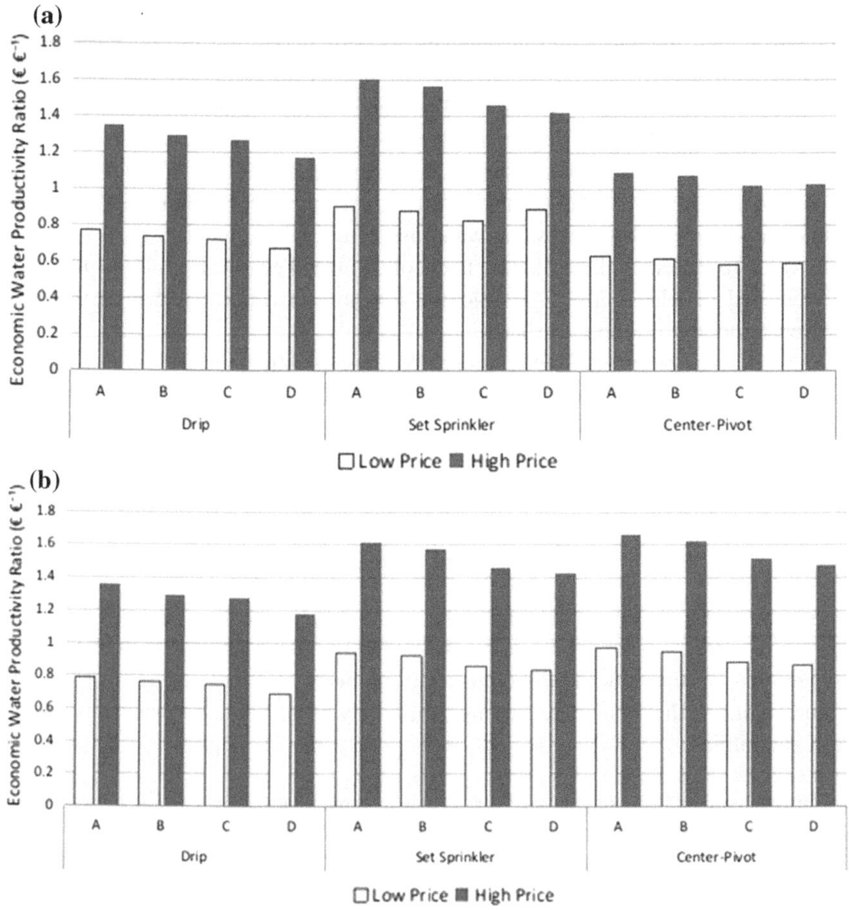

Fig. 6 Case study from the south of Portugal: Economic water productivity ratio (EWPR) for all deficit irrigation treatments and irrigation systems applied to **a** small farm, of 5 ha, and **b** large farm, of 32 ha (from Rodrigues et al. 2013)

does not allow the desired recovery of the investment costs, which also depend on the farm size. This may help policy makers to understand the contradictions between water saving and farm economic results.

Conclusion

The new water resources paradigm consequence of global change implies that new approaches must be adopted to water use by the agricultural sector. The allocation of water to non-agricultural users is progressively increasing. Therefore, water management needs to focus more than ever on the multiple water users. The problem is also impacted by the intensification of variability in climate regimes, which is expected to modify the volume as well as the temporal and spatial distribution of water storage and fluxes. The implications for the distribution, availability and sustainability of regional water resources are yet not clear.

A severe reduction of water supply for irrigated agriculture implies, in many situations that involve poor peasant farmers, serious social, economic and environmental impacts. This means that water use in agriculture requires the assessment of its value to understand the global effects of change. Technical solutions must also be identified in relation to agronomic and irrigation practices so that they can be adapted to cope with climate change.

The three case studies briefly presented here illustrate the question of water saving and irrigated agriculture development in different areas of the world where several new on-farm irrigation technologies are in use, in different decision-making contexts.

It is notable that the adaptation of irrigated agriculture requires the improvement of water productivity, through a change of technology and practices. The new technology available nowadays is a powerful resource to help cope with this problem of increasing agricultural water use. Adaptation requires micro- and macro-economic decisions that must be compatible with the farmers' technical know-how and their farms' economic sustainability. This process calls for society to intervene, specifically the policymakers, engineers and farmers' representatives, to find solutions that constitute a compromise between sustainable water saving and farm economics.

Acknowledgements The support of FCT, Portugal, through the research unit LEAF-Linking Landscape, Environment, Agriculture and Food (UID/AGR/04129/2013), and of FCT and FEDER, through the research project HIRT—"Modelling surface hydrologic processes based on infrared thermography at local and field scales" (PTDC/ECM-HID/4259/2014–POCI-01-0145-FEDER-016668), is acknowledged.

References

Alexandratos, N., & Bruinsma, J. (2012). *World agriculture towards 2030/2050: The 2012 revision* (ESA working paper No. 12-03). Rome: FAO.

Backeberg, G. R. (2014). Innovation through research and development for irrigation water management. *Irrigation and Drainage, 63,* 176–185.

Car, N. J., Christen, E. W., Hornbuckle, J. W., & Moore, G. A. (2012). Using a mobile phone short messaging service (SMS) for irrigation scheduling in Australia farmers' participation and utility evaluation. *Computers and Electronics in Agriculture, 84,* 132–143.

Darouich, H., Cameira, M. R., Gonçalves, J. M., Paredes, P., & Pereira, L. S. (2017). Comparing sprinkler and surface irrigation for wheat using multi-criteria analysis: Water saving vs. economic returns. *Water, 9,* 50. https://doi.org/10.3390/w9010050.

Gonçalves, J. M., & Pereira, L. S. (2009). A decision support system for surface irrigation design. *Journal of Irrigation and Drainage Engineering, 135*(3), 343–356.

Gonçalves, J. M., Horst, M. G., Pereira, L. S., & Muga, A. P. (2011). Furrow irrigation design with multicriteria analysis. *Biosystems Engineering, 109,* 266–275.

Horst, M. G., Shamutalov, S. S., Gonçalves, J. M., & Pereira, L. S. (2007). Assessing impacts of surge-flow irrigation on water saving and productivity of cotton. *Agricultural Water Management, 87,* 115–127.

IPCC (2014) In Core Writing Team, R. K. Pachauri, & L. A. Meyer (Eds.), *Climate change 2014: Synthesis report. Contribution of working groups I, II and III to the fifth assessment report of the intergovernmental panel on climate change* (p. 151). Geneva, Switzerland: IPCC.

Ishizaka, A., & Nemery, P. (2013). *Multi-criteria decision analysis: Methods and software.* Wiley. ISBN: 978-1-1199-7407-9.

Lenton, R. (2014). Irrigation in the twenty-first century: Reflections on science, policy and society. *Irrigation and Drainage, 63,* 154–157.

Lovejoy, S., & de Lima, M. I. P. (2015). The joint space-time statistics of macroweather precipitation, space-time statistical factorization and macroweather models. *Chaos, 25,* 075410. https://doi.org/10.1063/1.4927223.

Lovejoy, S., Del Rio Amador, L., & Hébert, R. (2017). Harnessing butterflies: Theory and practice of the stochastic seasonal to interannual prediction system (StocSIPS). In A. A. Tsonis (Ed.), *Nonlinear advances in geosciences.* Springer Nature (in press).

Masseroni, D., Ricart, S., Cartagena, R. F., Monserrat, J., Gonçalves, J. M., de Lima, I., et al. (2017). Prospects for improving gravity-fed surface irrigation systems in mediterranean European contexts. *Water, 9*(1), 20. https://doi.org/10.3390/w9010020.

Molden, D., Murray-Rust, H., Sakthivadivel, R., & Makin, I. (2003). A water-productivity framework for understanding and action. In J. W. Kijne, R. Barker, & D. Moldem (Eds.), *Water productivity in agriculture: Limits and opportunities for improvement.* Wallingford: CAB International; Colombo: IWMI.

Pedras, C. M. G., Pereira, L. S., & Gonçalves, J. M. (2009). MIRRIG: A decision support system for design and evaluation of microirrigation systems. *Agricultural Water Management, 96*(4), 691–701.

Pereira, L. S. (2011). Challenges on water resources management when searching for sustainable adaptation to climate change focusing agriculture. *European Water, 34,* 41–54.

Pereira, L. S., Cordery, I., & Iacovides, I. (2009). *Coping with water scarcity, addressing and challenges.* Springer.

Rinaldi, M., & He, Z. (2014). Decision support systems to manage irrigation in agriculture. *Advances in Agronomy, 123,* 229–279.

Rodrigues, G. C., Paredes, P., Gonçalves, J. M., Alves, I., & Pereira, L. S. (2013). Comparing sprinkler and drip irrigation systems for full and deficit irrigated maize using multicriteria analysis and simulation modelling: Ranking for water saving vs. farm economic returns. *Agricultural Water Management, 126,* 85–96.

Rodrigues, G.C., Sequeira, B., Paredes, P., & Pereira, L. S. (2010). PROASPER, um SAD para projeto de sistemas de rega por aspersão. In L. S. Pereira, F. R. B. Victória, P. Paredes, M. Garcia, E. Palácios, & A. Torrecillas (Eds.), *Tecnologias para o Uso Sustentável da Água em Rega. Edições Colibri e CEER* (pp. 15–19) (in Portuguese).

Rosenzweig, C., & Hillel, D. (1998). *Climate change and the global harvest. Potential impacts of the greenhouse effects on agriculture.* New York: Oxford University Press.

UNDP. (2006). *Human development report 2006. Beyond scarcity: Power, poverty and the global water crisis.* Houndmills, Basingstoke, Hampshire, New York: Palgrave Macmillan.

Promoting Sustainability Through Agro-industrial Waste Valorisation

Micael Silva, Patrícia Marques, Luís Coelho, Hartmut Nestler, Paula Castro and Cristina Galhano

Abstract Adaptation to climate change and sustainability encompasses several strategies as the agronomic by-products valorisation, transforming waste into resources, contributing to a most efficient use of land and resources and, simultaneously, to circular economy. This work aimed to study the effect of the rice husk (RH) incorporation on substrates in petunia, *Petunia x hybrid*, yield. The bioassay was carried out in 13 cm in diameter pots, randomly arranged, under greenhouse conditions, for 10 weeks. Four treatments were studied: (1) control, substrate without RH; (2) substrate with 10% RH; (3) substrate with 25% RH; and (4) substrate with 50% RH. There were four replicates, each consisting of five plants. Number of flower buds, aerial part diameter and length were evaluated weekly. At the end of the bioassay, aerial part and root weights, and substrate water retention capacity were evaluated. Results were submitted to One-Way ANOVA followed by the Tukey test, when significant differences were present. The obtained results suggest that the incorporation of a low-cost by-product of the rice processing industry—RH in soils with low drainage capacity may be a recommended practice promoting simultaneously the increase of crop production and the use of a low cost and widely available bio-waste, contributing this way to the circular economy premises, reducing wastes and targeting as well a more adapted agro-industry to a changing climate.

Keywords Circular economy · Climate change · Rice husk · *Petunia x hybrid* Sustainable agriculture

M. Silva · P. Marques · L. Coelho · C. Galhano (✉)
Coimbra College of Agriculture—Polytechnic Institute of Coimbra, Bencanta, 3045-601 Coimbra, Portugal
e-mail: cicgalhano@esac.pt

H. Nestler
Leal and Soares, S.A., Zona Industrial de Mira, 3071-909 Mira, Portugal

P. Castro · C. Galhano
Department of Life Sciences, CFE—Centre for Functional Ecology – Science for People & the Planet, University of Coimbra, Calçada Martim de Freitas, 3000-456 Coimbra, Portugal

F. Alves et al. (eds.), *Theory and Practice of Climate Adaptation*, Climate Change Management, https://doi.org/10.1007/978-3-319-72874-2_21

Introduction

The sustainable development of society has been one of the most debated subjects in the last few years. In fact, there is an evident concern on the part of politicians, managers, scientists, and society in general, when they realize that the markedly selfish way of over–exploitation of resources carried out by previous generations, the so-called "developed societies", is compromising our future. There are publication and legislation on issues such as climate change, global warming, loss of biodiversity, scarcity of natural resources, and actively pursuing ways to mitigate these ecological "disasters".

In agriculture, it has been sought to produce high nutritional quality food, free from residues, using the most environmentally harmless agricultural techniques, simultaneously assuring the required production and the desired Agroecosystem equilibrium. The significant increase of Organic Agriculture Production over the past few years is an evidence of this concern (Ghini and Bettiol 2000; IFOAM 2005). Actually, when compared to conventional agriculture, strongly reliant on several external factors, Organic Agriculture is considered a more sustainable agriculture production mode, as it uses techniques that promote soil productivity and the increase of available nutrients, avoiding and excluding the use of almost all synthetic chemical products, such as fertilizers, pesticides or growth regulators (FiBL 2000; Hole et al. 2005).

The 2030 Agenda for Sustainable Development (UN 2015) and its 17 Sustainable Development Goals (SDG) to end poverty, fight inequalities and block climate change, and ensure prosperity for all as part of a new sustainable development agenda, was unanimously approved by the 193 UN member states on the 25th September 2015, at a summit held at the UN headquarters in New York, and officially came into force on 1st January 2016. Each goal has specific targets to be achieved over the next 15 years. Among the specified goals, we would like to highlight the second, the forth, the twelfth, and the thirteenth. The second goal, *End hunger, achieve food security and improved nutrition and promote sustainable agriculture*, demonstrates that we should reconsider the way our food is grown, shared and consumed. Actually, nutritive food will be provide for entirely world population, generating decent profits, whereas supporting people–centred rural development and protecting the environment, if agriculture, forestry and fisheries were correctly done. Currently, terrestrial and water ecosystems, and consequently biodiversity, are being rapidly degraded. Furthermore, climate change is setting even more stress on the resources we depend on, increasing risks associated with disasters such as droughts and floods. The agrifood sector presents key solutions for development, being unquestionably important for hunger and poverty. By 2030, it is intended to guarantee sustainable food production systems and to implement resilient agricultural practices which will help to maintain Ecosystems, and that progressively improve the quality of soil. Goal 4, *Ensure inclusive and quality education for all and promote lifelong learning*, sets that a quality education is fundamental to accomplish people's lives improvement as well as sustainable development. Knowledge and skills necessary to promote sustainable development must be guaranteed to all students through education for sustainable development and sustainable lifestyles. Goal 12, *Ensure sustainable consumption and production patterns*, among other topics, includes sustainable consumption and production through resource and energy efficiency promotion, aiming at "doing more and better with less". Economic activities must increase net wellbeing gains through reduction of resource use, degradation, and pollution, along the whole lifecycle, whereas increasing quality of life. It obliges a systemic approach and cooperation among different stakeholders, involving consumers, researchers, scientists, among others, from producer to final consumer. Adequate information through standards and labels and engaging in sustainable public procurement should be provided to consumers to engage them through awareness-raising and education on sustainable consumption and lifestyles. Therefore, among others, by 2030: sustainable management and efficient use of natural resources should be guarantee; the environmentally sound management of chemicals and all waste throughout their life-cycle should be achieved to diminish their negative impacts on human health and the environment; considerable waste generation reduction through prevention, reduction, recycling and reuse is also desirable as well as the companies encouragement to implement sustainable practices and integrate sustainability information into activity reports. Goal 13, *Take urgent action to combat climate change and its impacts*, can be considered a global challenge with no borders, requiring international level coordinated solutions, taking into account that emissions anywhere affect people everywhere. Climate change is now affecting the whole world, and its impacts can be noticed in different ways, including changing weather patterns, rising sea level, and more extreme weather events. It is closely related to greenhouse gas emissions from human activities. Sustainable, scalable solutions are now available, permitting countries to catapult to cleaner, more resilient economies. Addressing the need to limit the rise of global temperatures, on the 4th of November 2016, the Paris Agreement on climate change entered into force, where all countries agreed to work to limit global temperature rise to well below 2 °C, and given the serious risks, to strive for 1.5 °C. For this, world's energy, industry, transport, food, agriculture and forestry systems must be transformed.

There is also an important issue that should be taken in consideration in this work. On 17th March 2016, the European Commission published a press release (EC 2016) entitled "Circular economy: New Regulation to boost the use of organic and waste based fertilisers" where the real importance of reuse raw materials currently considered waste comes out. This is one of the fundamental principles of the package of circular economy measures adopted in December 2015. The Commission has proposed a regulation that treats similarly the based on waste and the organic fertilizers, and the traditional non–organic fertilizers, simplifying the access to the EU single market for fertilizers of the first group. That regulation seeks to reduce the negative environmental impact by decreasing waste production and energy consumption, creating new market opportunities for innovative enterprises. Currently, only 5% of bio-waste is recycled. Nonetheless, if that percentage increases, bio-waste could replace up to 30% of non-organic fertilizers, which are made from imported resources or energy-intensive processes. The implementation of this directive will allow the conversion of problems into opportunities, i.e. to convert bio-waste into raw materials that can be used in fertilizers production, which can be a useful possibility for farmers and for businesses. By fulfilling all the Directive established common rules (safety, quality and labelling) for the production of bio-waste using fertilizers, they may be freely marketed throughout the EU, guaranteeing the highest level of soil protection as well as health and environmental risks. This regulation will also strongly contribute to the circular economy as value will be create from secondary raw materials that are fed back into the economy. For this, research, innovation and investment will have to develop promptly and the creation of jobs at the local level will be a must.

Rice husk (RH) a bio-waste which results from this crop production. Rice, *Oryza sativa* L., is the second most produced cereal in the world, after maize, constituting, therefore, one of the most relevant crops worldwide. It is consumed daily at least, by two-thirds of the world's population, with Asia, especially Vietnam and Thailand, the continent with the highest rice production, 92% of world production (http://www.ricehusk.com/).

The RH, the outer layer of the rice grain protects the endosperm and the embryo, is a by-product of the rice processing industry with major impact. In 2013, Vietnam alone produced 5.4 million tonnes of RH. As RH represents about 20% of the grain, it is estimated that 35,974 tonnes were produced (http://www.ricehusk.com/).

There are several possible uses of rice husk (Della et al. 2001; Foletto et al. 2005; Kumar et al. 2012; Reddy and Alvarez 2006; Rosa et al. 2009; Mistry 2016). It can be used for the production of energy, being direct combustion, gasification and pyrolysis the most used methods. The combustion of this biomass results in the use of released energy and ash. Ash can be used in the production of pure silica, cement and different types of silicates. It can also be applied as an adsorbent, as support for metal catalysts and in the synthesis of zeolites.

The use of RH in agriculture, as an horticultural conditioner for an organic soil, was pointed out in 1994 by Bellé and Kampf. Steffen et al. (2010) proposed the use of CA and bovine manure as a substrate for multiplication of earthworms and the production of tomato and lettuce seedlings.

Studies on the effect of ash from rice husk, at 2.5, 5, 7.5, 10, 12.5 and 15% concentrations, on okra germination were also carried out, being 2.5% the appropriate concentration to be used as fertilizer in agriculture (Rajor et al. 2011). According to Della et al. (2001), RH contains the main macronutrients P and K, the secondary macronutrients Ca and Mg, as well as Fe and Mn micronutrients and beneficial nutrients such as Na, Al and Si, constituting the latter 97.87%.

In Portugal, in 2014, the rice production was 162.100 tonnes (FAOSTAT 2016). Taking into account that in the central region, the RH is being sold at 60 €/t, its incorporation in organic substrates seems economically worthwhile. Still in Portugal, Atlantic Meals, a company dedicated to the drying, storage and commercialization of rice and maize, is developing a research project that aims to use RH as a renewable biological source with a chemical–energetic potential. The appearance of food additives, pesticides, as well as amorphous silica products, polymers and incorporation in mortars is expected. The intention is to pass RH from waste to raw material (Costa 2012).

The herein presented study came to light as a result of all the aforementioned and also the already existing partnership between a Higher Education Entity, Coimbra College of Agriculture—Polytechnic Institute of Coimbra (ESAC-IPC) and the company SIRO—Leal and Soares, SA. For its part, ESAC is a school integrated in the Polytechnic Institute of Coimbra, globally oriented to the pursuit of the objectives of polytechnic higher education in the scope of technologies, sciences and agricultural and related engineering and tourism and leisure, aiming at train qualified professionals, with a high level of demand in the technological, scientific, professional and cultural aspects, and to carry out research activities, providing also services to the community. The main activity of SIRO—Leal and Soares, SA is the composting of forest waste for the production of substrates and organic compounds, providing a variety of products, from substrates to fertilizers, thus offering integrated solutions, adapted to the established market needs.

The aim of this work was to study the effect of substrate RH incorporation on the production of an ornamental plant, *Petunia* x *hybrida* (common name petunia). It is also important to underline that this work intend to contribute to the achievement of the 2nd, 4th, 12th and, 14th sustainable goals of the 2030 while converting a problem (agro-industrial wastes) into a solution.

Materials and Methods

The experiment was carried out for 10 weeks, from the 21st of February to the 30th of April, 2015, under greenhouse conditions. One petunia plant was transplanted to 10-cm diameter pot with substrate. The study included four treatments: (1) control—substrate without rice husk; (2) substrate with 10% RH; (3) substrate with 25% RH, and (4) substrate with 50% RH. There were four replicates for each treatment and control. Each treatment included five pots which were randomly arranged in the greenhouse. The number of flower buds, and the aerial part diameter and length were weekly registered. Water retention capacity of each treatment substrate was also evaluated. At the end of the experiment, aerial part and root weights were determined. The results were submitted to a one-way analysis of variance (ANOVA), followed by the Tukey test, when significant differences were detected.

Results and Discussion

In a general way, considering the number of flower buds (Fig. 1), no significant differences among treatments were observed for all observation times. However, the number of flower buds greatly increased after the 8th week. It was in treatment 1 (control) that highest numbers of floral buds were produced, 42, followed by

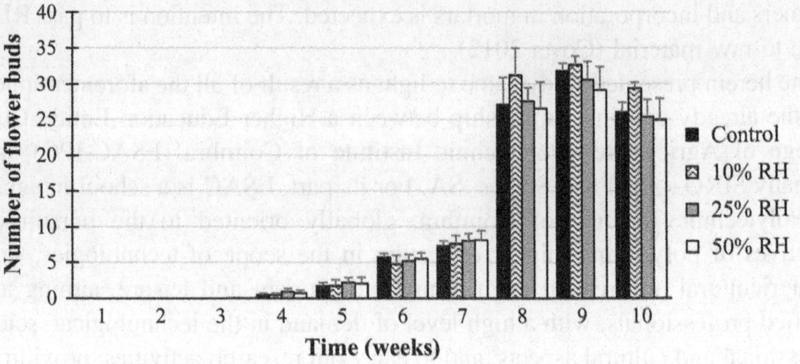

Fig. 1 Number of flower buds of *Petunia x hybrida* plants, over ten weeks, after transplantation to substrate with previous incorporation of rice husk (RH). Results correspond to the mean ± standard deviation of four replicates, each replicate consisting of five plants. Columns followed by different letters, for the same observation time, presented statistically significant differences ($p < 0.05$), according to Tukey test

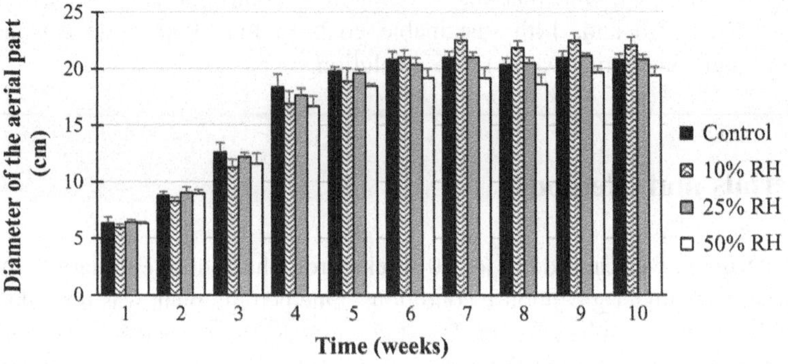

Fig. 2 Diameter of the aerial part of *Petunia x hybrida* plants, over ten weeks, after transplantation to substrate with previous incorporation of rice husk (RH). Results correspond to the mean ± standard deviation of four replicates, each replicate consisting of five plants. Columns followed by different letters, for the same observation time, presented statistically significant differences ($p < 0.05$), according to Tukey test

treatment 2 (substrate with 10% RH). However, considering mean values, treatment 2 was slightly superior (Fig. 1).

After the 5th week, it was also treatment 2 (10% RH) where plants with significantly larger diameters were obtained (Fig. 2), reaching at the end of the test 22.1 cm (mean value), compared to 20.8 and 19.4 cm measured in treatments 3 (25% of RH) and 1 (control), respectively.

Maximum values for aerial part growth were obtained after 10 weeks (Fig. 3). The longest length was once again observed in treatment 2 (10% RH), with 15.6 cm, followed by treatments 4 (50% RH) and control (15.1 cm), and treatment 3 (25% RH) with a similar value (15.0 cm)

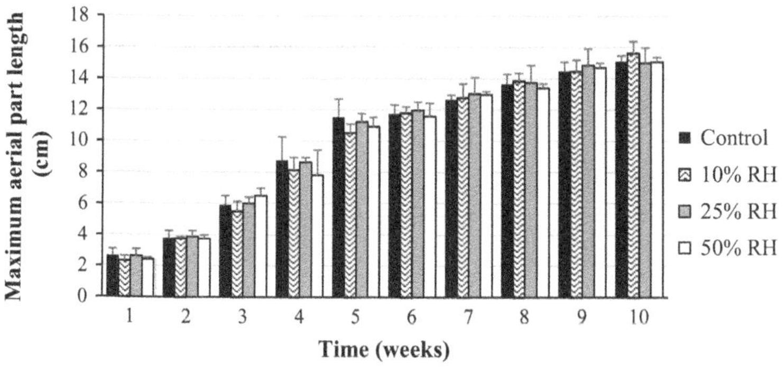

Fig. 3 Maximum length of the aerial part of *Petunia x hybrida* plants over ten weeks after transplantation to the substrate with previous incorporation of rice husk (RH). Results correspond to the mean ± standard deviation of four replicates, each replicate consisting of five plants. Columns followed by different letters, for the same observation time, presented statistically significant differences ($p < 0.05$), according to Tukey test

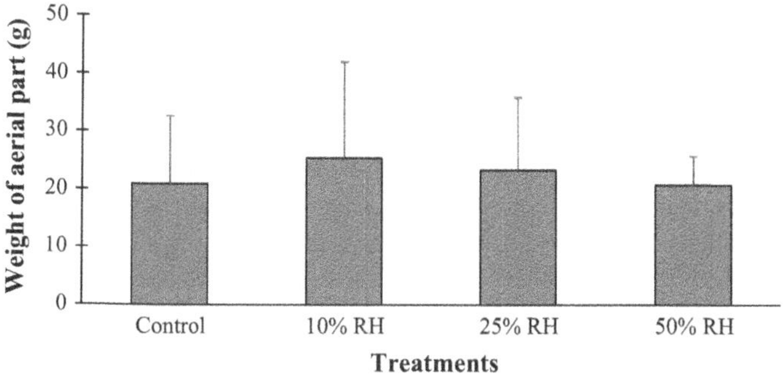

Fig. 4 Weight of the aerial part of *Petunia x hybrida* plants, 10 weeks after transplantation to substrate with previous incorporation of rice husk (RH). Results correspond to the mean ± standard deviation of four replicates, each replicate consisting of five plants. Columns followed by different letters, presented statistically significant differences ($p < 0.05$), according to Tukey test

A larger variability in aerial part of plants' weight was observed among treatments. Nevertheless, mean values were slightly higher in treatment 2 (10% in RH) (Fig. 4). Despite not statistically significant, higher percentages of RH in the substrate showed to affect aerial part of the plant, decreasing its weight. The highest weight value, of 124.7 g, was also verified in treatment 2, followed by control, treatment 3 (25% RH) and treatment 4 (50% RH), with maximum values of 86.5, 79.6 and 78.7 g, respectively.

It was treatment 4 (50% RH) that showed a significantly higher root weight, with a mean value of 20.1 g, (similar to control), when compared with the other substrates: 14.3 and 10.9 g, in treatments 2 (10% RH) and 3 (25% RH), respectively (Fig. 5).

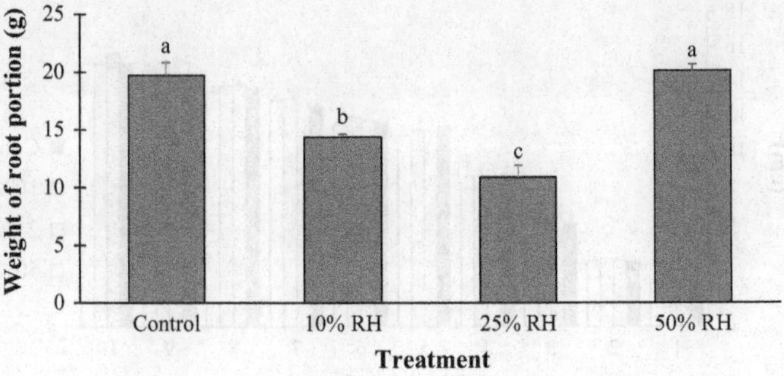

Fig. 5 Weight of the root section of *Petunia x hybrida* plants, ten weeks after transplantation to substrate with previous incorporation of rice husk (RH). Results correspond to the mean ± standard deviation of four replicates, each replicate consisting of five plants. Columns followed by different letters, presented statistically significant differences ($p < 0.05$), according to Tukey test

The volume of water drained from substrate was also analysed in this study (Fig. 6) and it was clearly observed that lower percentages of RH in plant substrate significantly increased water retention capacity. In contrast, treatments 3 (25% RH) and 4 (50% RH) showed significantly lower water retention capacity, 133.8 and 120.8 mL, respectively, when compared to control or substrate with only 10% of RH (Fig. 7).

It was verified that the incorporation of the bio-waste rice husk in substrates, in various percentages, promoted different water retention capacities, having this capacity diminished with the increase of the substrate incorporation. The increased porosity promoted by the RH has clearly led to a lower water holding capacity. In this preliminary study, it was possible to verify that treatment 2 (20% RH) obtained the best results, although no significant differences were detected between control. Most likely, the addition of RH improved the fertility of the substrate; however,

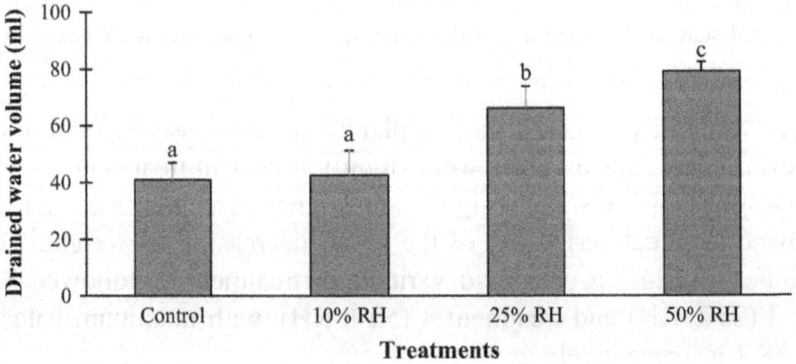

Fig. 6 Volume of water drained from substrate with previous incorporation of rice husk (RH). Results correspond to the mean ± standard deviation of four replicates, each replicate consisting of five plants. Columns followed by different letters, presented statistically significant differences ($p < 0.05$), according to Tukey test

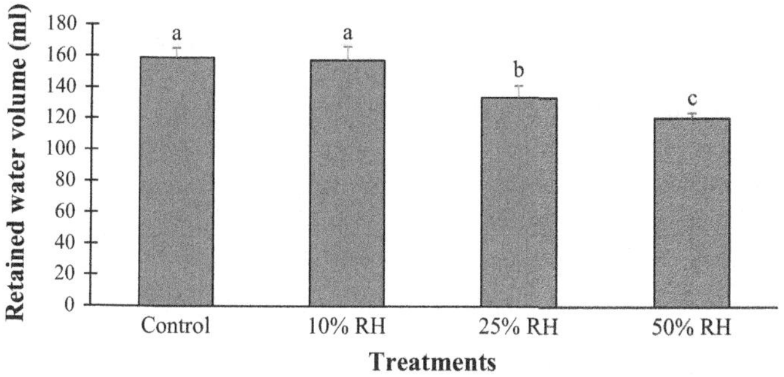

Fig. 7 Volume of water retained by substrate with previous incorporation of rice husk (RH). Results correspond to the mean ± standard deviation of four replicates, each replicate consisting of five plants. Columns followed by different letters, presented statistically significant differences ($p < 0.05$), according to Tukey test

taking into account that its incorporation changes substrate's structure this negatively influenced water retention capacity.

Despite the results documented for water retention capacity, the highest root weight was recorded in treatment 4 (50% RH). This effect can be explained by the fact that root system development in this treatment has been somewhat facilitated accounting the higher porosity present in the substrate. In another study with rice husk ash, Rajor et al. (2011), after testing several concentrations, reached the conclusion that the lowest concentration of 2.5% of the bio-waste, was also found to be more suitable to use as a fertilizer in agriculture.

Nevertheless, the similar values in treatments 4 and control are probably due to different causes. In treatment 4 the higher percentage of RH, causes a greater roots development in order to increase the volume of nutrient and water uptake. On the other hand, in control, as the availability of water and nutrients was higher, roots grew more. Root hair length may depend on soil porosity (Haling 2014) and higher root biomass obtained in treatment 4 (50% RH) may indicate excessive soil drainage of water. Under these conditions, the root system tends to be more developed to search for water (Aguiar 2014). Vergani and Graf (2016) also observed that an increase in root development correspondeds to an increase in soil permeability).

Aerial plant parameters (number of flower buds, diameter and length) seemed to be improved in treatment 2 (10% RH). The interaction dynamics between nutrients made available to the substrate and its porosity and consequent water retention, should be better studied in the future to improve knowledge on how to apply this type of bio-waste in different substrates and soils.

The porosity provided by the RH can be beneficial in soils with poor capacity to drain water. The weight of the root system indicates that the excessive addition of RH is not beneficial because it may lead to water stress when under water shortage events. However, small amounts seemed to be beneficial as they ameliorate aeration and fertility conditions of the substrate.

Conclusion

Climatic variability affects wordwild population and new approaches towards sustainable development must also take into account interactions amoung economic, agricultural and ecological systems. Agro-industrial waste valorisation is a desirable climate-resilient pathway that combine adaptation and mitigation to realize the goal of sustainable development. Rice husk (RH) is a worlwide a bio-waste from rice production which showed a potencial for its incorporation in ornamental plant growth substrates.

Main results of this work suggested that plant benefit the most from an incorporation of 10% of the bio-waste into the substrate (treatment 2). The application of RH in soils with little drainage capacity may be a recommended practice, while promoting the increase of crop production and valuing a low cost and a widely available bio-waste such as rice husk.

Bibliography

Aguiar, C. (2014). *Arquitetura de Plantas*. Bragança: Instituto Politécnico de Bragança, Escola Superior Agrária.

Bellé, S., & Kãmpf, A. N. (1994). Utilização de casca de arroz carbonizada como condicionador hortícola para uso orgânico. *Pesquisa Agropecuária Brasileira, 29*(8), 1265–1271.

Costa, A. R. (2012). *Vida rural*. http://www.vidarural.pt/atlantic-meals-quer-valorizar-casca-do-arroz-portugues/. Accessed Feb 2016.

Della, V. P., Kuhn, I., & Hotza, D. (2001). Caracterização de cinza de casca de arroz para uso como matéria-prima na fabricação de refratário de sílica. *Química Nova, 24*(6), 778–782.

EC. (2016). *Circular economy: New regulation to boost the use of organic and waste based fertilisers*. Press release. European Commission, Bruxel. Available at http://europa.eu/rapid/press-release_MEMO-16–826_en.htm.

FAOSTAT. (2016). *Food and Agriculture Organisation of United Nations*. http://faostat3.fao.org/download/Q/QI/E. Accessed Jan 2016.

FiBL. (2000). *Organic farming enhances soil fertility and biodiversity. FiBL Dossier No 1*. https://shop.fibl.org/CHen/1090-doc.html?ref=1. Accessed Feb 2016.

Foletto, E. L., Hoffmann, R., Hoffmann, R. S., Portugal, U. L., Jr., & Jahn, S. L. (2005). Aplicabilidade das cinzas da casca de arroz. *Química Nova, 28*(6), 1055–1060.

Ghini, R., & Bettiol, W. (2000). Proteção de plantas na agricultura sustentável. *Cadernos de Ciência & Tecnologia, 17*(1), 61–70.

Haling, R. E., Brown, L. K., Bengough, A. G., Valentine, T. A., White, P. J., Young, I. M., et al. (2014). Root hair length and rhizosheath mass depend on soil porosity, strength and water content in barley genotypes. *Planta, 239*(3), 643–651.

Hole, D., Perkins, A., Wilson, J., Alexander, I., Grice, P., & Evans, A. (2005). Does organic farming benefit biodiversity? *Biological Conservation, 122*(1), 113–130.

IFOAM. (2005). *International federation of organic agriculture movements. Principles of organic agriculture—Preamble*. http://www.ifoam.bio/en/organic-landmarks/principles-organic-agriculture . Accessed Feb 2016.

Kumar, A., Mohanta, K., Kumar, D., & Parkash, O. (2012). Properties and industrial applications of rice husk: A review. *Internacional Journal of Emerging Technology and Advanced Engineering, 2*(10), 86–90.

Mistry, B. (2016). Properties and industrial applications of rice husk. *International Journal of Engineering Science, 6*(10), 2677–2679.
Rajor, A., Xaxa, M., Mehta, R., & Kunal. (2011). Use of rice husk ash fertilizer amendment to soil and its effect on crop germination. In *National Conference on emerging trends in chemistry-biology interface*. India, p. 34.
Reddy, D., & Alvarez, M. (2006). Marine durability characteristics of rice husk ash-modified reinforced concrete. In *Fourth LACCEI International Latin American and Caribbean Conference for Engineering and Technology (LACCET'2006). Breaking Frontiers and Barriers in Engineering: Education, Research and Practice.* June 2006, Mayagüez, Puerto Rico.
Rosa, S. M., Santos, E. F., Ferreira, C. A., & Nachitigall, S. M. (2009). Studies on the properties of rice-husk-filled-PP composites—Effect of maleated PP. *Materials Research, 12*(3), 333–338.
Steffen, G. P., Antoniolli, Z. I., Steffen, R. B., & Mchado, R. G. (2010). Cáscara de arroz y estiércol bovino como sustratos para la multiplicación de lombrices de tierra y la producción de plántulas de tomate y lechuga. *Acta zoológica mexicana, 26*(SPE. 2), 333–343.
UN. (2015). *Transforming our world: The 2030 Agenda for Sustainable*. United Nations General Assembly resolution A/RES/70/1.
Vergani, C., & Graf, F. (2016). Soil permeability, aggregate stability and root growth: a pot experiment from a soil bioengineering perspective. *Ecohydrology, 9*(5), 830–842.

Are Increases in Maize Production in Malawi Due to Favourable Climate or the Farm Input Subsidy Program (FISP)?

Floney P. Kawaye and Michael F. Hutchinson

Abstract The impacts of climate change and climate variability on the water and agriculture sectors in Malawi prompted the Government to implement in 2005 an adaptation program, the Farm Input Subsidy Program (FISP), to support the production of maize, the staple food crop. This program has subsidised maize seed and fertilizer to support dryland smallholder maize production. This paper provides a robust, spatially explicit, analysis of the dependence of observed maize yield on climate and uses it to assess the significance of the impact of FISP on yields in Malawi. The analysis was conducted across three maize varieties by applying the GROWEST plant growth index model to monthly climate for eight agricultural production zones. The analysis shows that FISP has made a significant contribution to increased yield beyond the impact of variations in climate. This addresses the misconception that the increases in maize yield after 2005 were due only to favourable climate conditions. This misconception has fuelled debate about the efficacy of FISP and its eligibility for continued government financial support. It is envisaged that this analysis could assist the Government of Malawi in assessing the impacts of projected climate change on maize production and the viability of ongoing investment in FISP to support improvement in food security under a changing climate.

Keywords Adaptation · Maize · Food security · Spatially explicit yield modelling GROWEST

F. P. Kawaye (✉) · M. F. Hutchinson
Fenner School of Environment and Society, The Australian National University, ACTON ACT 2601, Australia
e-mail: floney.kawaye@anu.edu.au

M. F. Hutchinson
e-mail: michael.hutchinson@anu.edu.au

F. Alves et al. (eds.), *Theory and Practice of Climate Adaptation*,
Climate Change Management, https://doi.org/10.1007/978-3-319-72874-2_22

Introduction

Climate change and climate variability are anticipated to adversely affect agricultural production, food security and nutrition in many African countries including Malawi. These countries experience multiple stresses and have low adaptive capacity (IPCC 2014). Malawi has experienced increasing and intense climate change-related hazards over the recent decades. For instance Malawi experienced more droughts, floods and erratic rainfall patterns between 2000 and 2010 than between 1970 and 2000 (GoM 2011a; Matiya et al. 2011). Such climate events have impacted the water and agricultural sectors (Nangoma 2008; GoM 2006). This is of major national concern because Malawi's agriculture is largely rainfed (Mwase et al. 2014; Oxfam International 2011), and therefore climate change adaptation is inevitable (Matiya et al. 2011).

Malawi experienced hunger catastrophes due to severe droughts in 2001/02 and 2004/05 that affected over 3.2 million and 4 million people respectively (Chinsinga 2004). The issue of food security has since been debated extensively by politicians, policy makers, government organisations, village chiefs and in national newspapers (Sahely et al. 2005). Consequently, the Government of Malawi has implemented several policies and interventions to improve agriculture production and enhance food security under the changing climate. These include the Malawi Growth and Development Strategy II (MGDS) (GoM 2011a) and the Malawi Agricultural Sector Wide Approach (ASWAp), both of which promote adaptation in agriculture (GoM 2011b) and subsidy programs (Levy 2003). The Government of Malawi introduced the Farm Input Subsidy Program (FISP) in 2005 under the initiative of late President Dr Bingu wa Mutharika (De Schutter 2013). FISP is a large scale national program subsidising agricultural inputs, mainly fertilisers and seeds for maize production. The government provides the poorest 50 per cent of Malawian farmers with coupons to buy fertiliser and grain seed at a reduced price (Chinsinga 2007; De Schutter 2013). Before FISP, Malawi had faced persistent food insecurity for almost two decades (Chinsinga 2004), during which time Malawi depended on food aid and imports. Devaluation of the local currency partly also contributed to the adaption of FISP since farm inputs became unaffordable for most smallholder farmers (Chinsinga 2004; Devereux 2002; Stambuli 2002).

The aim of FISP is to overcome the country's continuing food shortages that have been exacerbated by extreme climate events. It is envisaged that FISP will increase farm production and incomes that would reduce the future demand for food hand-outs and dependence on food imports (De Schutter 2013; Chinsinga 2007). After the introduction of FISP Malawi experienced successions of plentiful maize production and harvests and succeeded in improving food security (Dorward and Chirwa 2011). Dorward et al. (2008) indicated that maize output increased from an average of 1.06 ton/ha in 2000–2005 to 2.27 ton/ha in 2009/2010. This allowed Malawi to meet its national food requirements which are estimated at 2.1 million metric tonnes of maize per year. The Gross Domestic Product (GDP) of Malawi increased between 2005 and 2010 by an average of 7.4%, higher than the

World Bank target rate of 6% for Sub-Saharan Africa. Furthermore, inflation decreased and the standard interest rates dropped from around 40 to 25% for the first time in two decades. Moreover, the number of Malawians living below the poverty line fell from 60% in 2004, to less than 45% in 2009. Denning et al. (2009) reported that Malawi experienced a reversal of fortunes from a food deficit of 43% in 2005, to a 53% food surplus in 2007. This enabled Malawi to export food to its neighbouring countries including Zambia and Zimbabwe.

Additionally, Denning et al. (2009) reported that Malawi was buffered from the economically and socially negative impacts of the global food price increases of 2007 and 2008 because they had enough food through FISP and did not have to buy food. The late president Dr Bingu wa Mutharika's leadership in moving the country from a 'basket case' to a breadbasket for southern Africa, was internationally applauded (De Schutter 2013). This had a significant impact on policy dialogue in Africa and by 2010 at least nine other African governments had adopted input subsidy programs (Jayne and Rashid 2013).

The Problem Statement

Other studies have criticised FISP. Dorward and Chirwa (2011) stated that official figures for maize production increases due to FISP are likely to have been inflated. While Denning et al. (2009) indicated that the number of Malawians living below the poverty line fell to less than 45% in 2009, official rural counted poverty rates increased slightly from 52% in 2004 to 53% in 2010 (Ricker-Gilbert et al. 2011). Even if maize production figures were accurate, it cannot simply be concluded that investment in FISP is the most efficient way of using scarce resources without a holistic examination of its performance (De Schutter 2013; Chinsinga 2007). Dorward et al. (2008) further argued that the surplus production of maize recorded since the introduction of the FISP cannot totally be credited to FISP. Favourable weather and climate in the years following the commencement of the program may have contributed to the additional maize production. De Schutter (2013) also indicated that recent droughts and unpredictable weather patterns have challenged the effectiveness of FISP and questioned whether it is able to meet Malawi's long-term food demands.

This study aims to reduce uncertainty about the relative impacts of climate and FISP on maize yields in Malawi by quantifying the dependence of observed maize yields on climate using a robust spatially explicit plant growth model. As noted by Schlenker and Lobell (2010), spatially explicit statistical studies of the dependence of crop yield on climate have often been limited by poor quantity and quality of historical yield data, resulting in model estimates with wide confidence intervals. This makes the application of such models to management and policy decisions problematic. The robust model presented here offers an alternative approach that is able to provide a spatially explicit and statistically significant assessment of the impact of climate on year to year variability in crop yields and the impact of the introduction of the FISP program. The results could help determine the impacts of projected climate change on maize production in Malawi and whether FISP should be terminated to enable investment to focus on other programs that to support more sustainable and efficient farming. Appropriate investments are needed to support food security improvement for Malawi as climatic patterns become more unpredictable. De Schutter (2013) has recommended that such programs should support extension services and the growth of agro-forestry and legumes.

Maize Production in Malawi

Three main varieties of maize are cultivated in Malawi. These are: (i) local (traditional) varieties; (ii) composite varieties; and (iii) hybrid varieties. There are major differences between the yield potentials of these varieties (Giertz et al. 2015; Pauw et al. 2010; Denning et al. 2009; JICAF 2008; Heisey and Smale 1995; Ngwira and Sibale 1986). Local varieties have the lowest yield. They are not subject to yield improvement programs and harvested seed is recycled from year to year. Composite varieties have higher yields and are often more drought tolerant. They are subject to yield improvement programs but seed can be recycled. Hybrid varieties are the highest yielding and most expensive. They are subject to strictly controlled yield improvement programs and harvested seed cannot be recycled.

Maize is grown in all the five agro-climatic zones (ACZ) across Malawi. As listed in Table 1, the ACZ have different climates and soil types. The maize production areas are classified into eight agro-ecological zones (AEZ) according to altitude (low, medium and high altitude areas). Table 1 shows that three of these AEZ are included in more than one ACZ. The AEZ are also known as Agricultural Development Divisions (ADDs). The eight ADDs consist of Blantyre, Machinga and Shire Valley (in the southern region), Kasungu, Lilongwe and Salima (in the central region) and Mzuzu and Karonga (in the northern region). Maize production data are reported for each ADD by the agricultural extension department under the Ministry of Agriculture Irrigation and Water Development (MoAFS 2012).

Maize in Malawi is grown by both smallholder farmers and estate farmers, with smallholder farmers in the overwhelming majority (Ngwira and Sibale 1986). There are 2.5–3 million smallholder households farmers who have customary rights to the land they cultivate (Mwase et al. 2014; Frenken 2005). These farmers mainly grow local maize varieties primarily as a subsistence crop as only less than 20% of the maize they produce is marketed as surplus (Kachule 2011; GoM 2010; Chirwa 2010). More than 90% of farm households grow maize which occupies over half of the smallholder cultivated land (USAID 2013; IFPRI 2011; Chirwa 2010; Denning et al. 2009; Frenken 2005). There are about 30,000 estates, each cultivating between 10 and 500 ha (Kumwenda et al. 2015; Chirwa 2008). Estate farmers occupy leasehold or freehold land (Kachule 2011) and prefer to grow composite or hybrid maize varieties solely for commercial purposes (Ngwira and Sibale 1986).

Table 1 Agro-climatic Zones (ACZ) and Agro-ecological Zones (AEZ) in Malawi and their characteristics

ACZ	AEZ (ADD)	Altitude (masl)	Climate	Mean temperature (°C)	Mean annual rainfall (mm)	Major soil groups
Highlands	Mzuzu, Karonga, Blantyre	1600–3000	Subhumid	10–26	1600–2200	Lithic, leptosols, lixisols
Plateau	Lilongwe, Kasungu	900–1600	Semiarid and subhumid	16–26	900–1400	Ferrasols, luvisols, lixisols
Rift valley escarpment	Blantyre, Mzuzu	600–1500	Semiarid and subhumid	14–24	900–1200	Lithic, leptosols
Lakeshore plains	Karonga, Mzuzu, Salima, Machinga	300–600	Semiarid	20–29	800–1100	Greysoils, vertisols, fluvisols
Lower Shire plains	Shire Valley	33–300	Semiarid	21–33	<800	Greysoils, vertisols, fluvisols

Source Fiwa (2015), Gietz et al. (2015)

Maize is grown under both irrigation and rain fed farming systems (JICAF 2008). However, rainfed farming dominates Malawi's agriculture as it covers about 99% of agriculture, mainly under smallholder farms (Mwase et al. 2014; Oxfam International 2011). Rainfed maize is planted between November and December (the start of the rainy season). It grows rapidly during the high rainfall months of January and February. Maize matures by early April and is harvested in late April or early May (MoAFS 2012). Maize yield in Malawi ranges from less than 1000 kg to over 4000 kg per hectare, depending on various factors including climate, location, seed variety, fertilizer use and labour (Ngwira and Sibale 1986). Policy related factors that significantly influence maize production in Malawi include access to credit, input and output markets and extension services (Tchale and Sauer 2007).

Methodology

This study conducted maize growth modelling to clarify the relative contributions of climate and FISP to improved maize yields after 2005. A simple robust growth estimation model called GROWEST, as described by Nix (1981) and implemented by Hutchinson et al. (2004), was applied to monthly climate across all eight ADDs. The key output from the GROWEST model is a dimensionless growth index (GI) that is the product of three separate indices accounting for the impacts of temperature, solar radiation and soil moisture. The temperature index depends on a prescribed optimum temperature for plant growth and the soil moisture index is calculated from rainfall and potential evaporation. The GI varies between 0 (climate totally limiting for growth) to 1 (climate optimal for growth). It represents relative plant growth and it has been found that GI accumulated over the growing season can be well calibrated with total dry matter production via simple linear regression. Summary seasonal outputs from the GROWEST model have been used to underpin a global agroclimatic classification, producing classes that agree well with natural vegetation formations (Hutchinson et al. 1992, 2005).

The supporting monthly climate data were interpolated from observed point meteorological data for Malawi and neighbouring areas using elevation dependent thin plate smoothing splines, as implemented in the ANUSPLIN software (Hutchinson and Xu 2013). This method has been widely applied to the interpolation of monthly and daily climate data (Hijmans et al. 2005; Hutchinson et al. 2009; McKenney et al. 2011). A key factor in the accuracy of the method is its incorporation of the impact of topographic elevation on spatial climate patterns. GROWEST was run on the interpolated monthly climate at 74 points that approximately equi-sampled the areas making up the 8 ADDs. The output GIs were averaged across each ADD to match the reported average maize yields for each ADD.

The GROWEST model runs on a weekly time step but can be applied to monthly climate data by interpolating monthly data to weekly data "on the fly". Weekly outputs were used in this study to provide finer scale demarcation of the maize growing season. Critical GROWEST parameters were determined directly from the observed yield data by minimising the regression sum of squares of observed annual maize yields on GIs averaged over the growing season. Robust regressions were carried out using a common linear dependence on the average GIs across all 8 ADDs but using a separate intercept for each ADD to allow for different site conditions and growing practice between different ADDs. Allowing a different slope for each ADD for the regressions did not improve the fits of the models. Thus each regression had 9 parameters, 8 intercepts and 1 slope, enabling robust estimation of a common slope across the 8 ADDs. Since the 8 intercepts address the site to site variability in yield, the remaining slope parameter addresses year to year variability due to climate as quantified by the growing season average GIs. This represents an alternative, more spatially explicit, approach to identifying the impact of climate variability on crop yield than broad scale approaches based on country level yield data, such as those described by Schlenker and Lobell (2010) and El-Maayar and Lange (2013).

The optimised GROWEST parameters were the optimum temperature for maize growth and the growing season. The model was able to select from either 30 °C or 31 °C for each maize variety in view of the documented optimum temperature for maize whole plant growth of around 30.8 °C with a standard error of 2 °C (Sanchez et al. 2014). It was found that an optimum growth temperature of 30 °C was better for local and composite varieties while 31 °C was better for the hybrid variety. The latter is consistent with breeding programs that are producing maize varieties with higher heat tolerance (Cairns et al. 2013). The model found that the growing seasons that best matched the observed yield data were periods of around 14–15 weeks starting from around the last week of December and ending in early April, with minor differences for each variety. This is contained within the limits of the documented standard maize growing season described above.

Maize yield analyses were applied to 14 years of yield data from 1997 to 2012. This gave a similar number of years before and after the introduction of FISP for the 2006 growing season. The years 1999 and 2000 were excluded from the analyses as these had inconsistent growing seasons caused by major flooding. The growth analyses were applied to annual maize yields across all eight ADDs. Three periods were modelled for each maize variety as follows: (i) the years before FISP, 1997–2005; (ii) the years after FISP, 2006–2012; and (iii) all years, 1997–2012. In order to make consistent comparisons between the analyses before and after FISP, the fitted growing seasons were constrained to be the same before and after FISP for each variety. Thus the fitted growing season for local varieties was the first 15 weeks of each year, the fitted growing season for composite varieties was the 15 weeks starting with the last week of December and the fitted growing season for hybrid varieties was the 14 weeks also starting with the last week of December. The GROWEST parameters were optimised by automating the GROWEST runs and the regressions using FORTRAN code and standard LINPACK numerical analysis software (Dongarra et al. 1979). Once optimisations were complete, comprehensive statistical analyses were computed using the standard statistical modelling package within Excel software.

Results

Reported maize yields by ADD over the entire analysis period are shown in Fig. 1. The years with low yields correspond to major droughts in 1997 and 2005. As discussed above, the composite and hybrid varieties have significantly higher yields than the local varieties. Composite yield data were not available for 1997 and 1998 so the analysis period was reduced to 1999–2012 for the composite variety. There were still sufficient data to obtain meaningful results.

A modest proportion of large outliers were removed from the analyses. These were identified by examining large standardised residuals from the initial regression analyses. They revealed occasional accounting errors in the annual yield data and clear anomalies in yields or cropping areas reported for particular regions making up each ADD in particular years. No more than one or two outliers were removed for each ADD for each analysis. Once the outliers were removed the growing season average GIs explained the year to year variability in the reported maize yields reasonably well before and after FISP. Figure 2 shows the fitted model for hybrid maize from 1997 to 2005 as it applied to each ADD. These plots confirm the suitability of fitting a common slope across the ADDs in the regressions.

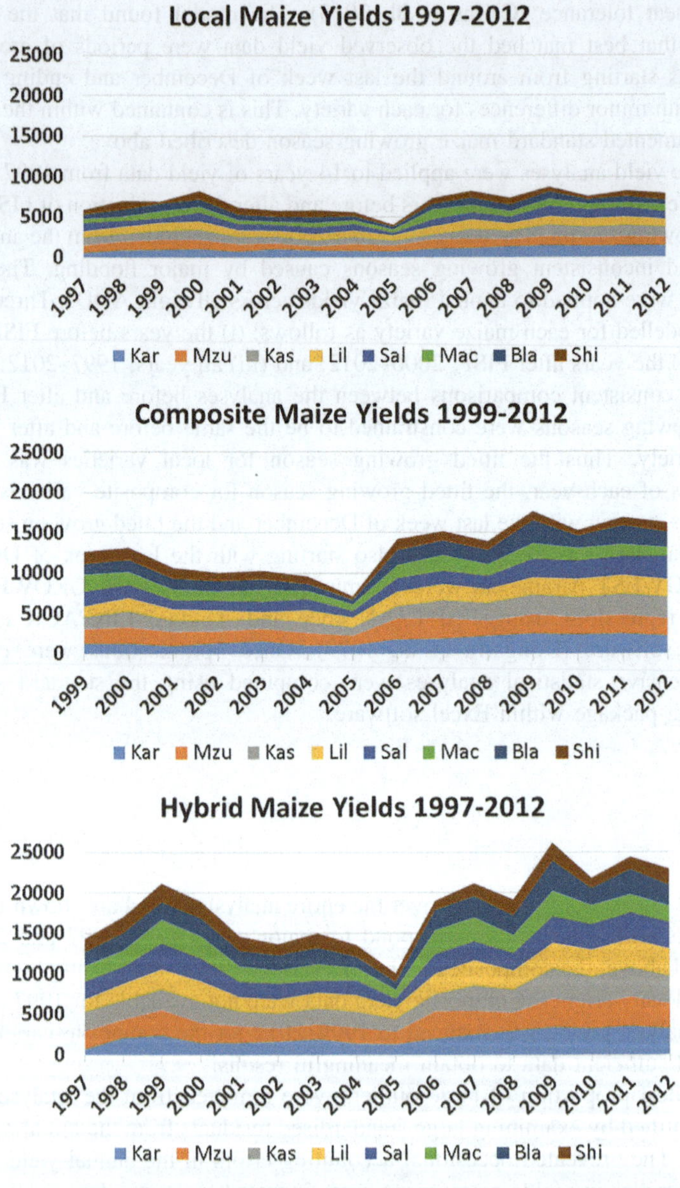

Fig. 1 Annual yields (kg/ha) for local, composite and hybrid maize over 8 ADDs from 1997 to 2012

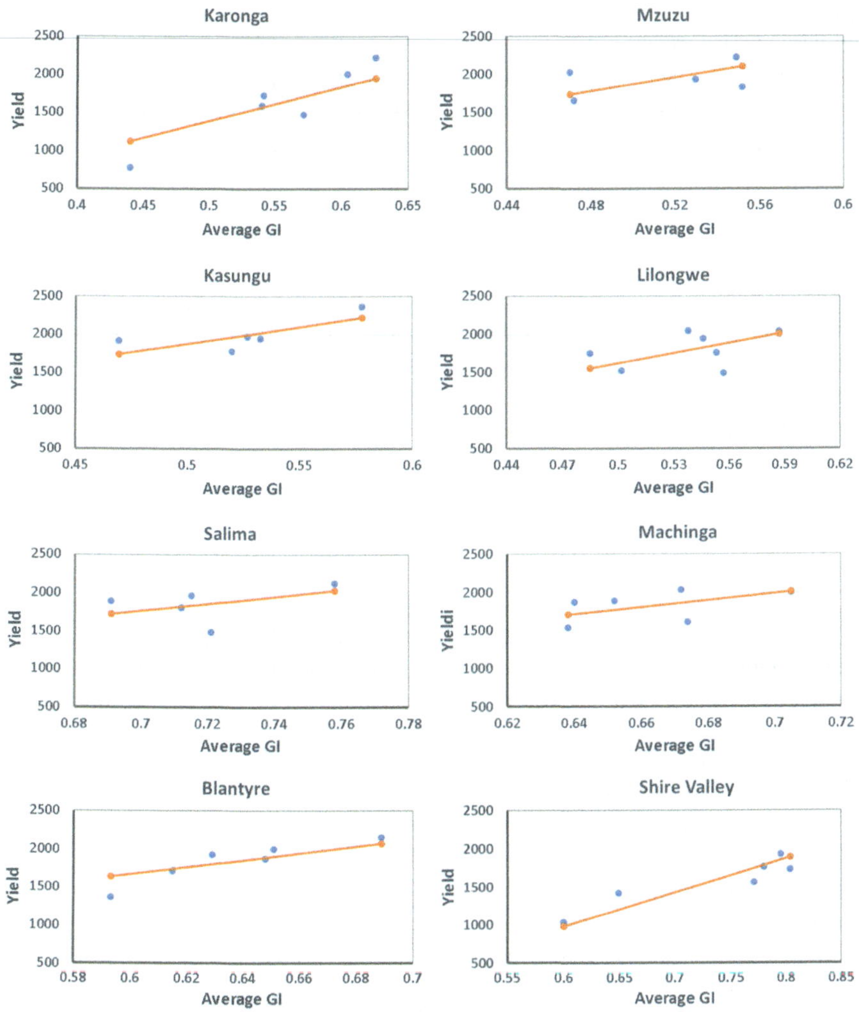

Fig. 2 Annual hybrid maize yields (kg/ha) vs growing season average GI for 1997 to 2005 and the fitted common slope regression model for each ADD

The results of the fitted regression models for the three varieties over the three periods are summarised in Table 2 and plots of the observed yield data versus the modelled yields before and after the introduction of FISP are shown in Fig. 3. All regressions for the two separate periods before and after the introduction of FISP provided reasonable explanation of the observed year to year variability in maize yields, particularly in the earlier period, with 38–55% of the observed year to year variation explained by the fitted models. The fits for the later period had somewhat less explanation, with 23–34% of year to year variation explained. This degree of explanation is modest but understandable given non-climatic contributors to year to

Table 2 Summary statistics for regressions of maize yield (kg/ha) on average growth index over the fitted growing seasons

Period	Number of data points	Slope of regression	Standard error of slope	Standard error of regression	Variance accounted for (%)
Local variety					
1997–2005	49	1697**	257	92	52
2006–2012	48	1920**	569	110	23
1997–2012	97	1520**	462	195	11
Composite variety					
1999–2005	37	2919**	708	200	38
2006–2012	47	5176**	1164	213	34
1999–2012	84	2896*	1254	434	7
Hybrid variety					
1997–2005	46	4464**	659	197	55
2006–2012	48	7363**	2199	252	25
1997–2012	94	2768	1844	652	2

Statistical significance of the fitted slopes at the 5 and 1% levels are denoted by * and ** respectively

year variability including significant changes in cropping areas and undocumented changes in management, such as delays in planting due to lack of soil moisture or flooding at normal planting times. Simple inconsistencies in reporting of yields are another potential contributor to year to year variability. The more variable climatic conditions observed in more recent years may have contributed to more variability in planting times after FISP and hence less explanation by a model with a fixed growing season. Nevertheless, the slopes of all regressions before and after FISP are all highly significant, with reasonably small standard errors. These coefficients could be used with reasonable confidence in estimating the impact of climate on maize yields, especially if these impacts were aggregated to a country level.

The impact of FISP on maize yields was assessed by examining the slopes of the fitted regression models before and after FISP and by fitting the model to the whole period from 1997 to 2012. As listed in Table 2, the slopes of the regressions for the local varieties before and after FISP are very similar, consistent with local varieties not being part of the FISP program. When the local maize model was fitted to all years the amount of explanation decreased but the fitted slope did not change significantly. This decline in overall fit does indicate that there have been some significant changes in management or reporting of local maize since the introduction of FISP, such as those listed above, but the net impact of climate on year to year variability has not changed.

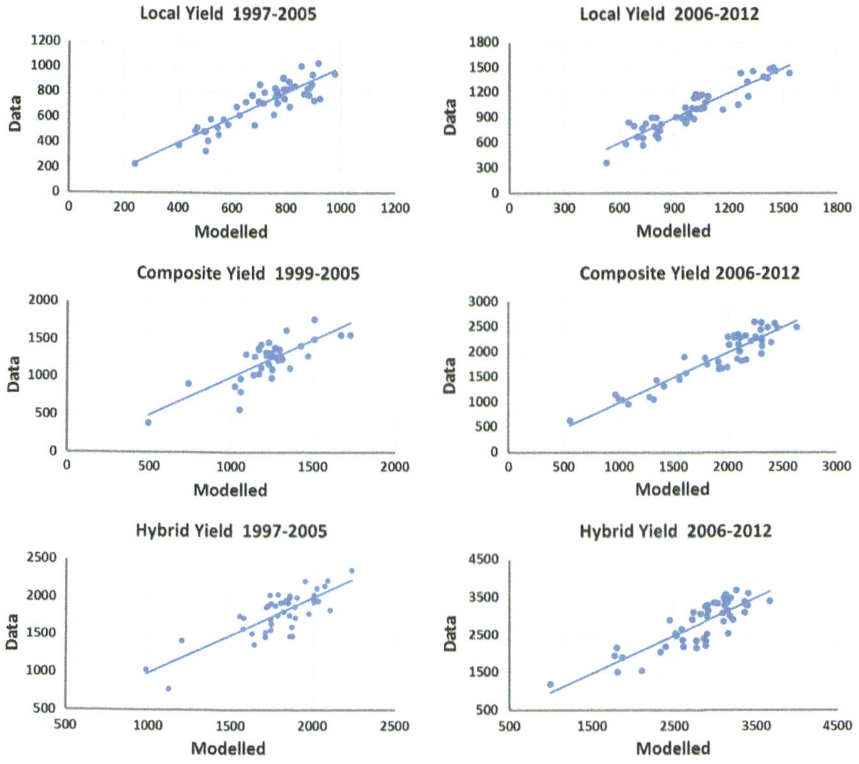

Fig. 3 Observed hybrid maize yield (kg/ha) versus modelled yield before and after FISP for local, composite and hybrid maize varieties

On the other hand, the slopes of the regression models for composite and hybrid varieties after FISP are nearly twice the slopes of the pre-FISP models. This is consistent with the near doubling of observed composite and hybrid yields after FISP. In statistical terms the differences between the two slopes are significant at the 5% level and 10% level for composite and hybrid yields respectively. The lower level of significance for hybrid yields reflects the somewhat poorer fit for hybrid yields after FISP. An additional factor not taken into account by the model is significant improvements in potential hybrid maize yield over the period 2006–2012. This is highly likely given the ongoing yield improvement programs being applied to hybrid maize. The fits to all years of data for composite and hybrid yields both significantly degraded the goodness of fits of the models, as confirmed by standard deviance analysis and the relative magnitudes of the standard errors of the regressions. This also confirms the significance of the impact of the introduction of FISP on composite and hybrid maize yields.

Discussion

The good fits of the models for the two separate periods indicate that climate via the GROWEST model can describe a significant fraction of the observed year to year variation in observed yields for all three maize varieties. The strength of this spatially explicit approach in obtaining statistically significant results, despite the difficulties noted by Schlenker and Lobell (2010) and others, is its basis on a simple process based plant growth model that can be robustly parameterised on minimal error prone data. The quality of the supporting monthly climate data is also likely to be a factor in its effectiveness. Particular confirming aspects of this approach are the ability of the model to reproduce known aspects of the maize cropping regimes directly from the yield data. These included the optimum temperatures for the three maize varieties and their normal growing periods. An additional confirmation was its ability to detect significant data errors in the reported yield data as outliers from the initial fitted models. In some cases the errors could be corrected and the data were retained while for others clear anomalies in the supporting data could be identified and the outliers were removed to improve the accuracy of the final fitted model. The reasonable goodness of fit of the model also confirms the overall accuracy of the reported maize yields, despite doubts expressed about their accuracy after the introduction of FISP.

The main limitation of the model is its dependence on an assumption that the many non-climatic factors that contribute to crop yield are constant over the calibration period. The relatively good fits before the introduction of FISP indicate that this assumption was well satisfied for all three varieties. The poorer fits after FISP, particularly for the hybrid varieties, is indicative of some variability in these factors after FISP. Continued improvement in potential yield due to ongoing scientific breeding programs is a clear likely contributor in the case of the hybrid varieties. A possible indirect climatic contributor is enhanced climatic variability leading to disruption of standard growing seasons. Major droughts and floods after the introduction of FISP may well have made such a contribution. Lesser contributors to model error include significant variation in cropping areas over time and mismatch in locations of supporting climate data and the locations of main production areas. It is possible that the model could be improved by applying it at a finer spatial level to the Rural Development Program areas that make up each ADD. But such a refinement is unlikely to improve the overall modelling of crop yields, particularly when they are aggregated to a country level.

Straightforward application of this model to assess the impacts of projected climate change on maize yields is also possible, perhaps with modification to take account of improvements in heat tolerance of maize varieties brought about by ongoing plant breeding programs, and to take account of carbon dioxide fertilization effects, although these are less significant for C4 plants such as maize (Berg et al. 2013). Inspection of the monthly climate data over the maize growing season for Malawi indicates that most temperatures are below the optimum maize growing temperature so a modest climate warming is likely to slightly increase maize yield. A more detailed assessment is required to assess the impact of the multi-degree warming projected for later this century.

Finally, the significant difference between the fitted models for composite and hybrid maize before and after the introduction of FISP indicates that FISP has made a very significant contribution to maize yields in Malawi. It is unlikely that possible improved management practice or any other non-climatic factors after FISP could have achieved such an impact. And the impact of climate on year to year variability on maize yield is well calibrated both before and after FISP so that climate is unlikely to have had a significant impact on the major increase in maize yields after the introduction of FISP.

Conclusion and Recommendations

This study has demonstrated the effectiveness of a robust, spatially explicit, process based statistical model of the impact of climate on year to year variability of reported yields maize for all three maize varieties grown in Malawi. The model is able to detect, and calibrate, significant aspects of the maize cropping regimes, including optimum growing temperatures and the normal growing season. It is amenable to application to assessing the impacts of projected climate change and has been able to demonstrate a significant maize growth response to the FISP program introduced in 2005.

It can be concluded that FISP has made a significant contribution towards the increased maize production and improved food security in Malawi. This underscores the importance of FISP as a climate change adaptation strategy in the Malawi agriculture sector for smallholder farmers. This suggests that FISP should not be terminated on the grounds that only a favourable climate after FISP was adopted contributed to the increased maize production. This finding should be taken into account by the Government of Malawi as it considers options for investing in food security through major irrigation schemes. Studies by de Fraiture et al. (2007) and Faurès et al. (2007) indicate that investment in both rain-fed and irrigated agriculture would be the best option to enhance food security and reduce people's vulnerability to external shocks and climate variability in Sub-Saharan Africa.

Acknowledgements The first author wishes to thank the Australian *Department of Foreign Affairs and Trade (DFAT), the Australian National University (ANU) and the GBI office in Malawi for the academic, material and financial support to carry out this study as part of her Ph.D. studies.*

References

Berg, A., de Noblet-Ducoudre, N., Sultan, B., Lengaigne, M., & Guimberteau, M. (2013). Projections of climate change impacts on potential C4 crop productivity over tropical regions. *Agricultural and Forest Meteorology, 170,* 89–102.

Cairns, J. E., Hellin, J., Sonder, K., Araus, J. L., MacRobert, J. F., Thierfelder, C., et al. (2013). Adapting maize production to climate change in sub-Saharan Africa. *Food Security, 5,* 345–360.

Chinsinga, B. (2004). Poverty and food security in Malawi: Some policy reflections on the context of crumbling traditional support systems. *Canadian Journal of Development Studies, 25*(2), 403–347.

Chinsinga, B. (2007). *Reclaiming policy space: Lessons from Malawi's fertilizer subsidy program.* Paper presented at the World development Report Workshop, Institute of Development Studies, 21–24 Jan 2007.

Chirwa, T. G. (2010). *Program evaluation of agricultural input subsidies in Malawi using treatment effects: methods and practicability based on propensity scores.* MPRA Paper No. 21236. Accessed 20th Dec 2016. http://mpra.ub.uni-muenchen.de/21236/.

Chirwa, E. (2008). Land tenure, farm investments and food production in Malawi. In *Institutions and Pro-Poor Growth (IPPG) research programme.* Discussion Paper No. 18. University of Manchester, Manchester.

De Fraiture, C., Wichelns, D., Rockstrom, J., Kemp-Benedict, E., Eriyagama, N., Gordon, L. J., et al. (2007). Looking ahead to 2050: Scenarios of alternative investment approaches. In David Molden (Ed.), *Water for food, water for life: A comprehensive assessment of water management in agriculture* (pp. 91–145). London, UK/ Colombo, Sri Lanka: Earthscan/IWMI.

Denning, G., Kabembe, P., Sanchez, P., Malik, A., Flor, R., Harawa, R., et al. (2009). Input subsidies to improve smallholder maize productivity in Malawi: Toward an African green revolution. *PLoS Biology, 17*(1), e23.

De Schutter, O. (2013). *Subsidies in the global food system II: Malawi's FISP—Success or Stumbling Block? United Nations special report.* Accessed 18 Nov 2016. http://futuredirections.org.au/publications/food-and-water-crises/28-global-food-and-water-crises-swa/1299-subsidies-in-the-global-food-system-ii-malawi-s-fisp-success-or-stumbling-block.html.

Devereux, S. (2002). Can social safety nets reduce Chronic Poverty. *Development Policy Review, 20*(5), 657–675.

Dongarra, J. J., Moler, C. B., Bunch, J. R., & Stewart, G. W. (1979). *LINPACK Users' Guide.* SIAM.

Dorward, A., & Chirwa, E. (2011). The Malawi agricultural input subsidy program: 2005/06–2008/09. *International Journal of Agricultural Sustainability, 16,* 232–247.

Dorward, A., Chirwa, E., Slater, R., Jayne, T., Boughton, D., & Valerie, K. (2008). *Evaluation of the 2006/07 agricultural input subsidy program,* Malawi: Final Report, London: School of Oriental and African Studies.

El-Maayar, M., & Lange, M. A. (2013). A methodology to infer crop yield response to climate variability and change using long-term observations. *Atmosphere, 4,* 365–382.

Faurès, J., Svendsen, M., & Turral, H. (2007). Reinventing irrigation. In D. Molden (Ed.), *Water for food, water for life: A comprehensive assessment of Water Management in Agriculture.* Earthscan and International Water Management Institute: London and Colombo.

Fiwa, L. (2015). *Improving rainfed cereal production and water productivity in Malawi: Modelling field management options in response to current and future climatic conditions.* Accessed 5th Feb 2017. https://lirias.kuleuven.be/bitstream/123456789/493765/1/Lameck_FIWA.pdf.

Frenken, K. (2005). *Irrigation in Africa in Figures: AQUAStAt Survey—2005* (Vol. 29).

Giertz, A., Caballero, J., Galperin, D., Makoka, D., Olson, J., & German, G. (2015). Malawi—Agricultural sector risk assessment. In *Agriculture global practice technical assistance paper.*

Washington, D.C.: World Bank Group, Accessed 10th Dec 2016. http://documents.worldbank.org/curated/en/2016/01/25799223/malawi-agricultural-sector-risk-assessment.

Government of Malawi. (2006). *National Adaptation Program of Action (NAPA)*. Ministry of Mines, Natural Resources and Environment.

Government of Malawi (GoM). (2010). *Malawi CAADP compact to support the successful implementation of the Agricultural Sector Wide Approach*. Malawi: Lilongwe.

Government of Malawi (GoM). (2011a). *Malawi Growth and Development Strategy II (MGDS II): From poverty to prosperity. 2011–2016*. Ministry of Economic Planning and Development.

Government of Malawi (GoM). (2011b). *Malawi agricultural sector wide approach (ASWAP): A prioritised and harmonised Agricultural Development Agenda. 2011–2015*. Ministry of Agriculture and Food Security.

Heisey, P. W., & Smale M. (1995). *Maize Technology in Malawi: A green revolution in the making?* In CIMMYT Research Report No.4. Mexico, D.F.: CIMMYT.

Hijmans. (2005). Very high resolution interpolated climate surfaces for global land areas. *International Journal of Climatology, 25*, 1965–1978.

Hutchinson, M. F., Nix, H. A., & McTaggart, C. (2004). *GROWEST version 2.0*. Fenner School of Environment and Society, Australian National University, Canberra.

Hutchinson, M. F., McIntyre, S., Hobbs, R. J., Stein, J. L., Garnett, S., & Kinloch, J. (2005). Integrating a global agro-climatic classification with bioregional boundaries in Australia. *Global Ecology and Biogeography, 14*, 197–211.

Hutchinson, M. F., Mckenney, D. W., Lawrence, K., Pedlar, J., Hopkinson, R., Milewska, E., et al. (2009). Development and testing of Canada-wide interpolated spatial models of daily minimum/maximum temperature and precipitation for 1961–2003. *Journal of Applied Meteorology and Climatology, 48*, 725–741.

Hutchinson, M. F., Nix, H. A., & McMahon, J. P. (1992). Climate constraints on cropping systems. In C. J. Pearson (Ed.), *Field crop ecosystems of the world* (pp. 37–58). Amsterdam: Elsevier.

Hutchinson, M. F., & Xu, T. (2013). *ANUSPLIN versions 4.4 user guide*. Fenner School of Environment and Society, The Australian National University, Canberra, Australia.

Intergovernmental Panel on Climate Change (IPCC). (2014). Climate change 2014: Synthesis report. In Core Writing Team, R. K. Pachauri & L.A. Meyer (Eds.), *Contribution of Working Groups I, II and III to the Fifth Assessment Report of the Intergovernmental Panel on Climate Change*. IPCC, Geneva, Switzerland, 151 pp.

International Food Policy Research Institute (IFPRI). (2011). *Could low adoption of modern maize varieties in Malawi be explained by farmers' interest in diverse seed characteristics? Policy Note 7*. Washington, DC: IFPRI.

Japan Association for International Collaboration of Agriculture and Forestry (JICAF). (2008). *The Maize in Zambia and Malawi*. Accessed 10th Mar 2017. http://www.apip-apec.com/ja/good-practices/files/The_Maize_in_Zambia_and_Malawi.pdf.

Jayne, T. S., & Rashid, S. (2013). Input subsidy programs in sub-Saharan Africa: A synthesis of recent evidence. *Agriculture Economics, 44*(6), 1–16.

Kachule, R. N. (2011). *Performance of the Agricultural Sector in Malawi*. Paper presented to Malawi Agricultural Sector Investment Program (MASIP) Secretariat on Priority Setting. Bunda College of Agriculture, Lilongwe, Malawi.

Kumwenda, I., van Koppen, B., Matete, M., & Nhamo, L. (2015). Trends and outlook: Agricultural water management in southern Africa. Country report—Malawi. In *Project report submitted to United States Agency for International Development's (USAID's) Feed the Future Program*. Pretoria, South Africa: International Water Management Institute (IWMI).

Levy, S. (2003). Starter packs and hunger crises: A briefing for policy makers on food security in Malawi. In *Key findings from the evaluation of Starter Pack/Targeted Inputs Program (TIP)*. Department for International Development, London, UK

Matiya, G., Lunduka, R., & Sikwese, M. (2011). *Planning and costing agricultural adaptation to climate change in the small-scale maize production system of Malawi*. International Institute for Environment and Development (IIED). London, UK

McKenney, D. W., Hutchinson, M. F., Papadopol, P., Lawrence, K., Pedlar, J., Campbell, K., et al. (2011). Customized spatial climate models for North America. *Bulletin of the American Meteorological Society,* 1611–1622. https://doi.org/10.1175/bams-d-10-3132.

Ministry of Agriculture and Food Security (MoAFS). (2012). *Guide to agricultural production and natural resources management in Malawi.* Ministry of Agriculture and Food Security, Department of Agricultural Extension Services, Lilongwe Malawi.

Mwase, W. F., Jumbe, C., Gasc, F., Owiyo, T., Manduwa, D., Nyaika, J., et al. (2014). Assessment of agricultural sector policies and climate change in Malawi-The Nexus between climate change related policies, research and practice. *Journal of Sustainable Development, 7*(6), 195–203.

Nangoma, E. (2008). *National adaptation strategy to climate change impacts: A case study of Malawi.* United Nation development Program (UNDP): Human Development Report for Malawi.

Ngwira, L. D. M., & Sibale, E. M. (1986). Maize research and production in Malawi. In B. Gelaw (Ed.), *To Feed Ourselves: Proceedings of the First Eastern, Central and Southern African Regional Maize Workshop* (pp. 50–56). CIMMYT: Mexico.

Nix, H. A. (1981). Simplified simulation models based on specified minimum data sets: The CROPEVAL concept. In A. Berg (Ed.), *Application of remote sensing to agricultural production forecasting* (pp. 151–169). Rotterdam: Commission of the European Communities.

Pauw, P., Thurlow, J., & van Seventer, D. (2010). Droughts and floods in Malawi. In *IFPRI Discussion Paper 00962*. International Food Policy Research Institute.

Oxfam International. (2011). *Supporting irrigation for food security in Malawi.* Oxford: Oxfam International.

Ricker-Gilbert, J., Jayne, T. S., & Chirwa, E. (2011). Subsidies and crowding out: A double-hurdle model of fertilizer demand in Malawi. *American Journal of Agricultural Economics, 93*(1), 26–42.

Sahely, C., Groelsema, B., Marchione, T., & Nelson, D. (2005). *The governance dimension of food security in Malawi.* A Report Prepared for USAID.

Sanchez, B., Rasmussen, A., & Porter, J. R. (2014). Temperatures and the growth and development of maize and rice: A review. *Global Change Biology, 20,* 408–417.

Schlenker, W., & Lobell, D. (2010). Robust negative impacts of climate change on African agriculture. *Environmental Research Letters, 5*(1), 014010.

Stambuli, K. (2002). *Long-run food security of a top-down agricultural strategy in Malawi.* Surrey, UK: Surrey Institute for Global Economic Research.

Tchale, H., & Sauer, J. (2007). The efficiency of maize farming in Malawi. A bootstrapped translog frontier. *Cahiers D?économie et Sociologie Rurales,* 82–83.

United States Agency for International Development (USAID). (2013). *Malawi climate change vulnerability assessment: African and Latin American Resilience to Climate Change (ARCC) program.*

Would Rainfed Agriculture Be the Right Option Under Climate Change Scenarios? A Case Study from Centro Region of Portugal

Albano Figueiredo, Carolina Alves, Joaquim Patriarca, Andreia Saavedra Cardoso, Paula Castro and João Loureiro

Abstract Agriculture has changed significantly during the last 30 years in Portugal. One of the main changes is related to agricultural abandonment, mainly driven by demographic dynamics on areas that are marginal in terms of productivity. Such trend affects mainly rainfed agriculture, and was significant in the inland Norte and Centro regions of Portugal between 1990 and 2010. By contrast, irrigated areas dedicated to agriculture are increasing. Considering a predicted reduction on water availability under future climatic scenarios, determined by the decrease in the amount of annual precipitation, it is necessary to set strategies to adapt agriculture to a new climatic context, considering both production and consumption/dietary trends. To do so, and using as case study the Centro Region of Portugal, this work aims to evaluate how the suitable area for agriculture might change under future climatic scenarios (RCP 4.5 and 8.5 scenarios for the two time-windows of 2041–2070 and 2071–2100), and identify measures that contribute to adapt agriculture to a new context. Such assessment is based on a

A. Figueiredo (✉)
Department of Geography and Tourism, University of Coimbra,
Colégio de São Jerónimo, 3004-530 Coimbra, Portugal
e-mail: geofiguc@gmail.com

A. Figueiredo · C. Alves · J. Patriarca
Centre for Studies in Geography and Spatial Planning (CEGOT),
University of Coimbra, Coimbra, Portugal

J. Patriarca
Institute for Systems Engineering and Computers Coimbra,
University of Coimbra, Coimbra, Portugal

A. S. Cardoso
Research Unit in Materials, Energy and Environment for Sustainability,
Agrarian School of Ponte de Lima (ESA), Viana do
Castelo Polytechnic Institute (IPVC), Viana do Castelo, Portugal

P. Castro · J. Loureiro
CFE-Centre for Functional Ecology - Science for People & the Planet,
Department of Life Sciences, University of Coimbra, Coimbra, Portugal

F. Alves et al. (eds.), *Theory and Practice of Climate Adaptation*,
Climate Change Management, https://doi.org/10.1007/978-3-319-72874-2_23

modelling approach that aims to evaluate suitability to agriculture, which is set from soil properties (soil type and texture), topographic parameters (such as slope and land morphology), and climatic conditions (water deficit). The expected reduction on water availability under future climatic scenarios, combined with recent trends on agriculture, namely the reduction of rainfed agriculture and the increase of irrigated agriculture areas, points to an unsustainable situation. This is of great concern, once there is a match between areas where water deficit is predicted to increase more and areas where irrigated area is expanding today. Thus, specific adaptation strategies/policies are needed to revert/cope with such trends, which must be spatially explicit and locally meaningful. The implementation of such approaches might be oriented by results from assessment of predicted changes in terms of suitable area for agriculture, but also consider economic and dietary aspects, an exercise that we try to validate based on the conditions of the Centro Region of Portugal.

Keywords Rainfed agriculture · Edaphoclimatic suitability model
Water deficit · Adaptation strategies · Climate change

Introduction

Adaptation or Mitigation?

Climate change is now considered to be a major threat to natural ecosystems and human activities, with great environmental and economic costs (Pimentel et al. 2005). Considering the predicted impacts, it is necessary to set priorities and dedicated measures in order to adapt to such a new context, a decision that should be based on the attributes of the territories. That condition demands that susceptibility, vulnerability or exposure should be considered on a first stage, in order to support the identification of adaptation measures that encompass the main vulnerabilities for a specific territory.

Considering the magnitude of foreseen climatic changes, validated by recent trends and observed facts (IPCC 2014), the focus seems to change from *mitigation*, where action is focused on reducing impacts, suggesting at some extent a capacity to *control*, to *adaptation*, indicating that action are oriented by measures focused on adapting to a new context. Despite the prevailing idea that adaptation measures could be also accepted as strategies to reduce impacts from climate change, in the sense of *mitigation,* evidences suggest that such actions must be focused on preparing for climate change impacts (Noble et al. 2014). Nevertheless, when considering farming systems and food systems at large, there is confirmation that a considerable amount of categories of adaptation have positive impacts respecting mitigation (Smith and Olesen 2010; Niles et al. 2017).

Climate Change Is Not in Question

It is not new to science that climate changed in the past, being the magnitude of change dependent on the time-scale considered. Considering large time-windows, several proxies point to different cycles of glaciation during the last 650,000, with climatic changes being attributed to very small variations in the Earth's orbit (Hays et al. 1976; Maslin 2016). But changes can also be detected for smaller time-windows and for recent past periods, such as the century scale (e.g. Little Ice Age) (Oerlemans 2005; Mann et al. 2008), or even at the decade scale (Beniston et al. 1994; Miranda et al. 2006). As expected, changes for small time-window periods are less evident, but can be detected by trends on databases for climatic records (Beniston et al. 2003) or from different proxies (Walther et al. 2002; Lloyd and Fastie 2003; Menzel et al. 2006; Torriani et al. 2007; Schleip et al. 2008; Sorg et al. 2012).

The evidence for climate change is convincing since the late 19th century, and has become an important topic of the 21st century. Current warming cycle is of particular importance since it is extremely likely (greater than 95% probability) that most of it results from human activity since the mid-20th century. Also, the current warming trend is evolving at a rate that is unprecedented over decades to millennia (Santer et al. 1996; Ramaswamy et al. 2006; Santer et al. 2003).

Climate is changing across the globe, modifying weather patterns, and increasing the vulnerability of many countries, regions, economic sectors and populations (IPCC 2014; EEA 2013). According to the records, temperatures warmed almost 0.75 °C from 1906 to 2005, becoming successively warmer than any preceding decade since 1850 (IPCC 2007, 2014). In Europe, from 2006 to 2015, the global average annual temperature was 0.83–0.89 °C higher than the pre-industrial period, with recent years becoming more and more known as the warmest years (EEA 2017a, b).

Climatic changes, including extreme events, have been sufficient to disturb, in different degrees, many of the planet's ecosystems, likely causing the loss of human lives, population displacement, and economic activities, including farming and agroforestry systems (WEF 2016; IPCC 2014).

Climate Change Impacts on Agriculture

Climatic conditions are the primary determinant of agricultural productivity. The water availability in the soil, conditioned by seasonal variation and spatial distribution of rainfall, together with the thermic conditions in each vegetative growing stage, govern, not only the spatial distribution of each type of crop, but also its productive capability, the carbon balance and ecosystem services (Pinto et al. 2006; Chang et al. 2016; Liang et al. 2017).

Therefore, it is highly expectable that climate change will affect the agriculture sector by causing impacts on the suitability and productivity of crops, conditioning the agricultural production (Olesen 2016). In particular, a higher variability in crop production is expected, with loss of productivity in certain crops and increase in others, accompanied by changes in the spatial distribution (Fellmann 2012), raising also concerns on food security and nutritional goals (FAO 2016).

Based on this evidence, agriculture is seriously threatened, forcing agricultural production strategies and design to be reassessed aiming its effective sustainability and resilience under foreseen climatic scenarios. This could entail the recovery of traditional management systems combined with the use of agroecological methods (Altieri and Nicholls 2017).

In Europe, climate change scenarios point to an increase in soil productivity in Northern Europe, favoured by increases in the period of crop's vegetative growth, and reversely by decreases in Southern Europe (Ciscar 2009). Some of the studies developed in Europe, point the Mediterranean Region as one of the most affected areas by climate change, especially in the warmer season, due to decreases in rainfall (ranging from 25 to 30%) and increases in temperature (ranging from 4 to 5 °C) (Skuras and Psaltopoulos 2012). Mediterranean agriculture is particularly vulnerable to the impact of climate change. Chiefly, changes in rainfall patterns deeply impact rainfed agriculture, limiting soil moisture and crop productivity, and new economic models of water use must be taken into account (Calzadilla et al. 2013; La Jeunesse et al. 2016).

For Skuras and Psaltopoulos (2012) the major threats to agriculture in Mediterranean countries, including those derived from climatic constraints, are related with:

- water availability and irrigation demand;
- fertility, salinity and soil erosion;
- conditions of growth, productivity and distribution of crops;
- soil use;
- conditions for livestock production;
- pests and diseases;
- adaptation/recovery in the face of extreme weather events.

As an European Southern country, Portugal is particularly sensitive to current and future climatic conditions shifting patterns. The National Climate Change Adaptation Strategy for Agriculture and Forests (APA 2013) highlighted some problems that will greatly affect the agricultural sector adaptive capacity and building resilience:

- increased impacts on the areas occupied by pasture land, permanent crops and temporary rainfed crops (cereals, in particular);
- water availability, significantly affecting temporary rainfed systems and pastures associated with extensive livestock farming;
- extreme weather events with aggravated impacts on horticulture and agricultural infrastructures;

- pests and diseases with changes in its dynamics and patterns, and new ones will emerge;
- increased susceptibility to desertification, with consequences on soil fertility and erosion.

Adding to these problems, the ageing of the active population in the agricultural and productive sectors, associated with low levels of education, is considered an important barrier to resilience, as a consequence of a greater resistance to innovation (introduction of new technologies and production systems) and weaker investment capacity.

Climate Change, Changing Food Systems and Diets

In a recent study on the vulnerability of the food system in Europe, conducted through consultation with actors and experts, the multidimensional nature of vulnerability was often linked to food system activities such as agriculture, post-production and food consumption (Paloviita et al. 2016).

According to the IPCC, vulnerability is "the degree to which a system is susceptible to, and unable to cope with, adverse effects of climate change, including climate variability and extremes"; and it is commonly agreed to be function of its exposure, sensitivity, and capacity to adapt (Engle et al. 2014).

The concept of vulnerability is frequently addressed in the field of food planning studies, since the negative results of vulnerability affect food security (Paloviita et al. 2016). The Food and Agriculture Organization (FAO) of the United Nations (UN) defined food security in 1996, as: "when all people, at all times have physical and economic access to sufficient, safe and nutritious food to meet their dietary needs and food preferences for an active and healthy life" (FAO 1996).

Food and agriculture are significant contributors to and heavily impacted by climate change. Concerning food system vulnerability, the impact of climate change on the sustainability of the food supply is a priority area, frequently addressed with agricultural suitability and crop yield assessments.

Even if sustainability considerations have largely been absent for years from food security assessments (Gustafson et al. 2016), there is a growing awareness that the food system's contribution to total anthropogenic environmental pressures and CC is considerable (Tukker et al. 2006; Garnett 2011).

Changes in Food Systems are essential to mitigate and adapt to climate change and meet food security demands. While it is estimated that agriculture contributes to 14–24% of global greenhouse gas emissions (GHG), estimates of emissions from agriculture and all the other food system activities are up to nearly 19–29% of all global GHG emissions (Vermeulen et al. 2012). Although agriculture makes the greatest contribution to total food system emissions, a food system approach enables a greater opportunity for adaptation and mitigation to CC, while sustaining food security (Campbell et al. 2016).

In particular, food consumption is a major area for climate change mitigation and adaptation, and there is a growing widespread support for a "demand restraint perspective". This perspective on sustainable food security proposes a dietary shift as a means of tackling the environmental impacts of the food system, as well as any food-related disease burden (Garnett 2014). Animal protein-intensive diets are at the forefront in terms of GHG emissions (Garnett 2011) and water consumption, as evidenced by recent studies on water consumption related to different diets (Vanham et al. 2017).

In Portugal, there are significant apparent food consumption deviations from the recommended dietary requirements, and food-related disease incidence is the main factor responsible for the years of life prematurely lost (Graça et al. 2016; Pinho et al. 2016). In a recent study for the Mediterranean, Galli et al. (2017) showed that Portugal had the highest *per capita* ecological footprint for apparent food consumption, mostly because of an animal protein-intensive diet (Galli et al. 2017).

The resolution for these problematic trends requires policies and actions to shift eating patterns, which need to target health, and sustainability-relevant consumption practices (Garnett et al. 2015). However, the National Climate Change Adaptation Strategy (APA 2013), does not consider the food system, either from the perspective of food supply or from a demand restraint perspective. Only recently, the food system was included as a key sector in a climate change adaptation plan developed for a group of municipalities, and strategies for joint action on adaptation, mitigation and food security were identified (Saavedra Cardoso et al. 2017a).

The linkages between the critical domains of water and food security require integrated approaches in the context of climate change. Therefore, sustainable food security should be pursued by balancing agricultural specialization, with protecting or promoting diversification of production from city region food systems (CRFS) (FAO 2012). The potential of CRFS to build regional food self-reliance has also been considered as a strategy to enhance resilience and promote sustainable land use and food and nutrition security (Ruhf 2015; Paci-Green and Berardi 2015).

Given the fundamental role of agriculture in human and animal welfare, there is a worldwide growing concern with the need to adapt agriculture and agroforestry (and forest systems) in order to better handle changing climate conditions and to preserve their multifunctional services. Rapid changes and transitions to climate change-resilient farming systems increasingly call for effective national and regional governance systems, financial support and investment in the sector as to promote innovation for sustainable agriculture, rural prosperity and food security (FAO 2017).

Measures to adapt agriculture, or other sectors, to a new climatic context demand the existence of studies that evaluate exposure, vulnerability and susceptibility to climate change. Considering all this, the main objective of this study was to produce spatial outputs that might help on the definition of adaptation measures focused on improving the capacity of agriculture to adapt to a new climatic context at a regional level, i.e., in the Centro Region of Portugal. Last, but not the least, this work also aimed to develop an algorithm that might be applied in other territories and for different resolutions.

Study Area

The Centro Region of Portugal, located at the central part of the Portuguese mainland, is a territory that encompasses very different environmental conditions, namely a significant difference between inland and coastal areas. While on low altitude coastal areas it is clear the dominance of gentle slopes, inland significant areas are associated to steep slopes, namely on "Cordilheira Central" (Fig. 1), one of the most important mountain massifs of the country.

In particular, only 34% of the region presents very gentle slopes (<5% of slope), and only 4% presents soils associated with high productivity (e.g., fluvisols), which are associated to alluvial plains mainly located at the end sections of the main rivers, mostly on western areas, used by intensive and irrigated agriculture. Only 9% of the area used by agriculture in the region is associated to such conditions, namely very gentle slopes and soils with high productivity. In the inland, despite the existence of areas with gentle slopes, those are very often associated to poor soils, namely on granitic substrates.

In terms of climate, the region is associated to a Mediterranean pattern, with significant differences in terms of rainfall and temperature regimes, determined by the distance to the Atlantic Ocean and altitude. The higher values of precipitation are associated with the mountains, whose summits are among the wettest in Portugal, while lower values are registered on easternmost inland areas, especially those located southeast and leeward to "Cordilheira Central". Such dryer areas are also those with higher variation in temperature along the year. Winter frosts are common, and hot and dry summers are particularly long. On the opposite, coastal areas have higher values of precipitation and lower temperature range, a fact associated to the higher influence of the Atlantic Ocean. In fact, frosts are very unusual on these areas. According to projected climatic scenarios, average annual precipitation will decrease in the region, mainly on southern and inland areas (Fig. 2), and average temperatures will increase; a spatial pattern that is clearly detected on results for water deficit (Fig. 11). In terms of extreme weather events, projections predict an increase in frequency, intensity and duration. It is expected an increase in the number of days with average maximum temperatures higher than 30 °C/35 °C, and in the number of days with precipitation higher than 20 mm. Such changes are also more pronounced on the southern and inland areas. The only change that benefits agriculture is related with the decrease in the number of frost days, associated to a projected rise on average minimum temperatures[1].

---

[1]Projected scenarios produced by IPMA—Portuguese Institute for Sea and Atmosphere, 2017. An online version can be accessed on http://portaldoclima.pt/.

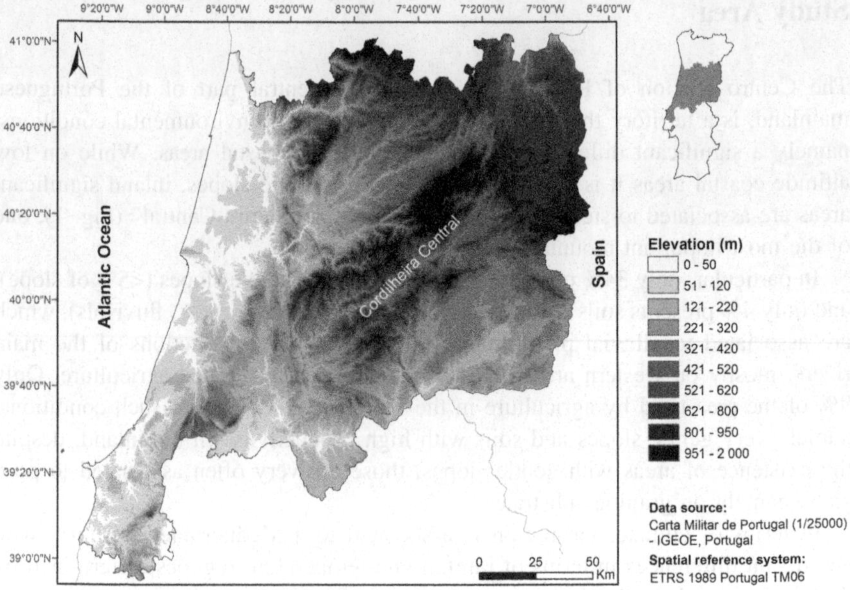

Fig. 1 Topography for the Centro Region of Portugal

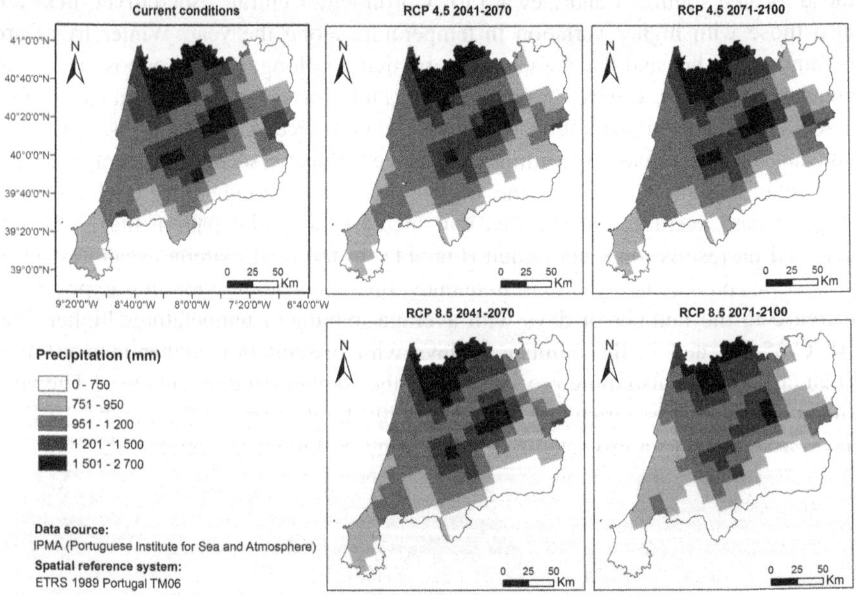

Fig. 2 Spatial distribution for precipitation for current climatic conditions and projected scenarios

Land Use

Topography and soil conditions are important to understand current land use patterns in the region. For the period under study (1990–2010), *forest and semi-natural areas* are the land use class with higher representation in the Centro Region, and registered an increase on importance of about 6% along this period (60–66%) (Fig. 3). Such evolution is mainly supported by the abandonment of agriculture, allowing the conversion of such areas to forest. In fact, from 1990 to 2010, the area dedicated to agriculture decreased about 7%.

As expected, the geographic pattern exhibited by agriculture is deeply associated to topographic and soil conditions. The areas where agriculture assumes higher importance at the landscape level are associated to gentle slopes, namely on the south-eastern sector and on the western areas (Fig. 4).

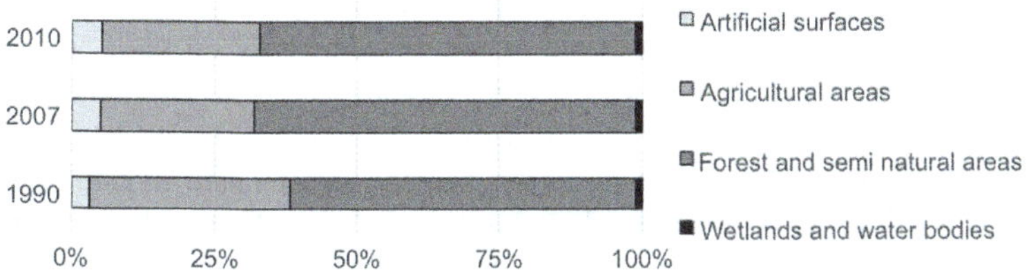

Fig. 3 Land use changes in Centro Region of Portugal between 1990 and 2010 (level 1). *Source* Carta de Ocupação do Solo, DGT (1990, 2007, 2010) (Land use and cover)

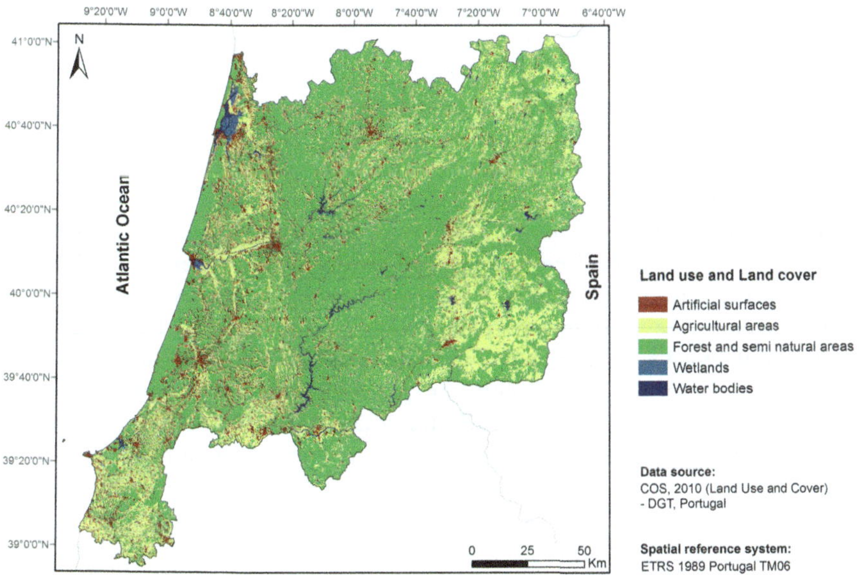

Fig. 4 Land use in Centro Region Portugal in 2010

While in the eastern sector rainfed agriculture is dominant, in western areas practices related to irrigation are significant, being very often associated with alluvial plains, where hydro-agricultural plans were installed. In fact, the potential area for irrigation is higher on the western part (Fig. 5—left), where irrigated area registered a higher increase between 1989 and 2009 (Fig. 5—right).

Such trend is clearly associated with investments on agriculture to ensure higher productivity, but increases water dependency. Moreover, it is opposite to the dominant pattern on inland areas, where it is clear an increasing conversion to extensive practices, namely to pastures or agroforestry systems on areas with poor soils, low availability of water, slopes unsuitable for the introduction of technology, and the prevalence of aged people, configuring very often a pre-stage to abandonment.

The loss in terms of area dedicated to agriculture is a common trend to inland Norte and Centro regions of Portugal, and is associated to important socio-demographic changes, namely associated to emigration to Europe (Almeida et al. 2009), promoting important changes on landscape (Almeida et al. 2012). Until the 1960s, agriculture was the most important activity on rural areas, a fact clearly identified by the dominance of area dedicated to such activity in the rural landscape. In fact, such activity played a major economic and cultural role, very often in combination with other activities, namely forestry and livestock production (Nunes 2008). Socio-demographic changes, in association with environmental constraints (steep slopes, nutrient-poor soils) and land holdings attributes (size, access), deeply contributed to abandonment, a trend that was reinforced after the implementation of new agricultural policies by the European Union (EU), namely the "set-aside" program, which promoted a breakdown in the productive system and the massive abandonment of agriculture in areas of low productivity. Such result is also determined by the lack of capacity to adapt to more competitive markets, contributing to the marginalization of such areas (Alves et al. 2016). In fact, 60 of the 100 municipalities of Centro Region are fully or partially classified as disadvantaged areas for agricultural activities (Portaria n.° 22/2015, de 5 de fevereiro).[2]

Non-irrigated land, which is associated to rainfed agriculture, registered a loss of about 20% between 1990 and 2010 (Fig. 6). In a more detailed analysis, it is possible to identify annual crops and olive groves as two types of rainfed agriculture that registered the highest decrease in terms of occupancy area (Fig. 7).

With an opposite trend, the area dedicated to irrigation increased, corresponding to 20% of the total area occupied by agriculture in 2010 (Fig. 8). Such increase in area is mostly promoted by an important expansion of irrigated annual crops, which increased 32% between 1990 and 2010. Because of their cultivation particularities, rice fields were the less important crop, concentrating essentially on the Lower Mondego valley.

---

[2]Portaria n.° 22/2015, 5th of february. Diário da República, Série I, n.° 25. Ministério da Agricultura e do Mar, 694–698.

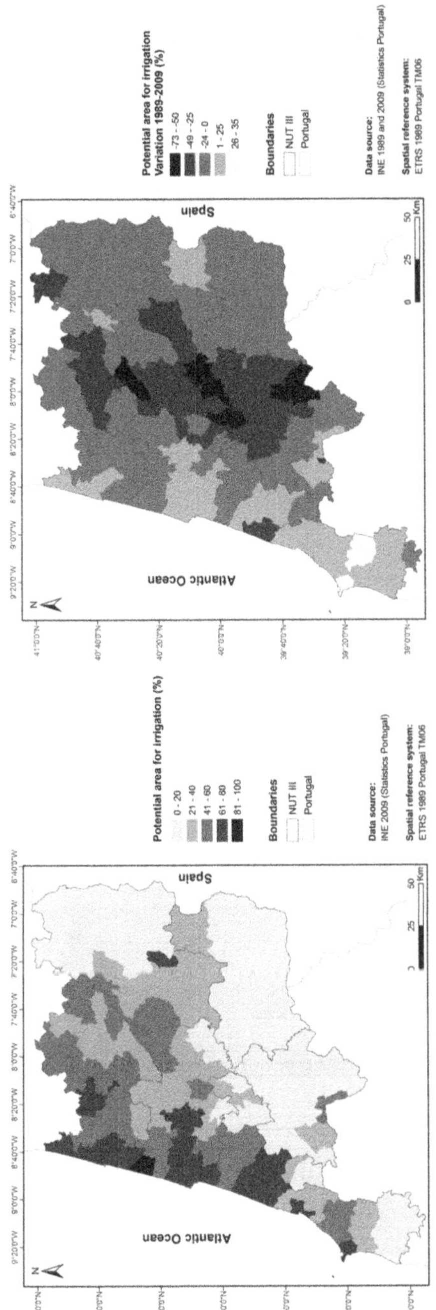

Fig. 5 Potential area for irrigation (2009) (left) and variation of irrigated area between 1989–2009 (right)

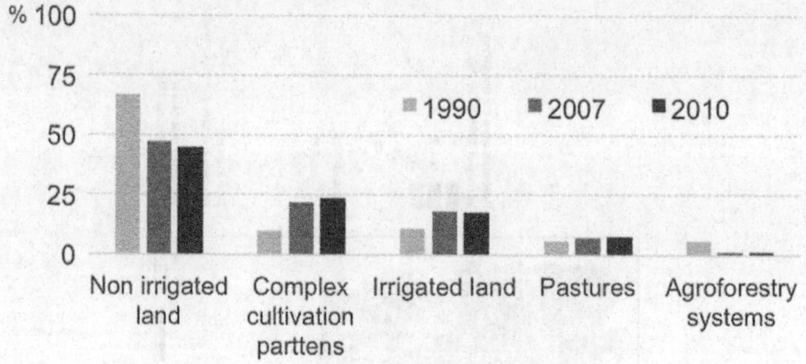

Fig. 6 Changes on area for different types of agriculture in the Centro Region (1990–2010)

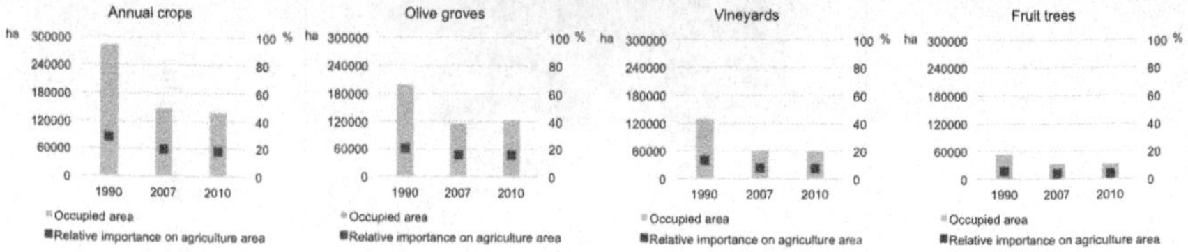

Fig. 7 Changes in area for different types of rainfed agriculture in the Centro Region (1990–2010)

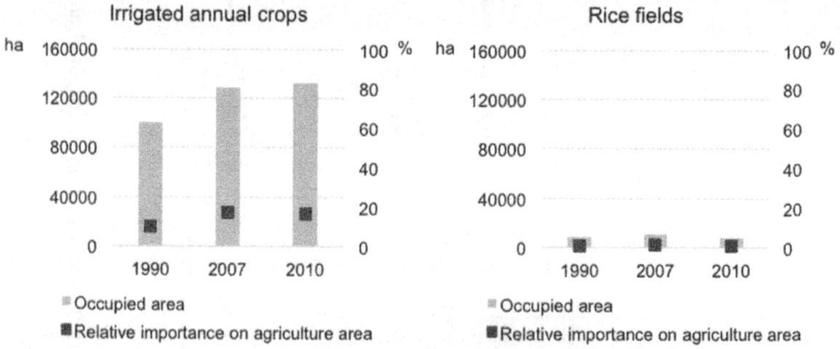

Fig. 8 Changes on area for different types of irrigated agriculture in the Centro Region (1990–2010)

Food Self-reliance Assessment

Food availability depends on production and food trade. To describe the current regional foodshed for the Centro Region, through the current production potential, we analyzed food consumption and regional production, i.e., a food self-reliance assessment. We used national data on food availability, the 2016 Portuguese Food Balance (INE 2017a), as a proxy for estimates of *per capita* food consumption. We considered these data to account for food consumption at a regional level, with

Table 1 Main crop and animal production (t), consumption and self-reliance: production in Portugal (1) and Centro Region (2); consumption (3) and self-reliance (4) in Centro Region

Plant crops and animal products	Production Portugal (t)	Production Centro Region (t)	Consumption Centro Region (t)	Self-reliance (%)
Cereal grains	1,138,086.00	308,234.00	242,994.34	126.85
Rice	169,289.00	33,307.00	45,370.27	73.41
Dried pulses	3603.00	1040.00	9515.86	10.93
Tubers	451,041.00	128,167.00	190,827.03	67.16
Vegetables	935,287.00	–	251 405.69	–
Fresh fruit	464,408.00	287,798.00	141,293.56	203.69
Citrus fruit	354,295.00	13,083.00	49,533.46	26.41
Nuts	40,130.00	4371.00	5607.56	77.95
Olive oil	69,526.84	7195.47	17,458.16	41.22
Wine	596,174.83	158,755.01	102,318.80	155.16
Beef meat	90,702,000.00	10,386,000.00	40,697.30	25,520.12
Pork meat	399,674,000.00	162,922,000.00	55,650.80	292,757.68
Poultry meat	369,119,000.00	278,045,000.00	72,388.52	384,100.93
Sheep and goat meat	18,242,000.00	3,506,000.00	3738.37	93,734.08
Eggs	134,576,000.00	116,713,000.00	19,796.39	589,567.02
Milk	1,849,375.00	219,757.00	165,508.04	132.78

Source Adapted from National Statistics of Food Availability (INE 2017a), Agricultural Statistics (INE 2017b) and Population Census (INE 2013)

reference to the total resident population (INE 2013), and compared the results with the regional production data to estimate the regional food self-reliance (RFSR). Our goal was to estimate the RFSR potential status by food agricultural products, i.e., the mass balance between the current production capacity and food consumption, for the resident population (number of inhabitants). The results of the food self-reliance assessment are shown in Table 1.

In 2016, the agricultural production of Centro Region met 127% of the annual required cereal grains, 204% of fresh fruits, 155% of wine, and 133% of milk consumption requirements. All the other food groups have self-reliances below 100% (Table 1), even though high for rice and nuts, respectively with 73 and 78% of group requirements attained. However, in what regards the animal production the RFSR exceeds the order of magnitude of consumption requirements by 200–6000 times, what evinces the regional profile of production specialization.

Consequently, some of these productions for which the region is self-reliant contribute heavily to domestic and international trade. These results enable us to identify areas of current food demand where increased regional production is necessary to meet food consumption requirements.

Approximately 50% of the virtual land import or "grabbed" by the EU is connected to soybean oil and feed products (Von Witzke and Noleppa 2010). This value increases if we consider the amounts of cereal grains used as feed. Intensive animal production, especially of granivores (pig and poultry farming), is largely independent on the utilized agricultural area (UAA) in the territories where this production takes place, although dependent on water for the production itself and other activities. However, the external dependence of cereals, soybeans and other raw materials used in feed, as well as fodder, can be considered as a vulnerability factor (Paloviita and Järvelä 2016). This is especially true in a context of uncertainty about maintaining global production potential, and the possible consequences of the fall in this potential on the price stability of these inputs.

Methodology

An edaphoclimatic suitability model focused on agriculture was created, based on a methodology structured by an ecological-landscape system planning perspective (Magalhães 2007; Saavedra Cardoso et al. 2017b). This spatial model relates edaphic (soil ecological value) (Cortez 2007; Cortez et al. 2013) and topographic components (slope, land morphology) (Cunha et al. 2013) with climatic variables. The variables for soil conditions and land morphology were used to produce edapho-morphological suitability, a procedure based on Multicriteria Decision Analysis (MCDA) (Munier 2011) that considers different forms of transforming variables to a common scale and assigns weights for the criteria used, according to the nature of the attributes and their relation with the MCDA goal (Satty 1988; Malczewski and Rinner 2015) (Fig. 9).

The estimation of the weight assigned to each factor was carried out considering the greater importance of the soil ecological value as a synthesis factor related to several qualities of FAO land evaluation methodology, namely: nutrient availability, nutrient retention capacity and rooting conditions (FAO 1976, 1988); according to Eliasson (2007), among the most commonly used qualities in assessing suitability for rainfed agriculture. According to Cortez (2007), the soil ecological value considers the intrinsic properties of the soil determining the biomass production capacity, namely: profile depth, nature of the originary material, clay and organic matter contents, structure, pH, cation exchange capacity and base saturation percentage. According to Eliasson (2007), slope is also among the qualities most used in assessments of agricultural suitability. Therefore, it was considered the second factor with the greatest weight due to its relationship with FAO qualities—potential for mechanization and erosion hazard (FAO 1988).

Land Morphology is a concept and a mapping method that represents landforms, categorised according to their hydrological position in the watershed (Magalhães 2001). This classification outlines two different systems, the wet and dry (concave-convex surfaces) in the hillslope profile, including valley bottoms, hilltops

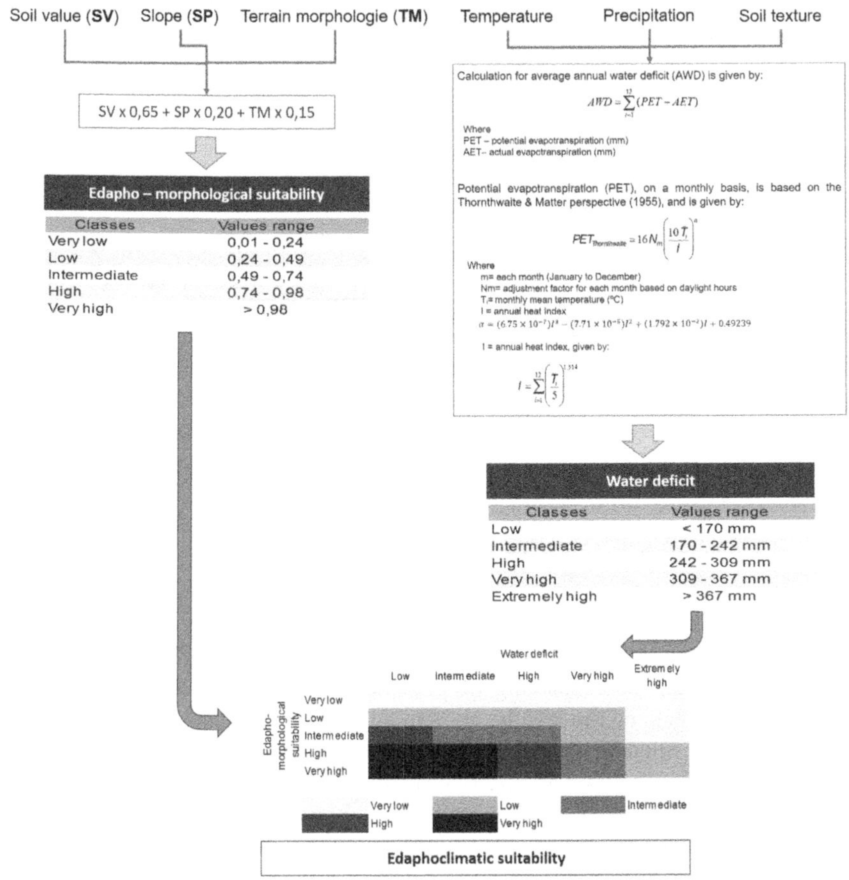

Fig. 9 Flow chart of the methodological approach for edapho-morphological suitability

and hillslope (Magalhães 2001, Cunha et al. 2017). As such, land morphology was considered as synthesis factor of water availability and nutrients availability, although as a factor it has no direct relation with the FAO qualities. For these reasons, and considering some redundancy with the other factors, its weight is lower.

The weight attributed to each class of the selected factors (from 1 to 5, increasing sense of class suitability) results from the literature review, from the definition of each class in the case of soil ecological value (Cortez 2007) and for slope the established bio-physical criteria to define natural constraints for agriculture (Orshoven et al. 2012).

The results, based on the weighted sum, were classified on different levels of edapho-morphological suitability, a classification procedure that relies on standard deviation (mean +/−1 standard deviation).

The integration of climatic variables, used to produce the edapho-climatic suitability model, was based on the use of water deficit, a calculation that follows the Thornthwaite method (Thornthwaite and Matter 1957; Fernández-García 1995). This method considers not only mean climatic conditions, based on mean temperature and precipitation for each month, but also different conditions in terms of soil water content that are available for plants/crops, considering data for soil texture (clayey: 100 mm; silty: 150 mm; sandy: 50 mm). Processes were generated automatically by an algorithm previously implemented in Python using the GDAL/OGR Python API (http://www.gdal.org/) and the package Numpy (http://www.numpy.org/). It calculates the water deficit for each pixel on the raster of the study area. Water deficit results were also classified into 5 classes based on quintiles.

The combination of both results, edapho-morphological suitability and water deficit, allowed the identification of different levels of edaphoclimatic suitability. Considering that previous variables were classified into 5 categories, the classification for suitability followed that condition. Thus, suitability level for each pixel was determined by the spatial coincidence of categories for each variable, as presented on Fig. 9.

As the main objective of this study was to produce a spatial output that contributes to a better understanding of potential impacts from climate change on agriculture, and help on the identification of adaptation measures, the calculation of edapho-climatic suitability was also carried out for future climatic scenarios. Two climatic scenarios (RCP 4.5 and RCP 8.5) and two time-windows (2041–2070 and 2071–2100) were considered in this study. The assessment of impacts from climate change was evaluated based on changes on average annual water deficit (AWD), using projections for mean monthly temperature and precipitation produced for the climatic projections. This was the only component that changed when assessing changes on edapho-climatic suitability under future climatic conditions, once soil conditions and land morphology were considered to be stable.

Results

Edapho-Morphological Suitability

Considering soil conditions and land morphology, it is clear that in the study area *very high* suitability is limited to small areas. Those areas are associated to alluvial plains located at the low section of the rivers, being the Baixo Mondego plain (lower Mondego) the most significant (Fig. 10). Also, only about 70% of the study area has low to very low edapho-morphological suitability considering agriculture.

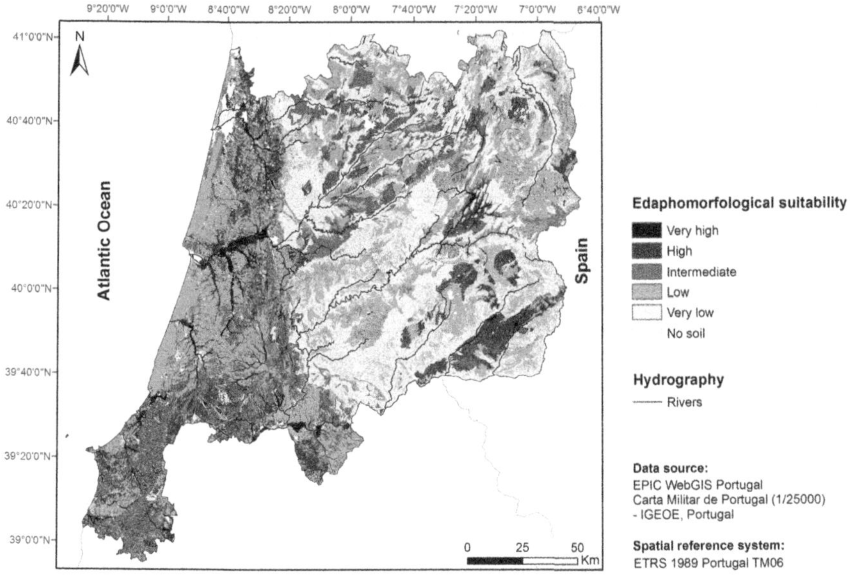

Fig. 10 Edapho-morphological suitability for the Centro Region of Portugal

It is also evident that western areas present higher suitability values, and only small patches have *intermediate* to *high* scores inland.

Water Deficit

In terms of water deficit, the lower values are identified on mountain areas, a fact determined by higher values of precipitation and lower mean temperatures, especially in the northwest and central areas (Fig. 11). The higher values of precipitation are greatly associated to the barrier effect associated with mountains considering prevalent western winds, responsible for the trajectory of low pressure systems (W-E) and associated fronts, which are responsible for a significant percentage of rain, reinforced by orography. Such context also explains the identification of higher values of water deficit in eastern areas, located on the leeward side of mountains. The higher values are also identified on southernmost areas, an expected pattern considering the increasing climatic dryness towards south (Fig. 11).

Considering climatic projections for both scenarios and the two time-windows considered, it is predicted an expansion of areas with *very high* and *extremely high* water deficit conditions, a result mostly dependent on the foreseen reduction of annual precipitation. Such increase on water deficit is expected to be higher for the RCP 8.5 scenario for the period 2071–2100, which is considered the "worst

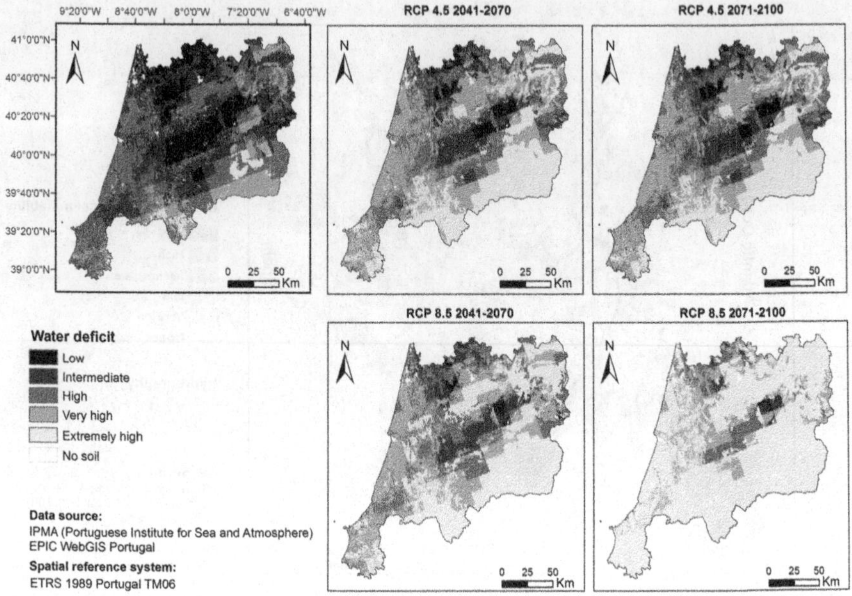

Fig. 11 Water deficit for current and future climatic conditions in the Centro Region of Portugal

scenario". Still, other outputs for water deficit also predict important changes, exhibiting a similar spatial pattern.

The areas that are more susceptible are those that already experience higher values of water deficit, namely on south and eastern sectors of the study area.

Edapho-Climatic Suitability

The combination of edapho-morphological suitability and water deficit reveals that a significant percentage of the study area has *low* to *very low* edapho-climatic suitability to agriculture under current conditions (Fig. 12). As expected, *very high* suitability is mainly associated to alluvial plains, being mostly concentrated on western areas. Although, considering the expected increase of water deficit under future climatic scenarios, even those areas might register a decrease on suitability in the future, if conversion to irrigated farming systems is not possible, e.g., as water supply diminishes and irrigation demands can not be satisfied. According to the results of each model, only very small areas would conserve *high* and *very high* suitability to rainfed agriculture under future climatic scenarios.

In fact, significant areas will register a significant loss of suitability, changing from *high* and *very high* to *low*, especially in the scenario RCP 8.5 for the period 2071–2100, considering current conditions as reference (Fig. 13).

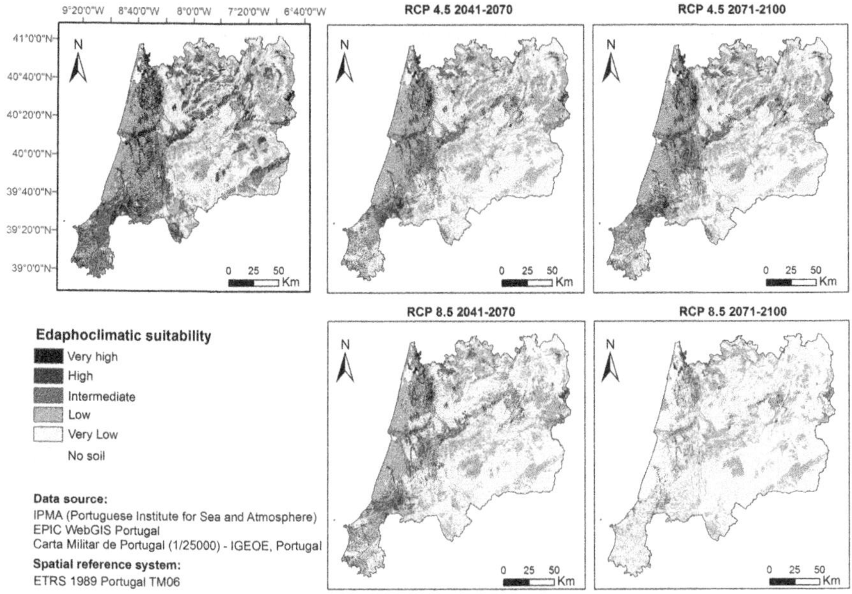

Fig. 12 Edaphoclimatic conditions for current and future climatic scenarios in the study area

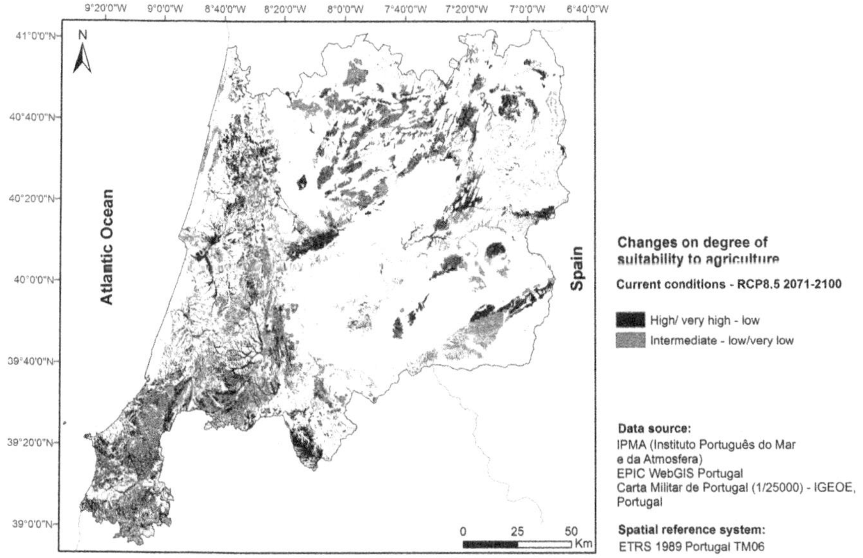

Fig. 13 Areas with higher loss on edaphoclimatic suitability in the study area for the scenario RCP8.5

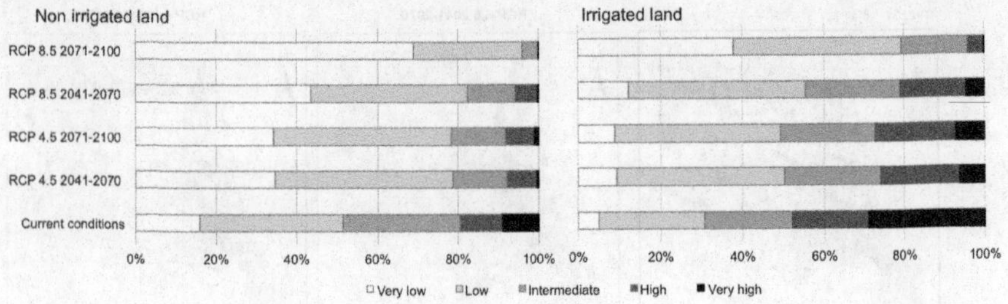

Fig. 14 Changes on edaphoclimatic suitability considering current area dedicated to agriculture

In fact, significant areas that are now occupied by agriculture (2010) will lose the capacity to support rainfed agriculture. Considering this type of land use, the loss of suitability is significant for both scenarios and time-windows, considering the important increase on water deficit. The model predicted that at least 40% of the area currently dedicated to rainfed agriculture will have *very low* suitability under future climatic scenarios, an important decrease considering that nowadays such value is around 15% of the study area. A prediction that is even higher under the scenario RCP8.5 for the period 2071–2100 (Fig. 14).

Discussion

Considering that is possible to produce outputs with higher resolution, the approach used in this study can be further applied to assess suitability at finer (local) scales, allowing the implementation of strategies for small territories or administrative areas. Such condition is of critical importance when considering the identification and application of adaptation measures at local scales. Although, validation based on current land use identified a very good match between land use and edapho-climatic suitability for current conditions, the validation of the model should be improved through the use of new strategies dedicated to the evaluation of models' results and by its application in different territories and at different scales.

For this study, a very clear mismatch between recent trends on agriculture and foreseen climatic scenarios was detected in the Centro Region of Portugal. Recent trends on agriculture identified an increase in irrigated agriculture and an important reduction in terms of rainfed agriculture, which reflects an increasing pressure on water resources. Such trend seems to be incompatible to climatic projections for future scenarios, considering that projections predict an important increase on water deficit, determined by a reduction in terms of annual average precipitation and an increase in mean temperature, forecasting a reduction in terms of available water resources.

The areas that are more susceptible to increasing water deficit under future climatic scenarios are those where agriculture has higher importance today. Also, even more relevant, is the increase of the area for irrigation on the southwest sector, predicted as highly susceptible to increasing water deficit.

Due to the increases in water deficit, models for edaphoclimatic suitability predicted an expansion of areas that have *low* to *very low* suitability. In many cases the loss of suitability is very relevant, and might affect significant areas, especially when considering scenario RCP8.5 for the period 2071–2100.

Considering the reduction on water availability, a critical resource for irrigation, which may contribute to the loss of suitability for rainfed agriculture, it is important to identify measures that could be implemented in the short-medium term, helping to adapt agriculture to a decrease in available water resources, which must be based on the following concepts: *assess* and *repair, convert* and *cooperate*. *Assess* and *repair* is mostly dedicated to the type of agriculture that is associated to intensive use of water, and depends on infrastructures to store and transport water—irrigated agriculture. In this specific case, and considering the importance in the western area of the region, it seems advisable to invest on reducing water wasting, mainly by monitoring and improving the quality of the infrastructures associated with water storage and distribution. For example, in the "Baixo Mondego" region, an important area associated with irrigation, there are evidences that point to significant waste because of delays in the system associated to human decision (Alves et al. 2017).

Repair and cooperation measures should not be only focused on the hard repair of systems, but should also encompass integrated water resources management, at river basin level, i.e. land use planning and decision-making processes, models and technologies for assessing the capacity of water resources to meet current and future water demands (UNESCO 2009). This should include *cooperation* among stakeholders and water management entities in order to coordinate and balance the various water-using sectors. It must include the assessment of water resources availability and future water demands based on climatic and socio-economic scenarios (Milano 2012). A monitoring system that considers, among others, a sustainable water allocation index (SWAI), based on the ratio of water supply to water demand should be implemented for this purpose.

However, the bigger effort should be focused on conversion in different domains, namely irrigation techniques, crop varieties, type of agriculture and even changes at the land use level. Related to irrigation, conversion should be dedicated to water saving by motivating/supporting farmers in updating irrigation techniques when that option is possible, giving preference to water-saving technologies. In addition, and despite the fact that such option very often implies a decrease in profit, it is necessary to use crop varieties that are lower water demanding, or even convert irrigated areas of lower productivity to rainfed agriculture.

Despite the fact that engineered and technological adaptation strategies are still the most common as adaptive responses, results from ecosystem-based solutions indicate that measures should not rely only on technology (Noble et al. 2014). For those areas currently associated to rainfed agriculture that are expected to lose suitability, it could be considered the conversion to agroforestry, silvopastoral and polycultural systems, advocated for their resilience to climate change (Altieri et al.

2015; Altieri and Nicholls 2017). Farming systems and agroecological practices related to the maintenance of *landscape, crop and genetic diversity* (polyculture and agro-silvo-pastoral systems, rotations, use of drought tolerant local varieties); *soil management* that, e.g., improves soil water holding capacity (cover cropping, green manures, no-till, organic farming) and *soil conservation* (contour farming, terracing, grass strips), have been considered relevant to enhance resilience to CC through various effects on soil quality and water conservation (Altieri et al. 2015). In particular, several studies on mitigation and adaptation in the agricultural sector have identified a range of practices that enhance soil carbon storage as relevant to reducing vulnerability to climate change (Smith and Olesen 2010). For instance, the adaptive capacity of soils can be enhanced through the increase of soil organic matter, which favors productivity over time, improves soil structure and water holding capacity and is therefore considered a win-win option, for mitigation and adaptation (Smith and Oleson 2010). However, considering the high level of abandonment of agriculture registered in the region on inland areas along decades (Almeida et al. 2009) and the risk of abandonment predicted for some municipalities on western areas (Alves et al. 2017), the conversion to forest seems to be a very good option, in particular for securing sustained water resources (Kohm and Franklin 1997). Moreover, agroforestry and forestry could give an important contribution to improve carbon sequestration and reduce soil erosion, a risk that will increase under future climatic scenarios, considering the increase in water deficit and the higher frequency and intensity of extreme rain episodes.

In what regards sustainable food security, the need to diversify sources of food supply should be acknowledged, by considering food system relocalization as a way to construct food system resilience. City region food systems (CRFS) can be advocated as an adaptation strategy in the context of CC for its capacity to endure and adjust to failures in the mainstream food system, in the event of natural or man-made disruptions in food supply or price volatility (Ruhf 2015).

Also, solutions to achieve food security under climate change consider currently "demand restraint" options in what regards food consumption, as measures for mitigation and adaptation. Animal protein-intensive diets are at the forefront in terms of GHG emissions and water consumption. This "climate friendly" dietary transition requires policies and actions to shift eating patterns, which need to target health, and sustainability-relevant consumption practices (Garnett et al. 2015).

However, the National Climate Change Adaptation Strategy, does not consider the food system, either from the perspective of diversifying food supply through CRFS or from a demand restraint perspective.

Final Remarks

Current challenges about climate change are related to the definition and implementation of measures that promote resilient societies concerning the new climatic context. Once agriculture provides basic services, namely food supply, it is important to conceive measures that contribute to adapt such sector to projected changes on climate, one of the main limiting factors for the activities in this sector.

Under this focus, and despite the fact that this is an exploratory study, the approach produced interesting results under the aim of assessing changes on suitability to agriculture in the Central Region of Portugal associated to projected climatic scenarios. The model, combining variables related to soil properties, topographic parameters and water deficit, revealed a worrying decrease on the edaphoclimatic suitability for rainfed agriculture under future climatic scenarios, especially for the RCP 8.5 scenario. Considering such predictions and increasing pressure on water resources, it is important to adopt a set of strategies that reinforce the capacity of agriculture to adapt, increasing resilience, and reduce vulnerability.

Although the fact that such approach does not consider requirements for specific crops, it can be helpful on the definition of context/structural measures, pointing out guidelines for action. But, if the model is produced for higher spatial resolution, spatially explicit results create the change to set measures adjusted to specific areas, and include other variables in the analysis.

References

Almeida, A. C., Nunes, A., & Figueiredo, A. (2009). *Mudanças no uso do solo no interior Centro e Norte de Portugal*. Coimbra: Imprensa da Universidade de Coimbra.

Almeida, A. C., Nunes, A., & Figueiredo, A. (2012). *Mudanças recentes na paisagem portuguesa*. In M. M. Passos, L. Cunha & R. Jacinto (Eds.), As Novas Geografias dos Países de Língua Portuguesa. Paisagens, Territórios e Políticas no Brasil e em Portugal (II) (pp. 113–135). São Paulo: Outras Expressões.

Altieri, M., Nicholls, C., Henao, A., & Lana, M. (2015). Agroecology and the design of climate change-resilient farming systems. *Agronomy for Sustainable Development, 35*(3), 869–890. https://doi.org/10.1007/s13593-015-0285-2.

Altieri, M. A., & Nicholls, C. I. (2017). The adaptation and mitigation potential of traditional agriculture in a changing climate. *Climatic Change, 140*(1), 33–45. https://doi.org/10.1007/s10584-013-0909-y.

Alves, C., Nunes, A., & Silva, J. (2016). Innovation towards more resilient territories: A case study from the Serra da Estrela, Portugal. In T. Noronha & H. Pinto (Eds.), *Innovation for resilience* (pp. 97–124). Universidade do Algarve.

Alves, C. D., Figueiredo, A., Patriarca, J., Castro, P., & Loureiro, J. (2017). Agricultura. *Plano de Adaptação às alterações climáticas da Comunidade Intermunicipal da Região de Coimbra* (pp. 149–236). Coimbra: Universidade de Coimbra and CIM-RC.

APA (Agência Portuguesa para o Ambiente). (2013). *Estratégia de Adaptação da Agricultura e das Florestas às Alterações Climáticas*. APA, Lisboa, Portugal.

Beniston, M., Keller, F., Koffi, B., & Goyette, S. (2003). Estimates of snow accumulation and volume in the Swiss Alps under changing climate conditions. *Theoretical and Applied Climatology, 76,* 125–140.

Beniston, M., Rebetez, M., Giorgi, F., & Marinucci, M. R. (1994). An analysis of regional climate change in Switzerland. *Theoretical and Applied Climatology, 49*(3), 135–159.

Calzadilla, A., Rehdanz, K., Betts, R., Falloon, P., Wiltshire, A., & Tol, R. S. (2013). Climate change impacts on global agriculture. *Climatic Change, 120*(1–2), 357–374. https://doi.org/10.1007/s10584-013-0822-4.

Campbell, B. M., Vermeulen, S. J., Aggarwal, P. K., Corner-Dolloff, C., Girvetz, E., Loboguerrero, A. M., et al. (2016). Reducing risks to food security from climate change. *Global Food Security, 11*, 34–43. https://doi.org/10.1016/j.gfs.2016.06.002.

Chang, J., Ciais, P., Viovy, N., Vuichard, N., Herrero, M., Havlík, P., … Soussana, J. F. (2016). Effect of climate change, CO_2 trends, nitrogen addition, and land cover and management intensity changes on the carbon balance of European grasslands. *Global Change Biology, 22* (1), 338–350. https://doi.org/10.1111/gcb.13050.

Ciscar, J. (Eds.). (2009). *Climate change impacts in Europe. Final report of the PESETA research project*. Luxemburg: Publications Office of the European Union.

Cortez, N. (2007). Sub-sistema Solo. *Estrutura Ecológica da Paisagem, conceitos e delimitação: escalas regional e municipal* (pp. 52–57). Lisboa: ISApress.

Cortez, N., Leitão, M., & Pena, S. (2013). Solo. In *Estrutura Ecológica da Paisagem, conceitos e delimitação: escalas regional e municipal* (pp. 83–104). Lisboa: ISApress. ISBN: 978-972-8669-53-9. Retrieved from http://epic-webgis-portugal.isa.ulisboa.pt/.

Cunha, N., Magalhães, M. R., Abreu, M., & Abreu, M. C. (2013). Morfologia do Terreno. In *Estrutura Ecológica Nacional. Uma proposta de delimitação e regulamentação* (pp. 51–66). ISAPress: Lisboa. ISBN: 978-972-8669-53-9. Retrieved from http://epic-webgis-portugal.isa.ulisboa.pt/.

Cunha, N. S., Magalhães, M. R., Domingos, T., Abreu, M. M., & Küpfer, C. (2017). The land morphology approach to flood risk mapping: An application to Portugal. *Journal of Environmental Management, 193*, 172–187. https://doi.org/10.1016/j.jenvman.2017.01.077.

EEA (European Environment Agency). (2013). *Adaptation in Europe. Addressing risks and opportunities from climate change in the context of socio-economic developments*. Publications Office of the European Union, Luxembourg. EEA Report No 3/2013.

EEA (European Environment Agency). (2017a). *Climate change, impacts and vulnerability in Europe 2016. An indicator-based report*. Publications Office of the European Union, Luxembourg. EEA Report No 1/2017.

EEA (European Environment Agency). (2017b). *Global and European temperature. Indicator assessment; data and maps*. Available at https://www.eea.europa.eu/data-and-maps/indicators/global-and-european-temperature-4/assessment.

Eliasson, Å., (2007). Review of land evaluation methods for quantifying natural constraints to agriculture. *The institute for environment and sustainability*, Joint Research Centre, Ispra (Italy). Luxembourg: Office for Official Publications of the European Communities. https/doi.org/10.2788/17545. Available online at: https://core.ac.uk/download/pdf/38609579.pdf.

Engle, N. L., de Bremond, A., Malone, E. L., & Moss, R. H. (2014). Towards a resilience indicator framework for making climate-change adaptation decisions. *Mitigation and Adaptation Strategies for Global Change, 19*, 1295–1312. https://doi.org/10.1007/s11027-013-9475-x.

Fernández-García, F. (1995). *Manual de climatología aplicada: clima, medio ambiente y planificación*. Madrid, España: Editorial Síntesis.

FAO—Food and Agriculture Organization of the United Nations. (1976). *A framework for land evaluation*. Soils bulletin n.° 32. FAO, Rome, Italy.

FAO—Food and Agriculture Organization of the United Nations. (1988). *Directives: Evaluation des Terres pour l'Agriculture Pluviale*. Bulletin Pédologique de la FAO 52. FAO, Rome, Italy.

FAO—Food and Agriculture Organization of the United Nations. (1996). *Rome declaration on world food security and world food summit plan of action*. World Food Summit, November 13–17. FAO, Rome, Italy.

FAO—Food and Agriculture Organization of the United Nations. (2012). *Food agriculture and cities: Challenges of food and nutrition security, agriculture and ecosystem management in an urbanizing world*. Rome, Italy: FAO.

FAO—Food and Agriculture Organization of the United Nations. (2016). *Climate change and food security: Risks and responses*. Rome, Italy: FAO.

FAO—Food and Agriculture Organization of the United Nations. (2017). *The future Trends of food and challenges agriculture*. Rome, Italy: FAO.

Fellmann, T. (2012). The assessment of climate change-related vulnerability in the agricultural sector: Reviewing conceptual frameworks. In A. Meybeck, J. Lankoski, S. Redferm, N. Azzu, & V. Gitz (Eds.), *Building resilience for adaptation to climate change in the agriculture sector* (pp. 37–61). Proceedings of a Joint FAO/OECD Workshop, April 23–24, 2012. FAO, OECD, Roma.

Galli, A., Iha, K., Halle, M., El Bilali, H., Grunewald, N., Eaton, D., et al. (2017). Mediterranean countries' food consumption and sourcing patterns. An Ecological Footprint viewpoint. *Science of the Total Environment, 578,* 383–391. https://doi.org/10.1016/j.scitotenv.2016.10.191.

Garnett, T. (2011). Where are the best opportunities for reducing greenhouse gas emissions in the food system? *Food Policy, 36*(S1), S23–S32. https://doi.org/10.1016/j.foodpol.2010.10.010.

Garnett, T. (2014). Three perspectives on sustainable food security: Efficiency, demand restraint, food system transformation. What role for life cycle assessment? *Journal of Cleaner Production, 73,* 10–18. https://doi.org/10.1016/j.jclepro.2013.07.045.

Garnett, T., Mathewson, S., Angelides, P., & Borthwick, F. (2015). *Policies and actions to shift eating patterns: What works?*. Oxford, UK: Food Climate Research Network; Chatham House, Royal Institute of International Affairs.

Graça, P., de Sousa, S. F., Correia, A., Salvador, C., Filipe, J., Carriço, J., et al. (2016). *Portugal. Alimentação Saudável em Números – 2015.* Lisboa, Portugal: Direção-Geral da Saúde. Retrieved from https://www.dgs.pt/em-destaque/portugal-alimentacao-saudavel-em-numeros-2015.aspx.

Gustafson, D., Gutman, A., Leet, W., Drewnowski, A., Fanzo, J., & Ingram, J. (2016). Seven food system metrics of sustainable nutrition security. *Sustainability, 8*(3), 196. https://doi.org/10.3390/su8030196.

Hays, J. D., Imbrie, J., & Shackleton, N. J. (1976). Variations in the Earth's orbit: Pacemaker of the ice ages. *Science, 194*(4270), 1121–1132.

Instituto Nacional de Estatística—INE. (2013). População residente em lugares censitários (N.º) por Local de residência. *Recenseamento da População e Habitação* – 2011. Instituto Nacional de Estatística, Lisboa, Portugal.

Instituto Nacional de Estatística—INE. (2017a). Capitação edível diária de produtos alimentares disponível para abastecimento (g/ hab.) por Tipo de produto alimentar. *Balança Aliment. Port.* – 2012–2016. Instituto Nacional de Estatística, Lisboa, Portugal.

Instituto Nacional de Estatística—INE. (2017b). *Produção Agrícola: Animal e vegetal – 2016.* Vários indicadores. Instituto Nacional de Estatística, Lisboa, Portugal.

IPCC (Intergovernmental Panel on Climate Change). (2007). *Climate change 2007: Synthesis report. Summary for policymakers.* Cambridge, UK and New York, NY, : Cambridge University Press.

IPCC (Intergovernmental Panel on Climate Change). (2014). *Climate change 2014. Impacts, adaptation, and vulnerability. Part A: Global and sectoral aspects. Contribution of working group II to the fifth assessment report of the Intergovernmental Panel on Climate Change.* In C. B. Field, V. R. Barros, D. J. Dokken, K. J. Mach, M. D. Mastrandrea, T. E. Bilir, M. Chatterjee, K. L. Ebi, Y. O. Estrada, R. C. Genova, B. Girma, E. S. Kissel, A. N. Levy, S. MacCracken, P. R. Mastrandrea, & L. L. White (Eds.). Cambridge, UK: Cambridge University Press.

Kohm, K. A., & Franklin, J. F. (1997). *Creating a forestry for the 21st century: The science of ecosystem management.* Washington, DC: Island Press.

La Jeunesse, I., Cirelli, C., Aubin, D., Larrue, C., Sellami, H., Afifi, S., … Dettori, M. (2016). Is climate change a threat for water uses in the Mediterranean region? Results from a survey at local scale. *Science of the Total Environment, 543,* 981–996. https://doi.org/10.1016/j.scitotenv.2015.04.062.

Liang, X. Z., Wu, Y., Chambers, R. G., Schmoldt, D. L., Gao, W., Liu, C., … Kennedy, J. A. (2017). Determining climate effects on US total agricultural productivity. *Proceedings of the National Academy of Sciences, 114*(12), E2285–E2292. https://doi.org/10.1073/pnas.1615922114.

Lloyd, A., & Fastie, C. (2003). Recent changes in treeline forest distribution and structure in interior Alaska. *Écoscience, 10*(1), 176–185.
Magalhães, M. R. (2001). *A Arquitectura Paisagista. Morfologia e complexidade*. Lisboa: Editorial Estampa. ISBN 972-33-1686-2.
Magalhães, M. R. (Coord.). (2007). *Estrutura ecológica da paisagem. Conceitos e delimitação – escalas regional e municipal*. Centro de Estudos de Arquitectura Paisagista "Prof. Caldeira Cabral". Lisboa: ISAPress, Instituto Superior de Agronomia.
Malczewski, J., & Rinner, C. (2015). *Multicriteria decision analysis in geographic information science*. Berlin: Springer. ISBN 978–3-540-74756-7.
Mann, M. E., Zhang, Z., Hughes, M. K., Bradley, R. S., Miller, S. K., Rutherford, S., et al. (2008). Proxy-based reconstructions of hemispheric and global surface temperature variations over the past two millennia. *Proceedings of the National Academy of Sciences, 105*(36), 13252–13257.
Maslin, M. (2016). In retrospect: Forty years of linking orbits to ice ages. *Nature, 540*(7632), 208–210.
Menzel, A., Sparks, T. H., Estrella, N., Koch, E., Aasa, A., Ahas, R., et al. (2006). European phenological response to climate change matches the warming pattern. *Global Change Biology, 12*(10), 1969–1976.
Milano, M. (2012). *Changements globaux en Méditerranée: impacts sur le stress hydrique et la capacité à satisfaire les demandes en eau*. PhD thesis. Université Montpellier 2, France.
Miranda, P. M. A., Valente, M. A., Tomé, A. R., Trigo, R., Coelho, M. F. E. S., Aguiar, A., et al. (2006). O clima de Portugal nos séculos XX e XXI. In F. D. Santos & P. Miranda (Eds.), *Alterações Climáticas em Portugal. Cenários, Impactos e Medidas de Adaptação. Projecto SIAM II* (pp. pp. 45–113). Lisboa: Gradiva.
Munier, N. (2011). *A strategy for using multicriteria analysis in decision-making*. London: Sringer.
Niles, M., Esquivel, J., Ahuja, R., & Mango, N. (2017). *Climate change and food systems: Assessing impacts and opportunities*. Washington, DC.: Meridian Institute.
Noble, I. R., Huq, S., Anokhin, Y. A., Carmin, J., Goudou, D., Lansigan, F. P., et al. (2014). Adaptation needs and options. In C. B. Field, V. R. Barros, D. J. Dokken, K. J. Mach, M. D. Mastrandrea, T. E. Bilir, M. Chatterjee, K. L. Ebi, Y. O. Estrada, R. C. Genova, B. Girma, E. S. Kissel, A. N. Levy, S. MacCracken, P. R. Mastrandrea, & L. L. White (Eds.), *Climate change 2014: Impacts, adaptation, and vulnerability. Part A: Global and sectoral aspects. Contribution of Working Group II to the Fifth Assessment Report of the Intergovernmental Panel on Climate Change* (pp. 833–868). Cambridge, UK and New York, NY: Cambridge University Press.
Nunes, A. (2008). *Abandono do espaço agrícola na "Beira Transmontana"*. Iberografias 13. Ed. Campo das Letras SA, Porto, Portugal.
Oerlemans, J. (2005). Extracting a climate signal from 169 glacier records. *Science, 308*(5722), 675–677.
Olesen, J. E. (2016). Socio-economic impacts—agricultural systems. In M. Quante & F. Colijn (Eds.), *North sea region climate change assessment* (pp. 397–407). Cham: Springer International Publishing.
Orshoven, J., Terres, J.-M., & Tóth, T. (2012). *Updated common bio-physical criteria to define natural constraints for agriculture in Europe. Definition and scientific justification for the common biophysical criteria*. Luxembourg: Publications Office of the European Union.
Paci-Green, R., & Berardi, G. (2015). Do global food systems have an Achilles heel? The potential for regional food systems to support resilience in regional disasters. *Journal of Environmental Studies and Sciences, 5*, 685–698. http://dx.doi.org/10.1007/s13412-873015-0342-9.
Paloviita, A., & Järvelä, M. (Eds.). (2016). *Climate change adaptation and food supply chain management*. New York, NY: Routledge. ISBN 9781315757728.
Paloviita, A., Kortetmäki, T., Puupponen, A., & Silvasti, T. (2016). Vulnerability matrix of the food system: Operationalizing vulnerability and addressing food security. *Journal of Cleaner Production, 135*, 1242–1255. https://doi.org/10.1016/j.jclepro.2016.07.018.

Pimentel, D., Zuniga, R., & Morrison, D. (2005). Update on the environmental and economic costs associated with alien-invasive species in the United States. *Ecological Economics, 52*, 273–288.

Pinho, I., Franchini, B., & Rodrigues, S. (2016). *Guia alimentar mediterrânico. Relatório justificativo do seu desenvolvimento*. Porto, Portugal: Direção-Geral da Saúde.

Pinto, P., Braga, R., & Brandão, A. (2006). Agricultura. In F. Santos & P. Miranda (Eds.), *Alterações climáticas em Portugal: cenários, impactos e medidas de adaptação, Projeto SIAM II* (pp. 213–231). Lisboa: Gradiva.

Ramaswamy, V., Schwarzkopf, M. D., Randel, W. J., Santer, B. D., Soden, B. J., & Stenchikov, G. L. (2006). Anthropogenic and natural influences in the evolution of lower stratospheric cooling. *Science, 311*(5764), 1138–1141.

Ruhf, K. Z. (2015). Regionalism: A New England recipe for a resilient food system. *Journal of Environmental Studies and Sciences, 5*, 650–868, 660. https://doi.org/10.1007/s13412-015-0324-y.

Saavedra Cardoso, A., Domingos, T., Magalhães, M. R., de Melo-Abreu, J., & Palma, J. (2017b). Mapping the Lisbon potential foodshed in Ribatejo e Oeste: A suitability and yield model for assessing the potential for localized food production. *Sustainability*, *9*(11), 2003. https://doi.org/10.3390/su9112003.

Saavedra Cardoso, A., Ramos, F., de Melo-Abreu, J., Loureiro, J., & Leston, S. (2017a). Alimentação. *Plano de Adaptação às alterações climáticas da Comunidade Intermunicipal da Região de Coimbra* (pp. 237–423). Coimbra: Universidade de Coimbra and CIM-RC.

Santer, B. D., Taylor, K. E., Wigley, T. M. L., & Johns, T. C. (1996). A search for human influences on the thermal structure of the atmosphere. *Nature, 382*(6586), 39.

Santer, B. D., Wehner, M. F., Wigley, T. M. L., Sausen, R., Meehl, G. A., Taylor, K. E., ... Brüggemann, W. (2003). Contributions of anthropogenic and natural forcing to recent tropopause height changes. *Science, 301*(5632), 479–483.

Saaty, T. (1988). *The analytic hierarchy process: Planning, priority setting, resource allocation*. New York: Mcgraw-Hill. ISBN 0070543712.

Schleip, C., Rutishauser, T., Luterbacher, J., & Menzel, A. (2008). Time series modeling and central European temperature impact assessment of phenological records over the last 250 years. *Journal Geophysical Research, 113*(G4), G04026.

Skuras, D., & Psaltopoulos, D. (2012). A broad overview of the main problems derived from climate change that will affect agricultural production in the Mediterranean area. In A. Meybeck, J. Lankoski, S. Redferm, N. Azzu, & V. Gitz (Eds.), *Building resilience for adaptation to climate change in the agriculture sector* (pp. 217–260). *Proceedings of a Joint FAO/OECD Workshop*. April 23–24, 2012. FAO, OECD, Roma.

Smith, P., & Olesen, J. E. (2010). Synergies between the mitigation of, and adaptation to, climate change in agriculture. *The Journal of Agricultural Science, 148*(5):543–552.

Sorg, A., Bolch, T., Stoffel, M., Solomina, O., & Beniston, M. (2012). Climate change impacts on glaciers and runoff in Tien Shan (Central Asia). *Nature Climate Change, 2*, 725–731.

Thornthwaite, C. W., & Mather, J. R. (1957). *Instructions and tables for computing potential evapotranspiration and the water balance* (Vol. 10, No. 3, pp. 185–311). Centerton, N.J.: Laboratory of Climatology, Publications in Climatology

Torriani, D. S., Calanca, P., Schmid, S., Beniston, M., & Fuhrer, J. (2007). Potential effects of changes in mean climate and climate variability on the yield of winter and spring crops in Switzerland. *Climate Research, 34*(1), 59–69.

Tukker, A., Huppes, G., Guinée, J., Heijungs, R., Koning, A., Van Oers, L., et al. (2006). *Environmental impact of products. Analysis of the life cycle environmental impacts related to the final consumption of the EU-25 European Commission*. Seville: European Commission Joint Research Centre Institute for Prospective Technological Studies.

United Nations Educational, Scientific and Cultural Organization (UNESCO). (2009). *Integrated water resources management guidelines at river basin level*. Retrieved from: http://www.unesco.org/water/.

Vanham, D., del Pozo, S., Pekcan, A. G., Keinan-Boker, L., Trichopoulou, A., & Gawlik, B. M. (2017). Water consumption related to different diets in Mediterranean cities. *Science of the Total Environment, 573,* 96–105.

Vermeulen, S. J., Campbell, B. M., & Ingram, J. S. I. (2012). Climate change and food systems. *Annual Review of Environment and Resources, 37*(2012), 195–222. https://doi.org/10.1146/annurev-environ-020411-130608.

Von Witzke, H., & Noleppa, S. (2010). *EU agricultural production and trade: Can more efficiency prevent increasing "land grabbing" outside of Europe?* Research Report, Humboldt University, Berlin.

Walther, G.-R., Post, E., Convey, P., Menzel, A., Parmesan, C., Beebee, T. J. C., et al. (2002). Ecological responses to recent climate change. *Nature, 416*(6879), 389–395.

WEF (World Economic Forum). (2016). *The Global Risks Report 2016* (11th ed.). Geneva.

Impact of the Light Microclimate on Photosynthetic Activity of Grape Berry (*Vitis vinifera*): Insights for Radiation Absorption Mitigations' Measures

Andreia Garrido, Richard Breia, João Serôdio and Ana Cunha

Abstract IPCC's predicted rise in mean temperatures, increase in the frequency of summer heat waves and decrease in soil water availability for the Mediterranean regions will have an impact on foliar and fruit photosynthesis. But mitigation measures aiming reducing radiation absorption by the vine canopy may pose light limitations to grape berry photosynthesis. This work focused on the influence of the light level of the canopy microenvironment where clusters develop on the photosynthetic competence of grape berry tissues (exocarp and seed integument) throughout fruit growing season by imaging PAM fluorometry. Clusters from low (LL), medium (ML) and high light (HL) microclimates were sampled from green to mature stages. Both tissues showed high maximum quantum efficiency (F_v/F_m) and photosynthetic capacity (ETR_m) at the green stage, exocarp extending to mature stages while seed photosynthetic activity was more restricted to green stage. The light microclimate had a significant effect on the photosynthesis of both tissues but

A. Garrido · R. Breia · A. Cunha (✉)
Department of Biology, School of Sciences, CITAB, Centre for the Research and Technology of Agro-Environmental and Biological Sciences, University of Minho, Campus de Gualtar, 4710-057 Braga, Portugal
e-mail: accunha@bio.uminho.pt

A. Garrido
e-mail: andreia.garrido@bio.uminho.pt

R. Breia
e-mail: richardgoncalvesbreia@gmail.com

J. Serôdio
Department of Biology, CESAM, Centre for Environmental and Marine Studies, University of Aveiro, Campus de Santiago, 3810-193 Aveiro, Portugal
e-mail: jserodio@ua.pt

A. Cunha
CEB, Centre of Biological Engineering, University of Minho, Campus de Gualtar, 4710-057 Braga, Portugal

F. Alves et al. (eds.), *Theory and Practice of Climate Adaptation*, Climate Change Management, https://doi.org/10.1007/978-3-319-72874-2_24

also in their photosynthetic phenotypes along the season. In LL, both tissues showed lower activity in all stages, higher susceptibility to photoinhibition and lack of response to short-term light acclimation; ML and HL grapes adjust their activity peaking at different light intensities, were more responsive to light changing conditions, recover better from high light. Overall, our results suggest that not only light/temperature stress conditions imposed by climate changes but also viticulture practices causing changes in canopy light microclimates may have significant impacts on grape berry photosynthesis and hence in fruit development and quality.

Keywords Fruit photosynthesis · Exocarp · Seed integument · Chlorophyll fluorescence · Imaging PAM fluorometry

Introduction

Agricultural production is environmentally sensitive, being highly influenced by changes in climate, soil water and nutrition, and land use practices. In many regions worldwide, viticulture is a central agronomic practice representing a key economic activity. It was estimated that in Europe *Vitis vinifera* cultivars account to 60% of the world wine production in 2011 (OIV 2015). Portugal is the 11th wine producer in the world (OIV 2015), being the Minho Wine Region one of the most important wine growing areas (Fraga et al. 2014).

It has been recognized that the European southern regions, with Mediterranean climates (Peel et al. 2007), will be most affected by climate change (COM 2009). Until the end of the 21st century, climate variation will be especially pronounced in these regions, with a temperature increase of 2.2–5.1 °C (IPCC 2014). Projections point to the intensification of summer conditions, with lower water availability, elevated temperature and solar irradiance (visible and ultraviolet) what will affect viticulture (Fraga et al. 2012).

Grapevine physiology and development, as grapes' metabolite composition, are inherent to each variety, but also strongly dependent on climatic factors (e.g. temperatures, radiation intensities/duration, and water availability) (Jones and Davis 2000). The viticulture and winemaking will be significantly affected especially by high temperatures interacting with high radiation levels and strong water deficits (Leeuwen and Darriet 2016). High radiation and its interaction with high temperatures (e.g. 30–40 °C) can cause several impacts on grapevines and grape berries, such as: (1) sunburns in leaves and clusters (Chuine et al. 2004); (2) stomatal and mesophyll limitations of foliar photosynthesis (Moutinho-Pereira et al. 2004); (3) changes in transpiration, respiration and photorespiration rates, as well as photoinhibition and chlorosis (Moutinho-Pereira et al. 2003), mainly associated with the production of reactive oxygen species.

The evaluation of climate change effects on viticulture and increasing awareness of growers to these impacts is of the utmost importance, due to the high socioeconomic relevance of the sector. Indeed, growers are adopting economically and environmentally sustainable short- and long-term adaptation measures, to mitigate the negative impacts. Some short-term adaptation measures aiming to reduce radiation/heat absorption include the adjustment of pruning times and training systems (Reynolds and Heuvel 2009) and application of leaf sunscreens (Dinis et al. 2016) or vine shadings (Greer et al. 2011). Long-term adaptation measures mainly include varietal and land-allocation changes (Fraga et al. 2015).

Recently, it was demonstrated that the application of kaolin (a white, inert clay mineral, that acts as a sunscreen, absorbing mainly UV and infrared radiation) in vines of 'Touriga Nacional' variety, at the Douro Demarcated Region, have decreased leaf temperature and increased stomatal conductance, photosynthesis and water use efficiency of exposed leaves (Correia et al. 2014).

The structural organization of canopies extensively studied for viticulture (revised by Reynolds and Heuvel 2009), have been recently revisited in the context of climate changes in order to devise adequate exposure of the fruits to sunlight (Leeuwen and Darriet 2016). In fact, the structure of the vines can interfere with the grape berry productivity and quality, not only because the exposed leaf area affects the photosynthetic activity (Smart 1974), but also because it influences the microclimate of the fruits (Schultz 1995). The term "microclimate" was adopted by Smart et al. (1985) to define the environmental conditions in the vicinity of leaves and fruits (Haselgrove et al. 2000). The microclimate in the vineyard has a fundamental influence in growth and in the biosynthesis of metabolites.

In higher plants, besides leaves, other vegetative and reproductive structures can also perform photosynthesis (Aschan and Pfanz 2003). The photosynthetic activity was already studied in fleshy fruits (apple and grape berry), indehiscent fruits (cereals) and dehiscent fruits like pea pods (Blanke and Lenz 1989). Regarding grape berry *(Vitis vinifera* L.), Breia et al. (2013) proved by imaging PAM fluorometry that the exocarp (skin) and seed integuments are the most photosynthetic competent tissues. The role(s) of photosynthesis in sink organs, such as fruits, are still not fully disclosed. Some authors claim that it may contribute with additional organic compounds for the growth of fruit and seed (Aschan and Pfanz 2003), but the complexity of the photosynthetic mechanisms together with the heterogeneous distribution among grape berry tissues supports other functions, like oxygen production and secondary metabolites biosynthesis (Breia et al. 2013). It is thus reasonable to conceive that any effect on grape berry photosynthesis will have multiple impacts on grape berry physiology and development and, ultimately, on berry and wine quality.

The main objective of this work was to study the influence of cluster's light microclimate on the photosynthesis of grape berry tissues (exocarp and seed integument) throughout the fruit growing season and to contribute with information for short-term mitigation measures' plans where a reduction in radiation absorption by the vines is involved.

Materials and Methods

Biological Material and Sampling Procedure

Grape berries (*V. vinifera* L. 'Alvarinho') were sampled in 2015 from bards with North-South orientation in a vineyard located in Barcelos in northern Portugal (+41° 29′58.4″ N, −8° 33′45.1″ O). In order to study the effect of average light intensity of the microenvironments where clusters typically develop in the canopy on grape berry photosynthetic profile, three light microclimates were selected: LL (low light), ML (medium light) and HL (high light). LL clusters grow in shaded inner zones of the canopy exposed to diffuse and transmitted light; ML clusters were less shaded but rarely receiving direct sunlight; and HL clusters were exposed to direct or reflected sunlight most of the day. Grape berry clusters (six per microclimate) were collected randomly in the morning (9–10 h) from the young green to the mature stage: GE—Green Early (6 weeks after anthesis, WAA); GL—Green Later (9 WAA); V—*Véraison* (14 WAA) and M—Mature (17 WAA). Green, *véraison* and mature stages were selected because they represent the main developmental stages of the grape berry (Kennedy 2002), at the cell and tissue growth pattern level but also at the physiological and metabolic levels. Two sampling times in the green phase were included to better characterize grape berry photosynthetic profile evolution and response to short-time acclimation. The material was transported in refrigerated boxes to the laboratory and used within 2–6 h for imaging fluorometry experiments. For other assays, berries were immediately frozen with liquid nitrogen and stored at −80 °C.

Characterization of the microclimates: average light intensities and temperatures

In order to characterize the selected microclimates throughout the developing season, light intensities and temperatures were registered in clear days near sampling days. The light intensity (μmol m^{-2} s^{-1}), was measured with a radiometer (LI-COR, LI-250 Light Meter) at five different points in each cluster, and the temperature (°C) was measured with an infrared digital thermometer (Infrared, DT8380), in the frontal and bottom regions of the cluster (Fig. 1). Six independent replicates per microclimate were considered by selecting clusters and bards randomly. The average light intensity of each microclimate is represented by the mean of the six replicates, after pooling the 5 individual measurements, with the respective standard error of the mean (SEM).

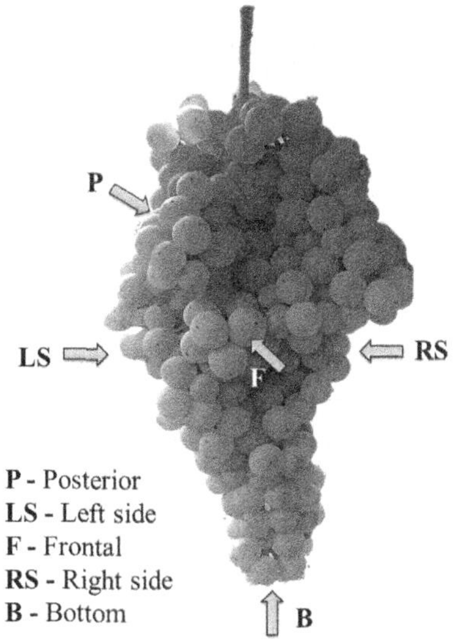

Fig. 1 'Alvarinho' cluster—arrows indicate the regions where light intensity and temperature were measured. The light sensor was positioned against the cluster facing outwards. The temperatures were measured at **F** and **B** at a distance of 10 cm. (Photograph obtained by the author)

Extraction and quantification of photosynthetic pigments

Small pieces of exocarp and seed integument (about 1 cm^2 or equivalent) were detached from different grape berries and placed in pre-weighed eppendorf tubes for fresh weight measurement. For each sampling time and light microclimate (LL, ML and HL) four independent replicates were prepared. One ml of 80% acetone (v/v) was added and samples were maintained at 4 °C, in the dark, for about 5 h with periodic agitation. The samples were then centrifuged at 15,000 rpm for 6 min and the absorbance (A) of the extracts read at 470, 647 and 663 nm. The concentrations of chlorophylls and carotenoids (µg/ml) were calculated using the equations described by Lichtenthaler (1987) and the results expressed in *µg* of photosynthetic pigment per *g* of tissue fresh weight.

Chlorophyll Fluorescence Analysis

For chlorophyll fluorescence experiments an imaging chlorophyll fluorometer was used (Open FluorCAM 800 MF; PSI, Czech Republic) comprised by four 13 × 13 cm LED panels emitting red light (emission peak at 621 nm, 40-nm bandwidth) and a 2/3 inch CCD camera (CCD381) with an F1.2 (2.8–6 mm) objective. Two of the LED panels provided modulated measuring light (<0.1 µmol m^{-2} s^{-1}), and the other two provided saturating pulses (>7500 µmol m^{-2} s^{-1}, 0.8 s). Chlorophyll fluorescence images were captured and processed using FluorCam7 software (PSI).

Before chlorophyll fluorescence experiments, intact grape berries were dark-adapted or short-term acclimated to different light intensities (50, 100, 200 and 450 μmol m^{-2} s^{-1}) for at least 30 min, selected on the basis of the average light intensities of the microclimates (Fig. 2). Homogeneous white light was provided by halogen lamps (Leica and Olympus), coupled to an external source (intralux, 5000-1, Volpi, Switzerland). Subsequently, exocarps and seeds were removed and arranged on upside-down multiwell plates.

The maximum quantum yield of PSII [$F_v/F_m = (F_m - F_0)/F_m$] was computed following a saturation pulse (SP) of 0.8 s and 3000 μmol m^{-2} s^{-1} as previously optimized (Breia et al. 2013). Three 24-well plates were prepared for each tissue type.

For rapid light curve (RLC) experiments, exocarps and seeds obtained from short-term light acclimated berries were exposed to nine incremental steps of actinic light (AL), each of 10 s, ranging from a photosynthetic photon fluence rate (PFR) of 0 to 234 μmol m^{-2} s^{-1}. Two 24-well plates were prepared per light-acclimated treatment. At the end of each AL step, a SP was emitted for determining the effective quantum yield of PSII [$\Phi_{II} = (F'_m - F_s)/F'_m$], a parameter correlated with the quantum yield of CO_2 fixation in a wide range of physiological conditions (Genty et al. 1989). Relative rates of electron transport through PSII ($ETR = \Phi_{II} \times PFR$) were calculated. RLC replicates ($3 \leq n \leq 12$) were fitted with empirical model proposed by Platt et al. (1980), and the maximum relative electron transport rate (ETR_m) were estimated.

To evaluate PSII susceptibility to the light conditions experienced during the short-term light acclimation period plus the RLC (AL series, and repetitive SP), F_v/F_m was measured after a dark-relaxation period of 15 min and compared with control values (obtained from dark-adapted grape berries).

Statistical Analysis

The results were statistically analyzed running Analysis of Variances tests (two-way and three-way ANOVA). Post hoc multiple comparisons were performed using the HSD Tukey (Statistica vs. 8, StatSoft, Inc.) or Bonferroni (Prism vs. 5, Software GraphPad, Inc.) tests. To meet variances' homocedasticity, a square root transformation of the variable 'light intensity' was done before running statistical analyses. The results were included in graphs following ns/asterisk notation: ns—not significant ($p > 0.05$); *—significant ($0.01 < p \leq 0.05$); **—very significant ($0.001 < p \leq 0.01$); ***—highly significant result ($p \leq 0.001$); and lettering notation, where significant differences between mean values were identified with different letters.

Results

The Selected Cluster-Growing Zones in the Canopy Represent Distinct Light Microclimates

The light intensities amongst the 5 measuring points varied and the range widens from LL to HL (LL: 5–11 μmol m^{-2} s^{-1}; ML: 11–60 μmol m^{-2} s^{-1}; HL: 60–1200 μmol m^{-2} s^{-1}). In order to characterize each light microclimate at each developmental stage, an average value was calculated (Fig. 2). Comparing the different developmental stages for each microclimate it is possible to observe that there was no relevant differences for LL and ML, but the average light intensity of HL microenvironment significantly increased from GE to *véraison* and then declined to mature stage. The differences between the 3 light microclimates were, however, very significant in all developmental stages, with ML (mean value considering all the stages: 28 μmol m^{-2} s^{-1}) receiving almost 4x more light that LL (7 μmol m^{-2} s^{-1}) and HL (258 μmol m^{-2} s^{-1}) 9x more than ML (Fig. 2). Together these results reveal that there is a high heterogeneity in the light received not only by the different grapes of a cluster but also by clusters from different microclimates. The variation both in space (at the cluster level) and time (during development) was much smaller in LL, showing that globally it represented a more homogeneous light microclimate. HL exhibited the largest variation, being a more heterogeneous and changing light environment, where grapes and clusters can experience very different light conditions.

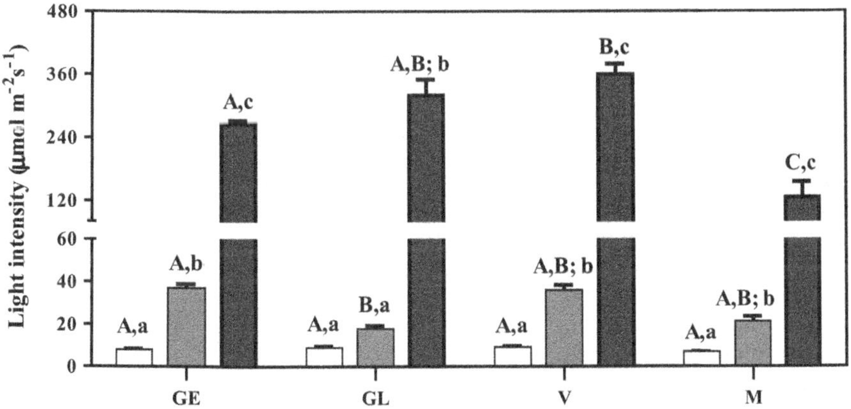

Fig. 2 Average light intensities (n_{pooled} = 6, +SEM) for each microclimate (LL ☐, ML ■ and HL ■) at the 4 fruit developmental stages: green early (GE); green later (GL); *véraison* (V) and mature (M). Statistical notation: capital letters represent the differences between stages, for the same microclimate; and lowercase letters represent the differences between microclimates for the same stage (mean values with a common letter were not statistically different).

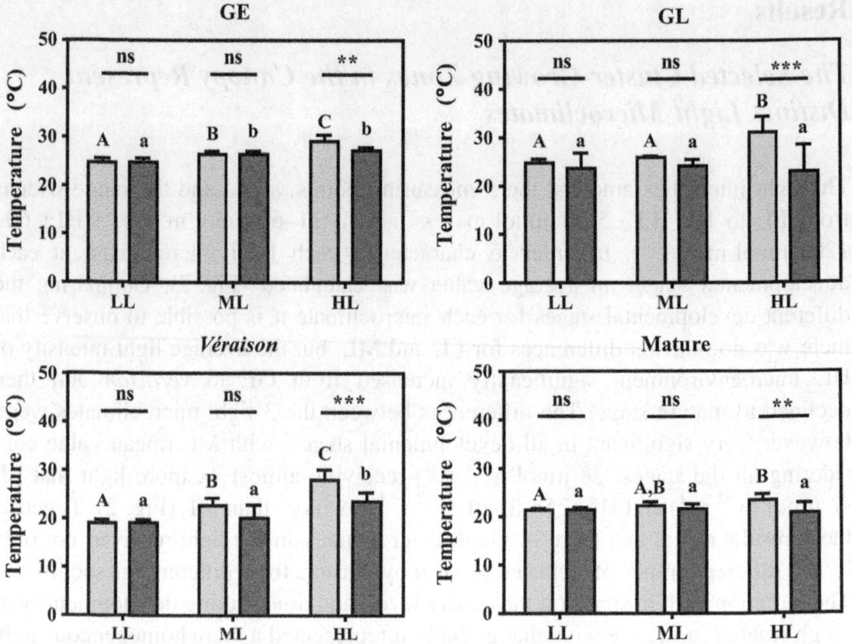

Fig. 3 Average temperatures measured in the morning (n = 6, +SD) at two cluster points, **frontal** (▒) and **bottom** (■), in the 3 light microclimates (LL, ML and HL) and the 4 developmental stages (GE, GL, V and M). Statistical notation: for each stage, the differences between microclimates are denoted in capital letters for frontal, and in lowercase letters for bottom (mean values with a common letter were not statistically different)

With respect to the temperature and when comparing microclimates the results varied in line with those obtained for light intensity (Fig. 3). LL clusters exhibited the lowest mean temperatures and the smallest variations along the season (19–25 °C), and HL the highest and most variable temperatures (20–31 °C). Comparing the two reading points in the cluster, only HL clusters exhibited significant differences, being the temperature of the frontal region higher than that of the bottom suggesting that, at least at the sampling hour, cluster temperature depends more from the incident/diffuse radiation and less from the reflected/emitted from the soil (Fig. 3).

Tissue's content in photosynthetic pigments varied with the light microclimate and along the season

The influence of the light microclimate on the content of photosynthetic pigments of both the exocarp and seed integument, and their variation throughout the developmental stages, is shown in Fig. 4. The content of total chlorophylls was

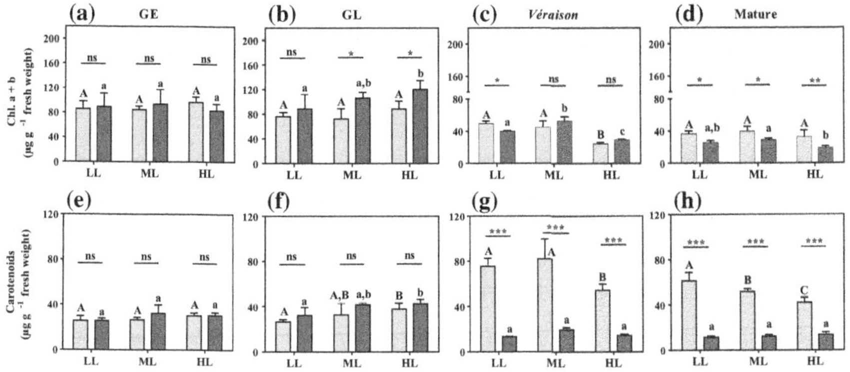

Fig. 4 Mean values (n = 4, +SD) of photosynthetic pigments' contents (µg/g fresh weight)—chlorophylls (**a–d**) and carotenoids (**e–h**)—in seed integuments (░) and exocarps (■), of clusters growing in the 3 light microclimates (LL, ML and HL) and at the 4 developmental stages (GE, GL, V and M). Statistical notation: for each stage, differences between microclimates are denoted in capital letters for seeds, and in lowercase letters for exocarps (mean values with a common letter were not significantly different)

similar in both tissues (ca. 80 µg/g) at the green early phase (Fig. 4a), being higher at the green phases (GE and GL) and suffering a progressive decrease till the mature phase (Fig. 4a–d). If at GE no differences were detected in chlorophyll content between seed and exocarp, neither between light microclimates, at the green later phase (GL), a significant increase of total chlorophylls in the exocarp with increasing microclimate luminosity was observed (Fig. 4a, b). HL exocarps had almost 50% more chlorophyll content than LL ones (Fig. 4b). For subsequent fruit development stages (*véraison* and mature), the chlorophyll content decreased progressively especially in exocarp and in a microclimate's light level-dependent manner (Fig. 4b–d). In fact, while in seed integuments chlorophylls content decreased from 80 to 40 µg/g, in exocarps it decreased from 120 to 20 µg/g (Fig. 4b–d).

Regarding the carotenoid content, and like what was observed for the chlorophyll content, at the green early stage (GE) there were no significant differences between tissues or light microclimates (Fig. 4e). Also, a significant increase was observed from GE to GL stage in response to the increase in microclimates' light level (Fig. 4f), however, in the case of carotenoids both tissues exhibit this increment (Fig. 4f, b). For subsequent developmental stages the carotenoid content of exocarps decreased two-fold, being similar between microclimates (Fig. 4g, h). In a very different manner, the carotenoid content of seed integuments increased sharply from GL to *véraison* particularly in the most shaded microenvironments (LL and ML) where the value more than doubled (Fig. 4f, g). From *véraison* to mature stage, a decrease was observed that was inversely related of the microclimate luminosity (Fig. 4g, h).

Photosynthesis of grape berry tissues under different light conditions: effect of long-term adaptation and response to short-term acclimation

Several photochemical parameters determined by imaging PAM fluorometry were used to characterize the photosynthetic activity of exocarps and seed integuments of grape berries located in different canopy light microclimates during fruit development. The response of grape berry tissues with different light-life histories to short-term light acclimation challenges was also assessed.

1. *Maximum quantum yield of PSII decreases after the green stage but only in seed integuments*

The maximum quantum yield of PSII (F_v/F_m), that reflects the probability of electrons being transferred from PSII reaction center for the transport chain of electrons by quanta absorbed (Baker 2008), was measured in grape berries as soon as arrived from the field (Fig. 5). In the green stage of development (GE and GL), F_v/F_m were relatively high (ca. 0.7) and no significant differences between light microclimates were detected. Corroborating previous results obtained with grape berry discs (Breia et al. 2013), integuments of isolated seeds presented generally significantly higher values than the exocarps (Fig. 5). However, for more advanced fruit developmental stages, while exocarps maintained F_v/F_m values, those of seed integuments decreased significantly, reaching values below 0.4 at mature.

2. *Grape berries adapted to higher light microclimates have tissues with higher photosynthetic capacities (ETR_m) that responds better to short-term light challenges but the exocarp is more sensitive to the light environment*

The effective quantum yield of PSII (Φ_{II}) measures the actual efficiency with which electrons are transferred from the PSII to the electron transport chain (Genty et al. 1989), being related therefore to the relative rate of electron transport (ETR). The value depends on light intensity but also on the efficiency of the non-photochemical processes (Murchie and Lawson 2013). After fitting the rapid light curve (RLC) replicates ($3 \leq n \leq 12$), the maximum relative electron transport rate (ETR_m) was calculated. This parameter translates the actual photosynthetic capacity of the sample under light-saturation conditions (Schreiber 2004). As results have shown low F_v/F_m and ETR_m for seeds at *véraison*, short-term acclimation experiments were not performed at mature stage, so ETR_m statistical analysis (3-way) was performed considering only the other three developmental stages.

Although seeds presented higher F_v/F_m at the green stages (Fig. 5), exocarps exhibited globally higher photosynthetic capacity under all light microclimates (Fig. 6a). The light microclimate where clusters develop had a significant effect on grape berry photosynthetic competence, with deep-shaded clusters (LL) resulting in tissues with a significantly much lower photosynthetic capacity (Fig. 6a). The seed peaked its photosynthetic capacity in the green phase showing a dramatic decrease at *véraison* while the exocarp suffer a smaller variation with photosynthetic

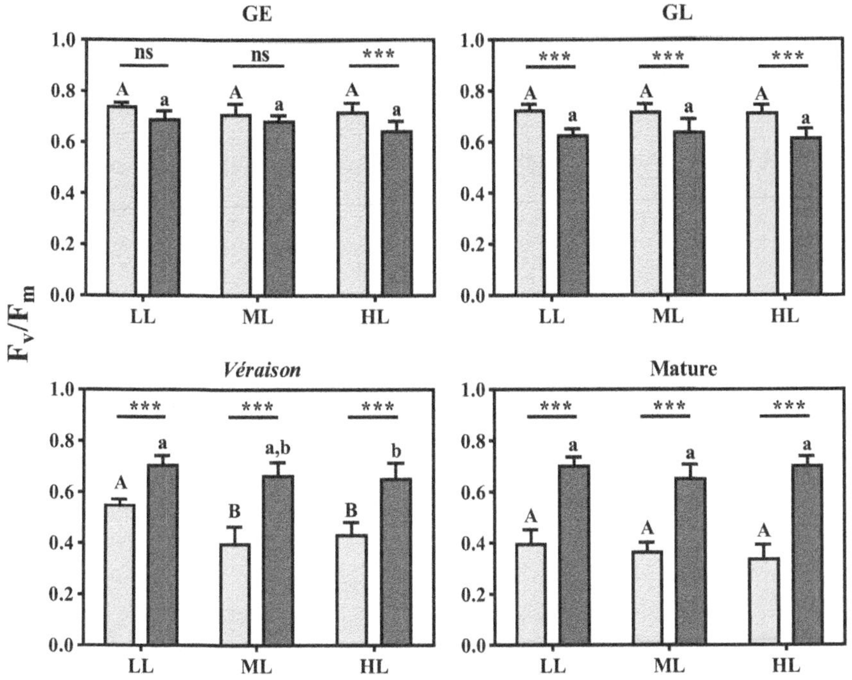

Fig. 5 F_v/F_m mean values (n = 18–22, +SD) in dark-adapted grape berries (control) for the seed integument (▒) and for exocarp (■), of clusters growing in the 3 light microclimates (LL, ML and HL) and the 4 development stages (GE, GL, V and M). Statistical notation: for each stage, differences between microclimates are denoted in capital letters for seeds, and in lowercase letters for exocarps (mean values with a common letter were not significantly different)

capacities at *véraison* similar to those at GL (Fig. 6b), what is in line with F_v/F_m results (Fig. 5).

The response to short-term acclimation light intensities was clearly dependent on the light microclimates, or light-life history of the grape berries, and type of tissue. With respect to the light-life history, LL tissues exhibited consistently the lowest ETR_m and did not respond to increasing acclimation light intensities (Fig. 7). The highest photosynthetic capacities values were registered in ML-grown grapes when acclimated at lower light intensities (50 and 100 μmol m^{-2} s^{-1}) and in HL-grown grapes when acclimated at the highest intensity tested (450 μmol m^{-2} s^{-1}) registering these grapes the widest variation in ETR_m (Fig. 7).

Regarding the photosynthetic tissue type, Fig. 8 shows that seed integuments of ML and HL grape berries were able to increase its photosynthetic capacity with increasing short-term acclimation light intensity, suggesting that being internal tissues the light transmitted by the exocarp and mesocarp was limiting for seed photosynthesis. The higher ETR_m exhibited by the exocarp when comparing to seeds (Fig. 6) is mainly explained by its better performance at lower acclimation

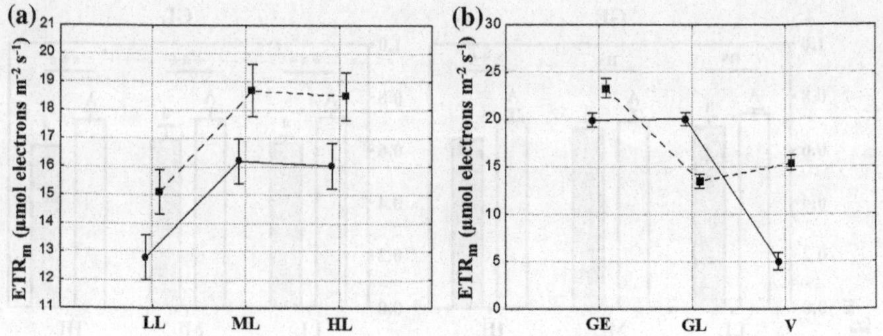

Fig. 6 Maximum relative electron transport rate (ETR_m) (µmol eletrons m^{-2} s^{-1}) of seed integument (—●—) and exocarp (···■···) of light-acclimated grape berries grown under 3 light microclimates (LL, ML and HL) (**a**), at 3 development stages (GE, GL, and V) (**b**)

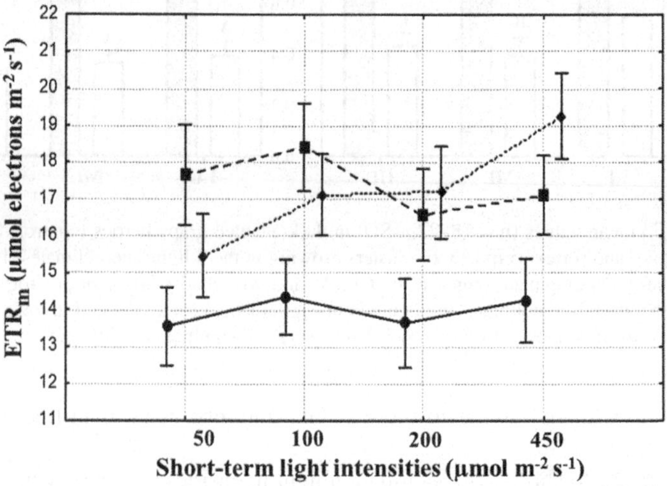

Fig. 7 Maximum relative electron transport rate (ETR_m) (µmol electrons m^{-2} s^{-1}) determined in tissues of grape berries grown under 3 light microclimates—LL (—●—), ML (--■--) and HL (····◆····) and short-term acclimated at 50, 100, 200 and 450 µmol m^{-2} s^{-1} before running RLC

light intensities (Fig. 8), but it has also a more plastic photosynthetic capacity probably due to a stronger interaction between short-term light acclimation and light-life history. Like seeds, LL exocarps did not respond to short-term acclimation light (Fig. 8). ML exocarps peak their photosynthetic capacity at lower light intensities (50 and 100 µmol m^{-2} s^{-1}), decreasing in a dose-dependent manner for higher intensities (200 and 450 µmol m^{-2} s^{-1}). On the other hand, HL grape berries only peaked at the highest short-term light intensity tested (450 µmol m^{-2} s^{-1}).

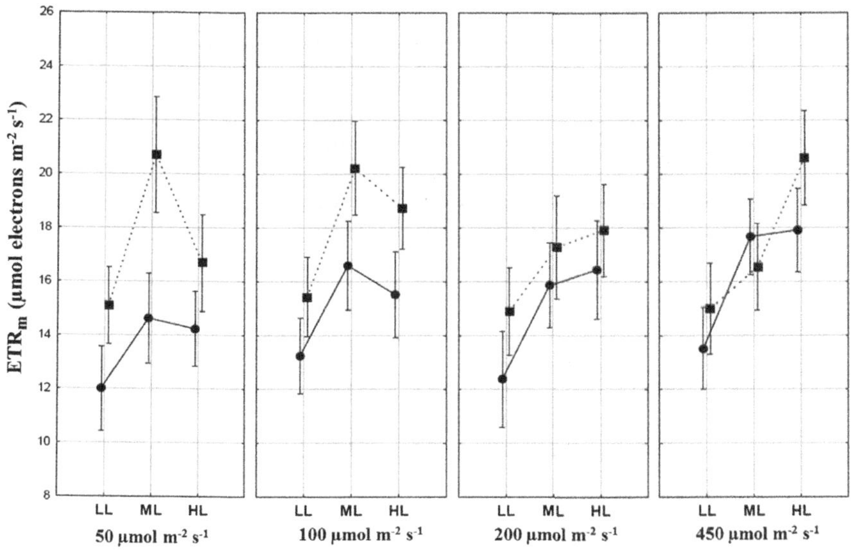

Fig. 8 Maximum relative electron transport rate (ETR_m) (μmol electrons m^{-2} s^{-1}) determined for the seed integument (—●—) and for the exocarp (····■····) of grape berries acclimated at 50, 100, 200 and 450 μmol m^{-2} s^{-1} before running RLC, and grown under 3 light microclimates (LL, ML and HL)

After a Light Challenge the Exocarp Recover F_v/F_m Values Better Than Seeds and LL Grape Berries Worse Than ML and HL Ones

To further evaluate potential negative effects on PSII due to the exposure to the light acclimation period and AL and repetitive SP during RLC, F_v/F_m was measured again after a 15 min. period of dark relaxation in grape berries grown under the three light microclimates in GL (data not shown) and *véraison* (Fig. 9). At the green stage (GL), both tissues fully recovered F_v/F_m values, only seeds of LL grapes acclimated at the highest intensities (200 and 450 μmol m^{-2} s^{-1}) didn't achieve the control values. However, at *véraison* while the exocarps roughly maintained their recovery profile (only LL grapes acclimated at 450 μmol m^{-2} s^{-1} exhibited an F_v/F_m significantly lower than the control), seeds integuments did not recover F_v/F_m for most conditions. Seeds from LL grapes presented significantly lower F_v/F_m than the controls for all short-term acclimation light intensities tested and in a dose-dependent manner. ML seeds generally recovered to control values but for HL seeds a dose-dependent inhibition pattern were again observed (Fig. 9). These results suggest that the seed integument but not the exocarp, after the green stage, not only loses photochemical efficiency and capacity, but also efficiency of its

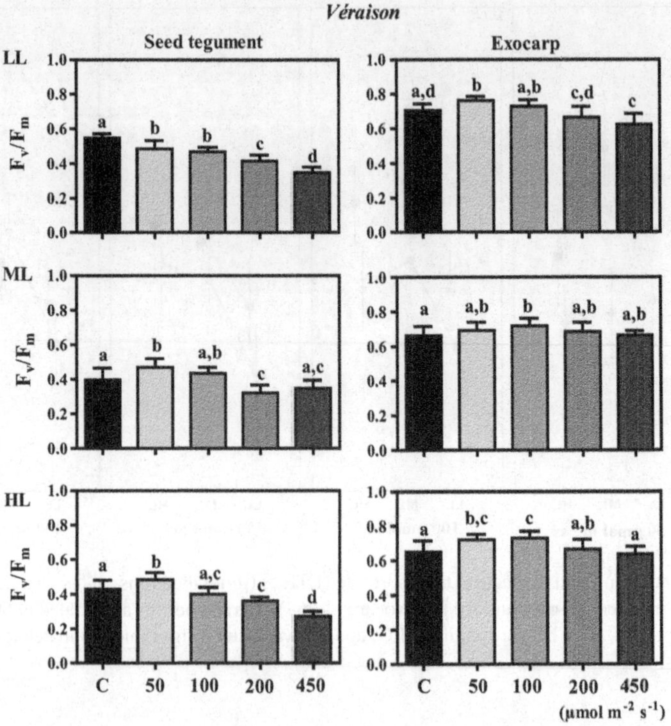

Fig. 9 Mean values of F_v/F_m ($12 \leq n \leq 18$, +SD) after 15 min of dark relaxation of seed integument and exocarp of grape berries grown under 3 light microclimates (LL, ML, HL) at *véraison*. The control (■) represents the initial F_v/F_m measured values. The increasing darker tones represent the increasing light intensities used for short-term acclimation (50, 100, 200 and 450 μmol m^{-2} s^{-1}) before the RLC, being the values of F_v/F_m obtained 15 min (in the dark) after the RLC. The letters above the bars correspond to statistical differences (two-way ANOVA)

photoinhibition repair mechanisms. The effect was dependent on short-term acclimation light intensity but also of the light microclimate.

These results are in accordance with those obtained for the other assessed photosynthetic parameters. In fact, and with the exception of LL grape berries clearly under a very limiting shaded environment and with very poor photosynthetic competence (Fig. 7), the exocarp has proved to be the tissue that better perform under a range of light microclimates and for most part of the developing season (Fig. 6), that better respond to short-term light changing conditions (Fig. 8) and better recover from light-induced inhibitory challenges (Fig. 9). On the other hand, seed integument is a more light-susceptible tissue, with its photosynthetic activity mainly restricted to green stages (Figs. 5, 6b and 9), but also responding to light microclimate (Fig. 6a) and short-term light changes (Fig. 8).

Discussion

Climate changes may have serious impacts in agriculture. The projections for grape-growing areas worldwide, including vineyards in the Mediterranean area like Douro Demarcated Region, point to severe changes in precipitation profile as well as elevated temperatures and solar irradiance (Fraga et al. 2012), affecting grape berry yield and quality. In this context, it is important to develop sustainable and rational short- and long-term adaptation measures, to mitigate these negative impacts in crop productivity and end-products quality. The aim of this work was to study the influence of the light microclimate on the photosynthetic capacity of grape berry tissues (exocarp and seed integument) throughout the fruit growing season but also on its capacity to respond to short-term light challenges and recover from light-induced inhibitory conditions. We also aimed to contribute with this knowledge for short-term mitigation measures programs where a reduction in radiation/heat absorption by the canopy is targeted.

The gross characterization of the microclimates in the vineyard clearly show that it was possible to identify three different microenvironments in the vicinity of the clusters, with distinct average temperatures but in particular with distinct average light intensities, setting the fundamental conditions for the experiment.

Does Grape Berries' Pigments Have Different Functions in Different Tissues and During Fruit Development?

With regard to photosynthetic pigments, the high contents registered for both tissues in the green stage that increase in response to the increasing luminosity of the microclimate for later green stages (Fig. 4) clearly suggest a photosynthetic function of these grape berry tissues as previously reported (Breia et al. 2013). After the green stage, the progressive decrease in chlorophyll content till mature, dependent on the light microclimate, suggest that the photosynthetic function is eventually losing relative importance in fruit physiology but also that the rate of decay could be explained by the developmental/maturation degree of the cluster or by direct light effects. Chlorophyll is constantly synthesized and destroyed (photo-oxidation) in the presence of light, but under very high light intensities, the degradation rate is higher. Hence it seems reasonable to conclude that the exocarp being an external tissue may experience a higher rate of degradation of chlorophyll than the seed integument, thus keeping this tissue relatively high contents until the final stages of development in all microclimates (Fig. 4). This decrease in chlorophyll content associated with fruit maturation has also been reported for other fleshy fruits (apple, mango and tomato), as the result of the substitution of chlorophyll by other colored pigments or due to the activity of chlorophyllase (reviewed by Seifert et al. 2014).

Interestingly, although the chlorophyll content of seed integuments roughly follows the trend of the exocarp during fruit development, the seed carotenoid content only parallels that of the exocarp during the green stage, supporting the view of a photosynthetic function. At *véraison,* seed carotenoids rise dramatically (Fig. 4). It is well documented that carotenoids, besides fueling photosystems' reaction centers with PAR energy, are involved in the protection of the photosynthetic apparatus against photoinhibitory damage by singlet oxygen ($^1O_2$), which is produced by the excited triplet state of chlorophyll (Young et al. 1997), and by superoxide radical. But carotenoid functions in seeds are less obvious than in other tissues and organs. Howitt and Pogson (2006) referred that the production of carotenoids in the seed could be important to produce abscisic acid (ABA) involved in seed dormancy (Maluf et al. 1997). It can also contribute to the antioxidant system in seeds, reducing the action of oxygen free radicals, which cause deterioration of membranes and premature aging of some seeds, such as wheat seeds (Calucci et al. 2004). Our results support the first hypothesis since the increase in seed carotenoids (Fig. 4) not associated with photosynthetic activity (Figs. 5 and 6b) occurs during seed maturation/dehydration phase (Ruiz-Sola and Rodríguez-Concepción 2012). A contribution to the antioxidant system could also be pictured but not to quench photochemical-derived ROS, but eventually by the high respiratory rates associated with seed metabolism (Pilati et al. 2007).

Grape Berry Tissues' Photosynthetic Activities are Differentially Coupled to the Microenvironment Being Differently Susceptible to Microclimate Changes: Are There Metabolic Consequences?

As shown above, there was a very significant reduction in chlorophyll content in exocarps and seed integuments during fruit ripening (Fig. 4a–d), however, the photochemical efficiency (F_v/F_m) varied differently in both tissues. The exocarp maintained the photosynthetic potential (ca. 0.7) until later stages of fruit maturation, while, seed integuments significantly reduced its photochemical efficiency. However, and like some species adapted to deep shade conditions (shade-lovers) (Puthur and Pardha Saradhi 2004), seeds exhibited higher F_v/F_m than exocarps at the green stages (Fig. 5). F_v/F_m values were similar to that previously described by Breia et al. (2013), but also to those of grapes berries of 'Bacchus' and 'Silvaner' variety from green later to *véraison* stages (Kolb et al. 2006). In tomato fruit peels F_v/F_m also decreased significantly during development reaching low values at the over-ripe stage (Carrara et al. 2001). A decrease in F_v/F_m was also observed for papaya due to chlorophyll degradation (Bron et al. 2004).

The results obtained for photosynthetic capacity (ETR_m) showed that the exocarps generally exhibited higher capacity than seeds (Fig. 6a) and extend its photosynthetic activity beyond the green stage (Fig. 6b), suggesting that exocarp photosynthesis may have a role during almost all fruit developmental stages. On the other hand, the high seed ETR_m values at ML and HL microclimates during the green stage (Fig. 6) support a role for seed photosynthesis in this period. In fruits of *Arum italicum*, the reduction of photosynthetic activity is correlated with a decrease of reaction centers and changes in the structural organization of PSII (Ferroni et al. 2013). A possible function could be fueling carbon and energy for seed development and metabolism. It was suggested that seed photosynthesis can contribute to compensate the deficit generated by some high demanding biosynthetic activities, such as the production of seed oils (Ruuska et al. 2004), as is the case of grape seeds.

The effect of light microclimate on the photosynthetic capacity of exocarps and seed integuments is well evident in our results. Grapes located in deep-shaded microclimates of the canopy (LL), exhibit very low photosynthetic capacity values throughout the developmental period (Fig. 6a) and are enable to respond to short-term variations in light intensity (Figs. 7 and 8). In this way, it seems reasonable to conclude that these grape berries (from LL) never develop a true photosynthetic chloroplast organization in both tissues probably due to grow in a microenvironment below its light compensation point, which is about 15–30 μmol m^{-2} s^{-1} for grapevines (Smart 1987), what constitutes a potential loss for grape berry carbon and energy budget.

The response to short-term acclimation light intensities was clearly dependent on the light-life history of the grape berries. The highest photosynthetic capacities values were registered in ML-grown grapes when acclimated at lower light intensities and in HL-grown grapes when acclimated at the highest intensity (Fig. 7), showing different light adaptation phenotypes and to be more linked to the microclimate. In addition, the wide variation in ETR_m in ML and HL grapes, proving to be responsive to a light-changing environment (Figs. 7 and 8), together with the low values in LL suggest that measures reducing or increasing light absorption may a have a huge impact on grape berry photosynthetic metabolism.

Does Grape Berry Photosynthetic Exocarp and Seed Integuments Exhibit Sun- and Shade-Tissues Traits?

Its is reported that sun and shade leaves differ in many aspects of their photosynthetic architecture, structure, machinery and activity (Lichtenthaler and Babani 2004). Sun leaves possess sun-type chloroplasts, that have a higher photosynthetic capacity on a leaf area basis, a higher content in total chlorophylls and carotenoids per leaf area (Lichtenthaler et al. 2007) and a greater capacity for photoprotective energy dissipation (Demmig-Adams and Adams 1992) than shade leaves, to name a few.

The lower values of total chlorophylls in GL together with lower ETR_m in seed integument when compared to exocarp in the photosynthetic green stage (Fig. 6b) is consistent with the profile of a shade tissue what are in accordance with some results in the literature. Aschan et al. (2005) reported the trend towards a lower ETR in stalks, nectaries, and anthers (in *Helleborus viridis* L.) indicating a more shade-type photosynthesis, what was explainable by the vertical orientation or hidden position of these plant parts. Another study showed that internal tomato tissues have lower ETR values compared with external ones (Smillie et al. 1999). In *Jatropha curcas* immature fruits maximum ETR was comparable to that of leaves of the same species, that decreased with the age of fruit (Ranjan et al. 2012). Chen and Cheng (2007) showed that the sun-exposed peel of apple fruit had higher PSII quantum efficiency than the shaded peel. The increase of total chlorophylls in the exocarp in response to increasing light level of the microenvironment (Fig. 4), together with the high increment in photosynthetic capacity of HL exocarps with increasing acclimation light intensities from 200 to 450 μmol m^{-2} s^{-1} (Figs. 7 and 8) is compatible with a more sun-type tissue. Also, the results for recovery of F_v/F_m values are in accordance with this view by showing a good recovery profile (Demmig-Adams and Adams 1992). Shade-adapted leaves of 'Cabernet Sauvignon' variety were more sensitive to photoinhibition than sun-adapted leaves (Iacono and Sommer 1996).

Potential Impacts of Reduction in Radiation/Heat Absorption Measures on Grape Berry Photosynthesis

Usually, in Douro Demarcated Region, leaves developed in typically moderate to severe abiotic stresses showed the lowest concentrations of photosynthetic pigments and lowest net photosynthetic rate (Moutinho-Pereira et al. 2007) with major impacts on productivity and quality. In this sense, mitigation measures are necessary, in order to reduce negative effects caused by high irradiance and extreme temperature in grapevine. Two short-term mitigation measures already under study are kaolin foliar application and vine shadings (Greer et al. 2011).

Recent studies had shown that kaolin foliar application on grapevine leaves of 'Touriga Nacional' avoid the reduction in the content in chlorophylls and carotenoids, F_v/F_m and ETR observed in control leaves from July to September (Correia et al. 2014; Dinis et al. 2016). On the other hand, experiments using vines shadings report a decrease in photosynthetic rates by 40% of 'Semillon' vines (Greer et al. 2011).

Although short-term mitigation measures are of paramount importance in some cases, our results show that a variation in the light microclimate of the clusters and also of short-term changes in their light environment had a major impact of grape berry photosynthesis.

Together these results call attention for two aspects regarding viticulture measures avoiding excess radiation absorption by the vines (leaves and fruits) but also in improving light absorption and productivity:

- Excess shading at the cluster level (either by mechanical shades or by kaolin or other chemicals applications) might dislocate fruit photosynthetic activity to lower levels, even below light compensation points.
- Excess exposure (e.g. more open conduction systems, defoliation, cluster thinning) may create microclimates that favour higher photoinhibition.

In fact, depending on the radiation level and acclimation status of the berries, measures at the canopy level may cause a decrease or increase in berry photosynthetic activity and consequently there may be effects on the secondary metabolism, which will affect the composition of the grape berries and therefore of the wine. Thus, a rational intervention accompanied by the monitorization of variations in light microclimate at the cluster level could optimize fruit photosynthesis and photosynthesis-related contributes to fruit physiology and quality.

Conclusions

The main objective of this work was to study the influence of the light microenvironment in the vicinity of developing clusters and also of short-term light acclimation challenges in the photosynthetic activity of grape berry tissues (exocarp and seed outer integument) throughout the growing season. From our results it was possible to conclude that exocarp's photosynthesis is maintained till the mature stage, suggesting participating in physiological processes that are present throughout fruit development and ripening, while seed photosynthesis is mainly active in the green stage. Tissue's photosynthetic pigments support the view that seeds photosynthetic activity could be related with seed differentiation and development, but the carotenoid content rise suggest a role in maturation/dehydration phase. Clusters may develop in very distinct canopy microclimates what dramatically affected grape berry tissues photosynthetic competence, performing very poorly under deep-shaded environments, but well under a range of medium (ML) to high light (HL) microclimates. Apparently, seed photosynthetic activity is optimized by higher acclimation light intensities, what is consistent of being an internal tissue. Overall, exocarps exhibit higher photosynthetic capacities than seeds in all microclimates tested, are more responsive to short-term light challenges and in a light microclimate-dependent manner and are less susceptible to light-induced photoinhibition.

Globally, results suggest that not only light/temperature stress conditions due to climate changes but also from viticulture practices that change cluster light microclimate may have significant impacts on the photosynthetic activity of both tissues with potential consequences on the grape berry metabolism, development and quality. To avoid fruit photosynthesis inhibitions or limitations it is necessary to program viticultural practices that allow adequate radiation/heat inputs both for leaves and fruits.

Acknowledgements This work is supported by: European Investment Funds by FEDER/COMPETE/POCI– Operacional Competitiveness and Internacionalization Programme, under Project POCI-01-0145-FEDER-006958 and National Funds by FCT - Portuguese Foundation for Science and Technology, under the project UID/AGR/04033/2013.

References

Aschan, G., & Pfanz, H. (2003). Non-foliar photosynthesis—A strategy of additional carbon acquisition. *Flora, 198,* 81–97. https://doi.org/10.1078/0367-2530-00080.

Aschan, G., Pfanz, H., Vodnik, D., & Batič, F. (2005). Photosynthetic performance of vegetative and reproductive structures of green hellebore (*Helleborus viridis* L. agg.). *Photosynthetica, 43* (1), 55–64. https://doi.org/10.1007/s11099-005-5064-x.

Baker, N. R. (2008). Chlorophyll fluorescence: A probe of photosynthesis in vivo. *Annual Review of Plant Biology, 59,* 89–113. https://doi.org/10.1146/annurev.arplant.59.032607.092759.

Blanke, M. M., & Lenz, F. (1989). Fruit photosynthesis. *Plant, Cell and Environment, 12,* 31–46. http://doi.org/doi:10.1111/j.1365-3040.1989.tb01914.x.

Breia, R., Vieira, S., Da Silva, J. M., Gerós, H., & Cunha, A. (2013). Mapping grape berry photosynthesis by chlorophyll fluorescence imaging: The effect of saturating pulse intensity in different tissues. *Photochemistry and Photobiology, 89*(3), 579–585. https://doi.org/10.1111/php.12046.

Bron, I. U., Ribeiro, R. V., Azzolini, M., Jacomino, A. P., & Machado, E. C. (2004). Chlorophyll fluorescence as a tool to evaluate the ripening of "Golden" papaya fruit. *Postharvest Biology and Technology, 33*(2), 163–173. https://doi.org/10.1016/j.postharvbio.2004.02.004.

Calucci, L., Capocchi, A., Galleschi, L., Ghiringhelli, S., Pinzino, C., Saviozzi, F., et al. (2004). Antioxidants, free radicals, storage proteins, puroindolines, and proteolytic activities in bread wheat (*Triticum aestivum*) seeds during accelerated aging. *Journal of Agricultural and Food Chemistry, 52*(13), 4274–4281. https://doi.org/10.1021/jf0353741.

Carrara, S., Pardossi, A., Soldatini, G. F., Tognoni, F., & Guidi, L. (2001). Photosynthetic activity of ripening tomato fruit. *Photosynthetica*. http://doi.org/10.1023/A:1012495903093.

Chen, L. S., & Cheng, L. (2007). The sun-exposed peel of apple fruit has a higher photosynthetic capacity than the shaded peel. *Functional Plant Biology, 34*(11), 1038–1048. https://doi.org/10.1071/FP07111.

Chuine, I., Yiou, P., Viovy, N., Seguin, B., Daux, V., & Ladurie, E. L. (2004). Historical phenology: Grape ripening as a past climate indicator. *Nature, 432,* 289–290.

COM. (2009). Comissão das Comunidade Europeias. *Adaptação Às Alterações Climáticas: Para Um Quadro de Acção Europeus Alterações Climáticas: Para Um Quadro de Acção Europeu.* Bruxelas.

Correia, C., Dinis, Lia-Tânia, Pinheiro, R., Fraga, H., Ferreira, H., Gonçalves, I., et al. (2014). Climate change and adaptation strategies for viticulture. *Journal of International Scientific Publications: Agriculture and Food, 2,* 424–429.

Demmig-Adams, B., & Adams, W. W. (1992). Photoprotection and other responses of plants to high light stress. *Annual Review of Plant Physiology and Plant Molecular Biology, 43*(1), 599–626. http://doi.org/10.1146/annurev.pp.43.060192.003123.

Dinis, L. T., Ferreira, H., Pinto, G., Bernardo, S., Correia, C. M., & Moutinho-Pereira, J. (2016). Kaolin-based, foliar reflective film protects photosystem II structure and function in grapevine leaves exposed to heat and high solar radiation. *Photosynthetica, 54*(1), 47–55. https://doi.org/10.1007/s11099-015-0156-8.

Ferroni, L., Pantaleoni, L., Baldisserotto, C., Aro, E. M., & Pancaldi, S. (2013). Low photosynthetic activity is linked to changes in the organization of photosystem II in the fruit of *Arum italicum*. *Plant Physiology and Biochemistry, 63,* 140–150. https://doi.org/10.1016/j.plaphy.2012.11.023.

Fraga, H., Malheiro, A. C., Moutinho-Pereira, J., & Santos, J. A. (2012). An overview of climate change impacts on European viticulture. *Food and Energy Security, 1*(2), 94–110. https://doi.org/10.1002/fes3.14.

Fraga, H., Malheiro, A. C., Moutinho-Pereira, J., & Santos, J. A. (2014). Climate factors driving wine production in the Portuguese Minho region. *Agricultural and Forest Meteorology, 185*, 26–36. https://doi.org/10.1016/j.agrformet.2013.11.003.

Fraga, H., Santos, J. A., Malheiro, A. C., Oliveira, A. A., Moutinho-Pereira, J., & Jones, G. V. (2015). Climatic suitability of Portuguese grapevine varieties and climate change adaptation. *International Journal of Climatology, 36*(1), 1–12. https://doi.org/10.1002/joc.4325.

Genty, B., Briantais, J.-M., & Baker, N. R. (1989). The relationship between the quantum yield of photosynthetic electron transport and quenching of chlorophyll fluorescence. *Biochimica et Biophysica Acta (BBA)—General Subjects, 990*(1), 87–92. http://doi.org/10.1016/S0304-4165(89)80016-9.

Greer, D. H., Weedon, M. M., & Weston, C. (2011). Reductions in biomass accumulation, photosynthesis in situ and net carbon balance are the costs of protecting *Vitis vinifera* "Semillon" grapevines from heat stress with shade covering. *AoB PLANTS, 11*(1), 1–13. https://doi.org/10.1093/aobpla/plr023.

Haselgrove, L., Botting, D., Heeswijck, R., Høj, P. B., Dry, P. R., Ford, C., et al. (2000). Canopy microclimate and berry composition: The effect of bunch exposure on the phenolic composition of *Vitis vinifera* L cv. Shiraz grape berries. *Australian Journal of Grape and Wine Research, 6*(2), 141–149. https://doi.org/10.1111/j.1755-0238.2000.tb00173.x.

Howitt, C. A., & Pogson, B. J. (2006). Carotenoid accumulation and function in seeds and non-green tissues. *Plant, Cell and Environment*. https://doi.org/10.1111/j.1365-3040.2005.01492.x.

Iacono, F., & Sommer, K. J. (1996). Photoinhibition of photosynthesis and photorespiration in *Vitis vinifera* L. under field conditions - effects of light climate and leaf position. *Australian Journal of Grape and Wine Research, 2*(1), 1–11. Retrieved from http://dx.doi.org/10.1111/j.1755-0238.1996.tb00089.x.

IPCC (International Panel on Climate Change). (2014). *Climate change 2014: Impacts, adaptation, and vulnerability. Part A: Global and sectoral aspects. Working Group II Contribution to the Fifth Assessment Report of the Intergovernmental Panel on Climate Change* (Cambridge).

Jones, G. V., & Davis, R. E. (2000). Climate influences on grapevine phenology, grape composition, and wine production and quality for Bordeaux, France. *American Journal of Enology and Viticulture, 51*(3), 249–261.

Kennedy, B. Y. J. (2002). Understanding grape berry development. *Practical Winery and Vineyard*, 1–5.

Kolb, C. A., Wirth, E., Kaiser, W. M., Meister, A., Riederer, M., & Pfündel, E. E. (2006). Noninvasive evaluation of the degree of ripeness in grape berries (*Vitis Vinifera* L Cv. Bacchus and Silvaner) by chlorophyll fluorescence. *Journal of Agricultural and Food Chemistry, 54* (September), 299–305. http://doi.org/10.1021/jf052128b.

Leeuwen, C. Van, & Darriet, P. (2016). The impact of climate change on viticulture and wine quality. *Journal of Wine Economics, 11*(1), 150–167. https://doi.org/10.1017/jwe.2015.21.

Lichtenthaler, H. K. (1987). Chlorophylls and carotenoids: Pigments of photosynthetic biomembranes. *Methods in Enzymology, 148*, 350–382. https://doi.org/10.1016/0076-6879(87)48036-1
.

Lichtenthaler, H. K., Ač, A., Marek, M. V., Kalina, J., & Urban, O. (2007). Differences in pigment composition, photosynthetic rates and chlorophyll fluorescence images of sun and shade leaves of four tree species. *Plant Physiology and Biochemistry, 45*(8), 577–588. https://doi.org/10.1016/j.plaphy.2007.04.006.

Lichtenthaler, H. K., & Babani, F. (2004). Light adaptation and senescence of the photosynthetic apparatus. Changes in pigment composition, chlorophyll fluorescence parameters and photosynthetic activity. In *Chlorophyll a Fluorescence: A Signature of Photosynthesis* (Papageorgi, pp. 713–736). Springer Netherlands.

Maluf, M. P., Saab, I. N., Wurtzel, E. T., & Sachs, M. M. (1997). The *viviparous12* maize mutant is deficient in abscisic acid, carotenoids, and chlorophyll synthesis. *Journal of Experimental Botany, 48*(311), 1259–1268. https://doi.org/10.1093/jxb/48.6.1259.

Moutinho-Pereira, J., Magalhães, N., Gonçalves, B., Bacelar, E., Brito, M., & Correia, C. (2007). Gas exchange and water relations of three *Vitis vinifera* L. cultivars growing under Mediterranean climate. *Photosynthetica, 45*(2), 202–207. https://doi.org/10.1007/s11099-007-0033-1.

Moutinho-Pereira, J. M., Correia, C. M., Gonçalves, B. M., Bacelar, E. A., & Torres-Pereira, J. M. (2004). Leaf gas exchange and water relations of grapevines grown in three different conditions. *Photosynthetica, 42*(1), 81–86. https://doi.org/10.1023/B:PHOT.0000040573.09614.1d.

Moutinho-Pereira, J., Magalhães, J. M., Correia, C., & Torres-Peireira, J. M. (2003). Effects of NW-SE row orientation on grapevine physiology under Mediterranean field conditions. In *Agricoltura Mediterranea* (Vol. 133, pp. 218–225). Pacini.

Murchie, E. H., & Lawson, T. (2013). Chlorophyll fluorescence analysis: A guide to good practice and understanding some new applications. *Journal of Experimental Botany, 64*(13), 3983–3998. https://doi.org/10.1093/jxb/ert208.

OIV. (2015). *International Organisation of Vine and Wine. Vine and Wine Outlook 2010–2011.*

Peel, B. L., Finlayson, B. L., & McMahon, T. A. (2007). Updated world map of the Köppen-Geiger climate classification.pdf. *Hydrology and Earth System Sciences, 11*, 1633–1644. http://doi.org/10.5194/hessd-4-439-2007.

Pilati, S., Perazzolli, M., Malossini, A., Cestaro, A., Demattè, L., Fontana, P., et al. (2007). Genome-wide transcriptional analysis of grapevine berry ripening reveals a set of genes similarly modulated during three seasons and the occurrence of an oxidative burst at vèraison. *BMC Genomics, 8*(1), 428. https://doi.org/10.1186/1471-2164-8-428.

Platt, T., Gallegos, C. L., & Harrison, W. G. (1980). Photoinibition of photosynthesis in natural assemblages of marine phytoplankton. *Journal of Marine Research, 38*(4), 687–701.

Puthur, J. T., & Pardha Saradhi, P. (2004). Developing embryos of *Sesbania sesban* have unique potential to photosynthesize under high osmotic environment. *Journal of Plant Physiology, 161*(10), 1107–1118. https://doi.org/10.1016/j.jplph.2004.03.002.

Ranjan, S., Singh, R., Soni, D. K., Pathre, U. V., & Shirke, P. A. (2012). Photosynthetic performance of *Jatropha curcas* fruits. *Plant Physiology and Biochemistry, 52*, 66–76. https://doi.org/10.1016/j.plaphy.2011.11.008.

Reynolds, A. G., & Heuvel, J. E. V. (2009). Influence of grapevine training systems on vine growth and fruit composition: A review. *American Journal of Enology and Viticulture, 60*(3), 251–268.

Ruiz-Sola, M. Á., & Rodríguez-Concepción, M. (2012). Carotenoid biosynthesis in arabidopsis: A colorful pathway. *The Arabidopsis Book, 10*. http://doi.org/10.1199/tab.0158.

Ruuska, S. A., Schwender, J., & Ohlrogge, J. B. (2004). The capacity of green oilseeds to utilize photosynthesis to drive biosynthetic processes. *Plant Physiology, 136*, 2700–2709.

Schreiber, U. (2004). Pulse-amplitude-modulation (PAM) fluorometry and saturation pulse method: An overview. In C. George (Ed.), *Chlorophyll a fluorescence: A signature of photosynthesis* (pp. 279–319). Dordrecht, The Netherlands: Kluwer Academic.

Schultz, H. R. (1995). Grape canopy structure, light microclimate and photosynthesis. I. A two-dimensional model of the spatial distribution of surface area densities and leaf ages in two canopy systems. *Vitis, 34*(4), 211–215.

Seifert, B., Pflanz, M., & Zude, M. (2014). Spectral shift as advanced index for fruit chlorophyll breakdown. *Food and Bioprocess Technology, 7*(7), 2050–2059. https://doi.org/10.1007/s11947-013-1218-1.

Smart, R. E. (1974). Photosynthesis by Grapevine Canopies. *Journal of Applied Ecology, 11*(3), 997–1006.

Smart, R. E. (1987). Influence of light on composition and quality of grapes. *Acta Horticulturae, 206*, 37–48. http://doi.org/10.17660/ActaHortic.1987.206.2.

Smart, R. E., Bobinson, J. B., Due, G. R., & Brien, C. (1985). Canopy microclimate modification for cultivar Shiraz I. Definition of canopy microclimate. *Vitis, 24,* 17–31.

Smillie, R. M., Hetherington, S. E., & Davies, W. J. (1999). Photosynthetic activity of the calyx, green shoulder, pericarp, and locular parenchyma of tomato fruit. *Journal of Experimental Botany, 50*(334), 707–718. https://doi.org/10.1093/jexbot/50.334.707.

Young, A. J., Phillip, D., & Savill, J. (1997). Carotenoids in higher plant photosynthesis. In M. Pessaraki (Ed.), *Handbook of photosynthesis* (pp. 575–596). New York: Marcel Dekker Inc.

Is It Possible to Completely Adapt Agriculture Production to the Effects of Climate Variability and Change in Central Argentina? New Approaches in Face of New Challenges

Mónica B. Wehbe, Roberto A. Seiler, Marta G. Vinocur and Ivan E. Tarasconi

Abstract The adaptation of agriculture has been a priority since the advent of studies on climate change at the end of the last century. Today agriculture faces two concurrent challenges: adapting to the impacts of climate variability and change and feeding a growing world population. But responses to these issues generally do not coincide with the goal of climate change mitigation at global scale and ecosystem sustainability efforts at the local scale. This is analyzed in this article through a case study on adaptation to droughts in one of the main agricultural areas of the world, central Pampas in Argentina. Results from the analysis show that current adaptation practices do not eliminate yield variability especially to climatic extremes. Therefore, extra measures for reducing the agricultural productivity gap is not always justified against the possible social and environmental costs of such actions, reinforcing the necessity for new approaches on the adaptation of agriculture to climate variability and change and to provide, at the same time, for healthy food and environments for humanity. Here, a focus on the resilience of the socio-ecological systems to climatic shocks is considered as complementary to more conventional adaptation practices when attempting to increase agricultural production but within the limits of the sustainable development goals.

Keywords Agriculture · Adaptations · Climate change · Droughts
Resilience · Socio-ecological systems

M. B. Wehbe (✉) · R. A. Seiler · M. G. Vinocur · I. E. Tarasconi
National University of Río Cuarto, Córdoba, Argentina
e-mail: mbwehbe@gmail.com; mwehbe@eco.unrc.edu.ar

R. A. Seiler
e-mail: roberto.seiler123@gmail.com

M. G. Vinocur
e-mail: mvinocur@ayv.unrc.edu.ar

I. E. Tarasconi
e-mail: tarasconi_ivan@yahoo.com.ar

F. Alves et al. (eds.), *Theory and Practice of Climate Adaptation*,
Climate Change Management, https://doi.org/10.1007/978-3-319-72874-2_25

Introduction

Globally, agriculture is being called to increase production to meet many different demands. It is expected from agriculture not only to feed a growing population, eradicate poverty and undernourishment and to produce renewable energy, but also to reduce greenhouse gases emissions in order to avoid being one of the main drivers of climate change. While trying to reach these multiple objectives, agriculture must also adapt to that changing climate, increased weather variability, and weather extremes which are expected to intensify in a warming world. A warmer atmosphere holds more moisture; precipitation can be heavier and cause more extreme flooding and on the other hand, higher temperatures, increase evaporation and can intensify droughts (Hansen et al. 2012). Impacts from climate change raise concerns on food availability (Vergara et al. 2014; FAO 2008), and on food prices in the coming years (OCDE/FAO 2011; CEPAL/FAO/IICA 2011). Preserving and enhancing food security thus requires agricultural production systems to increase productivity and, at the same time, to diminish the variability of crop production in the face of risks of climate and other risks of an agro-ecological and socio-economic nature (Beddington et al. 2012; Antle and Capalbo 2010; FAO and DWFI 2015).

Farmers as well as technology developers have been struggling to develop and to incorporate new agriculture practices and technological options to be able to cope with socio-economic changes as well as weather and climatic hazards (e.g. no tilling and other soil protection practices, long and short season seed varieties, genetically modified seeds, control of pests and diseases, fertilizers, and irrigation). These technologies are aimed at increasing yields and reducing variability to face extreme weather events. However, it will be argued here that, despite efforts, crop yield variability has not diminished, particularly in case of droughts (Lobell et al. 2014). Moreover, many of these practices, associated with intensive farming, are being criticized for the negative impacts on human health and on ecosystems (Carpenter et al. 2015; Viglizzo et al. 2011). This is why it is necessary to be able to determine the causes and extent of such variability in yields and to find advanced and new practices if further stabilizing food production, improving food quality and preserving agriculture ecosystems health should be achieved.

To stabilize outputs, production systems must become more resilient—i.e. capable of performing well in face of disruptive events. More productive and resilient agriculture requires transformations in the management of natural resources and higher efficiency in the use of these resources and other inputs for production (FAO 2010), as well as for preserving agro-ecosystems' functions and services. Moreover, this is not only true at a field scale. A resilience framework implies that all systems are *panarchically* articulated, meaning that structures and processes are linked across spatial and temporal scales based on both top-down and bottom-up interactions (Walker et al. 2006), and that adaptation refers to a process of managing resilience (Nelson 2011). Therefore, managing resources to adapt to changes (namely, climate variability and change) should make a system (an agriculture field) overcome that change, but in a way that fosters resilience not only of that particular system but also for the entire hierarchy in which the system is embedded (being the latest a plot, a landscape, a region or the entire planet). Under a socio-ecological system resilience approach, agriculture adaptations to climate variability and change should take into account not only what is needed at the field scale but also at larger scales and for longer periods (Eakin and Wehbe 2009); it is necessary to secure a "resilient Earth system" (Rockström et al. 2017: 7).

This study aims to illustrate the impacts of droughts on the agriculture system of central Argentina, exploring the hypothesis that despite advances in agricultural practices and technology, the variability of the yields of main crops has not diminished for any crop, or within any of the different cultivated areas. Possible reasons for this behavior were explored in order to improve adaptation and to stabilize crops yield based on the approach of socio-ecological systems' resilience.

Applying the resilience of socio-ecological systems approach to one of the most important agriculture and food producing areas in the world will probably provide an understanding on which are the possibilities to adapt to a changing climate and to increase agricultural production within the limits of the sustainable development goals.

Agriculture in the Argentinean *Pampas*

In Argentina, rainfed crop production is concentrated in the *pampas* region (Center-East of the country), which has always been the source of main food staples, the engine for the economy, and the source for foreign currency through exports in the country (Obschatko 2013). This agricultural region, one of the highest potentials for crop and livestock production in the world, is nowadays mainly devoted to soybean production, which is being exported as grain, oil, animal feed, and biodiesel. Corn represents the second main crop product among agriculture total exports; it is used to feed cattle and to produce bioethanol. Over the last two decades, land area devoted to cash crops in Argentina has increased by 90% by displacing livestock or other traditional crops and by incorporating new land through deforestation in the northwest of the country (INTA 2011).

Therefore, soybean expansion represents a process that not only involved an intensification of cash crop production but also the introduction of cash cropping into new agricultural land. This was possible due to its better adaptation capacity to different weather and soil conditions, and also due to lower investment requirements for planting compared to that of other crops. Agriculture expansion from the *pampas* to the North and West has meant: (a) a significant reduction in natural forests; (b) increased application of external inputs; (c) loss of biodiversity in particular ecosystems; (d) displacement of native population; (d) increasing risks of chemical contamination as well as wind and water soil erosion; (e) reductions in greenhouse gasses emissions with the expansion of no-tilling practices, although offset by increased emissions due to deforestation; and, (f) reduction in water-use efficiency due to land cover changes (Viglizzo et al. 2011).

Methodology

A comparative analysis of yield variability during the last four decades in relation to identified droughts events in two study sites located in the province of Córdoba (Argentina) characterized by different soils and climate conditions and for two cash crops was conducted. In the East, the area of the department[1] Marcos Juárez (MJ) is representative of the weather and cropping conditions of the *pampas*. The West, the department of Río Cuarto (RC), represents the characteristics of new agricultural areas or the extra-*pampas*. In both areas crop mix, production, yield variability, and yield trends of main crops at aggregated level were analyzed. Geographic, climatic, and socio-economic circumstances in both areas may contribute to the understanding of the challenge (i.e. opportunities and constraints) of adapting to climate variability and change. Prior research conducted in these areas allowed for the identification of main adaptation strategies taken by farmers to cope with economic and environmental changes (AIACC Final Report 2006). Time series (1975–2014 period) for soybean and corn crops—i.e. cultivated area, crop production and not harvested plots—available for both areas were used being SIIA[2] the crop data source (SIIA 2014).

To identify and characterize drought events during the last 40 years (1975–2014), the Standardized Precipitation Index (SPI) at different time scales (3, 12, 24 months) was used (McKee et al. 1993, 1995). The computation of the SPI requires monthly precipitation data from Río Cuarto and Marcos Juárez which were obtained from the Agro Meteorological Service of the University of Río Cuarto and from the National Institute for Agricultural Technology (INTA) data base, respectively. SPI values range between 2.0+ (extremely wet) and −2.0 and less (extremely dry). A drought event occurs any time the SPI is continuously negative and reaches an intensity of −1.0 or less (WMO 2012).

Based on drought occurrences, corn and soybean yield variability was analyzed in both Departments. Yield trends of corn and soybean crops were determined using the 5-year moving average and used as an indicator of the effects of the introduction of new and improved technology and the increased water availability for crops (Barros 2008). This is a conservative methodology since average values rather than maximum yield potentials were used to calculate negative yield variations. This may imply an underestimation of losses, especially when a series of "bad years" causes the 5-years moving average to decrease compared with maximum potential production of a crop. Moreover, plots that are reported as "not harvested"—i.e. those that agriculture producers decide not to harvest because yields are so low that the cost of harvesting them is higher than the expected income—were also not considered to calculate yield reductions. Although not harvesting a plot can have a high impact on an individual farmer, not harvested hectares are not significant in relation to total hectares allocated to a particular crop. Nevertheless when necessary

[1]Departments are the political divisions of the province.

[2]SIIA is the crop reporting division of the Ministry of Agriculture, Argentina.

they were pointed out in the analysis. Finally, additional information related to the studied area was obtained from previous research at the National University of Río Cuarto and from experimental and extension fields belonging to the National Institute of Agriculture Technology.

Main Agriculture Features

In this section, a description of the main agriculture features of Río Cuarto and Marcos Juárez Departments is presented including: (a) the evolution of total planted hectares and production of corn and soybean in the period 1975–2014; (b) main extreme weather events and farmers' responses to them such as technology adoption and new practices; and, (c) the identification of major drought events.

(a) Main soybean and corn production trends

Río Cuarto department in the West of Córdoba province is one of the regions where agriculture has been extended far away from the pampas. The area is characterized as humid to sub-humid, with decreasing precipitation patterns from the East to the West. The region's agro-food-based economy has contributed to the replacement of the natural land cover, suggesting that dynamics of this system over the years has been defined by rainfall, demography, productive activities and the consequent change in land use. Originally, agriculture included both livestock and crop activities but nowadays cash crop production prevails.

Soybean cultivation started during the 1990s, but it was not until 2003 that hectares cultivated with soybeans became higher than hectares cultivated with corn. From then on, soybean cultivation started to grow and in 2013 the total area cultivated with soybean had more than doubled. Meanwhile, land area with corn, although variable, never reached again the amount of hectares devoted to this crop in the 1980s.

Although total production of corn in the West is associated to a certain extent to cultivated area, variation in total corn production is higher than for soybean, suggesting a stronger dependence on interannual yield variations. It is important to note that during "bad years" the amount of hectares not harvested relative to planted area also affects total production beyond the solely impacts from decreasing yields. But above all, an increasing total production trend for the whole analyzed period reflects technological enhancement and adoption.

Even when an increasing productivity trend started after the incorporation of the whole soybean technological package (i.e. genetically modified soybean seed, no till farming), increases in soybean production were associated to a greater extent to the expansion of cultivated area. However, variability in total soybean production is, as in the case of corn, also related to yield variability as well as the decision of the farmers not to harvest plots when very low yields are evident.

Marcos Juárez department. To the East, soybean crop was introduced earlier, at the beginning of the 1970s, and, it was early in the 1980s that soybean hectares surpassed those of corn. Corn hectares show a more stable trend in an area characterized by the presence of a main winter crop like wheat. When short-cycle soybean appeared, double-cropping wheat/soybean was possible, rendering more income to farmers. To the end of the studied period, when short cycle corn was available, double cropping wheat/corn became another alternative for farmers (Calzada and Corina 2016). Total corn production started to diverge from other areas devoted to this crop during the 1990s, based on the introduction of technological change, and its consequent yield increase. Corn yields per hectare are highly variable and also affected by plots that were not harvested. Nevertheless, not harvested plots have been decreasing during the analyzed period. Total soybean production in this area is also highly variable due to changes in total cultivated area, yields per hectare, not harvested plots, and a preeminence of soybean or corn as a second crop within the double-cropping practice.

In both areas, years with low yields and the presence of plots that were not harvested, therefore low production, are mainly associated with droughts, as will be shown further along the study. In fact, the most worrisome climatic event highlighted by farmers from the south of Córdoba was drought (AIACC Final Report 2006). Hail storms, especially to the West, and floods to the South, were also mentioned as important events. The latest is particularly the case of farmers located in more topographically depressed areas, where production loss is more frequent since it is not only the deficit but also the excess of precipitation that may affect their production activities year after year.

(b) Extreme weather events and farmers responses

Both, in the West and East of Córdoba, thermal and water conditions are important variables causing crop yield variability, with soil moisture content being the most limiting factor. Winters are mild and short and are characterized by frost events and soil moisture deficits. Although there is a surplus in the average water balance of the region, mainly in the East, the interannual variability of precipitation can generate droughts or floods of different frequency and severity. In the SE of the region three major flood episodes occurred during 1979, 1986 and 1998, affecting agricultural production and the economy of the area for several years after each episode (Seiler et al. 2002; Wehbe et al. 2008).

Droughts are expected to grow in frequency and severity as a consequence of climate change together with increasing temperatures, both responsible for important crop losses. Future climate projections under moderate and high-emission green house gases' scenarios show that the occurrence of short term and long term droughts will be more frequent in the 21st century, with shorter duration but greater severity over agriculture areas of Argentina (Penalba and Rivera 2013).

Most of the available technology and management techniques are focused on the amelioration of drought impacts. Varieties and hybrids genetically adapted, use of chemicals and fertilizers, adjusting planting dates, soil water conservation practices, and special tilling systems are, among others, the main resources for adaptation to drought in the region. Not less important is the practice of supplemental irrigation that has been expanding in the last decades. However, the required investment and operation costs for introducing irrigation compared with crops market prices make extensive agriculture in the region mostly rainfed (Wehbe et al. 2011). New technologies in irrigation like sub-surface drip irrigation in extensive agriculture may become more applicable in the future to contribute to stabilize yields, together with increased green water use efficiency (INTA, Personal communication; Rockström et al. 2009). Impacts from hail storms are localized and crop insurance and cropping in different areas are usual adaptation options to cope or adapt to this kind of events. In the case of floods larger infrastructure works will be needed but are out of the hands of farmers and depend on political decisions which have not been manifested yet.

In general and from survey information, most common agronomic practices incorporated by farmers in response to climate variability were adjusting planting dates (36%); spatially distributing risk through planting or using geographically separated plots (52%); changing crops (12%); accumulating commodities as an economic reserve (85%); maintaining a livestock herd (70%), among others. Weather and climate information, when available, was also mentioned as been used to make decisions each year (Wehbe and Maurutto 2005; Wehbe et al. 2008). There have also been a number of other adopted technologies and practices, however, not all of them were direct responses to weather conditions, but rather general changes in production strategies for both economic and environmental reasons, including weather, which facilitated agriculture production growth.

New technology, in general, and climate adaptations have improved yields to a certain extent, but growing production trend, especially soybeans, has been the result of increased quantity of land devoted to this crop, whether replacing other existing crops and livestock production in the Pampas or incorporating new agriculture land.

(c) Major drought events in the period 1975/2014

The duration and magnitude of the major drought episodes in Marcos Juárez and Río Cuarto from 1975 were assessed using the SPI for three and six months (SPI 3, SPI 6), both periods ending in January for each year of the series. Significant drought occurred during the crop seasons 1988/89 and 1989/90 (Figs. 1 and 2). These droughts intensified their severity during both summers and lasted for almost 3 years. Moderate to severe drought happened during spring and summer of 1996–1997. Severe droughts during the summer occurred during 2000/01 and 2001/02 crop seasons although the 2000/01 drought was also present in the spring. An unusual short summer drought occurred during 2011/12 crop season where the lack of rain and a concurrent heat wave provoke very severe effects, which was devastating for the crops in the region. Sometimes droughts may be shorter in duration or lesser in magnitude but with differential impacts on the crops, according to soil moisture conditions before and after each one of the events.

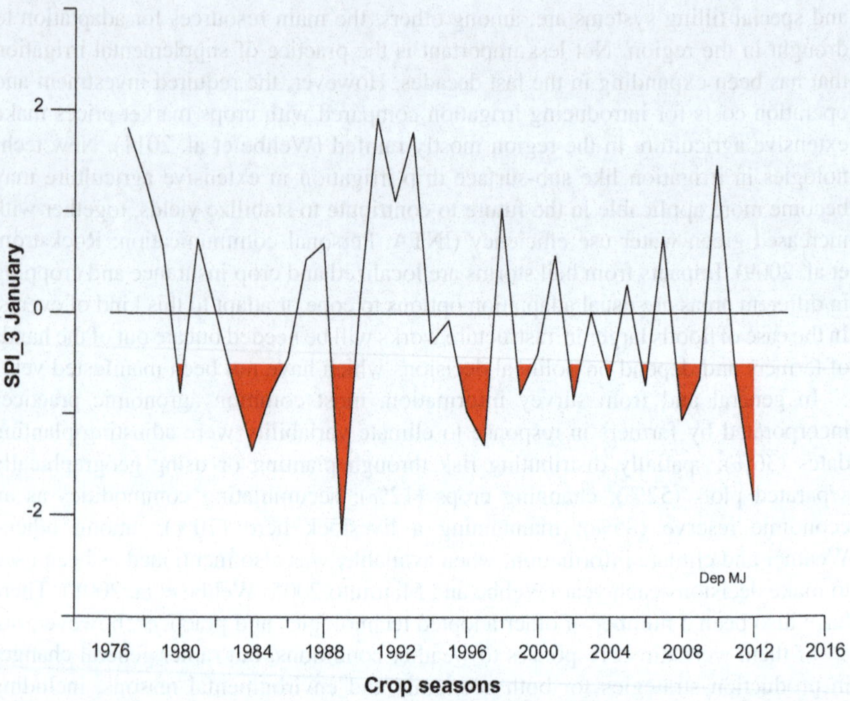

Fig. 1 Seasonal droughts in Marcos Juárez as indicated by the SPI_3 at the end of January

Results

Results from a comparative analysis of drought impacts on yield variability for corn and soybean in both cropping areas during the analyzed period are presented:

(a) *Corn yield trends*. Corn yields in the East (MJ—*pampas*) and West (RC—extra-*pampas*) of the province of Córdoba show a similar trend pattern (Figs. 3 and 4, top). The yield average for MJ remained constant from the beginning of the series up to the start of the 1990s. From there on the trend shows a constant increase reaching a plateau during the last years of the series. Similar performance is observed in RC but with an increase in the trend later in the nineties than in MJ and showing a slight decrease in the averages at present, due to more severe droughts than in MJ. The absolute yield values for both periods indicate the different carrying capacity of both areas. The East shows an average yield of about 4000 kg/ha before the nineties, and exceeding 8000 kg/ha at present. In the West area, the average yield was about 2500 kg/ha and 6000 kg/ha for the two periods of the series, respectively.

In the East, the analysis of corn yield as deviations from the trend (in kg/ha) (Figs. 3 and 4, middle) shows a significant number of events in which anomalies

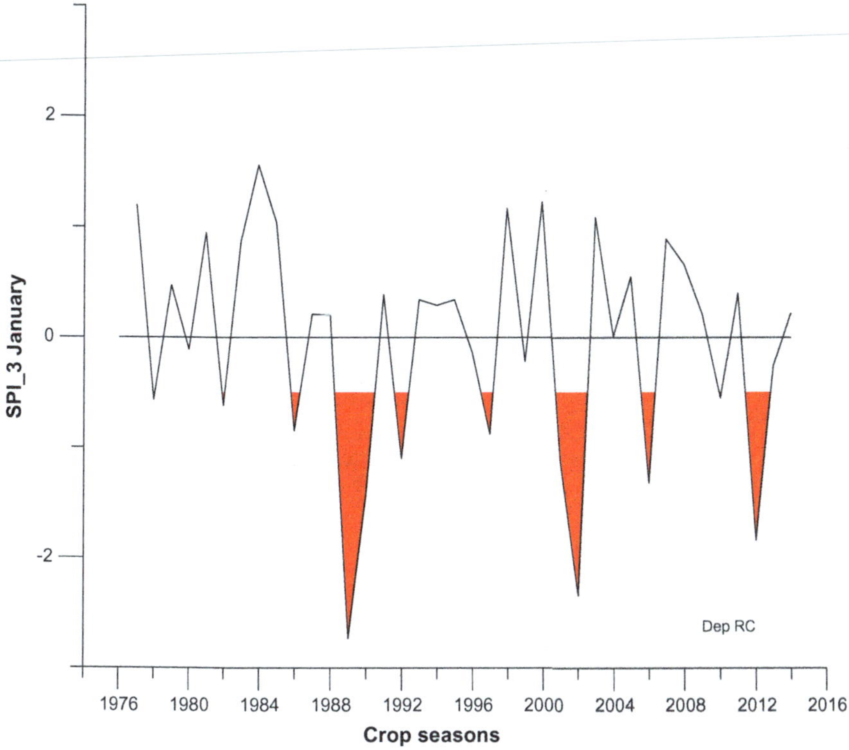

Fig. 2 Seasonal droughts in Río Cuarto as indicated by the SPI_3 at the end of January

occurred (n = 13 events), ranging from −500 to −2500 kg/ha, with three major decreases during the droughts 1989/90, 1996/97 and 2011/12 (−2500, −2000, and −2500 kg/ha, respectively). In the West, 10 anomalous seasons were recorded with yields deviations ranging from −500 to −2000 kg/ha with respect to trend of the area: −2000 in 1996/97, −1500 kg/ha in 2005–06, and −2000 kg/ha in 2011/12. Therefore, corn crop losses, measured as kilograms per hectare, seem to have not varied in absolute terms under drought conditions in 15 years despite technological changes and new agronomic practices.

(b) *Soybean yield trends.* The regional average yields for soybean increased throughout the series both in the East and in the West, with yields showing a recent plateau (Figs. 5 and 6, top). Differences are again in the absolute average values of the trend where in the East they started above 1800 kg/ha up to more than 3200 kg/ha. In the West maximum yield average in the series is about 2300 kg/ha.

The number of negative deviations from the trend of soybean yields (kg/ha) (Figs. 5 and 6, middle) was also significant in the East (n = 6 events) with variations in yield ranging from −500 to −1250 kg/ha, the latest recorded during the

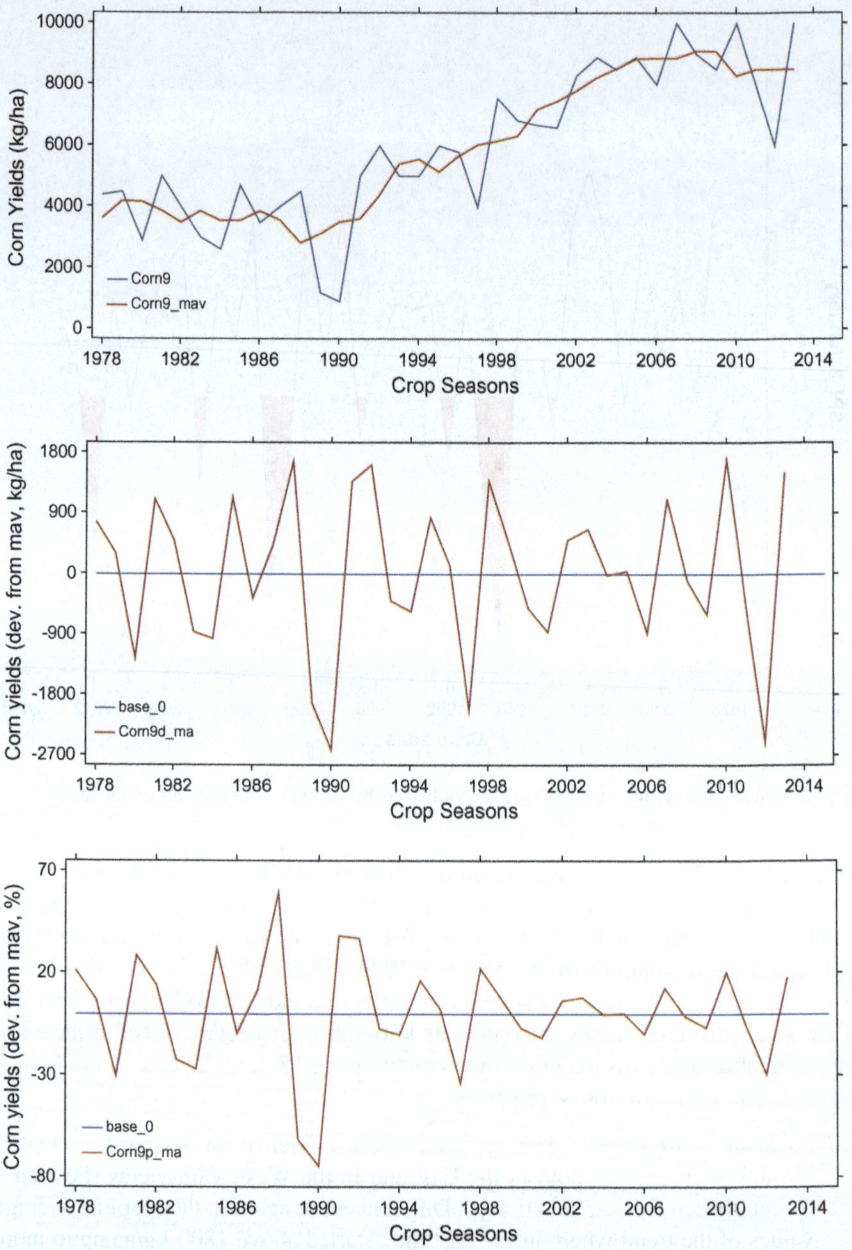

Fig. 3 Corn yield interseasonal variations in the wet region (Department of Marcos Juarez, Pampas). Top: yield variation (Corn9) and trend (Corn9_mav) as the 5 years moving average; middle: deviations from trend in kg ha^{-1} (Corn9d_ma); bottom: deviations as percentage of the trend (Corn9p_ma)

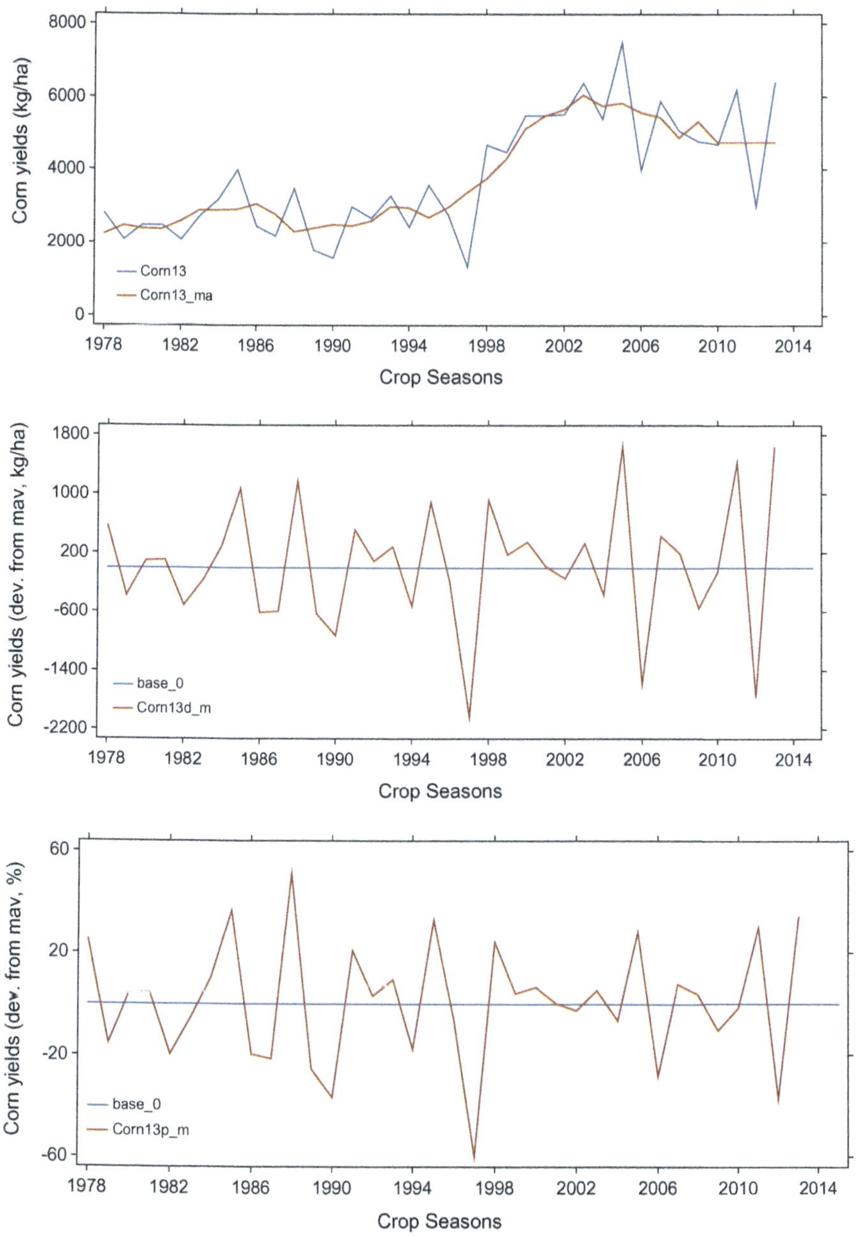

Fig. 4 Corn yield interseasonal variations in the sub-humid region (Department of Rio Cuarto, Extra Pampas). Top: yield variation (Corn13) and trend as the 5 years moving average (Corn13_ma); middle: deviations from trend in kg ha^{-1} (Corn13d_m); bottom: deviations as percentage of the trend (Corn13p_m)

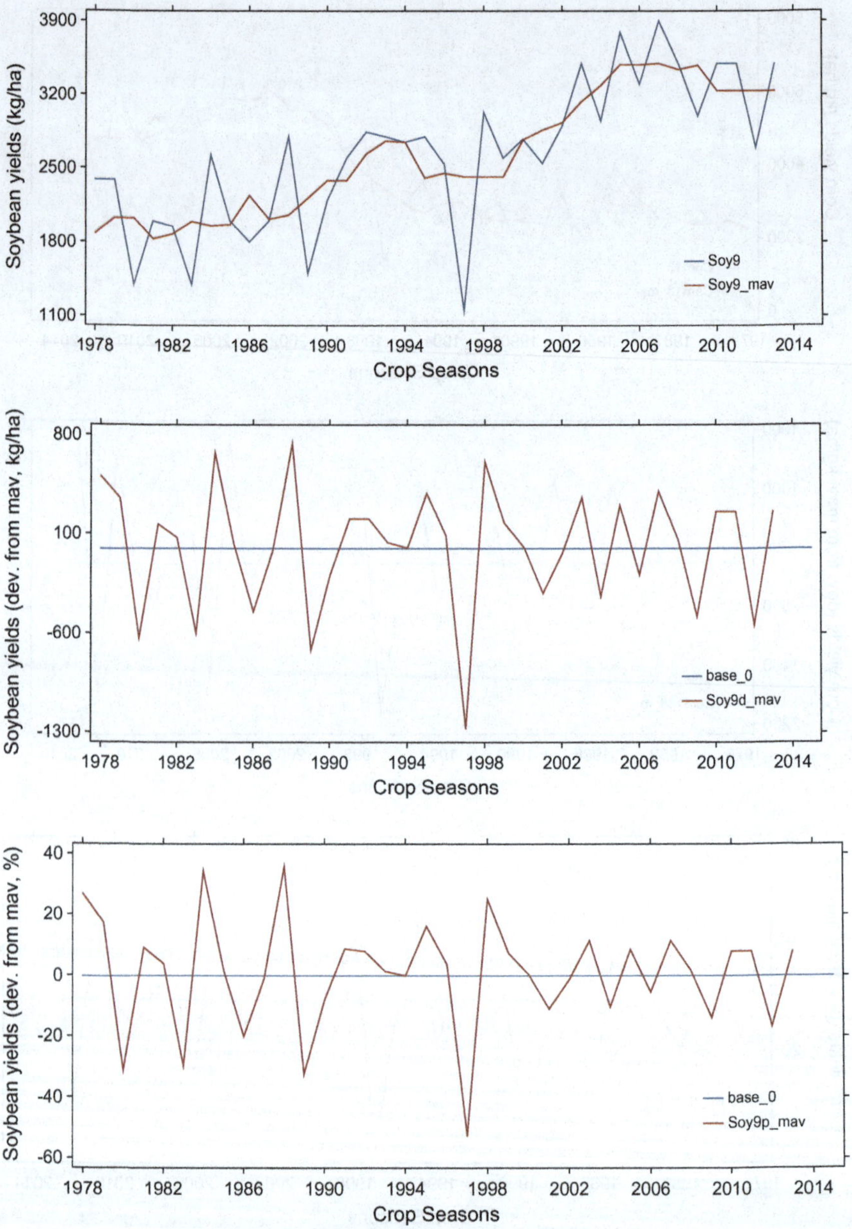

Fig. 5 Soybean yield interseasonal variations in the wet region (Department of Marcos Juarez, pampas), Top: yield variation (Soy9) and trend as the 5 years moving average (Soy9_mav); middle: deviations from trend in kg ha^{-1} (Soy9d_mav); bottom: deviations as percentage of the trend (Soy9p_mav)

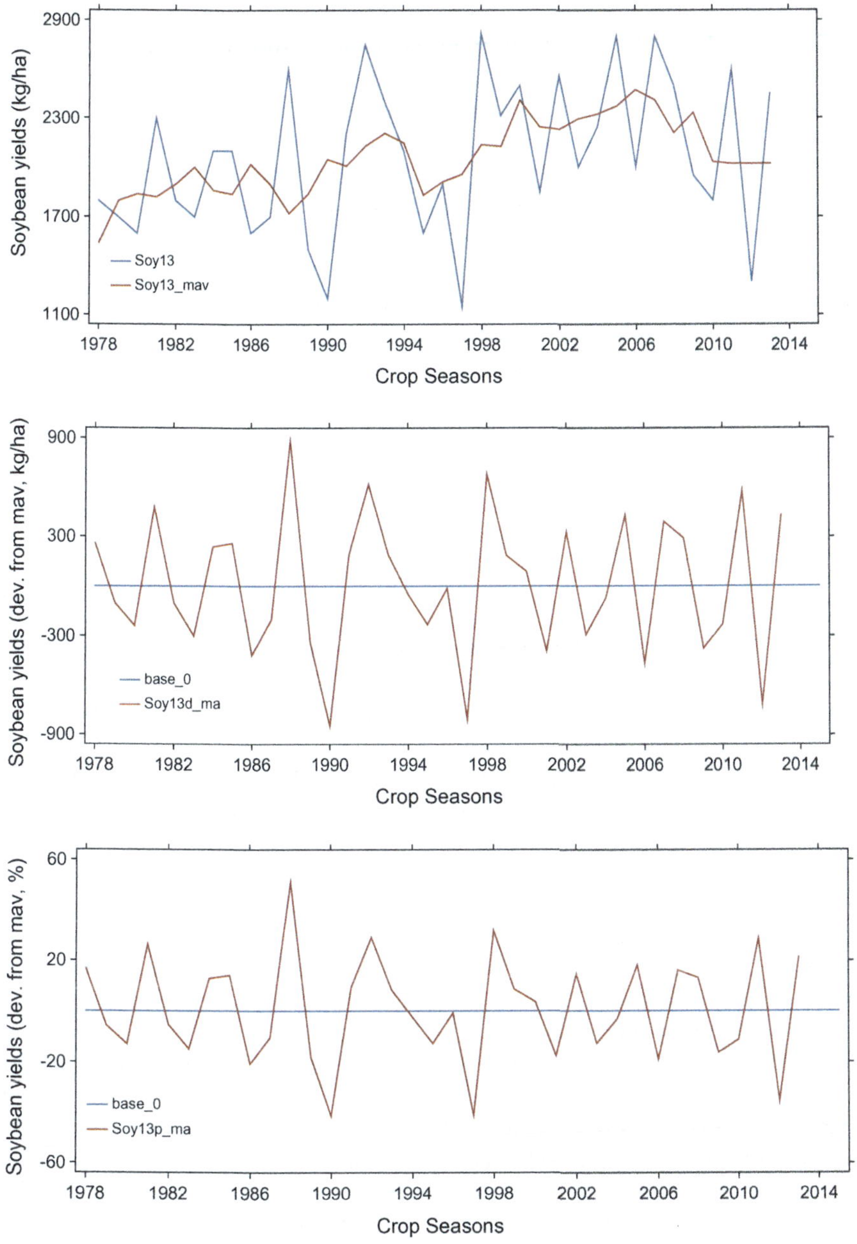

Fig. 6 Soybean yield interseasonal variations in the sub-humid region (Department of Rio Cuarto, extra pampas). Top: yield variation (Soy13) and trend as the 5 years moving average (Soy13_mav); middle: deviations from trend in kg ha^{-1} (Soy13d_ma); bottom: deviations as percentage of the trend (Soy13p_ma)

1996–97 drought episode. In the West, three cropping seasons were recorded as highly anomalous with deviations from the trend of −800 kg/ha in 1989–90 and 1996/97; and −600 kg/ha in 2011/12. This means a slight diminishing trend of negative outcomes in soybean production during droughts.

It seems possible to sustain the hypothesis that crop yield variability is not diminishing, at least not enough, in the case of corn production in both cultivated areas. For soybean yields a reduction in negative outcomes is more evident. However, it is worth to recall the characteristics of the summer crop season 2011/12 drought. Although precipitation during Dec–Jan were extremely low, the most damaging condition, especially for corn, was the heat wave that simultaneously affected the area, with average temperatures 5 °C above normal (Seiler 2012). This weather condition was not so evident before along the analyzed time series and was also unexpected due to the weak negative ENSO signal, surprising farmers who were already well aware about drought impacts on rainfed agriculture.

Add-on Results

Not all previous assertions are so straightforward. On the one hand, average yields for both crops and within different areas are not equal. Maximum corn crop yields are about 40% higher in the East compared with those in the West. And, in case of soybeans, maximum yields in the East are about 35% higher than in the West. This makes negative yield deviations different for each crop in each area when they are analyzed as percentages of average values. For both areas and for both crops, technological change and advanced agricultural practices have enabled increasing yields per hectare as seen in the regional averages. Nevertheless, under conditions of water deficit, crop losses do not appear to change much in terms of kilograms per hectare (or they are now even larger, depending on drought severity) during the second half of the analyzed period. Impacts measured by yield deviations as percentage of average values have been decreasing, but it does not mean that total production variability is diminishing neither in the *pampas* nor in extra-*pampas* regions, particularly during the latest years.

Furthermore, information related to yields per hectare only takes into account those hectares that have been harvested. But when a weather event strikes a region, there usually exist a varying quantity of "not harvested hectares" which should be accounted for into yield variability analysis. Finally, climate is not the only source of stress for the farmers to deal with, as prices and market variability are also important for them. Usually called "double exposure" (O'Brien and Leichenko 2000), this is a situation where selection among adaptive options do not necessarily fit with both, market and climate stresses. Farmers modify inputs if the ratios of the product prices to input prices are not favorable (analysis not incorporated in this study). Interviewed farmers have expressed that "*variability in governmental policies is more dangerous for us than climate variability... (sic)*" (Wehbe and Maurutto 2005).

It is evident from this study that there has been an adaptation process based on the incorporation of available technologies and management practices, but it has not been homogeneous for the different areas and among farmers. Since they prioritize returns to investments, and especially in the East of the studied region, they assign the best land to corn and the rest for soybeans; this may explain why although negative impacts on soybeans were reduced during the second half of the analyzed period, reductions were not as important as those achieved in the case of corn. Furthermore, economies of scale that characterizes this industrial-type of agriculture production and the lack of governmental aid means those smallholders have less capacity to incorporate costly options. This is particularly the case of supplemental irrigation which is not widely adopted, even when enough quantity and quality of ground water is available (Blarasín and Cabrera 2005).

Discussions

Total production variability in soybean and corn in both Departments cannot be explained only by total hectares allocated to each crop. Such variability is explained by a number of factors, mainly climate and available technologies, market prices and the political context among others. It has been shown that corn and soybean share similar patterns of increasing productivity for both analyzed areas (Río Cuarto, representing the extra-*pampas* areas, and, Marcos Juárez representing the core agricultural area of the *pampas* region). But absolute values of average yields differ as a consequence of differences in soil and climate between them, and between both cultivated crops. The increase of average yields result from a combination of the incorporation of improved technology and some specific adaptations to climate variability. In the case of corn, hybridization and the associated package of chemical inputs, soil management, and seeds adapted to local pests and diseases were firstly introduced; later short cycle crop season varieties and genetically modified corn seeds started to be useful. Not less important is the increasing availability of more accurate climatic information that contributes to the decision-making process, especially at the beginning of the cropping season. In spite of all the possible adaptations, analysis allows to assert that current adaptations taken by farmers have been not sufficient to decouple yields from drought events. Therefore, yearly availability of corn and soybean production is still highly variable.

Previous analysis provides the basis on which to suggest the following arguments:

(a) Some seed varieties, especially those of higher yield potential under more stable weather conditions, may have relatively higher sensitivity under drought conditions and other weather extremes; however, even if it is possible to acquire information about how weather will perform over a cropping season, it is still highly uncertain how changes in temperature or precipitation will affect a crop during its critical phenological stages;

(b) The latest drought episode recorded within the studied area has been characterized by the concurrence of a heat wave and winds; this kind of extreme weather event may start to be increasingly recurrent, affecting mainly corn production (IPCC 2014);

(c) Price fluctuations or the macro-economic policy context may cause uncertainties and time delays in the farmers' decisions making process. And, after the decision is taken it usually happens that not all farmers are able to meet their total demand of a particular seed variety or other inputs because the supply is short of;

(d) Farmers face variable weather conditions with frequent extremes—namely drought, floods, strong winds, hail storms or heat waves—which continue to exert a toll on total production, suggesting that farmers have not yet managed to ameliorate crop sensitivity to climate risks.

The selection of seed varieties together with weather condition may constitute a source of yield variability. In this sense, if there is no evidence of a proximate dry weather condition, farmers would choose a specific variety of higher yield potential, but if a drought does occur, negative impacts over yields may be larger than if the farmer had chosen an alternative lower-yielding variety (Lobell et al. 2014). But, even when a perspective of a drought exists, if the farmer prioritizes market situations when taking decisions, he may select a specific crop like corn, which is more sensitive to droughts but represents more kilograms per hectare at harvest. Oftentimes farmers are forced to sow corn even under unfavorable conditions for soil conservation reasons. Finally, crop yields may also be influenced by the resulting state of soils in the aftermath of fires which become strengthened by scarce precipitation, heat and strong winds during spring months, especially in northwestern areas. During late winter and early spring, fires have shown to be increasing in frequency and in area extension (Fisher et al. 2012).

The expected impacts of climate change are not the only threats to agriculture, namely food production for fulfilling the demand of an increasing population and to secure food access to the undernourished. Biofuel and bioplastic production based on crops (OCDE 2013), and increasing loss of soils due to monoculture will further contribute to declining availability of feed and food, independently of possibilities for declining yield variability due to a changing climate. In such a future, adaptation will not only need to improve yields, but it will also be necessary for total agriculture production to increase. After expanding agriculture into natural areas, and because of impacts on ecosystems it will be a whole range of different environmental changes that it will be necessary to adapt to.

Adaptations to Climate Change Under a Resilience of Socio-ecological System Approach

A question to be answered is if it is possible to reduce production variability due to climate extremes through types of adaptations which are not dangerous to and could enhance ecosystems capacities to provide for human services? First, it will be necessary to diminish variability and stagnated productivity growth, both consequences resulting from lack of new technological advances or from increasing frequency of droughts. Second, adaptations aimed at diminishing yields losses must not only be appropriate in terms of the environmental characteristics where agriculture is being practiced—like to keep buffer natural areas—but also in relation to human health security. The latest refers to the possibility to ameliorate negative impacts on humans and ecosystems characteristic of high industrial input based agriculture production. In these senses and based on the concept of resilience of socio-ecological systems—the capacity to buffer change, learn and develop—as a framework for understanding how to sustain and enhance adaptive capacity in a complex world of rapid transformations (Folke et al. 2002) some possibilities based on current and more widely acknowledged situations are identified in relation to the studied area:

- Appropriate agrochemicals management will be necessary to reduce yield losses and at the same time to avoid human health risks, pests and diseases resistance as well as the development of new more tolerant crop varieties;
- To improve harvesting, stocking and transport technology to avoid grain losses in each of these processes;
- To incorporate supplemental drip irrigation systems to enhance fertilization outcomes but always within certain criteria consistent with local ecosystems and human population needs of freshwater and equity on access to costly irrigation technologies (Wehbe et al. 2011);
- In areas where freshwater is scarce and must be shared with other uses and users, new technologies like enhancing management of green water need to be developed and incorporated. This is also true where groundwater quality would result detrimental for agriculture in the medium to long run (Rockström et al. 2009);
- To return to a more diversified agriculture in terms of variety of crops, livestock and production technologies for securing quantitative and qualitatively food and ecosystems which is one of the major concerns for human health (Cabell and Oelofse 2012);
- Reforestation of degraded land (due to mono cropping systems) with native species, and the introduction of cattle grazed woodland models instead of increasing feed-lot practices;
- Diversity in farm size and intensity may help reduce yield variability in more risky new agriculture areas in face of higher temperatures and droughts (Reidsma and Ewert 2008).

From the perspective of the resilience of socio-ecological systems, it will be necessary to increase diversity—diversity of responses and of functions (Walker et al. 2006)—within agro-ecosystems to be able to maintain the natural-social relationships within a desirable state. Local, small-scale, short run decisions accumulate through space and time to determine more or less resilient socio-ecological systems to current and future shocks at higher spatial and time scales (Wehbe 2014; Eakin and Wehbe 2009). Climate change entails a more uncertain future in relation to extreme climate events like droughts endangering the state of these systems. Therefore, it will be necessary to be creative and adaptive, or even more, to have the capacity to transform current state of agricultural systems to increase food production in order to meet current and future food security both in quantitative and qualitative terms.

Conclusions

Despite changes in agricultural technology, concerns persist over both the availability of enough food and of food of quality at global level and in face of climate change impacts. Argentina, one of the main food producer countries, exemplifies current difficulties encountered by agriculture to cope with weather and climate extremes such as droughts. In this paper, results from research in the Argentinean *pampas* that analyze yield variability within the last four decades in which significant technological changes in production have occurred were presented. It was found that while yield variability has slightly declined, production is still significantly affected by climatic variability and other sources of stressors. The analysis underscores the necessity for advanced and new practices if further stabilizing food production, improving food quality and preserving agriculture ecosystems health are to be achieved.

The possibilities for feeding a growing population by only closing yield gaps seem worldwide very unlikely under current climate variability, climate change impacts and current technologies. It has been shown that, even in one of the main agriculture areas in the world with access to most of the technologies and practices, yield reductions in face of climatic stresses are always possible due to: (a) a climate behavior highly uncertain; (b) the confluence of different weather patterns at one location which make current technologies unable to stabilize crop yields; (c) adoption of particular technologies and practices depends not only on weather or climate predictions but also on the contextual parameters to the process of decision making, namely markets, policies, etc. Furthermore, if these adaptations should maintain ecosystems integrity and human health at the same time, they not only should be confined to ways of sustainably intensification of agriculture but also they should include improvements in food production and distribution together with the promotion of profound changes in consumption patterns under a socio-ecological resilience approach.

References

AIACC Final Report. (2006). *Vulnerability and adaptation to climate variability and change: The case of farmers in Mexico and Argentina*. Project No. LA29. Washington, DC, USA: The International START Secretariat.

Antle, J. M., & Capalbo, S. M. (2010). Adaptation of agricultural and food systems to climate change: An economic and policy perspective. *Applied Economic Perspectives and Policy, 32*(3), 386–416.

Barros, V. (2008). Adaptation to climate trends: Lessons from the argentine experience. In *Climatic Change and Adaptation* (Chapter 17, p. 296). UK: Earthscan.

Beddington, J., Asaduzzaman, M., Clark, M., Fernández, A. et al. (2012). *Achieving food security in the face of climate change: Final report from the commission on sustainable agriculture and climate change*. CGIAR research program on climate change, agriculture and food security (CCAFS), Copenhagen, Denmark. Available online at www.ccafs.cgiar.org/commission.

Blarasín, M., & Cabrera. (2005). Agua subterránea y ambiente. Área de Promoción Científica de la Agencia Córdoba Ciencias S.E. Available at www.cba.gov.ar/wp-content/.../03/Agua+subterranea+y+ambiente.desbloqueado.pdf.

Cabell, J. F., & Oelofse, M. (2012). An indicator framework for assessing agroecosystem resilience. *Ecology and Society, 17*(1), 18.

Calzada, J., & Corina, S. (2016). *¿Qué conviene sembrar: soja o maíz de segunda? Una pelea pareja*. Bolsa de Comercio de Rosario, Informativo Semanal, No. 1781, Available at http://www.bcr.com.ar/Publicaciones/Informativo%20semanal/bcr2016_10_21.pdf.

Carpenter, S. R., Booth, E. G., Gillon, S., Kucharik, C. J., Loheide, S., Mase, A. S., et al. (2015). Plausible futures of a social-ecological system: Yahara watershed, Wisconsin, USA. *Ecology and Society, 20*(2), 10.

CEPAL/FAO/IICA. (2011). Volatilidad de precios en los mercados agrícolas (2000–2010): implicaciones para América Latina y opciones de políticas. Boletín No. 1, p. 36.

Eakin, H. C., & Wehbe, M. B. (2009). Linking local vulnerability to system sustainability in a resilience framework. *Climate Change, 93*(3–4), 355–377.

FAO. (2008). *The state of food insecurity in the world 2008. High food prices and food security—Threats and opportunities* (p. 56). Rome.

FAO. (2010). *"Climate smart" agriculture, policies, practices and financing for food security, adaptation and mitigation* (p. 41). Rome, Italy: Food and Agriculture Organization of the United Nations.

FAO, & DWFI. (2015). Yield gap analysis of field crops—Methods and case studies. In V. O. Sadras, K. G. G. Cassman, P. Grassini, A. J. Hall, W. G. M. Bastiaanssen, A. G. Laborte, A. E. Milne, G. Sileshi, & P. Steduto. FAO Water Reports No. 41, Rome, Italy.

Fisher, M. A., Di Bella, C. M., & Jobbagy, E. G. (2012). Fire patterns in central semiarid Argentina. *Journal of Arid Environment, 78,* 161–168.

Folke, C., Carpenter, S., Elmqvist, T., Gunderson, L., Holling, C., & Walker, B. (2002). Resilience and sustainable development: Building adaptive capacity in a world of transformations. *Ambio, 31*(5), 437–440.

Hansen, J., Sato, M., & Ruedy, R. (2012). Perception of climate change. *PNAS* E2415–E2423.

INTA. (2011). PRECOP - *Mayor valor agregado en origen*. INTA - Ministerio de Agricultura, Ganadería y Pesca, Presidencia de la Nación.

IPCC. (2014). Summary for policymakers. In: C. B. Field, V.R. Barros, D. J. Dokken, K. J. Mach, M. D. Mastrandrea, T. E. Bilir, M. Chatterjee, K. L. Ebi, Y. O. Estrada, R. C. Genova, B. Girma, E. S. Kissel, A. N. Levy, S. MacCracken, P. R. Mastrandrea, & L. L. White (Eds.), *Climate change 2014: impacts, adaptation, and vulnerability. Part A: Global and sectoral aspects. Contribution of working group II to the fifth assessment report of the Intergovernmental Panel on Climate Change* (pp. 17–18). Cambridge, UK: Cambridge University Press.

Lobell, D., Roberts, M. J., Schlenker, W., Braun, N., Braun, N., Little, B., et al. (2014). Greater sensitivity to drought accompanies Maize yield increase in the U.S. Midwest. *Science, 344,* 516–519.

McKee, T. B., Doesken, N. J., & Kleist, J. (1993). The relationship of drought frequency and duration to time scales. In *Preprints, Eighth Conference on Applied Climatology* (pp. 174–184). Anaheim, CA: American Meteorological Society.

McKee, T. B., Doesken, N. J., & Kleist, J. (1995). Drought monitoring with multiple time scales. In *Preprints, Ninth Conference on Applied Climatology* (pp. 233–236). Dallas, TX: American Meteorological Society.

Nelson, D. R. (2011). Adaptation and resilience: Responding to a changing climate. *WIREs Climate Change, 2,* 113–120. https://doi.org/10.1002/wcc.91Vol.2.

O'Brien, K. L., & Leichenko, R. M. (2000). Double exposure: Assessing the impacts of climate change within the context of economic globalization. *Global Environmental Change, 10,* 221–232.

Obschatko, E. (2013). El sector agroalimentario argentino como motor del crecimiento. In *Alimentar el futuro. Argentina proveedora de Alimentos para el mundo* (pp. 54–67). Buenos Aires: Cámara de Comercio Argentino-Holandesa.

OCDE/FAO. (2011). OCDE-FAO *Perspectivas Agrícolas 2011–2020.* OECD Publishing y FAO. http://dx.doi.org/10.1787/agr_outlook-2011-es.

OECD. (2013). *Policies for bioplastics in the context of a bioeconomy.* OECD Science, Technology and Industry Policy Papers, No. 10, OECD Publishing.

Penalba, O. C., & Rivera, J. A. (2013). Future changes in drought characteristics over Southern South America projected by a CMIP5 Multi-model Ensemble. *American Journal of Climate Change, 2,* 173–182.

Reidsma, P., & Ewert, F. (2008). Regional farm diversity can reduce vulnerability of food production to climate change. *Ecology and Society, 13*(1), 38.

Rockström, J., Falkenmark, M., Karlberg, L., Hoff, H., Rost, S., & Gerten, D. (2009). Future water availability for global food production: The potential of green water for increasing resilience to global change. *Water Resources Research, 45*, W00A12, https://doi.org/10.1029/2007wr006767.

Rockström, J., Williams, J., Daily, G., Noble, A., Matthews, N., Gordon, L., et al. (2017). Sustainable intensification of agriculture for human prosperity and global sustainability. *Ambio* 1–14. (First Online) ISSN: 0044-7447.

Seiler, R. A. (2012). *La sequía como fenómeno climático*. Presentación realizada en el marco del "Seminario: Sequía 2011–2012, Un análisis multidisciplinario". Res CD. No. 036/2012 (April 03, 2012), Universidad Nacional de Río Cuarto.

Seiler, R. A., Hayes, M., & Bressan, L. (2002). Using the standardized precipitation index for flood risk monitoring. *International Journal of Climatology, 22,* 1365–1376.

SIIA. (2014). *Sistema Integrado de Información Agropecuaria.* Secretaría de Agricultura, Ganadería y Pesca. Presidencia de la Nación. http://www.siia.gov.ar/.

Vergara, W., Ríos, A. R., Trapido, P., & Malarín, H. (2014). *Agricultura y Clima Futuro en América Latina y el Caribe: Impactos Sistémicos y Posibles Respuestas.* Banco Interamericano de Desarrollo, Documento de debate; No. IDB-DP-329. Available at www.iadb.org/cambioclimatico.

Viglizzo, E. F., Frank, F. C., Carreño, L. V., Jobbagy, E., Pereyra, H., Clatt, J., et al. (2011). Ecological and environmental footprint of 50 years of agricultural expansion in Argentina. *Global Change Biology, 17,* 959–973.

Walker, B. H., Gunderson, L. H., Kinzig, A. P., Folke, C., Carpenter, S. R., & Schultz, L. (2006). A handful of heuristics and some propositions for understanding resilience in social-ecological systems. *Ecology and Society, 11*(1), 13.

Wehbe, M. B. (2014). Needs for institutional transformations for managing freshwater and agriculture in South of Cordoba (Argentina). In *Third International Science and Policy Conference on the Resilience of Social & Ecological Systems: "Resilience and Development: Mobilizing for Transformation*". Montpellier, France, 4–8 de Mayo.

Wehbe, M. B., Bosch, E. A., Granda, J. A., & Tarasconi, I. E. (2011). *Riesgo sequía, impactos y manejo de riesgo en la agricultura del Sur de Córdoba*. III Congreso Regional de Economía Agraria-XVI Congreso de Economistas Agrarios de Chile-XLII Reunión Anual Asociación Argentina de Economía Agraria, Valdivia, Chile, 9–11 Noviembre.

Wehbe, M. B., Eakin, H.C., Seiler, R. A., et al. (2008). *Local perspectives on adaptation to climate change. Lessons from Mexico and Argentina*. In Climatic Change and Adaptation (Chapter 18, 315–331) Earthscan, UK.

Wehbe, M. B., & Maurutto, M. C. (2005). *Participatory local development: Importance of incorporating agriculture producers into public policy formulations towards reducing their vulnerability to climate variability and change*. 6th Open Meeting of the Human Dimensions of Global Environmental Change Research Community. BONN.

WMO. (2012). *Standardized precipitation index user guide*. WMO-No: 1090. p. 24.

Coffee Responses to Drought, Warming and High [CO_2] in a Context of Future Climate Change Scenarios

José Nobre Semedo, Weverton P. Rodrigues, Danielly Dubberstein, Madlles Q. Martins, Lima D. Martins, Isabel P. Pais, Ana P. Rodrigues, António E. Leitão, Fábio L. Partelli, Eliemar Campostrini, Marcelo A. Tomaz, Fernando H. Reboredo, Paula Scotti-Campos, Ana I. Ribeiro-Barros, Fernando C. Lidon, Fábio M. DaMatta and José C. Ramalho

Abstract Climate variability strongly determines agricultural productivity, further causing important economic and social impacts. In a context of global climate changes, the continuous enhancement of agricultural production in the coming years is a major challenge for plant science research. Coffee, one of the most important agricultural commodities worldwide, is grown in more than 80 countries in the tropical region. Several estimates point to a strong reduction on both coffee

J. N. Semedo · I. P. Pais · P. Scotti-Campos
Unidade de Investigação em Biotecnologia e Recursos Genéticos (UIBRG), Instituto Nacional de Investigação Agrária e Veterinária, I.P. (INIAV), Quinta do Marquês, Av. da República, 2784-505 Oeiras, Portugal

J. N. Semedo · A. E. Leitão · F. H. Reboredo · P. Scotti-Campos · A. I. Ribeiro-Barros
F. C. Lidon · J. C. Ramalho (✉)
GeoBioTec, Dept. Ciências da Terra (DCT), Faculdade de Ciências e Tecnologia (FCT), Universidade NOVA de Lisboa (UNL), 2829-516 Monte de Caparica, Portugal
e-mail: cochichor@isa.ulisboa.pt; cochichor@mail.telepac.pt

W. P. Rodrigues · E. Campostrini
Setor Fisiologia Vegetal, Centro de Ciências e Tecnologias Agropecuárias, Universidade Estadual Norte Fluminense - Darcy Ribeiro (UENF), Campos dos Goitacazes, Campos dos Goitacazes, RJ, Brazil

D. Dubberstein · M. Q. Martins · F. L. Partelli
Dept. Ciências Agrárias e Biológicas (DCAB), Centro Universitário Norte do Espírito Santo (CEUNES), Universidade Federal Espírito Santo (UFES), Rodovia BR 101 Norte, Km. 60, Bairro Litorâneo, São Mateus, ES CEP: 29932-540, Brazil

L. D. Martins · M. A. Tomaz
Centro de Ciências Agrárias e Engenharias (CCAE), Universidade Federal do Espírito Santo (UFES), Alto Universitário, Alegre, ES CEP: 29500-000, Brazil

A. P. Rodrigues
Forest Studies Center (CEF), Instituto Superior de Agronomia (ISA), Universidade de Lisboa (ULisboa), Tapada da Ajuda, 1349-017 Lisbon, Portugal

F. Alves et al. (eds.), *Theory and Practice of Climate Adaptation*, Climate Change Management, https://doi.org/10.1007/978-3-319-72874-2_26

yields and suitable areas in a near future, mostly related to predicted rising temperature, but also due to changes in intra- and inter-annual rainfall amounts and distributions. Nonetheless, recent findings from our team has shown that the coffee plant is more resilient that usually accepted, and that the negative impacts of rising temperature, at physiological and biochemical levels, were strongly mitigated by enhanced air [CO_2], which is considered one of the promoting agents of temperature rise. Also, the identification of ecophysiological and molecular traits that can promote plant acclimation to warming, in particular those related to the C-assimilation pathway, would foster the selection of more adapted/tolerant genotypes. In this context, this work aims at envisage leaf physiological responses in *Coffea* spp. subjected to supra-optimal temperatures, increased [CO_2], and water shortage conditions, contributing to this crop sustainability.

Keywords Agriculture · Climate changes · CO_2 · Coffee crop
Physiological heat impact · Warming · Water shortage

Introduction

Global climate change is one of the most notorious events of this century. Impacts from recent climate-related extremes, such as heat and cold waves, droughts and strong rainfall events, exposed the remarkable vulnerability of agricultural systems. Estimated future climate changes are expected to further amplify the existing climate-related risks and create new ones (IPCC 2013; Martins et al. 2017). There are several factors that contribute to this disturbing scenario, partly of which are still poorly known. The concentration increase of different greenhouse gases in the atmosphere, particularly carbon dioxide (CO_2), in among then. The air [CO_2] increase has been mainly caused by anthropogenic activities, such as the increased consumption of fossil fuels, changes in land use, linked to a global population rise, which has in turn contributed to a steady increase of air temperature (Buckeridge et al. 2007; ECDRP 2009; IPCC 2014). In fact, air [CO_2] increased from ca. 280 to 400 $\mu L\ L^{-1}$ from the beginning of the pre-industrial period to the present. Additionally, depending on the extent of measures to control future greenhouse

A. E. Leitão · A. I. Ribeiro-Barros · J. C. Ramalho
Plant-Environment Interactions and Biodiversity Lab (PlantStress&Biodiversity),
Linking Landscape, Environment, Agriculture and Food (LEAF), Instituto Superior
de Agronomia (ISA), Universidade de Lisboa (ULisboa), Quinta do Marquês,
Av. da República, 2784-505 Oeiras, Portugal

F. M. DaMatta
Departamento de. Biologia Vegetal, Universidade Federal Viçosa (UFV),
Viçosa, MG CEP: 36570-090, Brazil

gas emission scenarios, [CO_2] might reach between 421 and 936 $\mu L\ L^{-1}$ by 2100, with estimated parallel increases in global warming from 0.3–1.7 °C (best scenario) to 2.6–4.8 °C (worst case scenario without additional efforts to mitigate the increase of CO_2), when compared to the values observed by 1986–2005 (IPCC 2013, 2014). Irrespective of the cause/effect of increased [CO_2] in warming, it is indisputable that both air [CO_2] and temperature have been reaching new records every year. Such air temperature increase will have a widespread impact, leading to drastic changes in the air humidity, contributing as well to modify intra- and inter-annual rainfall regimes. These changes are based on more active hydrological cycles, changes in atmospheric and oceanic circulation, as well as increased water vapour retention capacity by the air (Buckeridge et al. 2007; CSIRO 2016). Altogether, these circumstances may lead to more frequent and extreme drought events.

In parallel, changes in air [CO_2] affects fundamental plant processes, with the potential to alter agronomic yields and the quality of products (Drake et al. 1997; Idso and Kimball 1997; Luo et al. 1999), with a positive role in crop production related to significant increases in net C-assimilation (Ainsworth and Rogers 2007; Kirschbaum 2011). These positive impacts are expected if the plants are capable of maintaining a sufficient strong sink strength to use the overproduced carbohydrates under high [CO_2], as observed in several plants, as loblolly pine (LaDeau and Clark 2001), orange sour (Idso and Kimball 1997) and grapevine (Moutinho-Pereira et al. 2009). Furthermore, by reducing photorespiration, [CO_2] increase will enhance net photosynthesis more at high than at low temperatures, thus with a possible offsetting of the negative impact of supra-optimal temperatures on yield (Polley 2002; DaMatta et al. 2010). Additionally, it has been reported clear mitigation effects of enhanced [CO_2] on the impact of supra-optimal temperatures in a few plant species, increasing tree growth and forest productivity (Boisvenue and Running 2006), as well as reinforcing plant functioning and protective mechanisms in coffee (Martins et al. 2016; Rodrigues et al. 2016).

Water deficits, often imposed concomitantly with increased temperature and high irradiation levels, is a strong limiting factor to plant growth, survival and productivity worldwide, causing significant constraints to the agricultural sector (Chaves et al. 2003). A conceptual definition of agricultural drought can be described as an extended period of deficient precipitation resulting in extensive damage to crops and yield losses (NDMC 2017). It is well established that, depending on stress duration and intensity, as well as on species/genotype tolerance, reduced water availability impairs a large number of physiological and metabolic processes in plants. Among them, the photosynthetic pathway is frequently affected, firstly due to stomatal closure (even at quite mild water deficits), and afterwards by non-stomatal impacts that arise with more severe stress imposition (Kaiser 1987; Chaves et al. 2003; Matos et al. 2010; Ramalho et al. 2014b). Growth and development of any culture depends directly on the performance of the photosynthetic metabolism. The decrease of C-assimilation will reduce sugar (and, therefore, energy) availability to sustain cellular metabolic and to achieve high productivity, with estimates of yield reductions as high as 50% for most major crop plants when subjected to important water restriction conditions (Wang et al. 2003). Moreover, in view of the predicted scenarios for climate changes, lower water availability conditions are likely to become even more important in the future in many world regions (Ramalho et al. 2014b), due to alteration of intra- and inter-annual rainfall patterns.

Coffee is a specific tropical crop, distributed from South and Central America to África and Southwest Asia regions, being cultivated in more than 80 countries, and supporting the livelihoods of about 25 million small farmers (Waller et al. 2007), and the economy of many of those countries. The world coffee trade is largely supported by only two species, *Coffea arabica* L. (Arabica type of coffee) and *Coffea canephora* Pierre ex A. Froehner (Robusta type of coffee), which together account for nearly 99% of the yielded coffee bean (Partelli et al. 2011; Davis et al. 2012). World annual production has been above 8 million tons of green coffee beans in the last years, generating an income of ca. US$ 173,000 million for the entire value chain of coffee (ICO 2014).

Several studies using mathematical models, based largely on temperature increase and on *C. arabica* characteristics, have predicted extensive impacts on production (Gay et al. 2006; Craparo et al. 2015), extinction of wild populations (Davis et al. 2012), rise to the incidence of pests and diseases (Magrach and Ghazoul 2015), reduction up to 50% of traditionally *C. arabica* cultivated areas already in 2050, and the need to shift the crop to new areas (Bunn et al. 2015b; Magrach and Ghazoul 2015), with heavier loss of climatic suitability to this at lower altitudes and high latitudes (Ovallle-Rivera et al. 2015). These problems may be further aggravated by the fact that actual plantations will be increasingly exposed to such new environmental conditions, since coffee plantations have a productive life span from 20 up to 50 years (Bunn et al. 2015a), depending on the possibility crop management and renewal. However, it must be underlined that such modelling predictions of severe impacts were mostly based on rising temperatures, without considering the recently reported mitigation effects of enhanced air [CO_2] and the resilience of the coffee crop to acclimate to stressful conditions (Martins et al. 2016; Rodrigues et al. 2016).

Given the facts described above, it becomes imperative to study the main mechanisms (with emphasis on the physiological, biochemical and molecular levels) involved in the acclimation/adaptation of different coffee genotypes, allowing them to withstand the impact caused by a constantly changing environment. This knowledge, namely coming from several ecophysiological indicators for different genotypes and their interaction with the environment, will help the selection of more tolerant individuals and will serve as tools for select/obtain new improved plants. Moreover, the use of adapted varieties, coupled with adequate agronomic management techniques, namely regarding a better water use efficiency, may contribute to the mitigation of environmental impacts and preservation of agricultural productivity (Dodd et al. 2011).

Water and Warming Constraints

Coffee leaf hydraulic architecture imposes a major constraint on the maximisation of the photosynthetic gas exchange of leaves, with a reducing effect on C-assimilation due to low stomatal conductance, followed by similar limitations associated with the mesophyll and biochemical constraints, although not related to inefficient RuBisCO functioning. This will limit the access of CO_2 to the carboxylation sites, thus, favouring the occurrence of relatively high photorespiration rates, which ultimately leads to further limitations to C-assimilation (Martins et al. 2014a; Ramalho et al. 2014a). Additionally, upon the imposition of mild water deficit conditions, some coffee genotypes show a fast and efficient stomatal closure, which will restrict the transpiration flow, but will further limit CO_2 flux into the leaf, reducing the photosynthetic rates, and, therefore, significantly limiting plant growth and productivity (Ronchi and DaMatta 2007).

Changes in rainfall seasonality could affect the growth and productivity of coffee crop, mainly in the planting and pre-flowering phases, as well as during fruit set and expansion/filling. At planting, the seedlings have a reduced foliar index area associated to immature roots, regarding its adaptation to the new soil environment, making them more prone to quickly suffer from water deficit. Despite the triggering effect of floral bud dormancy break caused by the first rains, the maintenance of low water availability in the pre-flowering phase may induce the production of abnormal flower buds or can cause their abscission. Furthermore, water shortage in the stage of fruit expansion/filling can also lead to appreciable productivity loss of and decreased bean quality (Crisosto et al. 1992; Camargo and Camargo 2001; Custódio et al. 2014). Furthermore, low soil water availability reduces cell turgidity, causing decreases in leaf development, and promotes leaf shed. This will reduce the photosynthetically active area, which further impacts plant growth and productivity (DaMatta and Ramalho 2006). As an example, two consecutive years (2014 and 2015) of severe rainfall reduction in the Espírito Santo State, Brazil, promoted average reductions of 15% in the first year, and of 48% in the 2nd year as regards *Coffea canephora* yields. Notably, significant aftereffects in the next year were expected in some cropping areas, due to the fragile plant status after such prolonged water deficit periods and the maintenance of high air temperatures, and also because plants were just removed in some areas, therefore decreasing the number of coffee plants and the extention of cropped area, with a first estimate of a ca. 18% decrease in 2017 in that region (CONAB 2017). Additionally, leaf shed may expose the fruits to a greater amount of solar radiation, which can accelerate their maturation and reduce the quality of the produced bean.

Another consequence to yield and bean quality related to a strong stomatal closure and low transpiration water flow from soil to the leaves, is related to the restriction of mineral nutrient absorption by roots and its translocation to all plant organs. Knowing that fruit are strong sinks, namely for minerals, such uptake limitation can promote deficiencies at leaf level, what might increase leaf senescence, and, therefore, decrease the potential for C-assimilation (Martinez et al. 2014), regarding, both the actual and next cropping season, since a defoliated plant has a much lower production capability.

In most situations, drought impacts can be greatly aggravated by supra-optimal air temperatures. Drought and unfavourable temperatures (sub and supra-optimal) are considered the main climatic limitations for coffee cultivation (DaMatta and Ramalho 2006; Ramalho et al. 2014a), although there are some intra- and inter-specific variations in regarding the tolerance limits. In general, for the coffee plants, extreme temperatures will impair the synthesis of proteins, lipids and carbohydrates, the metabolism of enzymes and hormones, whereas it strongly reduce the photosynthetic performance. Moreover, impacts at the membrane level could arise, namely in the lipid matrix components, with implications to all membrane bound reactions, as well as to membrane selectivity and cell compartmentalization. These are frequently reflected in increased loss of ions (given by electrolyte leakage), and extended lipoperoxidation of polyunsaturated fatty acids (Alonso et al. 1997; Queiroz et al. 1998; Campos et al. 2003; Dias et al. 2010; Scotti-Campos et al. 2014). Furthermore, such peroxidative chain reaction can promote ethylene synthesis, a growth regulator responsible for leaf and flower senescence (Finger et al. 2006; Dias et al. 2010).

Warming can further reduce coffee tree growth due to physiological/biochemical changes, namely by decreasing the photosynthetic capacity and increasing the respiratory rate (DaMatta and Ramalho 2006). Effectively, heat can induce a loss of assimilated carbon due to a proportionally greater increase in respiration and photorespiration relatively to photosynthesis (DaMatta et al. 2016). Furthermore, when submitted to high temperatures coffee plants undergo profound changes in development, flowering and fruit set processes. In fact, the frequent exposure to temperatures above 35 °C often leads to the production of abnormal flowers, as well as the abortion of flowers and fruits, compromising yield and quality of the coffee bean obtained (Camargo 1985). Also, supra-optimal temperatures will accelerate the development and maturation of fruits, leading to negative consequences on the final bean quality, as well as on the obtained beverage (Camargo 1985).

Increase of Air CO_2 Concentration

Plant growth and biomass production are closely linked to the rate of CO_2 uptake and assimilation via the photosynthetic pathway. In this sense, an increase of atmospheric [CO_2] can promote the rate of photosynthesis by increasing the carboxylation rate due to a higher CO_2 availability as substrate for carboxylation and because that will simultaneously reduce the photorespiration rate due to the competitive inhibition with O_2 over ribulose-1,5-bisphosphate carboxylase/oxygenase (RuBisCO) (Ramalho et al. 2013b; Ghini et al. 2015; Martins et al. 2016; Rodrigues et al. 2016). This has been shown to be frequently accompanied by increased water use efficiencies due to the combination of a reduction/maintenance of stomatal conductance together to the referred increase of net photosynthesis (Ramalho et al. 2013b; Rodrigues et al. 2016).

In many plant species, the photosynthetic rate may increase ca. 30–60% under 600–700 μL CO_2 L^{-1} in air, as compared to the rates obtained between 370 and 390 μL CO_2 L^{-1} (Ainsworth and Rogers 2007; Kirschbaum 2011; Taiz and Zeiger 2013). In accordance, increases within this range were observed in both *C. arabica* and *C. canephora* genotypes (Ramalho et al. 2013b; Ghini et al. 2015; DaMatta et al. 2016). Furthermore, it was recently reported that coffee genotypes from both *C. arabica* and *C. canephora* species can maintain their photosynthetic performance under the exposure to 37 °C along the entire diurnal period (Rodrigues et al. 2016). These responses reflects a tolerance to temperatures well above what might be expected, considering classical reports of negative impacts on photosynthesis above 25 °C, but fully agrees with the field conditions frequently found in coffee tropical cultivation areas (see DaMatta and Ramalho 2006). Additionally, the photochemical and biochemical performances of the photosynthetic apparatus were enhanced at such supra-optimal temperatures in the plants grown at enhanced air $[CO_2]$, when compared to the plants grown at normal air $[CO_2]$. When exposed to 42 °C all studied coffee genotypes were significantly affected in most photosynthetic parameters (particularly at the enzyme level), extended as well to the enzymes from the respiration pathway. However, such metabolic/functional impairments on the photosynthetic components were significantly mitigated in the plants grown under high $[CO_2]$ (Rodrigues et al. 2016). This was particularly clear on the photosystems functioning that showed a high heat tolerance, both at physical (energy capture) and photochemical (electron transport) processes (Rodrigues et al. 2016). This better functioning ultimately helped to lower the photoinhibition status of the photosynthetic apparatus (Martins et al. 2016). The greater metabolic performance was likely related to the reinforcement of protective and antioxidative mechanisms in response to supra-optimal temperatures up to 37 °C and a higher prevalence of those mechanisms at 42 °C in the plants grown under high $[CO_2]$ (Martins et al. 2016). This is of upmost importance, since an effective control of oxidative stress has proven to be decisive to coffee plant acclimation to several environmental stresses, namely, cold, high irradiance, drought and nitrogen starvation (Ramalho et al. 1998, 2014a; Fortunato et al. 2010; Batista-Santos et al. 2011; Cavatte et al. 2012). Such control of oxidative conditions is at least in part associated with protective molecules, namely heat shock protein 70 (HSP70), α-tocopherol, neoxanthin, lutein, carotenes and of raffinose family oligosaccharides (RFOs), as well as with higher activities of antioxidant enzymes, such as, Cu, Zn-superoxide dismutase, ascorbate peroxidase, glutathione reductase, and catalase. This was further linked to a large upregulation of genes related to other protective molecules, as early-light induced proteins (*ELIPs*) and chaperonins (*Chape 20* and *60*) (Martins et al. 2016). Notably, such network of response is in accordance to the proposed cross-talk between plant stress response mechanisms under abiotic stress conditions, aiming at controlling ROS formation and promoting its scavenging (Wang et al. 2004).

Another important aspect, although frequently neglected, is related to the maintenance of a mineral homeostasis under stress conditions, that is, that adequate levels and balance of minerals are kept. In coffee plants under adequate temperature, high air [CO_2] promoted only a moderate mineral "dilution" effect within the range of 7–25%, depending on the considered mineral and genotype (Martins et al. 2014b). The decrease of the photosynthetic potential (down-regulation) can be attributable to an N "dilution" in leaf tissues, related to a lower N-allocation to photosynthetic components (Ainsworth and Rogers 2007; Bader et al. 2010). Still, the changes observed in coffee plants likely reflected qualitative physiological changes rather than a nutrient deprivation (Thiec et al. 1995), since these plants were capable of maintaining significant higher metabolic functioning, without any sign of photosynthetic down-regulation (Ramalho et al. 2013b). Additionally, upon supra-optimal temperatures almost all of the macro- and micro-nutrient levels increased, thus offsetting the previous "dilution" effect found under adequate temperature conditions (25 °C). Such maintenance of mineral homeostasis could in turn have positive implication in plant thermotolerance due to mineral roles in cell metabolism. In fact, the triggering of antioxidative mechanisms in coffee plants was shown to greatly depend on leaf availability of N (Ramalho et al. 1998, 2000), Cu, Zn, Fe and Mn (Ramalho et al. 2013a), with Mn having as well a key role in the photosystem II (PSII) functioning. Therefore, the maintenance or increase of PSII performance and antioxidant enzyme activities (Martins et al. 2016; Rodrigues et al. 2016) are in accordance with the observed mineral changes (Martins et al. 2014b). This further suggests that the coffee plant could cope, in some extent, with the global warming in what concerns its mineral dynamics, although additional attention should be paid to adequate fertilization management.

Altogether, this resulted in a higher photochemical energy use ability and a lower photorespiration in the plants grown with higher air CO_2 availability at any temperatures analysed, preventing an energy overcharge and decreasing the probability of the formation of highly reactive molecular species of chlorophyll and O_2. Such higher photosynthetic use of energy under enhanced atmospheric [CO_2] should, in fact, be regarded as a primary mean of protection, which, together with the reinforcement of several protective molecules, and the maintenance of adequate leaf mineral contents, allowed the plant to sustain higher metabolic performance even at temperatures as high as 42 °C (Martins et al. 2016; Rodrigues et al. 2016).

In summary, coffee plant can cope with global warming in a higher extent than previously expected, with air [CO_2] constituting a key factor to coffee physiological thermotolerance, what is clearly relevant in the context of predicted scenarios of increased [CO_2] and global warming. Nevertheless, impairments will be imposed on this crop by supra-optimal temperatures and altered water availability patterns, therefore several actions must be taken into account to mitigate climate changes impact, including the adoption of adaptation strategies aiming at obtaining new improved genotypes regarding heat and drought tolerance aspects while maintain bean (and cup) quality.

Final Considerations

The predicted scenarios of global warming and increasing water deficit have already showed alarming signs to coffee production (Bunn et al. 2015b; Magrach and Ghazoul 2015; Martins et al. 2015). In fact, projections pointed to 50% drop in world yielded coffee bean by the year 2050, raising important concerns regarding the sustainability of coffee crop in many of its traditional cultivation areas. However, recent findings have shown that increased air [CO_2] can potentially significantly mitigate heat impacts on this crop, by increasing plant vigour and defense mechanisms. Therefore, strong multi-disciplinary and multi-institutional efforts have to be made, in order to carry out integrative studies to identify and characterize plant response mechanisms to estimated environmental changes that are predicted to occur (often simultaneously), namely, increased air [CO_2], warming, and reduced water availability. This will allow the identification of key traits used in the selection of better performance plants, and to assist breeding programs. These, together with the implementation of mitigation measures will be indispensable for an adequate crop management under such new environmental conditions, ultimately aiming at contribute to guarantee this crop sustainability.

Acknowledgements This work was partly supported by Portuguese national funds from Fundação para a Ciência e a Tecnologia through the projects PTDC/AGR-PRO/3386/2012, the research units UID/AGR/04129/2013 (LEAF) and UID/GEO/04035/2013 (GeoBioTec). Brazilian funding from CAPES (grants PDSE: 000427/2014-04, W.P. Rodrigues; 0343/2014-05, M. Q. Martins), CNPq and Fapemig (fellowships to F. M. DaMatta, F. Partelli and E. Campostrini) are also greatly acknowledged.

References

Ainsworth, E. A., & Rogers, A. (2007). The response of photosynthesis and stomatal conductance to rising [CO_2]: Mechanisms and environmental interactions. *Plant, Cell and Environment, 30,* 258–270.

Alonso, A., Queiroz, C. S., & Magalhães, A. C. (1997). Chilling stress leads to increased cell membrane rigidity in roots of coffee (*Coffea arabica* L.) seedlings. *Biochimica et Biophysica Acta, 1323,* 75–84.

Bader, M. K.-F., Siegwolf, R., & Körner, C. (2010). Sustained enhancement of photosynthesis in mature deciduous forest trees after 8 years of free air CO_2 enrichment. *Planta, 232,* 1115–1125.

Batista-Santos, P., Lidon, F. C., Fortunato, A., Leitão, A. E., Lopes, E., Partelli, F., et al. (2011). The impact of cold on photosynthesis in genotypes of *Coffea* spp.—Photosystem sensitivity, photoprotective mechanisms and gene expression. *Journal of Plant Physiology, 168,* 792–806.

Boisvenue, C., & Running, S. W. (2006). Impacts of climate change on natural forest productivity—Evidence since the middle of the 20th century. *Global Change Biology, 12,* 862–882.

Buckeridge, M. S., Mortari, L. C., & Machado, M. R. (2007). Respostas fisiológicas às mudanças climáticas: Alterações no balanço de carbono nas plantas podem afetar o ecossistema? In Fenologia-Ferramenta para conservação e manejo de recursos vegetais arbóreos (Editores técnicos), Embrapa Floresta, pp. 1–13.

Bunn, C., Läderach, P., Pérez Jimenez, J. G., Montagnon, C., & Schilling, T. (2015a). Multiclass classification of agro-ecological zones for arabica coffee: An improved understanding of the impacts of climate change. *PLoS ONE, 10*(10), e0140490.

Bunn, C., Läderach, P., Rivera, O. O., & Kirschke, D. A. (2015b). Bitter cup: Climate change profile of global production of Arabica and Robusta coffee. *Climatic Change, 129,* 89–101.

Camargo, A. P. (1985). O clima e a cafeicultura no Brasil. *Informe Agropecuário, 11,* 13–26.

Camargo, A. P., & Camargo, M. B. P. (2001). Definition and outline for the phenological phases of arabic coffee under Brazilian tropical conditions. *Bragantia, 60,* 65–68.

Campos, P. S., Quartin, V., Ramalho, J. C., & Nunes, M. A. (2003). Electrolyte leakage and lipid degradation account for cold sensitivity in leaves of *Coffea* sp. plants. *Journal of Plant Physiology, 160,* 283–292.

Cavatte, P. C., Oliveira, A. A., Morais, L. E., Martins, S. C., Sanglard, L. M., & DaMatta, F. M. (2012). Could shading reduce the negative impacts of drought on coffee? A morphophysiological analysis. *Physiologia Plantarum, 114,* 111–122.

Chaves, M. M., Maroco, J. P., & Pereira, J. S. (2003). Understanding plant responses to drought—From genes to the whole plant. *Functional Plant Biology, 30,* 239–264.

CONAB (Companhia Nacional de Abastecimento). (2017). Acompanhamento da safra brasileira de café 2017. Primeiro Levantamento, observatório agrícola: 1–99.

Craparo, A. C. W., Van Asten, P. J. A., Läderach, P., Jassogne, L. T. P., & Grab, S. W. (2015). *Coffea arabica* yields decline in Tanzania due to climate change: Global implications. *Agricultural and Forest Meteorology, 207,* 1–10.

Crisosto, C. H., Grantz, D. A., & Meinzer, F. C. (1992). Effects of water deficit on flower opening in coffee (*Coffea arabica* L.). *Tree Physiology, 10,* 127–139.

CSIRO (Commonwealth Scientific and Industrial Research Organisation). (2016). State of the climate 2016. The biennial CSIRO and Bureau of Meteorology State of the Climate report. www.csiro.au. Accessed February 17, 2017.

Custódio, A. A. P., Lemos, L. B., Mingotte, F. L. C., Barbosa, J. C., Pollo, G. Z., & Santos, H. M. (2014). Florescimento de cafeeiros sob manejos de irrigação faces de exposição solar e posições na planta. *Coffee Science, 9,* 245–257.

DaMatta, F. M., Godoy, A. G., Menezes-Silva, P. E., Martins, S. C. V., Sanglard, L. M., Morais, L. E., et al. (2016). Sustained enhancement of photosynthesis in coffee trees grown under free-air CO_2 enrichment conditions: Disentangling the contributions of stomatal, mesophyll, and biochemical limitations. *Journal of Experimental Botany, 67,* 341–352.

DaMatta, F. M., Grandis, A., Arenque, B. C., & Buckeridge, M. S. (2010). Impacts of climate changes on crop physiology and food quality. *Food Research International, 43,* 1814–1823.

DaMatta, F. M., & Ramalho, J. D. C. (2006). Impacts of drought and temperature stress on coffee physiology and production: A review. *Brazilian Journal of Plant Physiology, 18,* 55–81.

Davis, A. P., Gole, T. W., Baena, S., & Moat, J. (2012). The impact of climate change on indigenous Arabica coffee (*Coffea arabica*): Predicting future trends and identifying priorities. *PLoS ONE, 7,* e47981.

Dias, A. S., Barreiro, M. G., Campos, P. S., Ramalho, J. C., & Lidon, F. C. (2010). Wheat cellular membrane thermotolerance under heat stress. *Journal of Agronomy and Crop Science, 196,* 100–108.

Dodd, I. C., Whalley, W. R., Ober, E. S., & Parry, M. A. J. (2011). Genetic and management approaches to boost UK wheat yields by ameliorating water deficits. *Journal of Experimental Botany, 62,* 5241–5248.

Drake, B. G., Gonzàlez-Meler, M. A., & Long, S. P. (1997). More efficient plants: A consequence of rising atmospheric CO_2? *Annual Review of Plant Physiology and Plant Molecular Biology, 48,* 609–639.

ECDRP (European Commission Directorate-General Regional Policy). (2009). *The climate change challenge for european regions—Regions 2020.* Brussels.

Finger, F. L., Santos, V. R., Barbosa, J. G., & Barros, R. S. (2006). Influence of temperature on respiration, ethylene production and longevity of *Consolida ajacis* inflorescences. *Bragantia, 65,* 363–368.

Fortunato, A., Lidon, F. C., Batista-Santos, P., Leitão, A. E., Pais, I. P., Ribeiro, A. I., et al. (2010). Biochemical and molecular characterization of the antioxidative system of *Coffea* sp. under cold conditions in genotypes with contrasting tolerance. *Journal of Plant Physiology, 167,* 333–342.

Gay, C., Estrada, F., Conde, C., Eakin, H., & Villers, L. (2006). Potential impacts of climate change on agriculture: A case of study of coffee production in Veracruz, Mexico. *Climatic Change, 79,* 259–288.

Ghini, R., Torre-Neto, A., Dentzien, A. F. M., Guerreiro-Filho, O., Lost, R., Patrício, F. R. A., et al. (2015). Coffee growth, pest and yield responses to free-air CO_2 enrichment. *Climatic Change, 132,* 307–320.

ICO. (2014). *World coffee trade (1963–2013): A review of the markets, challenges and opportunities facing the sector.* ICC (International Coffee Council), 111–115 Rev. 1, 29 pp. Available online at: http://www.ico.org/show_news.asp?id=361. Accessed July 2016.

Idso, S. B., & Kimball, B. A. (1997). Effects of long-term atmospheric CO_2 enrichment on the growth and fruit production of sour orange trees. *Global Change Biology, 3,* 89–96.

IPCC. (2013). Climate change 2013: The physical science basis. summary for policymakers, Technical summary and frequent asked questions. In T. F. Stocker, D. Qin, G.-K. Plattner, M. M. B. Tignor, S. K. Allen, J. Boschung, A. Nauels, Y. Xia, V. Bex, P. M. Midgley (Eds.), Part of the working group i contribution to the fifth assessment report of the Intergovernmental Panel on Climate Change (p. 203). ISBN 978-92-9169-138-8.

IPCC. (2014). Climate change 2014: Mitigation of climate change. In O. Edenhofer, R. Pichs-Madruga, Y. Sokona, E. Farahani, S. Kadner, K. Seyboth, A. Adler, I. Baum, S. Brunner, P. Eickemeier, B. Kriemann, J. Savolainen, S. Schlömer, C. von Stechow, T. Zwickel, & J. C. Minx (Eds.), Contribution of working group iii to the fifth assessment report of the Intergovernmental Panel on climate change (p. 1435).Cambridge, UK: Cambridge University Press. ISBN 978-1-107-65481-5.

Kaiser, W. M. (1987). Effect of water deficit on photosynthetic capacity. *Physiologia Plantarum, 71,* 142–149.

Kirschbaum, M. U. F. (2011). Does enhanced photosynthesis enhance growth? Lessons learned from CO_2 enrichment studies. *Plant Physiology, 155,* 117–124.

LaDeau, S. L., & Clark, J. S. (2001). Rising CO_2 levels and the fecundity of forest trees. *Science, 292,* 95–98.

Luo, Y., Reynolds, J., Wang, Y., & Wolfe, D. (1999). A search for predictive understanding of plant responses to elevated [CO_2]. *Global Change Biology, 5,* 143–156.

Magrach, A., & Ghazoul, J. (2015). Climate and pest-driven geographic shifts in global coffee production: Implications for forest cover, biodiversity and carbon storage. *PLoS ONE, 10,* e0133071.

Martinez, H. E. P., Clemente, J. M., Lacerda, J. S., Neves, Y. P., & Pedrosa, A. W. (2014). Nutrição mineral do cafeeiro e qualidade da bebida. *Revista Ceres, 61,* 838–848.

Martins, L. D., Eugenio, F. C., Rodrigues, W. N., Brinate, S. V. B., Colodetti, T. V., Amaral, J. F. T., et al. (2015). A bitter cup: The estimation of spatial distribution of carbon balance in *Coffea* spp. plantations reveals increased carbon footprint in tropical regions. *Plant Soil Environment, 61,* 544–552.

Martins, L. D., Tomaz, M. A., Lidon, F. C., DaMatta, F. M., & Ramalho, J. C. (2014a). Combined effects of elevated [CO_2] and high temperature on leaf mineral balance in *Coffea* spp. plants. *Climatic Change, 126,* 365–379.

Martins, M. Q., Fortunato, A. F., Rodrigues, W. P., Partelli, F. L., Campostrini, E., Lidon, F. C., et al. (2017). Selection and validation of reference genes for accurate RT-qPCR data normalization in *Coffea* spp. under a climate changes context of interacting elevated [CO_2] and temperature. *Frontiers in Plant Science, 8,* art 307.

Martins, M. Q., Rodrigues, W. P., Fortunato, A. S., Leitao, A. E., Rodrigues, A. P., Pais, I. P., et al. (2016). Protective response mechanisms to heat stress in interaction with high [CO_2] conditions in *Coffea* spp. *Frontiers in Plant Science, 7,* art 947.

Martins, S. V. C., Galmés, J., Cavatte, P. C., Pereira, L. F., Ventrella, M. C., & DaMatta, F. M. (2014b). Understanding the low photosynthetic rates of sun and shade coffee leaves: Bridging the gap on the relative roles of hydraulic, diffusive and biochemical constraints to photosynthesis. *PLoS ONE, 9,* e95571.

Matos, M. C., Campos, P. S., Passarinho, J. A., Semedo, J. N., Marques, N. M., Ramalho, J. C., et al. (2010). Drought effect on photosynthetic activity, osmolyte accumulation and membrane integrity of two *Cicer arietinum* genotypes. *Photosynthetica, 48,* 303–312.

Moutinho-Pereira, J., Gonçalves, B., Bacelar, E., Cunha, J. B., Coutinho, J., & Correia, C. M. (2009). Effects of elevated CO_2 on grapevine (*Vitis vinifera* L.): Physiological and yield attributes. *Vitis, 48,* 159–165.

NDMC (The National Drought Mitigation Center). (2017). Drought Impact Reporter. www.drought.unl.edu. Accessed March 9, 2017.

Ovalle-Rivera, O., Läderach, P., Bunn, C., Obersteiner, M., & Schroth, G. (2015). Projected shifts in *Coffea arabica* suitability among major global producing regions due to climate change. *PLoS ONE, 10,* e0124155.

Partelli, F. L., Batista-Santos, P., Scotti-Campos, P., Pais, I. P., Quantin, V. L., Vieira, H. D., et al. (2011). Characterization of the main lipid components of chloroplast membranes and cold induced changes in *Coffea* spp. *Environmental and Experimental Botany, 74,* 194–204.

Polley, H. W. (2002). Implications of atmospheric and climate change for crop yield. *Crop Science, 42,* 131–140.

Queiroz, C. G., Alonso, A., Mares-Guia, M., & Magalhães, A. C. (1998). Chilling induced changes in membrane fluidity and antioxidant enzyme activities in *Coffea arabica* L. roots. *Biologia Plantarum, 41,* 403–413.

Ramalho, J. C., Campos, P. S., Teixeira, M., & Nunes, M. A. (1998). Nitrogen dependent changes in antioxidant system and in fatty acid composition of chloroplast membranes from *Coffea arabica* L. plants submitted to high irradiance. *Plant Science, 135,* 115–124.

Ramalho, J. C., DaMatta, F. M., Rodrigues, A. P., Scotti-Campos, P., Pais, I., Batista-Santos, P., et al. (2014a). Cold impact and acclimation response of *Coffea* spp. plants. *Theoretical and Experimental Plant Physiology, 26,* 5–18.

Ramalho, J. C., Fortunato, A. S., Goulao, L. F., & Lidon, F. C. (2013a). Cold-induced changes in mineral content in *Coffea* spp. leaves—Identification of descriptors for tolerance assessment. *Biologia Plantarum, 57,* 495–506.

Ramalho, J. C., Pons, T., Groeneveld, H., Azinheira, H. G., & Nunes, M. A. (2000). Photosynthetic acclimation to high light conditions in mature leaves of *Coffea arabica* L.: Role of xanthophylls, quenching mechanisms and nitrogen nutrition. *Australian Journal of Plant Physiology, 27,* 43–51.

Ramalho, J. C., Rodrigues, A. P., Semedo, J. N., Pais, I. P., Martins, L. D., Simões-Costa, M. C., et al. (2013b). Sustained photosynthetic performance of *Coffea* spp. under long-term enhanced [CO_2]. *PLoS ONE, 8,* e82712.

Ramalho, J. C., Zlatev, Z. S., Leitão, A. E., Pais, I. P., Fortunato, A. S., & Lidon, F. C. (2014b). Moderate water stress causes different stomatal and non-stomatal changes in the photosynthetic functioning of *Phaseolus vulgaris* L. genotypes. *Plant Biology, 16,* 133–146.

Rodrigues, W. P., Martins, M. Q., Fortunato, A. S., Rodrigues, A. P., Semedo, J. N., Simoes-Costa, M. C., et al. (2016). Long-term elevated air [CO_2] strengthens photosynthetic functioning and mitigates the impact of supra-optimal temperatures in tropical *Coffea arabica* and *C. canephora* species. *Global Change Biology, 22,* 415–431.

Ronchi C. P., & DaMatta F. M. (2007). Aspectos fisiológicos do café conilon. In R. G. Ferrão, A. F. A.Fonseca, S. M. Bragança, M. A. G. Ferrão, & L. H. De Muner (Eds.), *Café Conilon* (pp. 95–115) Vitória: Seag/Incaper.

Scotti-Campos, P., Semedo, J. N., Pais, I., Oliveira, M., Passarinho, J., & Ramalho, J. C. (2014). Heat tolerance of Portuguese old bread wheat varieties. *Emirates Journal of Food and Agriculture, 26,* 170–179.

Taiz, L., & Zeiger, E. (2013). *Fisiologia Vegetal*. Porto Alegre, Brazil: Artmed.

Thiec, D. L., Dixon, M., Loosveldt, P., & Garrec, J. P. (1995). Seasonal and annual variations of phosphorus, calcium, potassium and manganese contents in different cross-sections of *Picea abies* (L.) Karst. needles and *Quercus rubra* L. leaves exposed to elevated CO_2. *Trees, 10,* 55–62.

Waller, J. M., Bigger, M., & Hillocks, R. J. (2007). Coffee pests, diseases and their management. In J. M. Waller, M. Bigger, & R. J. Hillocks (Eds.), *World coffee production* (pp. 17–40, Chapter 2) Egham, Surrey, BC: CAB International.

Wang, W., Vinocur, B., & Altman, A. (2003). Plant responses to drought, salinity and extreme temperatures: Towards genetic engineering for stress tolerance. *Planta, 218,* 1–14.

Wang, W., Vinocur, B., Shoseyov, O., & Altman, A. (2004). Role of plant heat-shock proteins and molecular chaperones in the abiotic stress response. *Trends in Plant Science, 9,* 244–252.

How Can Global Change Affect Insect Population Dynamics in Mediterranean Ecosystems? A Case Study with Pine Shoot Beetle and Pine Processionary Moth

Teresa Maria Vasconcelos and Isabel Maria Duarte

Abstract Many insect species are able to cause important damage in agriculture, forestry and human health due to their ability to feed on plants and animals tissues or transmit fungi and other diseases. Entomologists have long recognised that periods of atypical weather conditions, such as drought, excessive precipitation or unusually hot or cold weather, have great impact on insect development. Understanding the effects of climate on ecological processes has become increasingly important in entomological research as a result of global warming, which appears to be affecting their geographic ranges and population dynamics. In Mediterranean ecosystems, pine shoot beetle, *Tomicus destruens* Woll, and pine processionary moth, *Thaumetopoa pityocampa* (Denis & Schiff), are examples of species that origin problems to man. *Tomicus destruens* is a phytophagus insect responsible for several damages on maritime pine (*Pinus pinaster* Aiton) stands during its feeding on needles and its capacity to inoculate fungi and other decaying biotic agents. *Thaumetopoea pityocampa* is also responsible for damages in *P. pinaster* stands and can provoke strong allergic reactions due the presence of urticains hairs. Both species have a physiological response to the temperature variation, which is observed in the life cycle duration and in the mortality rate. The present study aims to highlight the effect of global change/global warming in the bioecology of those Mediterranean species and their implication in forest productivity and human health.

T. M. Vasconcelos (✉) · I. M. Duarte
College of Agriculture, Polytechnic of Coimbra, Bencanta,
3045-601 Coimbra, Portugal
e-mail: tvasconcelos@esac.pt

I. M. Duarte
e-mail: iduarte@esac.pt

T. M. Vasconcelos · I. M. Duarte
Research Center for Natural Resources, Environment and Society Linking Landscape
(CERNAS), Polytechnic of Coimbra, Coimbra, Portugal

F. Alves et al. (eds.), *Theory and Practice of Climate Adaptation*,
Climate Change Management, https://doi.org/10.1007/978-3-319-72874-2_27

Keywords Climate change · Mediterranean ecosystems · Insect dynamics
Tomicus piniperda · *T. destruens* · *Thaumetopoea pityocampa*

Introduction

The genus *Pinus,* family Pinaceae, is widely distributed in the Northern hemisphere. In Mediterranean ecosystems, characterised by hot and dry summers, warm and wet winters and low amount of nutrients (Fouseki and Margaris 1981; Rapp et al. 1999), pine trees play an important role owing their capacity to form the dominant vegetation cover due to their ability to prevail in areas with adverse climatic and edaphic conditions (Bond 1989; Reich et al. 1995; Barbéro et al. 1998). Pine timber is extensively used in the naval and furniture industries, saw milling, particleboard and general construction, conferring an economic importance to this genus.

Maritime pine, *P. pinaster* Ait., is a native species of Portugal (Capelo and Aguiar 2005) that spreads naturally in the Mediterranean regions of France, Spain and Italy (subspecies *P. pinaster mesogeensis)* and in the Atlantic influenced regions of Portugal, Spain and France with the subspecies *P. pinaster atlantica* (Barbéro et al. 1998).

The relationship between herbivorous insects and their host plants has captured human attention since antiquity. The fact that insects are the most abundant animal group on earth and that half of them are herbivorous explains the importance of plant-insect relationships in biology. Throughout history, certain herbivorous insects have attracted attention by periodically devastating the plants upon which human populations depend (Miller 1986).

In Portugal, *P. pinaster* has a numerous entomofauna associated being pine shoot beetles, *Tomicus piniperda* L./*T. destruens* Wolland, and pine processionary moth, *Thaumetopoea pityocampa* Schiff, a threat to timber production and public health, respectively.

The development of both trees and insects are dependent on several environmental factors, such as, light, humidity and temperature but the latter has a greatest effect on insect development rates, due to their cold-blooded nature, and on the tree growth. Each tree or insect developmental stage requires a certain amount of heat units that accumulate faster under warmer conditions.

The understanding of the insect development plays a crucial role in insect pest management, since the timing of decision is determinant for the success of the management.

The aim of this study is to highlight the effect of global change/global warming in the bioecology of those Mediterranean species and their implication in forest productivity and human health.

Bioecology of the Insects

Pine Shoot Beetles—Tomicus *spp.*

Pine shoot beetles are bark beetles, insects that belong to the order Coleoptera, family *Scolytinae* that can seriously damage pinewood mainly due to its localisation into the bark of the host tree. Damage is done during the oviposition and development of the larvae, which weaken trees and increase mortality (Chararas 1962; Langström et al. 1999), and also for their capacity to introduce associated blue-staining fungi in the host tree (Solheim et al. 1992; Lieutier et al. 1995) depreciating timber quality (Nylinder et al. 2000; Vasconcelos 2005).

As bark beetles attack weakened or dead trees, they play an important ecological role in forest ecosystems initiating wood decomposition. They are one of the very few insect families in which the adult can penetrate the protective outer bark of woody plants, facilitating the introduction of pathogenic and decay-causing fungi essential to recycle nutrients from these long-lived plants. Due to their capacity to kill living trees, bark beetles act as natural thinning elements, eliminating weak trees and thus promoting growth and vigour on the remaining ones (Berryman et al. 1984). In addition, as they influence age, size and species distribution, they are a significant factor in forest succession. It is believed that the interaction between these species and trees started 200 million years ago (Christiansen et al. 1987).

If a surplus of suitable non-resistant breeding substrate is available, often as a result of drought or wind, bark beetles outbreak may occur (Berryman 1988). In these conditions, populations can rapidly increase and reach the critical threshold of attack density of the trees, (Langström et al. 1992; Lieutier et al. 1995) becoming forest pests and able to kill healthy trees (Christiansen et al. 1987).

Among bark beetles the genus *Tomicus* Latreille 1802 (Coleoptera: Curculionidae: Scolytinae) contains six described species known to infest pine stands worldwide (Wood and Bright 1992): *T. brevipilosus* Eggers 1929, *T. puellus* Reitter 1894, *T. pilifer* Spessivtsev 1919, *T. minor* Harting 1834, *T. piniperda* L. 1758 and *T. destruens* Wollaston 1865 and a new undescribed species (*Tomicus* sp. nov.), which develops on *P. yunnanensis* Franchet (Duan et al. 2004). The first three species and *Tomicus* sp. nov. are only known in Asia (Wood and Bright 1992), *T. piniperda* and *T. minor* are found throughout Eurasia and the remaining species around the Mediterranean Basin (Carle 1974; Mendel et al. 1985; Guerrero et al. 1997; Kerdelhué et al. 2002). As the genus *Tomicus* includes seven species among which six are present in Asia and four probably restricted to that region, the specific diversity is highest in Asia, suggesting this region as the centre of origin for the genus (Duan et al. 2004). *Tomicus* species differ from most other bark beetles species because of their unique post-emergence behaviour, consisting in maturation-feeding inside live shoots of healthy pine trees (Lieutier et al. 2003), and their monovoltinisme (one generation/year).

Tomicus piniperda is known to cause economic damage to pine species in Europe (Langström and Hellqvist 1990; Lieutier 1991), America (Czokajlo et al.

1997) and in the Mediterranean area (Ferreira and Ferreira 1990; Ghaioule et al. 1998; Ben-Jamâa et al. 2000). They can be a serious pest problem when they: (i) feed on pine shoots causing growth losses and top damage (Langström and Hellqvist 1990; Czokajlo et al. 1997), (ii) kill healthy trees (Langström et al. 2001) and (iii) introduce associated blue-stain fungi (Solheim et al. 1992) that decrease the quality of timber (Nylinder et al. 2000).

Trunk attack occurs in early spring (Chararas 1962; Sauvard et al. 1987), when temperatures begin to exceed 12 °C where they mate and develops until April—May (Vasconcelos and Duarte 2015). After emergence progeny, adults remain in shoots (Chararas 1962; Langström 1983; Sauvard et al. 1987) until temperature drop below freezing (Petrice et al. 2002). Afterwards, the adults begin to exit the shoots and walk along the branches and trunk to the base of the pine tree (Petrice et al. 2002), where they tunnel into outer bark and spend the winter. Nevertheless, in mild climates as southern Europe where winter temperatures seldom drop below freezing, *T. piniperda* adults overwinter in the shoots (Chararas 1962; Langström 1980).

Tomicus destruens is reported mainly in Mediterranean areas, such as Algeria (Chakali 2005), Morocco (Ghaioule et al. 1998), Portugal (Vasconcelos et al. 2005), Spain (Gallego et al. 2004), southern France (Kerdelhué et al. 2002) and Italy (Faccoli et al. 2005). These species perpetrate the trunk attack between October and November or in January, depending on wheather conditions and the progeny emerges in January—February or in April (Vasconcelos and Duarte 2015). As far as it is known, this specie does not overwinter in shoots, remaining there until it starts the reproductive activity (Fig. 1).

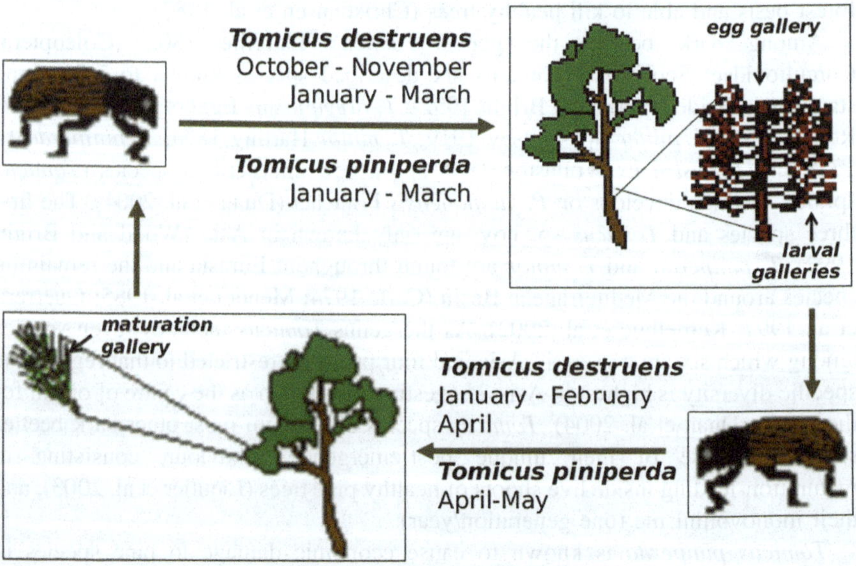

Fig. 1 Dispersal and sexual maturation flights of *Tomicus* spp.

Processionary Moth—Thaumetopoea Pityocampa

The major defoliator of *Pinus* genus inhabiting in Portugal and in the western Mediterranean basin is the pine processionary moth, *T. pityocampa* (Fig. 2). It severely defoliate, especially young stands, which can die if the attack occurs in successive years (Van Halder 2002). Accordingly, it is responsible for high productivity losses in Mediterranean stands (Arnaldo et al. 2010) and for the increase of host susceptibility towards the attack of another pests.

The life cycle of this defoliator is dependent on temperature and water stress (Arnaldo et al. 2011). The larval development typically occurs in Autumn-Winter, thus the caterpillars begin to feed on pine needles in September, undergoing the five stages of development. In late Autumn, they weave a nest to be shelted and survive the winter. Moreover, they produce small urticating hairs that are shed and can become airborne. Once in contact with the skin, the hairs can cause irritation, urticaria and papular rashes to humans, domestic animals and mammals in general. At the end of the winter season, that determines the end of larval development, the larvae burrow and leave the nests to pupate in the soil. In this trajectory they form a procession, after which the common name of this species was given. Pupae can diapause for variable period of time, namely from several months to several years (Devkota and Schmidt 1990). Under hot and dry summers, typical of Mediterranean regions, *T. pityocampa* has a long diapause and late emergencies, thus avoiding the excessively high summer temperatures. Conversely, in colder climates, the survival strategy is based on short diapauses and precocious emergencies, in this way completing the larval development before the temperature reaches negative values (Battisti et al. 2005).

Depending on the environmental conditions, adults hatch since late June to September (Pimentel et al. 2012).

Fig. 2 T*haumetopoea pityocampa* (processionary moth): **a** posture (cylindrical egg masses covered with scales, around the needles); **b** nest; **c** fifth-instar caterpillar in a 'procession' towards the soil; **d** male moth

Climate Alterations Pressure

The effects of temperature on the development time and brood development on insect populations has already been analysed (Chararas 1962; Sauvard 1989; Petrice et al. 2002; Régnière et al. 2012), which is crucial to understand the ecology of insect life histories. Temperature dependent phenology allows to be aware of the impacts of this factor on the geographical distribution, population dynamics and management of insects. The measurement of insect developmental, survival and reproductive responses to temperature poses practical challenges because of their modality, variability among individuals and high mortality near the lower and upper threshold temperatures. In the present work those aspects are analised in relevant Mediterranean insects (*Tomicus* spp. and *T. pityocampa*).

Pine Shoot Beetles—Tomicus *Spp.*

A study carried out in Portugal (Vasconcelos 2005), revealed that when temperature increase, the development rate of both *T. piniperda* and *T. destru*ens species diminish (Table 1).

Temperature had a significant positive effect on the survival rate of both *T. piniperda* and *T. destruens*, having a higher survival rate with lower variability at higher controlled temperatures (Vasconcelos 2005) (Fig. 3).

Under warmer conditions, such as those occurring in Portugal (Fig. 4), *T. destruens* replaces *T. piniperda* and became the most abundant and harmful species.

The same situation has also been reported in other Mediterranean areas (Faccoli et al. 2005; Gallego et al. 2004).

Processionary Moth—Thaumetopoea Pityocampa

The global warming registred over the past three decades affected the bioecology of this species, namely; (i) reduced larval mortality (Battisti et al. 2005); (ii) induced

Table 1 Effects of temperature in development rate of *Tomicus piniperda* and *T. destruens*

T. piniperda		*T. destruens*	
Temperature (°C)	Development (days)	Temperature (°C)	Development (days)
13	210 ± 5	11	210 ± 14
16	120 ± 4	15	150 ± 12
17	92 ± 12	17	112 ± 10
18	90 ± 5	20	90 ± 18
		22	61 ± 10

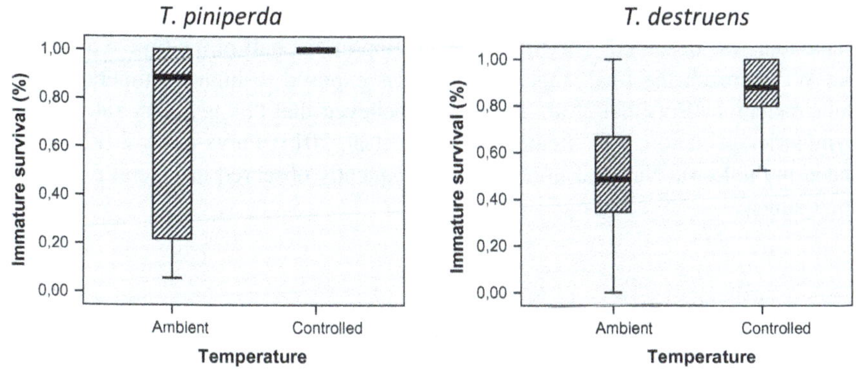

Fig. 3 Immature survival rate of *Tomicus piniperda and T. destruens* under ambient or controlled temperature

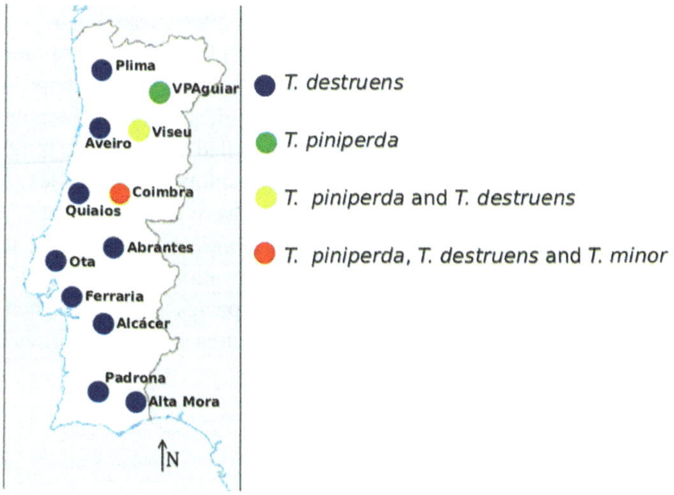

Fig. 4 Geographical distribution of pine shoot beetle species in Portugal (Vasconcelos and Duarte 2015)

behavioural changes, such as a long diapause and late emergencies protecting themselves from excessively high summer temperatures (Rousselet et al. 2010) and (iii) broadened geographical distribution, both in latitude and in altitude (Netherer and Schof 2010).

In the Portuguese National Forest of Leiria, a *T. pityocampa* population revealed an unprecedented behaviour. In 1997, a larval development was firstly recorded from June to September and pupae from September to May (Pimentel et al. 2006). This population was named as, summer population (SP); while the other, with the larval development in Autumn/Winter, as winter population (WP). Later, in 2009, in isolated trees of northern Portugal, under Atlantic influence, a similar deviant behaviour was observed. The SP larvae develop in just half of the time required by the WP to reach the final stage, because it is exposed to higher temperature and solar radiation (Pimentel et al. 2006). It is believed that this is a very rare case of sympatric speciation, still incipient (Santos et al. 2011; Paiva et al. 2012), originated in the Leiria National Forest and subsequently observed in several regions of the country.

Predictive Models

Synthesising, various factors, such as climate and host availability influence the geographical distribution of these species. As these pests are sensitive to tiny temperature variations, they are potentially able to undergo rapid distribution shifts. Consequently, as a result of current global warming, new economic threats are posed to pine stands in Europe and around the Mediterranean Basin.

In 2012, Horn et al. investigates the influence of climatic and host factors on the geographical distributions of both *Tomicus* species to establish maps of suitable areas, correlating these occurrence with meteorological records (variables describing climate). The bioclimatic variables were modelled and used to build potential species distribution maps. Within the climate descriptors, these species distribution were mostly correlated with temperature. They were shown to have parapatric distributions and opposite climate demands: *T. piniperda* occurring under cold climate and *T. destruens* in warmer temperature locations.

The predicted potential distribution of both species is useful when applying management strategies, and forecast the potential area infested as a result of temperature warming.

Discussion

Temperature is an important abiotic factor in the life cycle of *T. piniperda* and *T. destruens* because the effect on the development time and on the increase of survival rate (Bakke 1968; Vasconcelos 2005). At high temperatures, *T. destruens*, the most important pine shoot beetle in mild climates, presents a brood development time relatively short, which might allow the development of a second generation and consequently increase maritime pine damage. Moreover, the increase of survival in higher temperatures conditions can provide a greater populational density and a bigger pressure in healthy trees and subsequently decrease in timber production and increase economical losses.

The bigger tolerance to high temperatures detected in *T. pityocampa* summer populations, revealed by the longer diapause and the early larval development, might be explained by a response to the global warming (Santos et al. 2007) of this new population of processionary moth. In the other hand, the existence of two sympatric processionary populations has an immediate consequence to maritime pine stands because it increases pine defoliation, which weakens pines trees and favours the attack of other forestry pests, such as bark beetles in general and pine shoot beetles in particular.

Overall, climatic change, mainly global warming can act as important factors in insects biodynamic increasing (i) population density levels; (ii) number of generations per year of species that damages trees and (iii) healthy human and animal problems.

Acknowledgements This work is funded by National Funds through the FCT (Portuguese Foundation for Science and Technology) under the project PEst-OE/AGR/UI0681/2014.

References

Arnaldo, P. S., Chacim, S., & Lopes, D. (2010). Effects of defoliation by the pine processionary moth *Thaumetopoea pityocampa* on biomass growth of young stands of *Pinus pinaster* in northern Portugal. *Forest-Biogeosciences and Forestry, 3,* 159–162.

Arnaldo, P. S., Oliveira, I., Santos, J., Leite, S., Mendes Lopes, D. M., & Tome, M. (2011). Climate change and forest plagues: The case of the pine processionary moth in Northeastern Portugal. *Forest Systems, 20*(3), 508–515.

Bakke, A. (1968). Ecological studies of bark beetles (Col. Scolytidae) associated with Scots pine (*Pinus sylvestris* L.) in Norway with particular reference to the influence of temperature. *Meddelelser fra det Norske Skogsforsokvesen, 21*(83), 443–602.

Barbéro, M., Loisel, R., Quézel, P., Richardson, D. M., & Romane, F. (1998). *Pines of the mediterranean basin* (pp. 153–170). Ecology and Biogeography of Pinus: Cambridge University Press, Cambridge.

Battisti, A., Stastny, M., Netherer, S., Robinet, C., Schopf, A., Roques, A., et al. (2005). Expansion of geographic range in the pine processionary moth caused by increased winter temperatures. *Ecological Applications, 15*(6), 2084–2096.

Ben-Jamâa, M., Lieutier, F., & Jerraya, A. (2000). Les Scolytes ravageurs des pins en Tunisie. *Annales de L' inrgref, 4,* 27–39.

Berryman, A., Stenseth, N., & Wollkind, D. (1984). Metastability of forest ecosystem infested by bark beetles. *Researches on Population Ecology, 26,* 13–29.

Berryman, A. (1988). Pest and the stability of forest ecosystem. *Northwest Environmental Journal, 4,* 351–355.

Bond, W. J. (1989). The tortoise and the hare: Ecology of angiosperm dominance and gymnosterm persistence. *Biological Journal of the Linnean Society, 36,* 227–249.

Capelo, J., Aguiar, C. (2005). História Holocénica dos *Pinus* em Portugal. Uma interpretação geobotânica. In R. Silva & F. Páscoa (Eds.), *Actas do 5° Congresso Florestal Nacional*, SPCF-Edições, Lisboa, Portugal.

Carle, P. (1974). The decline of *Pinus pinaster* in Provence. Role of insects in changing the biological equilibrium of forests invaded by *Matsucoccus feytaudi. Annales des Sciences Forestières, 31*(1), 1–26.

Chakali, G. (2005). L' hilésine des pins, *Tomicus destruens* Wollaston 1865 (Coleoptera, Scolytidae) en zone semi-aride (Algérie). *Silva Lusitana, 13*(1), 113–124.

Chararas, C. (1962). *Scolytides des Conifères* (p. 556). Lechevalier. Paris: Encyclopédie Entomologique- XXXVIII, Éditions P.

Christiansen, E., Waring, R., & Berryman, A. (1987). Resistance of conifers to bark beetle attack: Searching for general relationship. *Forest Ecology and Management, 22,* 89–106.

Czokajlo, D., Wink, R., Warren, J., & Peale, S. (1997). Grow reduction of Scots pine, *Pinus sylvestris*, caused by the larger pine shoot beetle *Tomicus piniperda* (Coleoptera: Scolytidae), in the New York state. *Canadian Journal of Forest Research, 27,* 1394–1397.

Devkota, B., & Schmidt, G. H. (1990). Larval development of *Thaumetopoea pityocampa* (Den. & Schiff.) (Lepidoptera, Thaumetopoeidea) from Greece as influenced by different host plants under laboratory conditions. *Journal of Applied Entomology, 209,* 321–330.

Duan, Y., Kerdelhué, C., Ye, H., & Lieutier, F. (2004). Genetic study of the forest pest *Tomicus piniperda* (Col., Scolytinae) in Yunnan province (China) compared to Europe: new insights for the systematics and evolution of the genus *Tomicus. Heredity, 93,* 416–422.

Faccoli, M., Piscedda, A., Salvato, P., Simonato, M., Masutti, L., & Battisti, A. (2005). Genetic structure and phylogeography of pine shoot beetle populations (*Tomicus destruens* and *Tomicus piniperda*, Coleoptera, Scolytidae) in Italy. *Annals of Forest Science, 62,* 361–368.

Ferreira, M. C., Ferreira, G. W. (1990). *Pragas das resinosas - Guia de campo*. Série Divulgação, 3. Ministério da Agricultura, Lisboa.

Fouseki, E., & Margaris, N. (1981). Soil metabolism and decomposition in a phryganic (East Mediterranean) ecosystem. *Oecologia, 50,* 47–421.

Gallego, D., Canovas, F., Esteve, M. A., & Galian, J. (2004). Descriptive biogeography of *Tomicus* (Coleoptera: Scolytidae) species in Spain. *Journal of Biogeography, 31,* 2011–2024.

Ghaioule, D., Abourouh, M., Bakry, M., & Haddam, M. (1998). Insects ravageurs des forêts au Maroc. *Annales de la Recherche Forestière au Maroc, 31,* 129–156.

Guerrero, A., Feixas, J., Pajares, J., Wadhams, L. J., Pickett, J. A., & Woodcock, C. M. (1997). Semiochemical induced inhibition of behaviour of *Tomicus destruens* (Woll.) (Coleoptera: Scolytidae). *Naturwissenschaften, 84,* 155–157.

Horn, A., Kerdelhué, C., Lieutier, F., & Rossi, J. P. (2012). Predicting the distribution of the two bark beetles Tomicus destruens and Tomicus piniperda in Europe and the Mediterranean region. *Agricultural and Forest Entomology, 14*(4), 358–366.

Kerdelhué, C., Roux-Morabito, G., Forichon, J., Chambon, J.-M., Robert, A., & Lieutier, F. (2002). Population genetic structure of *Tomicus piniperda* L. (Coleoptera: Scolytidae) and validation of *T. destruens* (Woll.). *Molecular Ecology, 11,* 483–494.

Langström, B. (1980). Distribution of pine shoot beetle attacks within the crown Scots pine. *Studia Forestalia Suecica, 154.*

Langström, B. (1983). Life cycles and shoot-feeding of the pine shoot beetles. *Studia Forestalia Suecica, 163,* 1–29.

Langström, B., Annila, E., Hellqvist, C., Varama, M., & Niemela, P. (2001). Tree mortality, needle biomass recovery and growth losses in Scots pine following defoliation by *Diprion pini* and subsequent attack by *Tomicus piniperda. Scandinavian Journal of Forest Research, 16,* 342–353.

Langström, B., & Hellqvist, C. (1990). Spatial distribution of crown damage and growth losses caused by recurrent attacks of pine shoot beetles in pine stands surronding a pulp mill in Southern Sweden. *Journal of Applied Entomology, 110,* 261–269.

Långström, B., Hellqvist, C., Ehnström, B. (1999). Susceptibility of fire-damaged Scots pine (Pinus sylvestris L.) trees to attack by *Tomicus piniperda* L. In F. Lieutier, W. J. Mattson, M. R. Wagner (Eds.). *Physiology and genetics of tree-phytophage interactions. Colloques de l'INRA,* Versailles: INRA Editions, 90, 299–311.

Langström, B., Hellqvist, C., Ericsson, A., & Grefr, R. (1992). Induced defence reaction in Scots pine following stem attacks by *Tomicus piniperda. Ecography, 15,* 318–327.

Lieutier, F. (1991). Les coléopteres *Scolytidae* en France et les recherches menées par INRA. *Bulletin de la Societé Zoologique Française, 116,* 275–281.

Lieutier, F., Garcia, J., Yart, A., & Romary, P. (1995). Wound reactions of Scots pine (*Pinus sylvestris* L.) to attacks by *Tomicus piniperda* L. and *Ips sexdentatus* Boern. (Col., Scolytidae). *Journal of Applied Entomology, 119,* 591–600.

Lieutier, F., Ye, H., & Yart, A. (2003). Shoot damage by *Tomicus* sp. (Coleoptera: Scolytidae) and effect on *Pinus yunnanensis* resistance to subsequent reproductive attacks in the stem. *Agricultural and Forest Entomology, 5,* 227–233.

Mendel, Z., Madar, Z., & Golan, Y. (1985). Comparison of the seasonal occurrence and behaviour of seven pine bark beetles (Coleoptera: Scolytidae) in Israel. *Phytoparasitica, 13,* 21–32.

Miller, T. (1986). *Insect-Plant Interactions*. In J. Miller, T. Miller (Eds.), Springer.

Netherer, S., & Schopf, A. (2010). Potential effects of climate change on insect herbivores in European forests—general aspects and the pine processionary moth as specific example. *Forest Ecology and Management, 259*(4), 831–838.

Nylinder, M., Lunström, H., & Fryk, H. (2000). *Skador och fel pa tall-och grantimmer*. Sweden (in Swedish): Swedish University of Agricultural Sciences.

Paiva, M. R., Santos, H., Branco, M. (2012). Processionária do pinheiro na Mata Nacional de Leiria documenta caso muito raro de especiação simpátrica. *Notas Técnicas, 20.*

Petrice, T., Haack, A., & Poland, T. (2002). Selection of overwintering sites by *Tomicus piniperda* (Coleopter, Scolytidae) during fall shoot departure. *Journal of Entomological Science, 37,* 48–59.

Pimentel, C., Calvão, T., Santos, M., Ferreira, C., Neves, M., & Nilsson, J. Å. (2006). Establishment and expansion of a *Thaumetopoea pityocamp* (Den. & Schiff.) (Lep. Notodontidae) population with a shifted life cycle in a production pine forest, Central-Coastal Portugal. *Forest Ecology and Management, 233*(1), 108–115.

Pimentel, C., Santos, M., Ferreira, C., & Nilsson, J. Å. (2012). Temperature, size, reproductive allocation, and life-history evolution in a gregarious caterpillar. *Biological Journal of the Linnean Society, 105*(2), 340–349.

Rapp, M., Regina, I. S., Rico, M., & Gallego, H. A. (1999). Biomass nutrient content, litterfall and nutrient return to the soil in Mediterranean oak forests. *Forest Ecology and Management, 119,* 39–49.

Régnière, J., Powell, J., Bentz, B., & Nealis, V. (2012). Effects of temperature on development, survival and reproduction of insects: Experimental design, data analysis and modeling. *Journal of Insect Physiology, 58*(5), 634–647.

Reich, P. B., Kloeppel, B. D., Ellsworth, D. S., & Walters, M. B. (1995). Different photosynthesis-nitrogen relations in decidous hardwood and evergreen coniferous tree species. *Oecologia, 104,* 24–30.

Rousselet, J., Zhao, R., Argal, D., Simonato, M., Battisti, A., Roques, A., et al. (2010). The role of topography in structuring the demographic history of the pine processionary moth, *Thaumetopoea pityocampa* (Lepidoptera: Notodontidae). *Journal of Biogeography, 37*(8), 1478–1490.

Santos, H., Burban, C., Rousselet, J., Rossi, J. P., Branco, M., & Kerdelhué, C. (2011). Incipient allochronic speciation in the pine processionary moth (*Thaumetopoea pityocampa*, Lepidoptera, Notodontidae). *Journal of Evolutionary Biology, 24*(1), 146–158.

Santos, H., Rousselet, J., Magnoux, E., Paiva, M. R., Branco, M., & Kerdelhué, C. (2007). Genetic isolation through time: allochronic differentiation of a phenologically atypical population of the pine processionary moth. *Proceedings of the Royal Society B: Biological Sciences, 274*(1612), 935–941.

Sauvard, D., Lieutier, F., & Levieux, J. (1987). Répartition spatiale et dispersión de *Tomicus piniperda* L. (Coleoptera, Scolytidae) en foret d' Orleáns. *Annales des Sciences Forestières, 44,* 417–434.

Sauvard, S. (1989). Capacités de multiplication de *Tomicus piniperda* L. (Col. Scolytidae). *Journal of Applied Entomology, 108,* 164–181.

Solheim, H., Langström, B., & Hellqvist, C. (1992). Pathogenicity of the blue stain fungi *Leptographium wingfieldii* and *Ophiostoma minus* to Scots pine: Effect of tree pruning and inoculum density. *Canadian Journal of Forest Research, 23,* 1438–1443.

Van Halder, I. (2002). *Pragas e doenças das florestas do Sul da Europa*. Cestas: IEFC- Institut Européen de la Forêt Cultivée.

Vasconcelos, T. & Duarte, I. (2015). *Biotic/Abiotic Factors-Plant*. Chapter VI: 193–220. In E. Sousa, F. Vale & I. Abrantes (Eds), *Pine Wilt Disease in Europe. Biological Interactions and Integrated Management*. 328p. Edição FNAP, ISBN 978-989-99365-2-2.

Vasconcelos, T. M. (2005). Structuration des populations portugaises de *Tomicus* spp (Coleoptera: Scolytinae), aspects moléculaires et comportamentaux en liaison avec les espèces de pins hôtes. Dissertação de Doutoramento em Engenharia Florestal e em Biologie et Ecologie des populations. Instituto Superior de Agronomia, Universidade Técnica de Lisboa and École Doctorale de la Universidade de Orléans.

Vasconcelos T., Branco, M., Gonçalves M. & Cabral, T. (2005). Observation on flying activity of *Tomicus* spp. in Portugal during a long period. *In*. Lieutier F., Ghaioule D. (Eds). *Entomological research in mediterranean forest ecosystems*, INRA-Editions, Versailles, France.

Wood, S., Bright, D. (1992). A Catalog of Scolytidae and Platypodidae (Coleoptera), Part 2: Taxonomic Index Vol. A. *Great Basin Naturalist Memoirs*, 13, Brigham Young University, 1553 pp.

Author Biographies

Dr. Teresa Maria Vasconcelos Forestry Engineer, Instituto Superior Agronomia, Lisbon University. Associate professor at the College of Agriculture of Coimbra Polytechnic Institute. Researcher in the Centre for Natural Resources, Environment and Society (CERNAS), Portugal. Master Degree in Ecology, Coimbra University, Portugal and Ph.D. in cotutelle in Biologie des Populations et Eco-Ethologie/Forestry Engineering, Orléans University, France/ISA-Lisbon University. Specialization in Forest Pest and Management.

Dr. Isabel Maria Duarte Agronomic Engineer, Instituto Superior Agronomia, Lisbon University. Associate professor at the College of Agriculture of Coimbra Polytechnic Institute. Researcher in the Centre for Natural Resources, Environment and Society (CERNAS), in Portugal. Master Degree in Techonology of Crop Protection, Reading University, UK and Ph.D. in Science, Biology, Minho University, Portugal. Specialization in Crop Protection (Virology and Nematology) and recently in antibiotic resistance genes. Research Projects with Scottish Crop Research Institute.

Part V
Technological approaches to Climate Change Adaptation/Innovative Approaches towards Low Carbon Economics

Urban Agriculture Practices as Initiatives for Mitigation and Adaptation to Climate Change: Possibilities for Urban Farms in a South American City

Ana Valquiria Jonck, João Marcelo Pereira Ribeiro, Issa Ibrahim Berchin, Francesca Chaher Perini and José Baltazar Salgueirinho Osório de Andrade Guerra

Abstract Urbanization has become one of the main challenges of modern societies. Therefore, transforming urban areas to sustainable environments is strategic for mitigating and adapting to climate change. Florianópolis—the case of this study—is a South American city that in the next 30 years will double its population, the city already suffers from bad urban management, therefore solutions that combine social, environmental and economic aspects are necessary for achieving sustainable development. The city has begun some programs for urban agriculture and greater efforts are possible. Today, urban agriculture can provide ways of increasing community integration, decreasing greenhouse gases emissions, improving the urban waste management and revitalizing the local economy. This paper highlights urban agriculture practices applied in different cities of the world and how they relate with the Sustainable Development Goals, promoting an exchange of ideas and experiences among researchers, the community and the local government.

A. V. Jonck · J. M. P. Ribeiro · I. I. Berchin · F. C. Perini
J. B. S. O. de Andrade Guerra (✉)
Universidade do Sul de Santa Catarina, Rua Trajano,
219, Centro, Florianópolis, Santa Catarina 88010-010, Brazil
e-mail: baltazar.guerra@unisul.br

A. V. Jonck
e-mail: anajonck15@gmail.com

J. M. P. Ribeiro
e-mail: joaomarceloprdk@gmail.com

I. I. Berchin
e-mail: issaberchim@gmail.com

F. C. Perini
e-mail: franchaher@hotmail.com

F. Alves et al. (eds.), *Theory and Practice of Climate Adaptation*,
Climate Change Management, https://doi.org/10.1007/978-3-319-72874-2_28

Keywords Urban agriculture · Sustainable development · Cities
Sustainable development goals

Introduction

Urbanization has become one of the main challenges of modern society. Cities are centers of human interactions. Approximately 50% of the world's population lives in cities today, and by 2050, 75% of society will be living in urban centers (Buhaug and Urdal 2013; UN 2014).

Sustainable cities are means of action to assure opportunities for everyone and to reduce the impacts of climate change (UN 2017f; Satterthwaite 1997). Transforming urban areas in sustainable environments is strategic for mitigating and adapting to climate change. Today, urban agriculture, for example, presents a solution to aid the increase of communitarian integration, decrease greenhouse gas emissions, and improvement of urban waste management, while it also revitalizes the local economy (Brown and Jameton 2000; Smit and Nasr 1992).

The authors selected some SDGs that fit with urban agriculture. The main goals are: (1) eradication of poverty, (2) end hunger, (6) ensure availability and sustainable management of water and sanitation for all, (9) build resilient infrastructure, promote inclusive and sustainable industrialization and foster innovation, and (11) sustainable cities and communities.

Florianópolis—the case of this study—is a south American city that in the next 30 years will double its population, the city already deals with bad urban management, therefore solutions that comprehend social, environmental and economic aspects are necessary for achieving sustainable development. This study highlights urban agriculture practices applied in different cities worldwide and how they relate with the SDGs and how these practices can aid Florianópolis to deal with its' problems as well as to achieve a balanced and healthy urban development. That way, promoting an exchange of ideas and experiences among researchers, civil society and local governments.

Urban Agriculture

The world has been facing many challenges throughout the past few decades. The population growth and the unsustainable use of resources have become main threats to the planet (Deelstra and Girardet 2000). Cities are struggling to accommodate all the world´s inhabitants, since its majority lives in urban areas. Human activities are negatively impacting our cities and creating many social, economic and ecological problems—such as climate change, pollution, degradation of resources, unemployment and deficit on food supply (De Bon 2009).

The emergence of these problems demands that cities adopt more sustainable ideas and policies, establishing a healthier relationship between communities and the environment (Mersal 2016). One solution that has been receiving more attention is the production of food crops within cities or, urban agriculture (Van Veenhuizen 2006; Rakodi 1988).

Urban agriculture practices have turned cities into sustainable food producers (Smit and Nasr 1992), which is essential for: economic development, poverty alleviation, and social inclusion; nutrition and food security; preservation and public health (Brown and Jameton 2000).

Urban agriculture becomes essential due to the lack of opportunities in other economic sectors: creating jobs, providing and improving livelihoods and increasing food and nutrition security (World Bank 2007; Zezza and Tasciotti 2008). Urban agriculture is responsible for employing more than 200 million people (Zezza and Tasciotti 2008). Other studies have shone that urban agriculture helps to reduce poverty in cities (Ravallion et al. 2007) and that urban farms allow people to have access to and to consume a variety of healthy and nutritious food, like vegetables and fruits (Zezza and Tasciotti 2010).

Gardens already provide more than 40% of family food consumption needs (Niñez 1985). Urban agriculture helps save money on the purchase of food, allowing people to improve their buying power by saving around 20% of their income (Mougeot 2000; de Zeeuw and Drechsel 2015). With these savings, they are able to buy higher-value items like meat, and spend on other necessities such as health care (de Zeeuw and Drechsel 2015).

Urban farms as a source of income can improve farmer's profits by selling the surpluses of their food production. According to Mougeot (2000), urban agriculture is more profitable than rural-based production; therefore, aside from only helping to increase food security, it also helps to combat urban poverty.

In addition, urban agriculture improves public health in more than just nutritional ways. It adopts more organic and agro-ecological principles. Sometimes urban farmers reject the use of synthetic chemical pesticides and fertilizers in favor of more local, organic and locally generated soil nutrients (Nicholls et al. 2002), this way adapting to local resources. Moreover, under-utilized areas with poor sanitary conditions, such as garbage dumps and vacant lots, could transform into productive and healthy environments (Altieri et al. 1999).

Nutrients and pathogen pollution of the ecosystem are negatively affecting the health of the population and the ability of the environment to sustain future generations (Obeng and Wright 1987). Agriculture helps to improve the quality of urban environment through greening and reduction of pollution.

Urban agriculture helps to address a global waste management problem. Liquid and solid, organic and inorganic, waste can be recycled and transformed into resources for sustainable development (Armar-Klemesu 2000). It is possible to transform urban waste into organic matter for fresh food production, reducing costs, improving the quality of the food, conserving energy and natural resources, and improving the quality of the environment (Smit and Nasr 1992). Besides helping on waste disposal, that many times is not recycled and ends up on open dumps (Smit and Nasr 1992).

Urban farms have other positive effects on individual health and community improvement. For example, gardening helps with stress-reduction and relieves the urban environment. It also helps to develop fine motor skills and creates opportunities for leadership development and community organization (Brown and Jameton 2000). In cities, many vacant lots can be seen, some are vacant because they are not suited for constructions, such as steep slopes and flood plains (Altieri et al. 1999; Bryld 2003). These areas can be stabilized and protected from erosion through urban agricultural uses. This destination would also increase green areas in the city, helping to reduce dust and absorbing pollutants, also increasing humidity and reducing radiation heat (Bryld 2003).

Another problem that urban agriculture can potentially resolve is the current unsustainable agricultural production system. This system is requiring more and more land, therefore, the habitats of natural ecosystems are often destroyed in order to create new farmland, besides using methods that are harmful to the environment (Miccoli et al. 2016).

Urban farms also bring about the reduction of municipal management and city infrastructure costs, for example, by providing more green areas, less maintenance of damaged roads will be necessary, as it will not be as crucial to transport food from other areas if grows in the city. This way these funds can finance other sectors, like security (Smit and Nasr 1992). Despite being a practice that offers many benefits, urban agriculture, still experiences controversies. The main problems it faces are the stereotypes, mismanagement, contamination and illegality. There are some social insecurities connected to the practice of urban agriculture, like cultural taboos in relation to the use of waste in the farms (Smit and Nasr 1992) that worsen due to the mismanagement of those wastes, especially water, which offers a great risk of contamination to consumers and producers if mistreated (Bryld 2003). The vision of urban agriculture as a marginal phenomenon, (Miccoli et al. 2016) and the lack of investments in the sector, especially in developing countries (Mârza et al. 2015) is a deterrence to greater development of these practices.

Additional challenges faced by urban agriculture includes, lack of space in urban areas that result in high prices for the rent of lands (Deelstra and Girardet 2000; Smit and Nasr 1992). These problems of space and price depict the practice of a migratory nature, which only satisfies some needs for a short period, which is not sufficient to attend to and maintain the demands for food (Miccoli et al. 2016). In some countries urban agriculture is illegal, because some governments fear the possibility of food contamination (Bryld 2003).

Another challenge urban agriculture faces is the use of agrochemicals, which offer more risk in urban areas than in rural areas (De Bon et al. 2010; Mougeot 2000). Air pollution in urban areas, especially adjacent to roads and airports, threatens the reduction of yields and affects the quality of some crops. (Agrawal et al. 2003).

Methodology

This study presents a review of cases that were selected from literature that highlights best practices in urban agriculture. The first step allowed the alignment of the paper with the subject matter, and through a qualitative analysis, the selection of urban agriculture practices with its positive and negative aspects, that would further aid the execution of urban farm projects (Table 1).

The paper aligned the practices highlighted by the cases studied with the SDGs, whose objective is to mitigate the impacts of climate change, especially in developing countries. The main problems of urbanization in Florianópolis were identified. Based on select cases of urban agriculture, practices were selected on how they could aid in the achievement of the SDGs in Florianópolis, as well as the problems of the city.

Urban Agriculture Practices

By analyzing case studies, their process of implementation and their results positive and negative aspects of urban agriculture and the relations among the cases are reviewed.

In the study of Barthel and Isendahl (2013), about Constantinople in the 6th century B.C., show that urban agriculture, was achieved in gardens that showcased vegetable crops, orchards and livestock and potentialized by organic urban waste. Of special concern even then, and as said by Bryld (2003) and Mougeot (2000) before, was the careful management of waste, in order to avoid crop contamination,

Table 1 Chosen case studies for qualitative analysis

Author	City	Method
Barthel and Isendahl (2013)	Constantinople	Historical analysis; multi-scalar approach
Lupia and Pulighe (2015)	Rome	Google Earth photointerpretation
Tsuchiya et al. (2015)	Bangkok	Geographical approach analysis
Cahya (2016)	Jakarta	Multi Dimensional Scaling (MDS)
Nyantakyi-Frimpong et al. (2016)	Ashaiman	Qualitative field research; group meetings; interviews
Peng et al. (2015)	Beijing	Multifunctional quantitative research; index system; Analytic Hierarchy Process (AHP); triangular illustration method
Pribadi and Pauleit (2016)	Jabodetabek	Multivariate techniques, Geographically Weighted Regression
Pölling et al. (2016)	Ruhr	Geostatistics analysis to investigate the patterns of professional/commercial urban agriculture

as well as the zoonosis that can be transmitted to humans by urban animal husbandry. But it is largely believed that due to its vibrant urban agriculture practices, Constantinople was able to establish a successful, resilient, economically efficient society that was less dependent on commercial trade with other city/states and able to better sustain city sieges.

In Rome, in 2013 (Lupia and Pulighe 2015), urban agriculture also manifested itself in gardens, being in many cases residential, besides vacant lots and parks, all in the peri-urban zone. Rainwater was used, what optimizes the supply and the availability of potable water (Smit and Nasr 1992), the reuse of rainwater also contributes to reduce the water deficiency in Europe (Lupia and Pulighe 2015).

Besides Rome, Indonesia's massive capital, Jakarta also has residential farms, the majority of which, has been increasing domestic income, health, food security and community well-being (Cahya 2016). Unfortunately, Jakarta faces a common problem to the practice of urban agriculture in developing countries, as mentioned by Mârza et al. (2015), which is the lack of governmental incentives to support it.

The cities of Ruhr—Rhein-Ruhr is the most metropolitan region of Germany—and Bangkok—capital of Thailand—in turn, have a more commercial approach to urban agriculture. Tsuchiya et al. (2015) verified that in Bangkok, especially in peri-urban areas, urban agriculture generates income for many families who sell their products in the traditional street markets of the city. These street markets allow urban crop producers to contribute to a more decent life style for family members while contributing to sustainable and healthy urbanization.

In turn at Ruhr (Pölling et al. 2016), urban agriculture is more professional. It is represented by three economic models: low coast specialization, diversification and differentiation. Horticulture is part of the low coast specialization model, maintaining profit and competitiveness. The diversification model presents equine livestock and complements the other models. And the differentiation model covers traditional farms that are varying its productions and services, also business not related to agriculture but which are engaging in the sector (example.g. non-governmental organizations that run farms working as shelters, or designed for agritourism) (Pölling et al. 2016).

As in Bangkok (Tsuchiya et al. 2015), the farms in Ruhr are also located in peri-urban areas, forming a transition zone between these environments. Urban agriculture practices here contribute to the urban economy, generate more food and nutrition security, help in the adaptation of the environment as well as, with improvements in the agricultural sector through new experiences (Pölling et al. 2016).

Many cities have multifunctional systems of urban agriculture, like Beijing and Jabodetabek. According to Peng et al. (2015), in Beijing the agricultural system complements and is complemented by other systems including social, economic, and ecological in many dimensions and functions resulting in a multifunctional system.

Urban multifunctional agriculture is an effective option to promote a modern agricultural development that is sustainable and diversified, making itself present in urban progress and maintains the balance between resources, conservation, agricultural production, urban expansion and the natural environment (Peng et al. 2015).

At Jabodetabek Metropolitan Area, ten types of urban farms were registered specifically in peri-urban areas (Pribadi and Pauleit 2016). These multifunctional systems, as verified in the study of Beijing (Peng et al. 2015), are very efficient in the environmental sphere in conservation and augmentation of biodiversity (Bryld 2003; Deelstra and Girardet 2000). Depending on the region, the potential for food security varies, as well as the level of capital invested, these factors, on the other hand, depend on the connectivity within markets and urban zones (Pribadi and Pauleit 2016), as well as on the possible zoonosis transmission by urban livestock (Mougeot 2000).

The desakota phenomenon—according to Pribadi and Pauleit (2016): the interweaving of the urban, peri-urban and rural areas—of urban farms in Jabodetabek are considered of extreme importance to face poverty, food insecurity, and environmental degradation. They represent an option for sustainable urban development, reducing urban vulnerability to natural disasters, better adaptation to climate change and balance between different ecosystems. Besides providing new jobs, more income, inclusion of poor workers into the formal market and of peri-urban agriculture itself (Pribadi and Pauleit 2016).

One of the worst scenarios for urban agriculture within the cases studied was in the city of Ashaiman, Ghana by Nyantakyi-Frimpong et al. (2016). In this city, the exaggerated use of agrochemicals is a very common practice and one of the major complaints of producers, decision makers and consumers. Besides being extremely harmful to health, agrochemicals are a form of soil and atmospheric contamination (De Bon et al. 2010; Mougeot 2000). According to government officials, the regulation of the use, production and importation of agrochemicals is made, although the lack of information of the producers about these chemicals, is the main factor for the excessive use of them (Nyantakyi-Frimpong et al. 2016). Another issue is the use of non-treated waste water for crop irrigation (Bryld 2003), also noticed in Constantinople (Barthel and Isendahl 2013).

These two main issues, the excessive use of agrochemicals and the use of non-treated wastewater, are aggravated by the insecurity of land ownership that producers face. Many urban farmers rent land, but only for a brief period, resulting in a desperate use of agrochemicals in order to produce and harvest crops faster (Nyantakyi-Frimpong et al. 2016). The insecurity of land ownership is a common problem for modern urban agriculture (Smit and Nasr 1992), due to the lack of space in urban areas in general (Deelstra and Girardet 2000). Many farmers optimize city spaces with their productions, such as lands close to roadways and idle land.

Another issue in Ashaiman's urban farms is gender inequality in work division. Women perform most manual jobs while the men only plant and apply agrochemicals, this type of situation turns urban agriculture against one of its most important principles, which is the inclusion and empowerment of all members of the community (Mougeot 2000).

It is important to consider though, some factors that may justify all of these problems: the historical context of the city, the lack of information of the farmers about agrochemicals and non-treated waste water, the effects of climate change, the mutations/resistance of plagues to agrochemicals and some public policies (Nyantakyi-Frimpong et al. 2016).

Nyantakyi-Frimpong et al. (2016) propose some options that can improve urban agriculture in Ashaiman: an urban planning that involves urban agriculture and works with urban zoning and land owning; spreads more information about urban agriculture; encourages new methods that are not so dependent of agrochemicals and that are more equal in the division of work; strengthen methods of biological control of plagues and use composts, organic fertilizers and manure, as well as planting polycultures (Nyantakyi-Frimpong et al. 2016).

Urban Agriculture as a Strategy for Sustainable Development

By developing and implementing sustainable agriculture in urban areas, many issues approached by the SDG would be addressed at the same time. The main goals related to urban agriculture are: (1) eradication of poverty, (2) end hunger, (6) ensure availability and sustainable management of water and sanitation for all, (9) build resilient infrastructure, promote inclusive and sustainable industrialization and foster innovation, and (11) sustainable cities and communities.

The first goal to be analyzed is Goal 1, which aims to eradicate poverty by 2030. According to the United Nations, "The international poverty line is currently defined at $1.90 or below per person, using 2011 United States dollars purchasing power parity (ppp)" (UN 2017a). Since many of these families live with less than $2.00 a day, the idea is to reduce expenditures. By producing their own food, the families will not need to spend money buying the food they can produce (de Zeeuw and Drechsel 2015).

The next goal is known as Zero Hunger, which the main objective is to end hunger or malnutrition by 2030 (UN 2017b). Urban agriculture improves food security and reduces hunger through increasing availability. According to the Community Food Project, urban agriculture projects produced 18.7 million pounds of food in 2005, and made available more than 726,000 lb for community consumption (Kobayashi et al. 2010).

In addition, the target 2.4 of this Goal states that: "By 2030, ensure sustainable food production systems and implement resilient agricultural practices that increase productivity, help maintain ecosystems, and strengthen capacity for adaptation to climate change, extreme weather, drought, flooding and other disasters and that progressively improve land and soil quality" (UN 2017b).

Urban agriculture becomes essential to achieve this goal because it not only reduces hunger and malnutrition, but it is also a sustainable food production system that helps to maintain the ecosystem.

As for Goal 6 (UN 2017c), "ensure availability and sustainable management of water and sanitation for all", it fits in a more specific context, for example, in cities with poor sanitation and sewage treatment or only low availability of water, like the city of Florianópolis (FloripAmanhã 2016), and Rome (Lupia and Pulighe 2015). In cases like this, urban agriculture can be useful by utilizing wastewater (Armar-Klemesu 2000) or rainwater (Lupia and Pulighe 2015), decreasing the pressure in the water systems and increasing the availability of potable water for human consumption.

Goal 9 (UN 2017d), which aims to build a resilient infrastructure, promote sustainable industrialization and foster innovation. In this context urban agriculture creates green areas and relieves the urban environment (Altieri et al. 1999; Bryld 2003), as well as innovations in the way food is produced (Miccoli et al. 2016), and the use of local resources (Nicholls et al. 2002) and also, even reduces on municipal management expenditures (Smit and Nasr 1992).

The last Sustainable Development Goal related to urban agriculture is goal 11, sustainable cities and communities. The United Nations predicts that 6 out of 10 people will be living in urban centers by 2030 (UN 2017e). Urbanization brings many challenges, and one of them is the increasing demand for food. For this reason, urban farming plays an important role in contributing to the development of sustainable cities. According to Deelstra and Girardet (2000, p. 47), "urban farming can help to create an improved microclimate and to conserve soils, to minimize waste in cities and to improve nutrient recycling, improve water management, biodiversity, the O_2—CO_2 balance, and the environmental awareness of the city inhabitants".

Possible Practices for the City of Florianópolis

The city of Florianópolis is the capital of the State of Santa Catarina in Brazil. Florianópolis is an island, but also occupies a continental coast. Its ecosystem is composed by the Atlantic Rainforest, mangroves, restinga forest, dunes, and various beaches (Florianópolis 2017). Despite the increasing urbanization, Florianópolis has many preserved and idle areas, which present great potential to become sustainable urban farms.

Florianópolis does not have many projects for urban agriculture or in the sustainable development area. In the case of urban agriculture, all projects are community gardens, that occur in the North and Southern region of the island, as well as in the federal university, located in the central area of the island, in a municipal park and a horto forest, both in the center of the city, which is more urbanized (Campeche 2017; Floripa 2017).

Although, the city presents some challenges for the implementation of urban agriculture, the main issues of Florianópolis are related to the garbage collection system and sewage treatment network (Florianópolis 2008), irregular constructions (Florianópolis 2008; De Souza 2012), and urban planning. Urban farms could work out many of these problems, as well as contribute to achieve several SDG (United Nations 2017g). The main SDG in this context are: Goal number 6, assure basic sanitation and water in a sustainable way for all, and Goal 9, achieve resilient infrastructure, promote an inclusive and sustainable industrialization and foster innovation.

In the case of the waste collection system and the deficit of sewage treatment network, urban agriculture practices contribute through recycling waste that would become resources for urban farming (Armar-Klemesu 2000; Smit and Nasr 1992; Barthel and Isendahl 2013), as well as the practice of composting (Tognetti et al. 2007), therefore aiding Florianópolis in becoming more sustainable and contributing to achieve Sustainable Development Goal number 11, sustainable cities and communities. Besides contributing to decrease the pressure on the sewage treatment network, for reusing wastewater, also helping to achieve Sustainable Development Goal number 6. Although, it is important to remember that the waste mismanagement can lead to contamination of crops and damage producers and consumers (Bryld 2003).

Regarding the irregular buildings problem, urban farms can help to stabilize the land where these constructions are (if that is the case), like steep slopes, where these constructions are more expressive (Altieri et al. 1999; Bryld 2003). In 2007, Florianópolis' townhouse made a report about risk reduction, stating that the city presents high risk of land sliding in nine areas, what would affect directly 1.498 people, mostly of low income, who inhabit the mentioned area, this report was reevaluated in 2012, and 8 areas still present high risk of land sliding (Mathias 2011; Florianópolis 2008; De Souza 2012). Relocating these people to safest locations, would help improve their quality of life, and urban agriculture would make it possible for them to overcome poverty and hunger, which would also aid in the achievement of Sustainable Development Goal number 1, zero hunger.

Urban planning and infrastructure: urban agriculture would reduce the cost in infrastructure and municipal management, through creating multifunctional green areas, which would prevent the city from constructing and spending so much on the maintenance and treatment of these areas (Smit and Nasr 1992). This could also apply to spaces that would relief urban landscape (Brown and Jameton 2000), additionally using idle land or land not suited for construction (Altieri et al. 1999; Bryld 2003) which often function as garbage dumps, transmitting diseases (Altieri et al. 1999).

In practice, some case studies provide evidence that may contribute to the formation of an urban farm plan in Florianópolis. The city could successfully apply practices such as those accomplished in Rome (Lupia and Pulighe 2015); Jakarta (Cahya 2016) and Bangkok (Tsuchiya et al. 2015). Another important point is to keep the existing projects and expand them.

Rome (Lupia and Pulighe 2015) and Jakarta's (Cahya 2016) residential gardens would be well implemented in Florianópolis' neighborhoods that present a larger number of houses rather than buildings, for example, in beach neighborhoods, which are far from the largest markets downtown. Urban agriculture would increase nutrition and domestic income. Also in the case of Rome, it is important to highlight that they used rainwater to irrigate their crops. This practice would also be functional in Florianópolis, due to the sewage treatment network problem. The utilization of rainwater for irrigation can assure more food security and would decrease the pressure on water and on the sewage treatment network, increasing the availability of potable water.

In Bangkok (Tsuchiya et al. 2015) the use of urban agriculture as a form of income generation, would also be functional. Like Bangkok, Florianópolis also has many street markets, and the vendors who participate in them mostly come from other cities of Santa Catarina, therefore, their products are usually not fresh due to the long distances they travel to arrive in the city. In this case, if low-income families who live in Florianópolis were to engage in urban agriculture, they could increase domestic income through the sale and consumption of their agricultural products, besides providing fresh, organic and environmentally and socially sustainable crops, and contribute to a more sustainable urbanization (Tsuchiya et al. 2015).

Urban agriculture in Beijing (Peng et al. 2015) operates in a multifunctional way, including in preservation. Florianopolis' ecosystem, especially when it comes to the reduction of the destruction of the Atlantic Rainforest over time, could benefit substantially from this practice. Also, if applied in more parks or in idle land in the center of the city, urban farms would help to decrease the pollution levels. In social functions urban farms, could be used, for example, in schools as a form of environmental education and a source of healthy food for students, and also if applied in the penitentiary urban agriculture can be a leisure, learning and working activity, this would also be ways for achieving Sustainable Development Goal number 9.

Conclusion

By the results obtained with the revision of the urban agriculture case studies, the negative and positive aspects of this practice emphasized how they could contribute to climate change mitigation and implementation of the SDG, as well as more balanced urbanization. Sustainable cities, as urban centers can achieve future goals without compromising the present ones, being resilient and strategic environments for climate change mitigation. In this sense, the presented case studies depict the possibilities of urban agriculture application in many different realities, including in the city of Florianópolis, Brazil., which, as a developing city, presents some problems related to disordered and non-planned urbanization, entailing issues that additionally compromise sustainable development. Urban agriculture comes as an alternative to reduce the pressure on the ecosystems.

Some actions that fit with sustainable development and would help Florianópolis in the mitigation and adaptation to climate change are: (1) recycling of urban wastes, (2) urban agriculture as an activity that generates income to poor families, assuring food security, (3) Reduction of the agricultural pressure over native ecosystems, (4) the use of urban agriculture as a multifunctional method that would also act in preservation and environmental education, through expanding green spaces in cities and being pilot projects and studies. Although it is necessary to manage waste used by urban farms with responsibility, this is a fundamental point to avoid many negative aspects of urban agriculture.

Therefore, it is necessary that political engagement raise awareness and encouragement, through incentives, public policies and debate. The population, must also, seek alternative and healthier forms to produce food, this way helping the planet and reducing climate change and becoming more resilient towards these issues.

Acknowledgements The authors would like to thank Kirsten Ann Conrad e George Gabriel for their contributions in the revision of this paper This study was conducted by the Research Centre on Energy Efficiency and Sustainability (Greens), from the University of Southern Santa Catarina (Unisul), in the context of the projects: Building Resilience in a Dynamic Global Economy: Complexity across scales in the Brazilian Food-Water-Energy Nexus (BRIDGE)), funded by the Newton Fund, Fundação de Amparo à Pesquisa e Inovação do Estado de Santa Catarina (FAPESC) and the Research Councils United Kingdom (RCUK).

References

Agrawal, M., Singh, B., Rajput, M., Marshall, F., & Bell, J. N. B. (2003). Effect of air pollution on peri-urban agriculture: A case study. *Environmental Pollution, 126*(3), 323–329.

Altieri, M. A., Companioni, N., Cañizares, K., Murphy, C., Rosset, P., Bourque, M., et al. (1999). The greening of the "barrios": Urban agriculture for food security in Cuba. *Agriculture and Human Values, 16*(2), 131–140.

Armar-Klemesu, M. (2000). Urban agriculture and food security, nutrition and health. Growing cities, growing food. Urban agriculture on the policy agenda, 99–118.

Barthel, S., & Isendahl, C. (2013). Urban gardens, agriculture, and water management: Sources of resilience for long-term food security in cities. *Ecological Economics, 86,* 224–234.

Brown, K. H., & Jameton, A. L. (2000). Public health implications of urban agriculture. *Journal of Public Health Policy, 21*(1), 20–39.

Bryld, E. (2003). Potentials, problems, and policy implications for urban agriculture in developing countries. *Agriculture and Human Values, 20*(1), 79–86.

Buhaug, H., & Urdal, H. (2013). An urbanization bomb? Population growth and social disorder in cities. *Global Environmental Change, 23*(1), 1–10.

Cahya, D. L. (2016). Analysis of urban agriculture sustainability in Metropolitan Jakarta (Case study: Urban agriculture in Duri Kosambi). *Procedia-Social and Behavioral Sciences, 227,* 95–100.

Campeche, N. (2017). Mapa da Agricultura Urbana em Florianópolis. Retrieved April 01, 2017, from http://novash.com.br/agricultura-urbana-em-florianopolis/.

De Bon, H., Parrot, L. & Moustier, P. (2009). Sustainable urban agriculture in developing countries: A review. In *Sustainable Agriculture* (pp. 619–633). Springer Netherlands.

De Bon, H., Parrot, L., & Moustier, P. (2010). Sustainable urban agriculture in developing countries. A review. *Agronomy for sustainable development, 30*(1), 21–32.

Deelstra, T., & Girardet, H. (2000). Urban agriculture and sustainable cities. In N. Bakker, M. Dubbeling, S. Gündel, U. Sabel-Koshella, H. de Zeeuw (Eds.), Growing cities, growing food. Urban agriculture on the policy agenda. Feldafing, Germany: Zentralstelle für Ernährung und Landwirtschaft (ZEL), 43–66.

De Souza, L. A. (2012, October). Revisão do Plano Municipal de Redução de Risco - PMRR - Florianópolis - SC (Rep. No. 1). Retrieved April 12, 2017, from http://www.pmf.sc.gov.br/arquivos/arquivos/pdf/29_05_2013_14.03.33.119d4683dda798d135e95ce684a3d570.pdf.

de Zeeuw, H. & Drechsel, P. (Eds.) (2015). *Cities and agriculture: Developing resilient urban food systems*. Routledge.

Florianópolis, P. (2008, July). ELABORAÇÃO DO PLANO MUNICIPAL INTEGRADO DE SANEAMENTO BÁSICO DO MUNICÍPIO DE FLORIANÓPOLIS (Rep. No. 01). Retrieved April 12, 2017, from http://www.tratabrasil.org.br/novo_site/cms/files/governo.pdf.

Florianópolis, P. (2017). FLORAM. Retrieved April 01, 2017, from http://www.pmf.sc.gov.br/entidades/floram/index.php?cms=ecossistema&menu=0.

Floripa, Q. (2017). Hortas Comunitárias Florianopolis. Retrieved April 01, 2017, from https://www.google.com/maps/d/viewer?mid=11WeyKaMBbxWb-22TEcl93PQOyYs&ll=-27.4361365%2C-48.45840809999993&z=11.

FloripAmanhã. (2016, September 28). Pesquisa da FloripAmanhã revela quais as prioridades e problemas mais urgentes de Florianópolis. Retrieved May 28, 2017, from http://floripamanha.org/2016/09/pesquisa-da-floripamanha-revela-quais-as-prioridades-e-problemas-mais-urgentes-de-florianopolis/.

Kobayashi, M., Tyson, L. & Abi-Nader, J. (2010). *The activities and impacts of community food projects 2005–2009*. Report from the community food project competitive grants program, 1–28.

Lupia, F., & Pulighe, G. (2015). Water use and urban agriculture: Estimation and water saving scenarios for residential kitchen gardens. *Agriculture and Agricultural Science Procedia, 4,* 50–58.

Mârza, B., Angelescu, C., & Tindeche, C. (2015). Agricultural insurances and food security. The new climate change challenges. *Procedia Economics and Finance, 27,* 594–599.

Mathias, L. (2011, December 18). Florianópolis tem nove pontos de risco muito alto de deslizamentos envolvendo 1.498 moradores. Notícias do Dia. Retrieved April 12, 2017, from https://ndonline.com.br/florianopolis/noticias/florianopolis-tem-nove-pontos-de-risco-muito-alto-de-deslizamentos-envolvendo-1-498-moradores.

Mersal, Amira. (2016). Sustainable urban futures: Environmental planning for sustainable urban development. *Procedia Environmental Sciences, 34,* 49–61.

Miccoli, S., Finucci, F., & Murro, R. (2016). Feeding the cities through urban agriculture the community esteem value. *Agriculture and Agricultural Science Procedia, 8,* 128–134.

Mougeot, L. J. (2000). Urban agriculture: Definition, presence, potentials and risks. Growing cities, growing food: Urban agriculture on the policy agenda, 1–42.

Nicholls, C. I., Pérez, N., Vasquez, L., & Altieri, M. A. (2002). The development and status of biologically based integrated pest management in Cuba. *Integrated Pest Management Reviews, 7*(1), 1–16.

Niñez, V. K., (Ed.) (1985). *Household food production: Comparative perspectives*. International Potato Center.

Nyantakyi-Frimpong, H., Arku, G., & Inkoom, D. K. B. (2016). Urban agriculture and political ecology of health in municipal Ashaiman, Ghana. *Geoforum, 72,* 38–48.

Obeng, L. A. & Wright, F. W. (1987). *The co-composting of domestic solid and human wastes*.

Peng, J., Liu, Z., Liu, Y., Hu, X., & Wang, A. (2015). Multifunctionality assessment of urban agriculture in Beijing City, China. *Science of the Total Environment, 537,* 343–351.

Pölling, B., Mergenthaler, M., & Lorleberg, W. (2016). Professional urban agriculture and its characteristic business models in Metropolis Ruhr, Germany. *Land Use Policy, 58,* 366–379.

Pribadi, D. O., & Pauleit, S. (2016). Peri-urban agriculture in Jabodetabek Metropolitan Area and its relationship with the urban socioeconomic system. *Land Use Policy, 55,* 265–274.

Rakodi, Carole. (1988). Urban agriculture: Research questions and Zambian evidence. *The Journal of Modern African Studies, 26*(03), 495–515.

Ravallion, M., Chen, S., & Sangraula, P. (2007). New evidence on the urbanization of global poverty. *Population and Development Review, 33*(4), 667–701.

Satterthwaite, D. (1997). Sustainable cities or cities that contribute to sustainable development? *Urban studies, 34*(10), 1667–1691.

Smit, J., & Nasr, J. (1992). Urban agriculture for sustainable cities: Using wastes and idle land and water bodies as resources. *Environment and urbanization, 4*(2), 141–152.

Tognetti, C., Mazzarino, M. J., & Laos, F. (2007). Improving the quality of municipal organic waste compost. *Bioresource Technology, 98*(5), 1067–1076.

Tsuchiya, K., Hara, Y., & Thaitakoo, D. (2015). Linking food and land systems for sustainable peri-urban agriculture in Bangkok Metropolitan Region. *Landscape and urban planning, 143,* 192–204.

United Nations (UN). (2014). World urbanization prospects. Department of Economic and Social Affairs, from https://esa.un.org/unpd/wup/publications/files/wup2014-highlights.Pdf. Retrieved January 2017.

United Nations (UN). (2017a). Progress of Goal 1 (n.d.). Sustainable development knowledge platform. Retrieved from https://sustainabledevelopment.un.org/sdg1.

United Nations (UN). (2017b). Progress of Goal 2 (n.d.). Sustainable development knowledge platform. Retrieved from https://sustainabledevelopment.un.org/sdg2.

United Nations (UN). (2017c). Progress of Goal 6 (n.d.). Sustainable development knowledge platform. Retrieved from https://sustainabledevelopment.un.org/sdg6.

United Nations (UN). (2017d). Progress of Goal 9 (n.d.). Sustainable development knowledge platform. Retrieved from https://sustainabledevelopment.un.org/sdg9.

United Nations (UN). (2017e). Progress of Goal Eleven (n.d.). Sustainable development knowledge platform. Retrieved from https://sustainabledevelopment.un.org/sdg11.

United Nations (UN). (2017f). Sustainable cities: Why they matter? From http://www.un.org/sustainabledevelopment/wp-content/uploads/2016/08/16-00055K_Why-it-Matters_Goal-11_Cities_2p.pdf. Retrieved January 2017.

United Nations (UN). (2017g). SDGs. Sustainable development knowledge platform. Retrieved April 05, 2017, from https://sustainabledevelopment.un.org/sdgs.

Van Veenhuizen, R. (2006). *Cities farming for the future: Urban agriculture for green and productive cities*. IDRC.

World Bank. (2007). *Global economic prospects 2007: Managing the next wave of globalization*. Washington.

Zezza, A., & Tasciotti, L. (2008). Does urban agriculture enhance dietary diversity? Empirical evidence from a sample of developing countries. In *12th EAAE congress, people, food and environments: global trends and european strategies*, Vol. 2629. Ghent, Belgium.

Zezza, A., & Tasciotti, L. (2010). Urban agriculture, poverty, and food security: Empirical evidence from a sample of developing countries. *Food policy, 35.4*, 265–273.

Climate Adaptation Practices in Building Constructions: Progress and Limitations in Dar es Salaam, Tanzania

Elinorata Mbuya, Nathalie Jean-Baptiste and Alphonce G. Kyessi

Abstract Climate change and increased floods events are a major challenge for cities in the developing world. African cities face greater challenges having to deal with fewer management capacities simultaneously with increased intensity of climate related events. This paper presents an assessment of measures employed in construction of residential buildings and the impact of recurrent flooding events. Floods adaptation strategies take a central role here and they are assessed through empirical analysis of adaptation practices at household levels in Dar es Salaam, Tanzania. The emphasis is on key features of adaptation such as: knowledge of building technicians and technocrats, regulatory bodies, forms of buildings' contract and the relationship between different key actors. The paper highlights several recommendations to improve the quality of buildings in flood-prone areas, these include forming binding contracts among parties, recognition of informal actors and processes, appropriate building construction as well as storm water drainage systems, buildings' retrofitting as well as identifying maladaptive approaches. The paper concludes with the significance for urban design and policies in climate change adaptation in rapid growing urban environments and calls for guiding fragmented and *adhoc* practices to increase the robustness of different forms of settlements in East Africa.

Keywords Adaptation · Building construction technology · Climate change Urban floods · Tanzania

E. Mbuya (✉) · N. Jean-Baptiste · A. G. Kyessi
Institute of Human Settlements Studies, Ardhi University,
P. O. Box 35124, Dar es Salaam, Tanzania
e-mail: angelambuya@yahoo.co.uk

F. Alves et al. (eds.), *Theory and Practice of Climate Adaptation*,
Climate Change Management, https://doi.org/10.1007/978-3-319-72874-2_29

Introduction

The latest scientific research on climate change shows the world experiencing sea level rise, global temperature rise, warming oceans, and other extreme events (Rosenzweig et al. 2015; IPCC 2012, 2014; Kebede and Nicholls 2012; Thompson 2010). Other consequences are urban floods in riverine and coastal areas, affecting people and infrastructure in low-lying zones (IPCC 2014; De Risi et al. 2013a, b; Sakijege et al. 2012); and water stress linking to hydro-climatic changes, conflicts and human security (Zografos et al. 2013). The relationship between climate change and flood disasters is emphasised by several scholars (Cardona et al. 2012; De Risi et al. 2012). The Summary for Policy Makers (IPCC 2007) suggests that the costs to adapt to flooding could consume at least 5–10% of gross domestic products of respective countries.

Cities are presently experiencing urbanisation around the world, with the fastest urbanisation rates being in Asian and African cities (Seneviratne et al. 2012). The trend of population growth for selected African cities shows that the urban population in Africa is growing (UNHABITAT 2014; World Population Data Sheet 2013; WB 2009). By 2050, Africa's population is projected from 1.1 billion to 2.4 billion. This growth occurs at a pace which put existing infrastructure under significant pressure. Subsequently, settlements are formed informally partially due to lack of adequate surveyed plots for housing. In Tanzania, informal settlements occupy close to 80% of its largest city, Dar es Salaam. Makoba (2009) adds that these settlements have a higher occupancy ratio than those in planned settlements. Consequently, there is a higher risk of human losses from floods. This point is further stressed in the IPCC report, which reveals with high confidence that population growth and urbanisation increase human impacts in settlements, consequently, higher exposure to risks, particularly in the low lying coastal zones (IPCC 2014; CLUVA 2010). These impacts lead to vulnerability of people and built structures, especially in situations where poverty, high population densities and decaying and substandard physical infrastructure pervade (Cernea 2003). Furthermore, the impacts are projected to increase with expected climate change and continued urban development (Meixner et al. 2016; Dai 2012). Hence, there is a growing need to develop adaptation measures that enable the physical structures to withstand floods, rendering them and the urban populace safe.

The urban climate change vulnerability framework illustrated in Fig. 1 shows clear linkages between vulnerability, hazards and adaptive capacity (UNISDR 2004). These linkages are particularly evident in Tanzania where increasing floods are exacerbating existing challenges. However, there are opportunities to reduce disaster risk, and enhance community resilience through employing adaptation strategies. This paper presents a comprehensive vulnerability assessment on flood impacts on residential buildings with suggestions for adaptation.

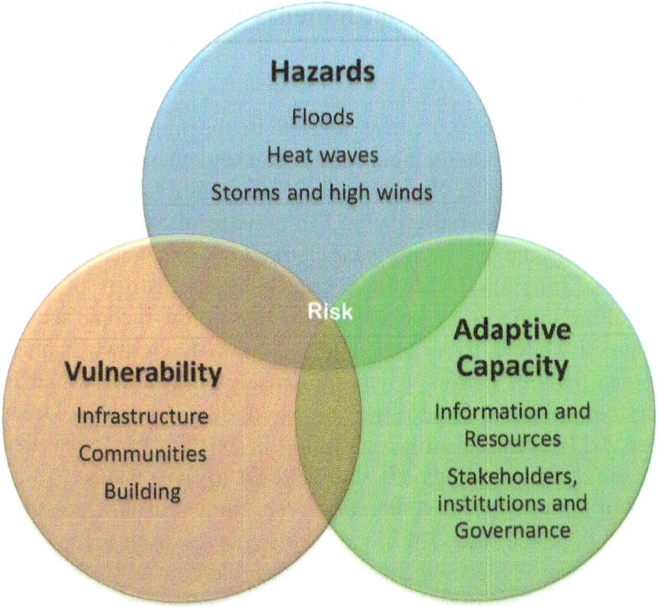

Fig. 1 The Climate Change vulnerability framework. *Source* UNISDR (2004)

Methodology

A case study approach was deployed for this research for a deeper exploration. The settlement of Sunna was selected following 3-stages to ensure we arrive at a case which is information rich. Units of analysis were selected based on stratified random sampling, whereas the strata were identified based on profile of building sites and family income. Then several data collection techniques were used. First, a literature review was conducted on buildings' vulnerability to floods. The documents' review found more insights on specific impacts and existing coping practices against floods. Second, household questionnaires were designed to collect quantitative and qualitative data on house conditions, socio-economic characteristics of exposed residents. Some questions also addressed the motivation and choices to settle in flood threatened areas and the buildings' construction processes. Third, key informant interviews and focus group discussions provided insights on the impacts on building structures. Key informants included local government officers such as the municipal engineer responsible for building inspections, the town planner, a municipal architect, building construction artisans, as well as residents living in the study areas and those who were relocated but returned back. Observations and onsite inspections complemented the information obtained from interviews. In addition, laboratory tests to authenticate the quality of existing building materials against standards, and, computer simulations were conducted.

Susceptibility of Informal Settlements to Flooding: The Case of Sunna

The Sunna sub-ward is a settlement located in the municipality of Kinondoni, Northern Dar es Salaam. On the South-West, the settlement is bordered by a major road whilst two rivers (River Ng'ombe and River Msimbazi) run along the Eastern and the Southern borders respectively. The altitude of Sunna is below the 10 m contour line. During the last twenty years, Sunna has expanded by about 30%; it's location in close proximity with the central business districts of Dar es Salaam (CBDs) and affordability of land made it attractive to low-income settlers. Sunna is partially planned while a larger and unplanned part has grown haphazardly. The unplanned area features houses which are densely built between narrow passages without surface drainages. Field observations conducted between December 2013 and October 2014 reveal constructions filling in existing tiny open areas between houses and some new buildings emerging closest to river banks.

Residents of Sunna experience flooding every rainy season. Occasionally, floods are a result of high sea tide. The flood depths ranges from a few centimetres to 0.72 m, while in some places the water reaches close to 2 m. Sunna experiences severe floods with approximately 10 years return cycles. In such cases, the flood height almost reaches the houses' rooftops. These extreme events have prompted a number of strategies most of which are geared towards protecting the houses while few are non-structural, such as cleaning the rivers of debris and solid waste to prevent the river overflowing their banks. In Sunna some houses are modest with a single room, others are large units with more than five bedrooms. The majority of residents are low-income house owners with a minority being financially privileged. Recurrent flood events combined with poor infrastructure and limited economic resources, makes this settlement particularly vulnerable.

Findings on Analysis of Vulnerability: Urban Buildings, Construction and Finances

This section explains the extent of the flood exposure in Dar es Salaam and further addresses the susceptibility of urban buildings by looking at two main aspects: quality and strength of building materials and construction processes; and the finances.

To begin, an estimation of flooding areas was calculated from exposed areas observed from an overlay of a sub-wards map with a topographic wetness index (TWI) map of Dar es Salaam. Figure 2 illustrates the extent of spatial flooding whereas approximately 10% of planned settlements and 26.8% of informal areas are flooding or prone to floods. It must be noted that approximately 80% of houses are

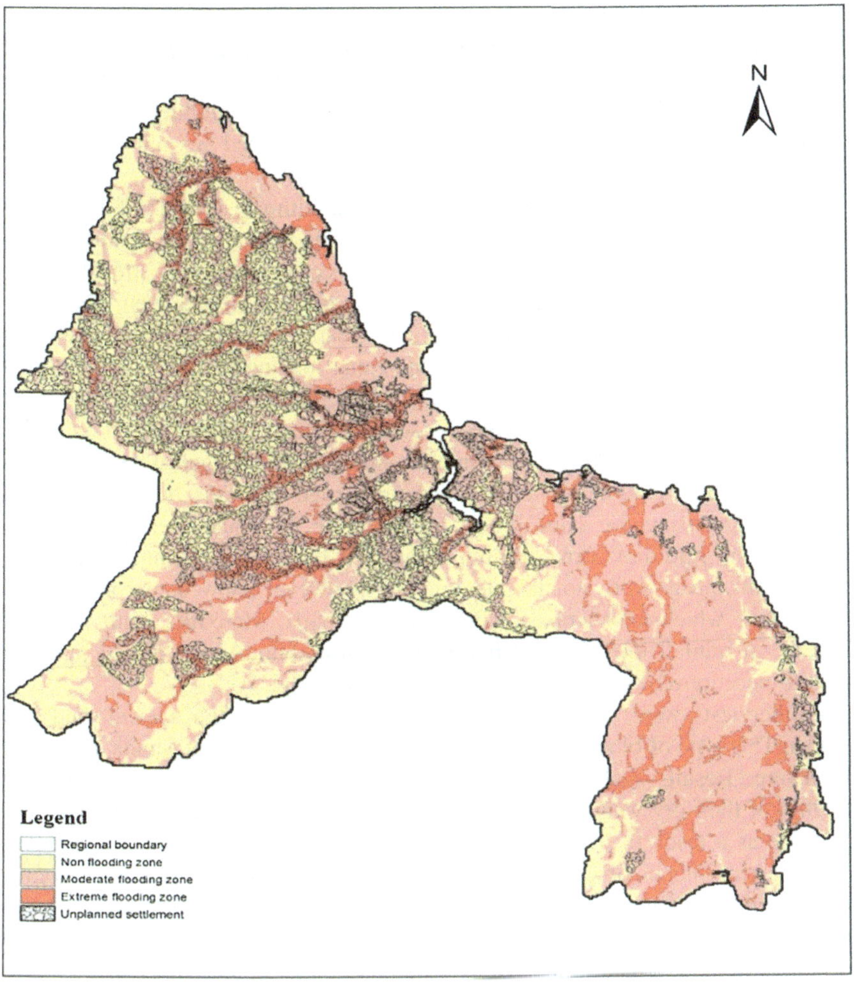

Fig. 2 Map of Flood prone areas in Dar es Salaam. *Source* Author, 2014

built in unplanned areas, and 20% are located in formally planned areas. Correspondingly, a larger proportion of people and assets are at risk, particularly those in flood-prone areas with highly densified buildings and a higher occupancy ratio.

Quality of Building Materials

Building materials tested for their strength were in the form of cement-sand blocks, and wall specimens built from the blocks. Test results showed that cement sand blocks used are weak because failure occurred at an average of 1.6 N/mm^2 whereas the recommended strength standard is 2.8 N/mm2. However, rare cases featured with very few blocks observed to be of standard compressive strength. Such cases were based on decision by building developers to comply with directives issued by building technicians, particularly in observing recommended mix ratios for cement and sand for blocks' making.

Axial load test of the wall specimen revealed an ultimate strength of 991 kN/m^2 while the actual strength (σ_a) was 28.8 kN/m^2. This is a desirable condition for a house to perform against axially applied forces; because the actual strength is less than the ultimate strength reached through testing of the wall specimen. The results imply that the capacity of the wall is sufficient to carry the service stress or self-load in the present condition. Furthermore, a diagonal loading of wall specimens was carried out to establish the wall's resistance. We found that the ultimate strength (σ_u = 423 $kN/m^{2)}$ was greater than the actual strength, (σ_a = 240 kN/m^2). This indicates a desirable conditions for a house while performing against applied forces. Following that a load-deformation test were carried on walls built with similar materials and using the same procedures as done in Sunna. Results showed a deflection of 1.30 mm corresponding to applied force of 9 kN/mm^2. The micro cracks that appear were not threatening, they however show signs of structural weaknesses which may worsen with repeated floods. Moreover, with projected increase in rainfall intensity and floods, the houses built with types of materials need to be improved if they are to adapt to floods (Fig. 3).

Further tests were carried out for water absorption of the cement-sand wall blocks. This type of blocks are used in the majority of houses in Sunna are porous by nature. Results showed that the blocks reached a maximum absorption at 41.90% of their dry weight, with a corresponding strength reduction of 54%. This could be as a result of using incorrect mix ratios of cement, sand and aggregates while making blocks or working with concrete. We found that the mix contained less cement than the required standard, consequently the bonding between cement and other particles (sand and aggregates) is inadequate. The strength reduction for

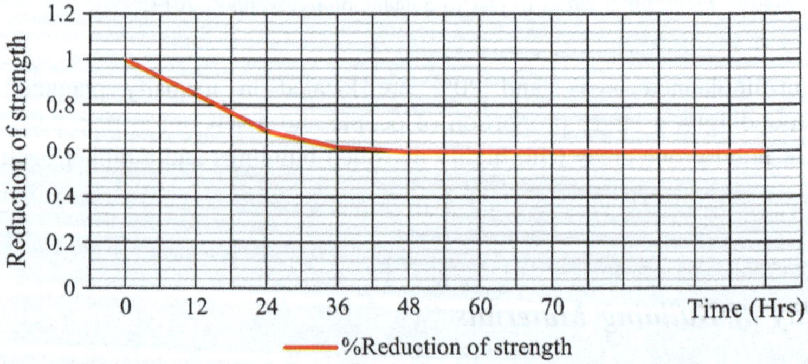

Fig. 3 Water absorption with relative percentage reduction of strength in block works. *Source* Author, 2016

the blocks mostly occur below ground level especially when the damp proof course for protecting the building against moisture is not properly fixed. This weakness is then transferred across the building and gradually weakening it structurally.

In Sunna, 4 per cent of the houses were built without concrete foundations. Indeed, a number of house developers claim that it is not important to build a house on a concrete foundation in Dar es Salaam where the subsoil is sandy, with a good bearing capacity. However, with such arguments, building site are found in building sites with different soil characteristics, including clay, and sandy-clay. In these cases, trenches are excavated from which buildings' are erected. The technology involves constructing a block wall straight from the bottom of the trench, whereas the appropriate method would be to construct a strip of concrete foundation, before construction a block wall.

It must be said, that constructions without foundations may significantly increase buildings' exposure to flood risks because foundations are a major building support. For the most part around the embankment area of the settlements where floods are characterised by a relatively high velocity, buildings with weak or absence of a concrete foundation are subjected to scouring effects and they can be being slightly uplifted by moving waters. In the lowest part (Sunna C) with a flat terrain, the ground is wet with floods retained for weeks long. Many houses have their plinths and foundations submerged in water. In this scenario, floods take a more serious state whereas water ooze from subfloors and external wall to cause indoors wetness and permanent wet walls. This naturally aggravated the conditions of houses which are poorly erected. In contrast, presence of houses built with better standard was apparent. As such, houses built on good foundations formed 68%; those with good quality of blocks for external wall are 40%; and use of strong mortar feature in 70.9% of the houses. On average, 59.3% of building structures used good quality materials and complied with local construction standards[1].

Household Finances

Household finances play a role in accessing a good quality house. In Sunna approximately 29% of households spent a substantial amount of financial resources in the construction and in protecting their houses in anticipation of floods. The table below shows the relationship between the quality of buildings and the households' income.

The provision of foundations, together with the type used was assessed in the Sunna settlement, in relation to household income levels reflecting financial ability. Different foundation types indicating different qualities, were found to be distributed across all income groups, with the biggest concentration being with the

[1]The procedures here means the guidelines and directives issued by building artisans and technicians.

Table 1 Relationship between family income and foundation quality

Provision of foundations for building structures			
Presence/type of foundation	Least income category (%)	Middle income earners (%)	Highest income households (%)
No foundations or compacted sand	7	31	2
Plain concrete foundations	4	25	2
Reinforced concrete foundations	3	15	8

middle income earners (see Table 1). This trend on relationship between household income and quality was also observed for the quality of walling and in the use of building technicians.

In a heterogeneous settlement such as Sunna, financial ability is a limiting factor towards accessing technically sound, durable and resilient houses because sometimes it limits developers' affordability in constructing good houses. However, it is not a very influential factor when considering buildings' strength on the facet of design considerations for construction in flood prone areas. The relationship between income and house quality was found to indicate a positive but not a very strong relationship ($r = 0.359$). This further implies that high income households do not necessarily construct houses with good quality walls. Houses owned both by the affluent or the poor were flood susceptible.

Landuse Planning and Development Control

Additional factors exacerbating floods include weak enforcement of the prepared plans, where the Kinondoni Municipal Council and the Ministry of Lands, Housing and Human Settlements Development issue development and rights of occupancy in areas declared unsuitable for house development without prior site drainage plans. Independent survey of land patches and construction of drainages without proper channelling to bigger drainage channels lead to localised floods, which add on the effects of the floods from upstream. This information is important to decision makers in addressing urban flood vulnerability concerns, which in turn affects buildings' stability. Authorities, especially municipalities, should take each flood event as a learning platform. Wisner et al. (2004) discusses the concept of adaptation to vulnerability being produced gradually.

Climate Adaptation and Coping Strategies for Protecting and Retrofitting Buildings in Sunna

Houses built in flood-prone areas are susceptible to various impacts, the most common being indoor and outdoor inaccessibility, micro-cracking, collapse of adobe housing structures, mud accumulations and instability against floods due to poor foundations design and construction. This implies that householders need to continuously invest in protecting and maintaining their properties.

One of the strategies used to protect buildings include raising toilets to a height presumed to be above the normal flood levels. Raising toilets represent 13% of the coping strategies implemented in Sunna. Typically, the floors for the main house and other outside structures including the toilets would be raised above the expected flood levels. However, residents mostly focused on raising foundations for pit latrines. This once again is based on limited resources; toilets cover a small surface area and consequently lesser amount of building materials than when raising the foundation of the whole house. Only 3% of Sunna residents raised their entire building units. Moreover, we observed that people considered toilets as critical during flooding, there is a general understanding that pit latrines should be protected from overflowing, however there have been instances where latrines are emptied into flowing storm water during heavy rainfalls, exacerbating already precarious sanitary issues in the settlements.

Another coping strategy practiced by 36% of houses involved constructing barrier walls at the front door of the house. This measure was proven to be ineffective in many houses. On one hand, knowledge about flood levels appeared to be scanty amongst builders and home developers. On the other hand, many walls were constructed with little consideration for the direction of the water flow, resulting in water and accumulation of mud on one or more sides of the building units. This caused floods going over the barrier walls in some situations (Fig. 4a).

Majority of buildings are constructed from cement-sand blocks, which are porous materials. Some of the house owners apply a cement lining finishing around the perimeter of the buildings to prevent percolation of floods into the buildings (Fig. 4b.). The efficiency of this mechanisms, which forms 3% of all coping strategies depends on the smoothness of the cement lining, to allow water slip over the building surface. However, in some cases limitation in construction knowledge are observed where inappropriate use of extra strong cement lining lead to shrinkage cracks, causing buildings to be structurally weak.

Coping measures differ depending on the location of the buildings within a settlement; the high elevation areas, the embankment, or lowest elevated zones. Some buildings located at the embankment have their external walls supported by buttress or retaining walls. We found 6% of houses in Sunna using this technique. However, such retaining walls were not built properly; some lacked continuity and were detached from the main houses and failed to properly function as structural supports, showing yet another knowledge limitation in construction in flood prone areas (Fig. 4c).

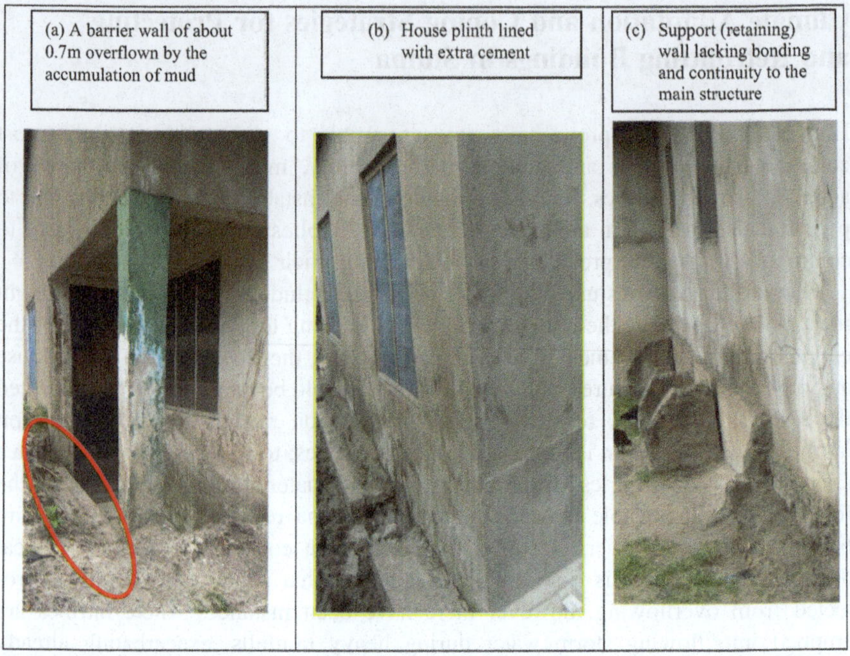

Fig. 4 Coping measures mostly found in building structures in Sunna. *Source* Author, 2014

The use of sand bags laid around houses is another technique used to protect housing structures. Sand bags are half filled and placed lengthwise around the buildings at the bottom of the walls. Positioned as such, they act as shock absorbers against flash floods and lower the impact of debris against the house. When placed appropriately, sand bags can serve as an effective protection measure. They should overlap similarly to a wall made of cement blocks. In Sunna, we found that sand bags were placed haphazardly in a single layer instead of overlapping at half-lengths. Moreover, the sacking materials deteriorated within a short period of time.

Another strategy against floods consists in placing rubble filling or concreting around houses. This aims at controlling water intrusion. This technique is expensive compared to the others, however is proven to be effective as it suppresses water infiltration both inside and around the external house walls.

The techniques mentioned above showed some advantages in strengthening the existing building structures in Sunna, yet some clear disadvantages were seen in particular in the effectiveness of controlling water intrusion in the houses amounting to maladaptation and misconceived safety. To further assess the efficacy of the above measures, a computer simulation through the visual risk and vulnerability (VISK) platform was employed to calculate the structural vulnerability for houses in informal settlements for Dar es Salaam. The simulation uses a methodology discussed in De Risi et al. (2013a) in which houses' risk levels are assessed against

Table 2 Assessment of efficacy of coping strategies

	Building component function in coping with floods	Risk, R (%)
1	80 cm platform	R = 9.28
2	Upgraded material properties	R = 22.81
3	Front door with 1 m barrier	R = 24.43
4	Default model, with a 40 cm platform	R = 24.54
5	(Appropriately) sealed doors and windows	R = 25.57
6	Upgraded vertical connections	R = 26.00
7	House without a platform, but with a 1 m barrier	R = 32.10

various coping strategies. Findings from the VISK analysis show a high risk level was found despite improved sealing of openings, use of good quality materials, good structural vertical connections and a barrier wall is as high as 1.00 m. The lowest risk level was found when all of the above occur and the house foundation is raised to 0.80 m high (see Table 2). The simulation supports the fact that the use of good quality materials and the services of perceived knowledgeable building artisans alone are not sufficient to protect the houses from flooding. A combination of measures are necessary to warrant safety of houses in flood-prone areas.

The Construction Process, Workmanship and Disaster Risk Reduction

Knowledge in building design and construction forms a significant contribution for buildings in informal areas to adapt to climate change induced floods and reduce flood risks. This is in the context of designs and specifications for construction that would reduce vulnerability of the buildings. Limitations in buildings' construction knowledge denies informal settlement dwellers access to flood safe shelter. The discussion on construction knowledge and workmanship is a further reflection on quality of building materials and how they are built into a system to form a complete, performing structure. Propositions of the Construction Management Theory suggests that building materials quality is determined by their sources and the processing methods and procedures (Windapo 2013).

The long term technical performance of building structures is largely determined by properties of materials, but also by the quality of the structure itself, including the specific construction knowledge and skills, as observed in several cases. Examples from Sunna, particularly, show that some home builders use own labour for simpler elements of house construction such as excavation of the foundation trenches, walling and fixing openings. The aim of using household labour force is to save cost. Building artisans are employed for specific building components which are complex and critical for houses' stability, such as setting out the houses, concreting for the foundations and roofing. Some works such as block making is done through family labour, which significantly compromises quality of blocks, for incorrect mix of materials and poor compaction; in such cases, blocks are hand-compacted with a light weight, flat headed tool such as a bottle or wood plank. However, some house owners, actually many, employ building artisans for complete construction works, while others contract registered building contractors, the major reasons being to seek safety of the houses.

Mechanical properties of building materials, the construction process and the general use of building technicians were the focus on an assessment on buildings' quality, which explained how building structures became vulnerable to floods. The technological ability in house construction in terms of floods' resistance is largely lacking, and where practiced, the designs and coping strategies meant to protect building structures are inefficient, making these buildings and people continually exposed to flood risks.

Conclusion

The state of the art of building constructions in flood prone areas is a concern in Dar es Salaam. Present house construction processes do not allow for feedback and accountability. This contributes to the perpetuation of poor construction practices and ultimately this contributes to various types of losses of building assets after each flood. The lack of technical knowledge and misconceptions about the properties of building materials put residents at risk. Local Government Authorities quality assurance in settlements such as Sunna is not guaranteed and homeowners often avoid using qualified building technicians which exacerbate their risk level. Capacity building of stakeholders is critical. We noted in this study that the inefficiency of coping and adaptation strategies is largely caused by insufficient knowledge and skills among all stakeholders. In addition, there are increasing concerns about storm water drainages, neighbourhoods are often poorly drained which is essential in flood prone areas. Also, inappropriate land use planning and building regulations contribute to worsen climate change induced floods through the effect from inappropriate human activity on the built environment.

Adaptation should be context specific, responding to specific climatic related events and multiple risks. In the case of Sunna, building protection knowledge is an essential part of the necessary adaptation strategies and implies the involvement of all actors and stakeholders. The way forward is a holistic approach which implies using a combination of techniques together with employment of skilled construction artisans. Given that urban systems are now integrated, the way towards reducing flood risk may well encompass a tripartite approach including well-constructed building structures, good neighbourhood surface drainage and informed communities.

Acknowledgements Acknowledgements go to the Climate Change and Urban Vulnerability (CLUVA) Project and Ardhi University (ARU) for financing this research. Thank you to the Officials in the Local and Central Government Authorities, residents and community leaders at Magomeni Sunna subward, and my mentors, Dr. N. Jean-Baptiste and Prof. A. G. Kyessi.

References

Cardona, O. D., Van Aalst, M. K., Birkmann, J., Fordham, M., McGregor, G., & Perez, R., et al. (2012). Determinants of risk: Exposure and vulnerability. In managing the risks of extreme events and disasters to advance climate change adaptation. In C. B. Field, V. Barros, T. F. Stucker, D. Qin, D. J. Dokker, K. L. Ebi, et al. (Eds.), *A special report of Working Groups I and II of the Intergovernmental Panel on Climate Change (IPCC)* (pp. 67–80). Cambridge University Press, Cambridge, UK and New York, NY, USA.

Cernea, M. M. (2003). *The urban environment and population relocation*. The World Bank, Washington, DC: World Bank Discussion Papersbreak.

CLUVA. (2010). Climate change and Urban Vulnerability in Africa, Assessing vulnerability of urban systems, populations and goods in relation to natural and man-made disasters in Africa, Seventh Framework Programme.

Dai, A. (2012). Increasing drought under global warming in observations and models. *Nature Climate Change, 3,* 52–58.

De Risi, R., Jalayer, F., De Paola, F., Iervolino, I., Giugni, M., Topa, M. E., et al. (2013a). Flood risk assessment for informal settlements. *Natural Hazards, 2013*(69), 1003–1032.

De Risi, R., Jalayer, F., Iervolino, I., Manfredi, G., & Carozza, S. (2013b). VISK: A GIS-compatible platform for micro-scale assessment of flooding risk in urban areas. In M. Papadrakakis, V. Papadopoulos, & V. Plevris (Eds.), *COMPDYN, 4th ECCOMAS thematic conference on computational methods in structural dynamics and earthquake engineering*. Greece: Kos Island.

De Risi, R., Jalayer, F., Iervolino, I., Kyessi, A., Mbuya E., & Yeshitela, K., et al. (2012) Guidelines for vulnerability assessment and reinforcement measures of adobe houses, CLUVA project deliverable D2.4. Available at http://www.cluva.eu/deliverables/CLUVA_D2.4.pdf. Accessed 24 Dec 2012.

IPCC (The Intergovernmental Panel on Climate Change). (2007). *Climate change 2007: Synthesis report synthesis report. An assessment of the intergovernmental panel on climate change.*

IPCC. (The Intergovernmental Panel on Climate Change). (2014). Climate change 2014: Impacts, adaptation, and vulnerability. Part A: Global and sectoral aspects. In C. B. Field, V. R. Barros, D. J. Dokken, K. J. Mach, M. D. Mastrandrea & T. E. Bilir et al. (eds.), Contribution of Working Group II to the Fifth Assessment Report of the Intergovernmental Panel on Climate Change (pp. 11–32). Cambridge University Press, Cambridge, United Kingdom and New York, NY, USA.

Makoba, N. D. (2009). *Enabling affordable housing for low income households in Tanzania: An integrated cost modelling approach*. Katholieke Universitat Leuven.

Meixner, T., Manning, A. H., Stonestrom, D. A., Allen, D. M., Ajami, H., Blasch, K. W., et al. (2016). Implications of projected climate change for groundwater recharge in the western United States. *Journal of Hydrology, 534*(2016), 124–138.

Nicholls, R. J., & Kebede, A. S. (2012). Indirect impacts of coastal climate change and sea-level rise: The UK example. Research article. *Climate Policy, 12,* S28–S52. Taylor & Francis. Available: www.tandfonline.com/tcpo.

Rosenzweig C. W., Solecki, P., Romero-Lankao, S., Mehrotra, S., Dhakal, T. B. & S. Ali Ibrahim. (2015). *ARC3.2 summary for city leaders. Urban Climate Change Research NetworkUCCRN*. New York: Columbia University.

Sakijege, T., Lupala, J., & Sheuya, S. (2012). Flooding, flood risks and coping strategies in urban informal residential areas: The case of Keko Machungwa, Dar es Salaam, Tanzania. *Jàmbá: Journal of Disaster Risk Studies, 4*(1), 10.

Seneviratne, S. I., Nicholls, N., Easterling, D., Goodess, C. M., Kanae, J. & Kossin, Y. L.et al. (2012). Changes in climate extremes and their impacts on the natural physical environment. In *Managing the risks of extreme events and disasters to advance climate change adaptation*.

Thompson, L. G. (2010). Climate change: The evidence and our options. *The Behavior Analyst, 33*(2), 153–170.

United Nations Human Settlements Programme. (2014). The State of African Cities 2014: Re-imagining sustainable urban transitions. UN-HABITAT.

UNISDR. (2004). Terminology, Basic terms of disaster risk reduction. Reduction Strategy International for Disaster, www.unisdr.org/files/isdrterminology11.pdf. Website visited June 20, 2013.

Windapo, A. (2013). Fundamentals of Construction Management. www.bookboon.com. Website visited November 2, 2014.

Wisner, B., Blaikie, P., Cannon, T., & Davis, I. (2004). *At risk: Natural hazards, people's vulnerability and disasters*. London: Routledge.

World Bank. (2009). *Making development climate resilient: A World bank strategy for sub-saharan Africa* (p. 46947). Report No: The World Bank.

World Population Data Sheet. (2013). Population reference Bureau. A report on world population highlights' focus on wealth and income inequality. www.prb.org.

Zografos, C., Goulden M. C., & Kallis, G. (2013). Sources of human insecurity in the face of hydro-climatic change. *Global Environmental Change, 1207*, 10.

Irrigation Water-Saving Technologies to Adapt to Global Changes in the Yellow River Basin, China: A Hetao Case Study

Qingfeng Miao, Haibin Shi, José Manuel Gonçalves and Luis Santos Pereira

Abstract Water resources management in the Yellow River basin, China, is facing a paradigmatic change in consequence of an unbalanced supply and demand due to an increased demand for water from non-agricultural sectors and a reduced supply due to climate change that reduced precipitation and increased climatic demand. The problem is aggravated by low equity of spatial water allocation in the basin. A supply reduction in the upstream basin area aims to control the water scarcity conditions occurring in the middle and lower reaches of the basin. Forecasted scenarios on water resources allocation and use for agriculture in the upper reaches of the Yellow River basin point out for the need to reduce irrigation water withdrawal and increasing land and water productivity. This paper focus on the Hetao Irrigation District, Inner Mongolia, in upper reaches of Yellow River, where sustainable water saving irrigation is being implemented in response to global changes occurring in the Yellow River basin. That implementation requires technological adaption referring to modernization of canal water conveyance and delivery, which refers to upgrading the hydraulic regulation and control structures, reducing operational runoff wastages, and improving system management. At field level, modern irrigation technologies adapted to local conditions are under implementation. The paper focus on an application to Dengkou area.

Q. Miao · H. Shi
College of Water Conservancy and Civil Engineering,
Inner Mongolia Agricultural University, Hohhot 010018, China
e-mail: imaumqf@imau.edu.cn

H. Shi
e-mail: shb@imau.edu.cn

J. M. Gonçalves (✉)
College of Agriculture, Polytechnic of Coimbra, Bencanta,
3045-601 Coimbra, Portugal
e-mail: jmmg@esac.pt

J. M. Gonçalves · L. S. Pereira
Centro de Investigação em Agronomia, Alimentos, Ambiente e Paisagem (LEAF),
Instituto Superior de Agronomia, Universidade de Lisboa, 1349-017 Lisbon, Portugal
e-mail: lspereira@isa.ulisboa.pt

F. Alves et al. (eds.), *Theory and Practice of Climate Adaptation*,
Climate Change Management, https://doi.org/10.1007/978-3-319-72874-2_30

Keywords Yellow River basin · Hetao irrigation district · Irrigation delivery scheduling · Irrigation demand simulation · Surface irrigation

Introduction

The Yellow River is the second largest river in China and the main source of water in the Northwest and North China, flowing across nine Provinces and supplies water to about 130 million people, mostly farmers and rural people. The Yellow River basin is located in the semi-arid and semi-humid regions which experiences severe water scarcity, with an average annual runoff of 58×10^9 m^3 (Li et al. 2004), and where the climate variability and change as well as human activities have been the two major driving factors that are influencing water resources availability and use (Liu et al. 2008). With the rapid development of social economy, unprecedented dramatic changes were happened in the Yellow River's water cycle in past 50 a multi-year cut-out happened in Yellow River in the 1990s shakes the world. Observed natural runoff decreased obviously, and water scarcity is becoming more and more serious (Liu 2006). In fact, water resources management in the Yellow River basin is facing a paradigmatic change in consequence of an unbalanced supply and demand due to reduced supply caused by reduced precipitation due to climate change and an increased demand for non-agricultural sectors, thus intensifying conflicts between water supply and water demand (Liu et al. 2017). The problem is aggravated by low equity of spatial water allocation in the basin. Supply reduction in upper area aims to control the water scarcity conditions occurring in the middle and lower reaches of the basin (Yu 2006) (Fig. 1).

The Hetao irrigation district (Hetao), located in the upper reaches of the Yellow River, is one of the three largest irrigation districts of China, with 570 000 ha of irrigated land, with 250 km long and more than 50 km wide. Hetao has a continental

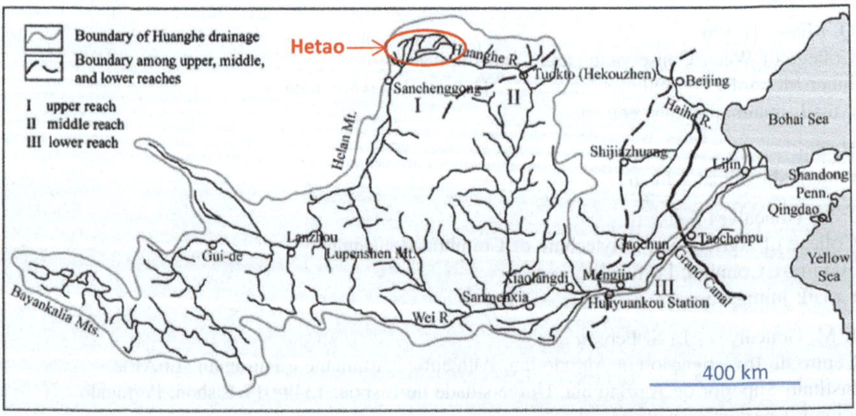

Fig. 1 Map of the Yellow River basin and location of Hetao Irrigation System (from Yu 2006)

monsoon climate, a semi-arid area with an average annual rainfall of about 200 mm, where only irrigated agriculture is feasible. The irrigation canal network is supplied directly from the Yellow River, with a complex conveyance system, being the irrigated area organized by 361 divisions. Each division comprises several sectors, each one supplied by a unique branch canal, with a relative independent operation. Each division has a Water Users Association (WUA), uncharged of water distribution operation and collective hydraulic structures maintenance.

Hetao traditionally uses 5.0×10^9 m^3 $year^{-1}$ of irrigation water. Nowadays, due to increased demand for non-agricultural sectors, the Yellow River Commission aims to reduce the Hetao demand to 4.0×10^9 m^3 $year^{-1}$ over the 2020s, that require to adapt, including the adoption of various water-saving technologies. The problem is aggravated during periods of drought and severe water scarcity if the water use priority is given to non-agricultural uses. However, a heavy reduction in water for agriculture may have very important social consequences, thus a more flexible water allocation policy is advocated, limiting restrictions to irrigation to no more than 10% of the average quota in dry years (Shao et al. 2009). Nevertheless, the need for sustainable use of water and land resources in the basin implies that water conservation and water-saving practices should be implemented for alleviation of water scarcity, however with consideration of impacts of/on groundwater and salinity (IWC-IM 1999; Cai et al. 2003; Pereira et al. 2003; Wang et al. 2005; Xu et al. 2010, 2011).

This paper focuses the main problems related to the adaptation of Upper Yellow River basin irrigation to global changes and water scarcity through the application of water-saving technologies and agricultural development, presenting an application to Dengkou area, Hetao District.

Literature

Hetao climate change was analysed by several authors that concluded (Zhang et al. 2013; Wang et al. 2014, 2016): (i) The annual precipitation has a stable behaviour (Fig. 2a); (ii) The annual mean temperature has a rising trend of 1.2 °C in 50 years, and the maximum and minimum air temperature also significantly increased (Fig. 2b), with a higher extreme temperature distribution with a longer and colder winter and a hotter and shorter summer season; (iii) The decrease of relative humidity, wind speed, sunshine hours and solar radiation, and the increase of the vapor pressure deficit mainly due to the increase of temperature and the decrease of air relative humidity, implying a stable reference evapotranspiration (Fig. 2c).

The adaptation of the irrigated agriculture includes the adoption of several technologies that focus on water saving, improved water use performance and higher water productivity (Pereira et al. 2009). Farmer practices should be adapted to consider changes in rainfall patterns and intensity call for improved water and soil conservation that improves infiltration, soil water storage and control of evaporation from the soil. Adaptation in agriculture needs to be considered and implemented jointly with adaptation in water resources planning and management because water is assuming a

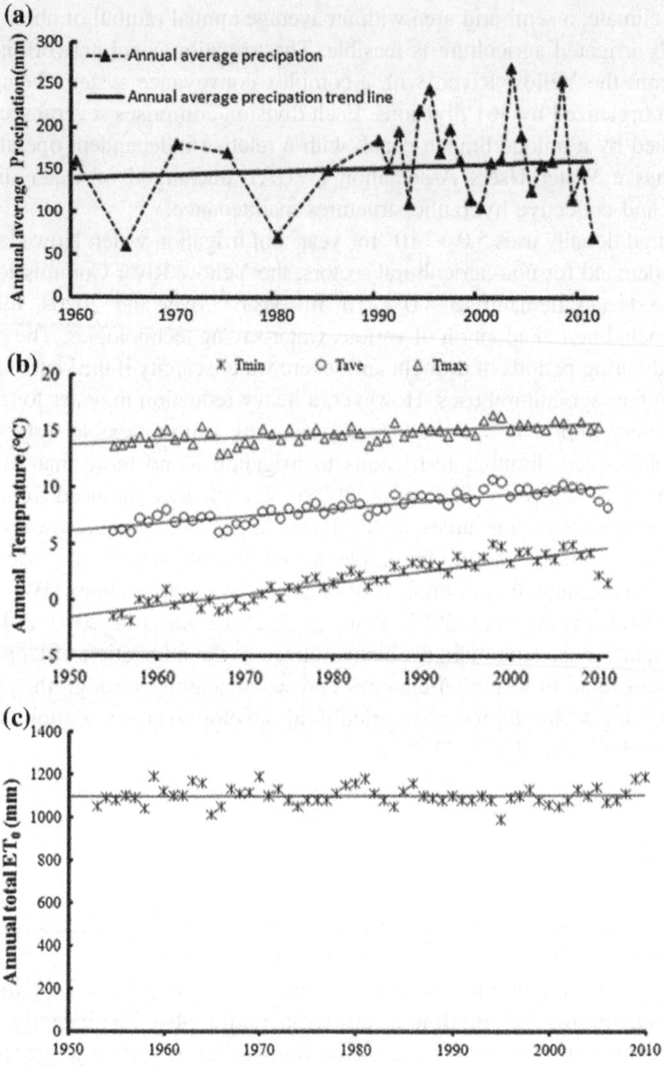

Fig. 2 Hetao variation of annual temperature, precipitation and reference evapotranspiration. Adapted from Wang et al. (2014, 2016)

growing importance when considering future scenarios of climate change. Adaptation does not resume to technologies but has a great relevance in terms of societal awareness, public participation, institutional building, and governance. Knowledge has to expand, to extend and to support improvements at all levels, namely the farm (Pereira 2011a, b), to succeed in food safety, environmental protection and upgrading, improved use of water and welfare of rural populations (Pereira 2017).

Water saving at irrigation district/division level requires irrigation modernization at several scales, namely on-farm, where the major responsibility is of the farmer, and off-farm, comprising the irrigation and drainage collective system. This process is particularly complex because it should be compatible with agricultural development and satisfying the criteria of sustainability. A variety of water saving technologies is considered by Hetao and Inner Mongolia water managers (IWC-IM 1999) aiming to reduce the agricultural demand for water, to improve environmental conditions, and to increase water productivity and farmers' incomes. These technologies interact with groundwater and salinity at various scales (Xu et al. 2010, 2011) and comprise: (a) the improvement of the water conveyance service, mainly through upgrading water delivery and reducing operational runoff wastages; (b) the improvement of farm water use when implementing improved crop irrigation schedules with low to moderate deficit irrigation; and (c) the betterment of irrigation performances through adopting precise land levelling and upgraded farm irrigation technologies for furrowed and flat level basin irrigation systems. Impacts of these technologies were analysed in previous studies applied to Huinong and Hetao irrigation systems (e.g., Gonçalves et al. 2007; Pereira et al. 2007; Xu et al. 2013).

The measures for improving water use at farm scale refer to upgrading crop irrigation scheduling, precise land levelling and water-saving irrigation technologies (Miao et al. 2015, 2016). The management of irrigated agriculture with less available water is a great challenge that implies the adoption of modern technologies that enable water saving, optimising water productivity and improving farmers' income (Pereira et al. 2003, 2007). This includes field irrigation modernization referring to modern technologies of surface irrigation well adapted to local conditions, including precise land levelling of furrowed and flat basin irrigation systems, and optimized crop irrigation scheduling (Gonçalves et al. 2007; Gonçalves and Pereira 2009). Pressurized systems could be economically feasible in the areas where water is pumped from groundwater, particularly for horticultural crops. However, the Autumn irrigation to refill soil root zone and leaching salts need that basin irrigation is applied.

Methods

The study focus on the Dongfeng sector, located in the Uulan Buh area, longitude 106° 44′E–107° 05′E, latitude 40° 17′N–40° 44′N, with total area of 16,314 ha, and supplied by the Dongfeng main canal. The Dongfeng main canal is the largest and the longest main canal in Dengkou county, with a history of more than 100 years. The Dongfeng canal is 63 km long, with a design discharge of 25 m^3/s. It is divided into four sections: the first is 22.5 km long, with a bottom width of 14 meters; the second section is 15 km long with 10 m bottom width; the third one is 8.3 km long with 8 m bottom width; and the fourth is 16.7 km long with a bottom 6 meters wide. These four sections are managed by four departments of Wulanbuhe irrigation area. The annual water supply to the Dongfeng canal is presented in Fig. 3 relative to the period 1988–2012.

The irrigation water reaches the fields through a hierarchical network of canals. Dongfeng comprises 480 sub-sectors, each one with an area ranging from 20 to

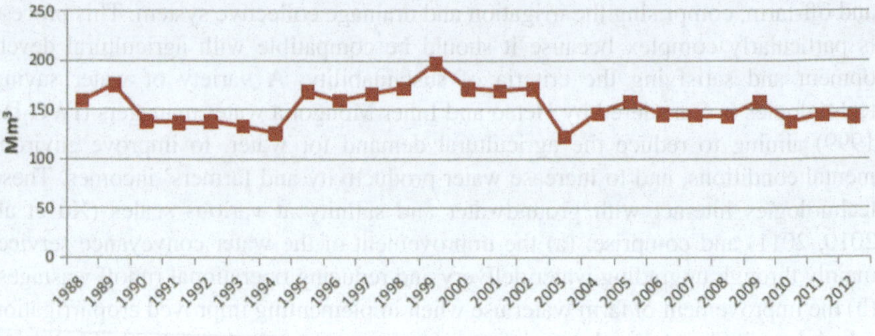

Fig. 3 Annual water supply to the Dongfeng main canal, diverted from the Yellow River during 1988–2012

40 ha. The irrigation system has the following components (Fig. 4): (a) the farm irrigated fields, (b) the unit irrigated area, comprising a variable number of fields, which is the area supplied by a distributor under the direct control of the farmers, (c) the sub-sector (Fig. 4b), which is the area served by a branch and several sub-branch canals and that supply a variable number of distributors, (d) the sector (Fig. 4a), grouping a number of sub-sectors served by branch canals supplied by the main canal.

Table 1 presents the characterization of sub-sectors in Dongfeng. The number of land parcels per sub-sector is 350–550 (median = 394), with an individual area ranging from 500 to 2500 m^2 (median = 750 m^2). Land parcels, which are quite small, are clustered according to a rotational delivery through farm distributors that deliver the water to the fields (typical layout is presented in Fig. 4b). Traditional basin irrigation is adopted. The typical field length is 50 m and field widths vary between 7 and 50 m. For each irrigation event, fields are supplied with a nearly constant discharge during the irrigation time established by the WUA according to the farmer's demand and water availability. The main crops are wheat, maize and sunflower, which relative distribution in 2010–2012 is presented in Table 2. The irrigation season starts by early April for wheat. Usually a maximum of 7 irrigation

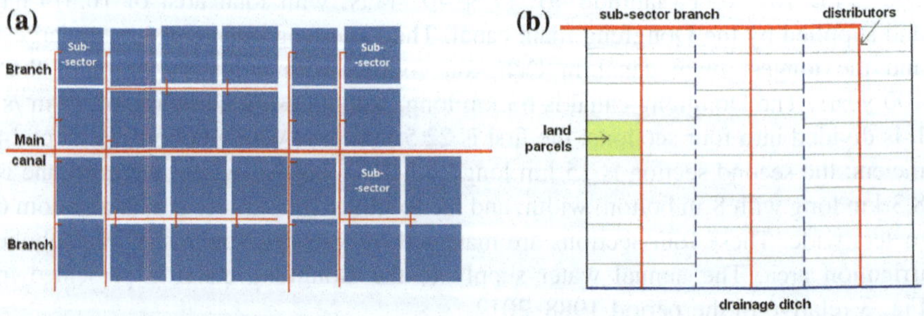

Fig. 4 Schematic representation of **a** a sector network and **b** of a single sub-sector, where red lines represent irrigation canals and dashed blue lines represent drainage ditches

Table 1 Dongfeng sub-sector characterization

Sub-sector area (ha)	Number of distributors	Sub-sector width (m)	Sub-sector length (m)	Number of sub-sectors
21	3	300	700	11
24	4	400	600	20
28	4	400	700	60
32	4	400	800	140
35	5	500	700	110
40	5	500	800	79
42	6	600	700	60

Table 2 Main crops distribution in Dongfeng sector (%)

	2010	2011	2012
Wheat	47	25	16
Maize	27	37	57
Sunflower	26	38	27

events are used: 1–3 events in the spring and 1–3 events in summer, with an interval between events of 10–20 days (Miao et al. 2015, 2016), and the autumn irrigation, after the harvest, to store water, leach salts and improve soil physical conditions through freezing and thawing (Li et al. 2010).

The performance of basin irrigation system depends on the design, land levelling and farmers' management, mainly the irrigation scheduling adopted, the inflow rate applied and the appropriateness of adopted cut-off time. Over-irrigation is often practiced for salts leaching and to avoid water deficits that may result from unexpected delays in water delivery, which are out of farmers' control. Because the water fee relates to the field size and does not depend upon the volume of water used, there is no incentive for water saving. However, adopting a limited deep percolation is desirable for maintaining the salts concentration at an appropriate level. Pereira et al. (2007) and Miao et al. (2015) reported on the limitations of traditional irrigation when aiming at modernization, achieving water saving and controlling salinity. Since farm operational losses are mainly due to deep percolation, it results clearly the need to control deep percolation, thus to provide for improved beneficial water use fractions (BWUF, Pereira et al. 2009) aimed at water saving. This improvement requires a combination of precise land levelling, aimed at uniform and controlled water infiltration while adopting appropriate inflow rates and better targeted time duration of irrigation events, i.e., well-adjusted cut-off times based on adopting an appropriate irrigation scheduling.

It happens, however, that the delivery schedules adopted by the WUA in charge of the canal system management often lead to delayed water applications relative to crop demand, which impedes the adoption of appropriate irrigation scheduling and strongly contributes to farmers' adoption of over-irrigation, particularly for maize since it has low tolerance to salinity and to water stress. The gross and net irrigation

Table 3 Gross and net irrigation depths (GID and NID) and beneficial water use fraction (BWUF) for present and improved systems (2010; Miao et al. 2015)

Crop	Number of irrigations	Present			Improved conditions		
		GID (mm)	BWUF (%)	NID (mm)	GID (mm)	BWUF (%)	NID (mm)
Maize	4	134	60	90	90	89	80
Wheat	3	145	60	87	90	89	80
Sunflower	3	145	60	87	90	89	80

Table 4 10-day delivery schedule schemes (e.g. 2010)

Scheme and crop		10-day periods													
		12	13	14	15	16	17	18	19	20	21	22	23	24	25
A	Maize							1		2		3			4
	Wheat		1		2			3							
	Sunflower				1			2		3					
B	Maize						1		2		3			4	
	Wheat	1		2			3								
	Sunflower						1		2		3				

depths (GID and NID) and the BWUF for present conditions and for estimated for improved ones are presented in Table 3.

Irrigation schedules are based upon Miao et al. (2015, 2016). In the following, 10-days irrigation schedules were organized adopting several schemes considering field data and after analyzing sector the observed inflow hydrographs as for the examples in Table 4.

To simulate the sector demand and delivery, the SEDAM model was applied. This model uses an up-scaling approach for modelling the demand and delivery, starting at the downstream units served by the distributor channels, and aggregating the demand to the sub-branch and branch canals and, finally, to the sector. It was developed for Yellow River basin context (Pereira et al. 2003) and is described by Gonçalves et al. (2007). Field irrigation evaluations and modelling were used for assessing the present situation of farm irrigation performance of wheat and maize, and generating alternatives for surface irrigation improvement (Miao et al. 2015, 2016). To test the demand and delivery simulation for the present scenario, data collected relative to the supply systems were used. The demand at the unit scale was built using results of the irrigation scheduling simulation SIMDualKc model (Rosa et al. 2012) and the surface irrigation simulation model SIRMOD (Walker 1998). The first generated the irrigation dates, depths, and impacts on yields, and the second produced the time duration of the irrigation and the respective performance indicators. The demand by each unit—discharge rate, duration and timing of deliveries—was then aggregated with those of other units supplied by the same sub-branch canal and, then, to the area served by the corresponding branch canal, so

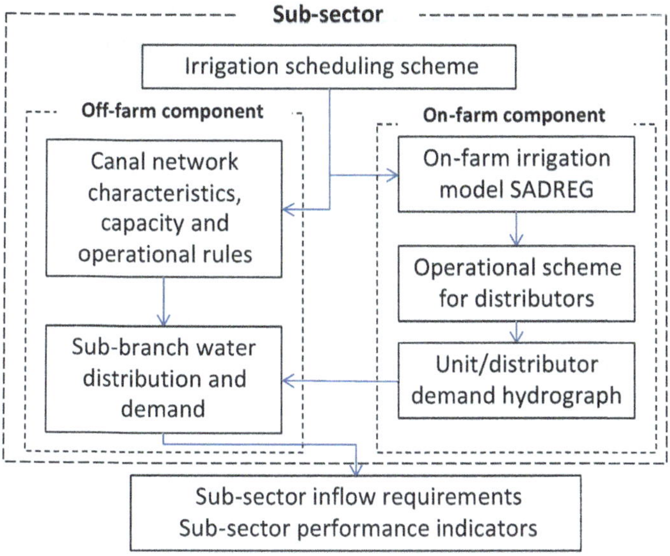

Fig. 5 Schema and components of SEDAM model

producing the sub-sector demand (Fig. 5). Finally, the generated sub-sector demands were aggregated at the sector and the division scales. The water demand at the distributors' level was built from that of the farm fields accounting for delivery losses.

The simulation of the demand at the farm (unit) scale requires information on: (i) the irrigation schedule, i.e. the irrigation timings, duration and depths (D, mm), each computed for a given combination climate—soil—groundwater depth—crop; (ii) the inflow rates into the distributors (Q_T) served by any sub-branch canal calculated from the area irrigated and the probability of occurrence of inflow rates; (iii) the seepage volumes estimated with an efficiency ratio relative to sub-branches and distributors; (iv) the inflow rate at the field level q_F (l/s) determined using field observations, which allows to estimate the number of fields irrigated simultaneously (N_F) with $q_F = Q_T/N_F$; (v) the inflow rates per unit width of basins, which refer to three conditions—low, medium and high, with 1.0, 2.0 and 3.0 l $s^{-1}m^{-1}$—as for the options previously considered in Huinong (Pereira et al. 2007). Irrigation schedule data were used with SIRMOD model, which is integrated in the SADREG model (Gonçalves et al. 2007), for each combination soil infiltration—field length—field width—inflow rate to determine the time duration of the irrigation. These models run for typical data sets and produce a database used as input to SEDAM according to the field characteristics considered, including demand hydrographs, and data on percolation, runoff return flows, labour requirements and crop yield attributes.

The water demand in each sub-sector is estimated with a daily time step. The model first computes the number of days of water distribution considering some flexibility in the deliveries and taking into consideration the current practices. When simulating improved scenarios, the current delivery procedures are modified, mainly those referring to the order of units' irrigation, day-time duration of irrigation, branch inflow rates, and non-used delivered water added to runoff. In addition, the model provides for adopting a fixed rotation water delivery or a more flexible rule in relation to crops demand.

Results

The performance of Hetao traditional and precise zero levelled basins, cropped with wheat, maize and sunflower, was assessed through field evaluation of basin irrigation (Miao et al. 2015). Present water use and water saving data of modernized basin irrigation are presented in Table 5. This assessment shows that better irrigation performance and water saving definitely require improved land levelling, inflow rates and cut-off times in combination with adequate irrigation scheduling practices. The field evaluations of precise land levelled basins showed quite good distribution uniformity (DU), above 90%. However high BWUF and low DP can be achieved when inflow rates are high enough to produce a quick advance, and when the cut-off times were adjusted to reduce application depths to the values required to refill the root zone at time of irrigation. It was concluded that good performance and high water saving may be achieved with basin irrigation in Hetao, nevertheless requiring improved design of farm systems including land levelling and inflow rates, and the implementation of irrigation scheduling to properly select the irrigation timings and the cut-off times (Miao et al. 2016). Moreover, achieving these requirements needs appropriate training and support is given to the farmers to change from traditional to modern irrigation management.

In Hetao, improvements are being implemented for water conveyance and distribution include lining of main, sub-main and distributor canals, and upgrading the respective hydraulic regulation and control structures. The effective system water saving requires the integration of off-farm canal distribution network technologies with the on-farm and field irrigation systems. This development requires the

Table 5 Water saving of modernized basin irrigation (from Miao et al. 2015)

Water use	Maize	Wheat	Sunflower
Present water use (PWU, mm)	484	434	232
Water use in precise zero levelled basins (mm)	340	360	220
Water Saving due to precise land levelling (mm)	144	74	12
Water use in levelled basins, with cut-off adjusted (mm)	313	276	161
Additional water saving (mm)	27	84	59
Total water saving (TWS, mm)	171	158	71
Relative water saving (TWS/PWU) (%)	35	36	31

monitoring of conveyance improvements and delivery schedules, discharges and hydraulic constraints, reducing operational runoff wastages, and improving system management in order to implement a delivery scheduling that matches farm demand, thus controlling operational water wastes and providing for improved land and water productivity, with particular attention to the peak period of delivery to farm distributors, and assessing impacts of delivery schedules on crop growth and yield.

The analysis of present and improved demand and its comparison with supply refers to the period 2010–2012 (Miao et al. 2017). The main canal conveyance ratio applied was 0.85, in agreement with the observed which ranged between 0.80 and 0.90. The applied non-beneficial water use ratio in the sub-sector distributors was 6–8%. Results from the application of SEDAM model are presented in Figs. 6 and 7 and refer to the 10-day time step. In Fig. 6, the simulated demand is compared with the recorded sector supply. For the current scenario and the three years period, results show that the demand was larger than the supply during most of the irrigation season (decades 12–21), which makes evident the occurrence of significant supply deficit. This fact agrees with the current irrigation practice, although over-irrigation at field level is practiced when water is available. During the initial period, from mid-April to mid-May (decades 11–14), the deficit was mild in 2010 and 2011, and there was a surplus in 2012. Water demand during this period is mainly for wheat irrigation, though in 2012 an irrigation was applied prior to sunflower sowing. The high surplus in decades 12–14 in 2012 is explained because the precipitation was relatively high during this period and the delivery was not adjusted to this condition, thus resulting in high delivery losses.

Deficit irrigation (Fig. 6) occurred due to delivery delays and insufficient water availability during the periods when irrigation was practiced. It may be noticed that the period from 10 June to 20 July (decades 17–20) is critical, corresponding to the last irrigation of wheat and the first irrigation of maize and sunflower (Miao et al. 2016). In turn at the late season period, after mid-August (decades 24 and 25), the deficit is less relevant. Results in Fig. 6 show the importance of improving the conveyance and delivery systems through upgrading the hydraulic regulation and control structures to improve water delivery and reduce operational runoff wastages, thus improving the system management in order to implement a delivery scheduling that matches farm demand. The improvement of water use at farm by upgrading and adopting appropriate crop irrigation scheduling, adopting precise land levelling to increase basins distribution uniformity and controlling percolation, are required to control the irrigation demand, so easing canal operation and management to make supply matching demand.

For the improved scenario, mainly resulting from the on-farm modernization (performance described in Table 3), the simulated demand was compared with the 10-day recorded supply (Fig. 7). Results show that there is potential to reduce the sub-sector water demand through on-farm water saving and, therefore, for supply to match demand.

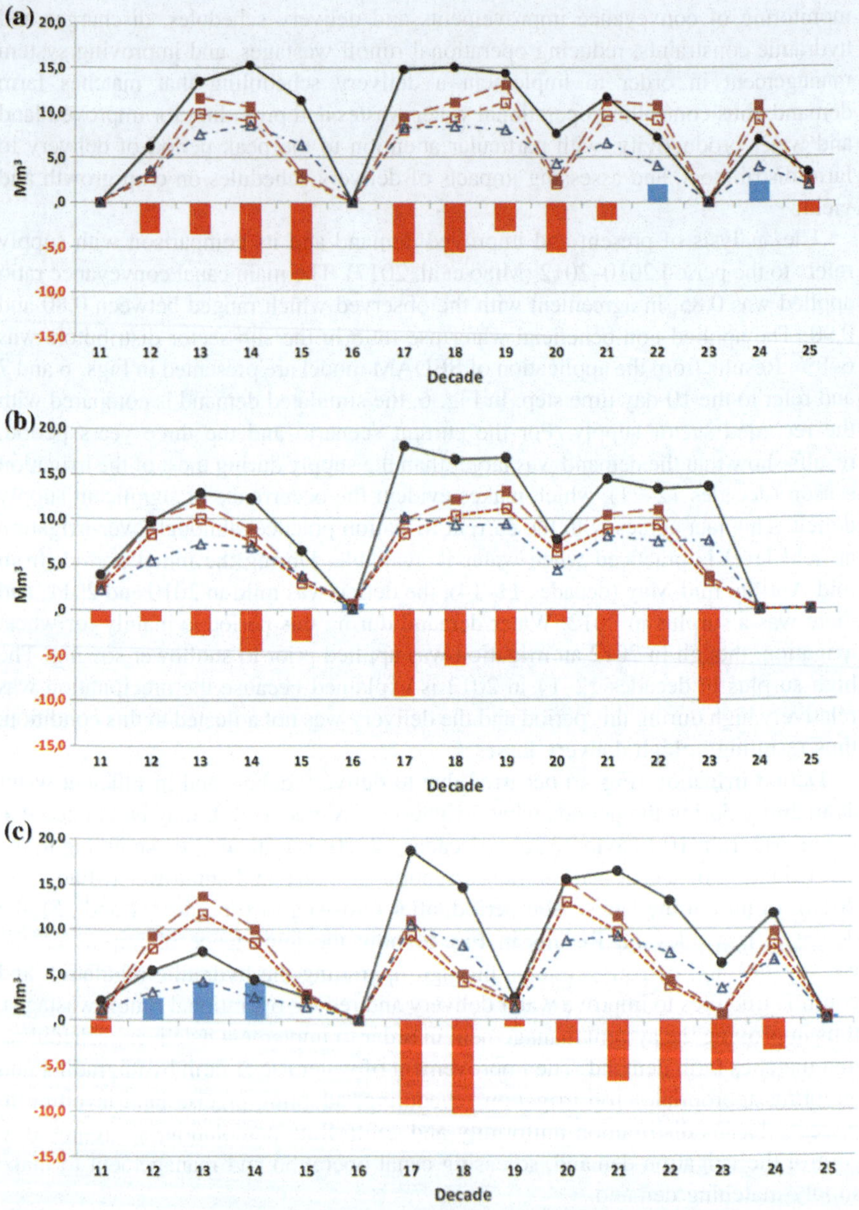

Fig. 6 10-day present sector recorded supply (—■—), sub-sector recorded supply (--□--), sub-sector simulated demand (—●—), and farm beneficial water use (– △–) for **a** 2010, **b** 2011, and **c** 2012. Also depicted the sub-sector supply and demand deficit (■) and surplus (■)

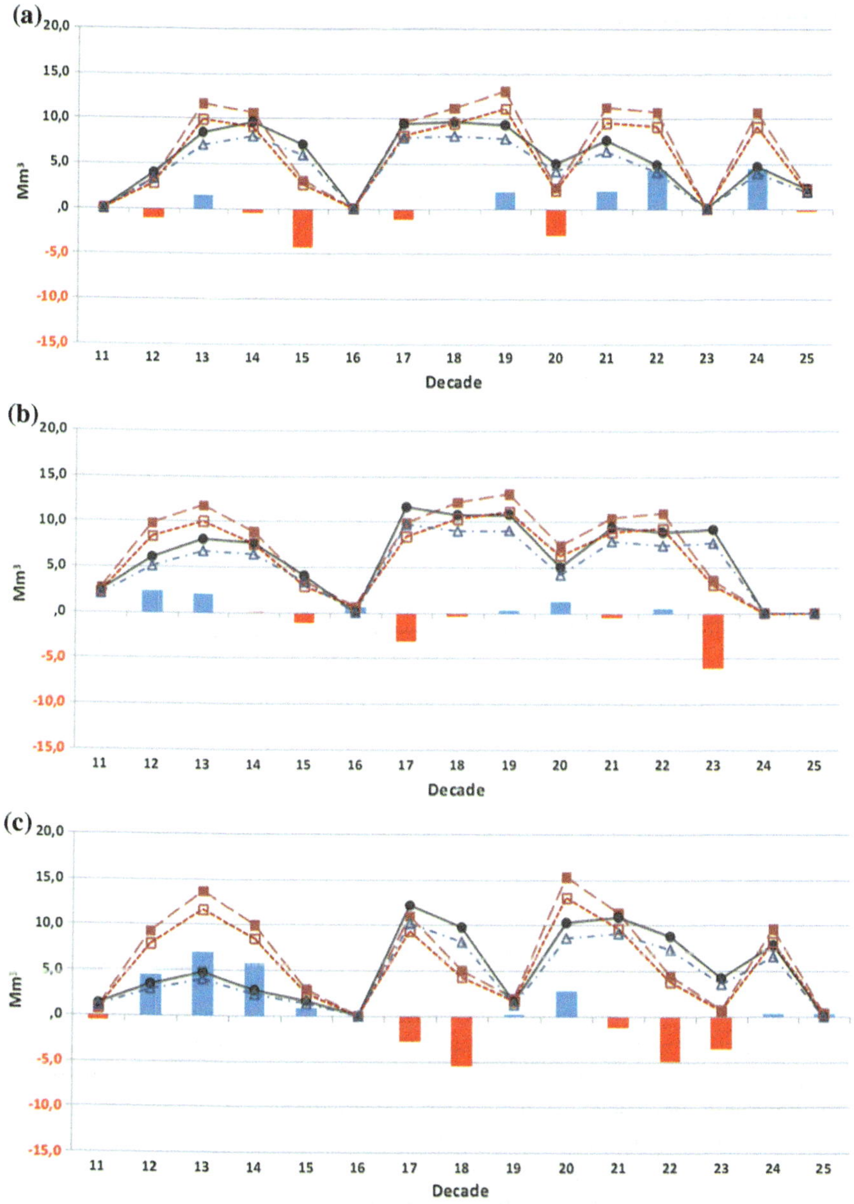

Fig. 7 10-day sector recorded supply (—■—), sub-sector recorded supply (--□--), sub-sector simulated improved demand (—●—), improved on-farm beneficial water use (– △–), sub-sector supply and demand deficit (■) and surplus (■) in years **a** 2010; **b** 2011; and **c** 2012

Table 6 Summary of annual supply-demand balance

	2010		2011		2012	
Supply (Mm^3)	84.7		88.6		82.0	
	Present	Improved	Present	Improved	Present	Improved
Demand (Mm^3)	126.8	81.3	144.5	93.1	120.2	78.9
Deficit (Mm^3)	−46.2	−10.6	−56.5	−11.5	−49.1	−18.5
Deficit (%)	−55	−12	−64	−13	−60	−23
Surplus (Mm^3)	4.2	14.0	0.6	6.9	11.0	21.6
Surplus (%)	4.9	16.5	0.6	7.8	13.4	26.3

Results show that it is likely feasible to adopt full irrigation with only mild irrigation deficits in a few decades (in 2010: 15, 18, 20; in 2011: 15, 17, 23; in 2012: 17, 18, 22, 23). The surplus of water supply in other decades (in 2010: 13, 19, 21, 24; in 2011: 12, 13, 20; in 2012: 12, 13, 14, 20) indicates that there is a significant potential for canal conveyance and delivery water saving. Sector water saving could also be optimized when improvements in the canal conveyance system would be enforced.

Table 6 presents a brief account of the annual supply-demand water balance, so comparing current with improved scenarios for each of the three years analyzed. The present deficit corresponds to a fraction of 50%, that could be reduced to 13% with on-farm improvements. In turn the surplus for improved scenario is identical to the deficit, which point out to the requirement to adjust the delivery to the on-farm demand.

Discussion

Previous studies in Yellow River basin, particularly in Hetao, focusing off- and on-farm irrigation systems, allow the use of consistent modelling tools and knowledge to support planning and managing irrigation modernization aiming to water-saving and adapt to global changes (e.g., Xu et al. 2010, 2013; Miao et al. 2015, 2016). The present study follows the related methodologies applied to Dengkou area, focusing on integrated water-saving technologies to improve off-farm canal distribution network and on-farm irrigation systems. It considered the following steps: (1) assessment of canal conveyance and distribution network at sector level through monitoring the delivery schedules, discharges and hydraulic constraints, with particular attention to the peak period of deliveries to farm distributors; (2) evaluation of water conveyance improvements according to Yellow River official guidelines, namely canals lining, upgrading the hydraulic regulation and control structures, reducing operational runoff wastages, and improving system management in order to implement a delivery scheduling that matches farm demand, thus controlling operational water wastes and providing for improved land

and water productivity; (3) characterization of the field irrigation modernization referring to modern technologies of surface irrigation well adapted to local conditions, including precise land levelling of furrowed and flat basin irrigation systems, and optimized crop irrigation scheduling; (4) assessing impacts of delivery schedules on crop growth and yield; and (5) assessment of delivery and field irrigation improvements including in economic and social terms.

Results show that actual supply is not enough to guarantee full irrigation. However, they also show that a significant reduction of sub-sector demand would occur, which would change the actual situation of deficit irrigation to nearly full irrigation with only mild deficits in peak period. The surplus of water supply in some decades allows significant delivery water saving, that could be optimized when improvements in the canal conveyance system would be enforced. Deficit irrigation is not well controlled by farmers, which implies income losses. Solutions include adopting water saving through improved on-farm irrigation, as demonstrated in previous studies. Achieving these requirements is only possible if appropriate training and support are provided to the farmers, namely to help them changing from traditional to modern irrigation management, and through a good interaction between farmers and WUA that likely allows to implement an appropriate irrigation scheduling program. Nevertheless, improving the delivery schedules is definitely required to base farm betterments. This fact implies revising the needs for upgrading the hydraulic structures and respective operation and management. The use of model SEDAM, combined with models SADREG and SIMDualKc, should be useful to support related assessments and supporting decisions.

Conclusion

Irrigation water management in the Hetao Irrigation District, in the upper reaches of the Yellow River, is facing the challenge of adapting to climate change and to the increased water demand due to the competition from non-agricultural users. Sustainable water saving irrigation in Hetao requires the modernization of conveyance, delivery and on-farm systems. Adopting a decision support system methodology to model the irrigation demand and delivery at sector levels has shown to provide a useful analysis, and that the decision tool SEDAM responded well to analyze the available data and coherently integrate data of multiple scale and multiple sources. The adaptation to climate change aims at sustainable irrigation, thus referring to modernization of canal water conveyance and delivery through upgrading the hydraulic regulation and control structures, reducing operational runoff wastages, and improving system management. At farm level, it requires the implementation of modern irrigation technologies adapted to local conditions, which are under implementation.

The current study case of Dengkou area have shown that a good performance and high water saving may be achieved with basin irrigation if appropriate training and support is given to the farmers to change from traditional to modern irrigation management. Moreover, and a good interaction between farmers and WUA are required to implement appropriate delivery schedules needed to support farm betterments.

Acknowledgements This study was funded by the Key project of National Natural Science Foundation, No. 51539005; and the Key Project of National Thirteenth Five-year Scientific and Technical Support Plan, No. 2016YFC0400205, contracted with the Ministry of Science and Technology, China. The support of FCT, Portugal, through the research unit LEAF (UID/AGR/04129/2013) is acknowledged.

References

Cai, L. G., Mao, Z., Fang, S. X., & Liu, H. S. (2003). The Yellow River basin and case study areas. In L. S. Pereira, L. G. Cai, A. Musy, & P. S. Minhas (Eds.), *Water saving in the Yellow River Basin: Issues and decision support tools in irrigation* (pp. 13–34). Beijing: China Agricultural Press.

Gonçalves, J. M., & Pereira, L. S. (2009). A decision support system for surface irrigation design. *Journal of Irrigation and Drainage Engineering, 135,* 343–356.

Gonçalves, J. M., Pereira, L. S., Fang, S. X., & Dong, B. (2007). Modelling and multicriteria analysis of water saving scenarios for an irrigation district in the Upper Yellow River basin. *Agricultural Water Management, 94*(13), 93–108.

IWC-IM. (1999). *Construction and rehabilitation planning project for water-saving in Hetao Irrigation District of the Yellow River basin. Inner Mongolia.* Hohhot: Institute of Water Conservancy and Hydropower of Inner Mongolia (in Chinese).

Li, C. H., Yang, Z. F., & Wang, X. (2004). Trends of annual natural runoff in the Yellow River basin. *Water International, 29*(4), 447–454.

Li, R. P., Shi, H., Takeo, A., Zhang, Y. Q., Zhang, X. H., & Flerchinger, G. N. (2010). Scheme of water saving irrigation in autumn based on SHAW model in Inner Mongolia Hetao irrigation district. *Transactions of CSAE, 26*(2), 31–36. (in Chinese).

Liu, C. M. (2006). A study of evolutionary laws and maintaining mechanism of renewable capacity of the Yellow River's water resources. *Advances in Earth Science, 10,* 991–998. (in Chinese).

Liu, L., Ma, J., Luo, Y., He, C., & Liu, T. (2017). Hydrologic simulation of a winter wheat–summer maize cropping system in an irrigation district of the Lower Yellow River Basin, China. *Water, 9,* 7. https://doi.org/10.3390/w9010007.

Liu, Q., Yang, Z., & Cui, B. (2008). Spatial and temporal variability of annual precipitation during 1961–2006 in Yellow River Basin, China. *Journal of Hydrology, 361,* 330–338.

Miao, Q., Shi, H., Gonçalves, J. M., & Pereira, L. S. (2015). Field assessment of basin irrigation performance and water saving in Hetao, Yellow River basin: issues to support irrigation systems modernisation. *Biosystems Engineering, 136*(60), 102–116.

Miao, Q., Rosa, R. D., Shi, H., Paredes, P., Zhu, L., Dai, J., et al. (2016). Modeling water use, transpiration and soil evaporation of spring wheat–maize and spring wheat–sunflower relay intercropping using the dual crop coefficient approach. *Agricultural Water Management, 165,* 211–229.

Miao, Q., Shi, H., Huo, Y., Gonçalves, J. M., Paredes, P., & Pereira, L. S. (2017). Assessing improvements at irrigation sector level aimed at water saving. Application of model SEDAM to Dongfeng, Hetao, Yellow River basin. In: *4th Inter-Regional Conference on Land and Water Challenges "Sustainable Innovation in Irrigation", Texcoco, Mexico.*

Pereira, L. S. (2011a). Challenges on water resources management when searching for sustainable adaptation to climate change focusing agriculture. *European Water, 34,* 41–54.

Pereira, L. S. (2011b). Using MODFLOW and GIS to assess changes in groundwater dynamics in response to water saving measures in irrigation districts of the upper Yellow River basin. *Water Resources Management, 25,* 2035–2059.

Pereira, L. S. (2017). Water, agriculture and food: challenges and issues. *Water Resources Management, 31,* 2985–2999.

Pereira, L. S., Cai, L. G., Musy, A., & Minhas, P. S. (Eds.). (2003). *Water savings in the Yellow River Basin, issues and decision support tools in irrigation* (p. 290). Beijing: China Agriculture Press.

Pereira, L. S., Cordery, I., & Iacovides, I. (2009). *Coping with water scarcity. Addressing the Challenges* (p. 382). Dordrecht: Springer.

Pereira, L. S., Gonçalves, J. M., Dong, B., Mao, Z., & Fang, S. X. (2007). Assessing basin irrigation and scheduling strategies for saving irrigation water and controlling salinity in the Upper Yellow River basin, China. *Agricultural Water Management, 93*(3), 109–122.

Rosa, R. D., Paredes, P., Rodrigues, G. C., Alves, I., Allen, R. G., & Pereira, L. S. (2012). Implementing the dual crop coefficient approach in interactive software. 1. Background and computational strategy. *Agricultural Water Management, 103,* 8–24.

Shao, W. W., Yang, D. W., Hu, H. P., & Sanbongi, K. (2009). Water resources allocation considering the water use flexible limit to water shortage—A case study in the Yellow River Basin of China. *Water Resources Management, 23,* 869–880.

Walker, W. (1998). *SIRMOD—Surface irrigation modeling software.* Logan: Utah State University.

Wang, L., Liu, T., Ding, Y., Wang, G., & Liu, X. (2016). Characteristics and tendency of climate change in the Hetao irrigation District in the past 50 years. *Journal of Beijing Normal University, 52*(3), 402–407. (in Chinese).

Wang, X. Z., Gao, Q. Z., & Lu, Q. (2005). Effective use of water resources, and salinity and waterlogging control in the Hetao Irrigation Area of Inner Mongolia. *Journal of Arid Land Resources Environment, 19*(6), 118–123. (in Chinese).

Wang, X., Liu, H., Zhang, L., & Zhang, R. (2014). Climate change trend and its effects on reference evapotranspiration at Linhe Station, Hetao Irrigation District. *Water Science and Engineering, 7*(3), 250–266.

Xu, X., Huang, G. H., Qu, Z. Y., & Pereira, L.S. (2010). Assessing the groundwater dynamics and impacts of water saving in the Hetao Irrigation District, Yellow River basin. *Agricultural Water Management, 98*(2), 301–313.

Xu, X., Huang, G. H., Qu, Z. Y., Pereira, L. S. (2011). Using MODFLOW and GIS to assess changes in groundwater dynamics in response to water saving measures in irrigation districts of the upper Yellow River basin. *Water Resources Management, 25,* 2035–2059.

Xu, X., Huang, G. H., Qu, Z. Y., Xu, X., Huang, G. H., Sun, C., Pereira, L. S., Ramos, T. B., Huang, Q. Z., Hao, Y.Y. (2013). Assessing the effects of water table depth on water use, soil salinity and wheat yield: Searching for a target depth for irrigated areas in the upper Yellow River basin. *Agricultural Water Management, 125,* 46–60.

Yu, L. (2006). The Huanghe (Yellow) River: Recent changes and its countermeasures. *Continental Shelf Research, 26,* 2281–2298.

Zhang, D., Liu, X., & Hong, H. (2013). Assessing the effect of climate change on reference evapotranspiration in China. *Stochastic Environmental Research and Risk Assessment, 27,* 1871–1881.

Emission Trading Schemes and Carbon Markets in the NDCs: Their Contribution to the Paris Agreement

Veronica Caciagli

Abstract At present there are 18 emission trading schemes (ETS) operating in the world. At the COP21 in Paris a clear path for climate and energy policies has been outlined over the coming decades: states have committed themselves to reach zero net emissions by the second half of this century and to maintain the average global temperature rise well below 2 °C, with an additional effort to stay within 1.5 °C. 191 countries, whose emissions cover 98.9% of global greenhouse gas emissions, have already submitted their INDCs/NDCs. The contribution of this paper is twofold. First, it presents a review of the existing ETS and the incumbent ones, like the Chinese national scheme. The ETS currently existing are summarized in a table and than described. Moreover, an overview of the countries which are planning to start new cap-and-trade systems is provided, as well as a table showing which countries included a reference to an ETS in their INDC/NDC. Second, it offers a comprehensive overview about what the countries included in their NDCs about national ETS and international carbon markets. In fact, many developing countries are now considering to access carbon finance to fulfil part of their commitments. But in order for them to receive this kind of finance, and therefore to be able to achieve their climate targets, there has to be a demand for credits. For this reason, understanding whether there will be this match between demand and offer for carbon credits will lead to more efficient choices in policy making, for both sides of the market and thus contribute to the achievement of the Paris Agreement.

Keywords Climate change · Carbon markets · Emission trading
Paris agreement · ETS · NDC · NDCs · COP21 · Cap-and-trade
Climate mitigation · NDCs implementation

V. Caciagli (✉)
Independent Consultant and Journalist, Piazzale Caduti Della Montagnola, 48, Rome, Italy
e-mail: veronica.caciagli@gmail.com

F. Alves et al. (eds.), *Theory and Practice of Climate Adaptation*,
Climate Change Management, https://doi.org/10.1007/978-3-319-72874-2_31

Introduction: Emission Trading Schemes

At the 21° Conference of Parties of the UNFCCC in December 2015, the Paris Agreement was signed. The Agreement contains a pathway to a low carbon future: the Parties have agreed the ultimate goal to maintain the average global temperature rise well below +2 °C, with an additional effort to stay within +1.5 °C. They committed to reach the peak of greenhouse gas emissions (GHGs) as soon as possible and to reduce global net emissions to zero by the second half of this century.

The Paris Agreement is a milestone in the climate negotiations, but studies show that to achieve a maximum increase in temperature of +1.5 °C, more drastic reductions of GHGs are needed. According to Rogelj et al. (2016), the world should reach zero net emissions already between 2045 and 2060.

On the contrary, mitigation pledges submitted voluntarily by the countries with their Intended Nationally Determined Contributions (INDCs) in the lead up to the COP21 and beyond, are not even sufficient to have a reasonable good chance to stay within +2 °C. The world, in the case that the pledges are fulfilled *in toto*, is projected to stabilize the temperature at about +3 °C (Kitous and Keramidas 2015). This is why the Paris Agreement contains also a long-term plan, with a process of review of the nationally determined contributions every five years. If the countries will be determinate in achieving the ultimate goal in terms of temperature increase, they will have to raise their climate change mitigation commitments.

The Paris Agreement leaves the Parties free to accomplish their commitments in the modalities that suit them best: it does not prescribe, for example, the quota of energy efficiency, nor of renewable energy, that should be achieved. It does not describe whether emission trading schemes (ETS) or carbon market should be used.

Nevertheless, energy will play an important role in the decarbonisation process. Energy use accounted for 68% of the total of GHG emissions in 2015 (IEA 2016); also, the demand for energy is foreseen to grow at least in the first half of the century due to the increase of population and well being in some countries and to the necessity to provide access to electricity in some others. This make energy the key sector to reach the targets set in the Paris Agreement about climate change mitigation.

One of the most effective way to reduce GHG emissions is carbon pricing (OECD 2016). This can be done though a tax or through carbon markets: while only pricing the carbon is not enough to fight climate change, as a combination of several other actions is necessary (on forestry, consumers' behaviours, production patterns, etc.), it is part of the solution.

In this document, the focus is on carbon markets and emission trading schemes.

The contribution of this paper is twofold. First, it presents a review of the existing ETS and the incumbent ones, like the Chinese scheme, proving an easy-to-read summary of the current policies about emission trading. The ETS currently existing and scheduled to start are summarized in a table and then described. Moreover, an overview of the countries which are planning to start a new ETS is provided.

Second, it offers a comprehensive overview about what the countries include in their INDCs/NDCs, specifically about emissions trading and generally about carbon markets. The method chosen for this second part was to report all the wording and sentences that the countries used in their INDCs/NDCs, so that the reader can get a quick but comprehensive overview of how carbon markets are included. It follows a brief summary of the main findings.

Information gathered in this paper will be useful to policy makers in the planning of ETS as part of their climate change policies, as well as planning their actions to take part to international carbon markets to access additional carbon finance.

In fact, many countries are now in the phase of planning their policies about carbon markets. Developing countries are planning financial options to receive funding to fulfil part of their commitments. But in order for them to receive this kind of finance, there has to be a demand for credits, coming from industrialised countries. At the same time, industrialised countries are planning whether to buy carbon credits to achieve part of their commitments. For this reason, understanding whether there will be this match between demand and offer for carbon credits, will lead to more efficient choices in policy making, for both sides of the market.

Emission Trading System in the World

An Emissions Trading System (ETS) is an artificial market developed with the aim of limiting the total emissions of certain substances or gases. The main benefit for using carbon markets among the measures to reduce GHG emissions is that they stimulate to achieve emission reductions where they are cheapest, and therefore reaching the targets at a lower economic cost.

ETS are usually developed in the form of cap-and-trade systems: the ceiling (or cap) is based on policy decisions and generally following a downward trend over the years. This cap, made of emission credits, is then distributed among the parties concerned by the regulations, which, at the beginning of the reference period, are assigned through a current account to them (allocation phase). The basic unit of account is 1 ton of CO_2 equivalent (CO_2e). At the end of each period, participants must 'surrender' the amount of emissions they have been allocated and the units are cancelled from their account. Those who have reduced emissions more than the amount of emission credits allocated will have the opportunity the surplus. Those who are in deficit, or who have issued more tonnes of CO_2 equivalent than allowed, will have to fill in the difference by purchasing emission rights. This is how the market is generated.

The cap ensures that climate targets are met. This is the main advantage of this system: the climate policies and the reduction trajectory are decided at the beginning of the period. However, this does not mean that an ETS can deliver by itself the necessary GHG reductions needed: experience has shown that in order to reach long-term climate goals, the reduction trajectory needs to be designed to work together with other policies (ICAP 2017a).

The alternative form of ETS is baseline-and-credit. In this case, there is a certain performance, or baseline, and the amount of credits is calculated in respect to the difference with the baseline. It is the kind of scheme at the base of some project-based crediting systems, like the Clean Development Mechanisms of the Kyoto Protocol and the REDD+.

The national ETS that are currently operating are of the cap-and-trade form.

The emission market instrument is therefore an alternative to the more classical "command and control" policy, through which the public authority sets limits and then checks the compliance. From an economic point of view, an emissions trading mechanism allows to reduce emissions at the lowest overall cost for the economy: in fact, subjects with an abatement cost lower than the price of CO_2e will be incentivized to reduce their emissions, whereas subjects with higher abatement costs will find more economic to buy credits. Thus, according to economists, if global emission systems were to be established, or to link the various national systems, emissions would be reduced with a lower overall cost (Sato et al. 2016).

Emission Trading Schemes Already in Force

At present (20th July 2017) there are 18 emission trading schemes operating in the world; Table 1 that summarizes their main characteristics.

European Emission Trading Scheme
The European Emissions Trading Scheme (EU ETS) is the largest greenhouse gas market currently existent. The EU ETS is a cornerstone of the EU's policy about climate change. It covers more than 11,000 installations in 31 countries (28 EU Member States, plus Norway, Iceland and Lichtenstein) and limits emissions from more than 11,000 heavy energy-using installations (power stations and industrial plants) and airlines operating between these countries. It covers around 45% of the EU's GHG emissions.

It has been operating for more than ten years: the first phase started in 2005. The price of the units (AAUs, Assigned Amount Units), has been volatile through the years and, for long periods, too low: there was a first fall in prices in 2006, due to the initial over-allocation of credits from Eastern European countries, followed to a phase of high prices up to 30 EUR (Ellerman et al. 2010). In the third phase, from 2012, there was a second fall, with 1 ton of CO_2 traded at less than EUR 4: this decrease was mainly attributed to the economic crisis and the reduction in industrial production (Koch et al. 2014; De Perthuis et al. 2014). The market was indeed in a situation of oversupply (Fig. 1).

"With such a low price signal and strong uncertainties regarding the evolution of the carbon price, the 'market design' of the current EU ETS is not sufficient to trigger the necessary investments—commented Cécile Goubet and Anaïs Maillet, of the French Ministry of the Environment, Energy and Marine Affairs. The current

Table 1 Emission trading schemes in the world

No.	Countries	Name of the scheme	Status	Emission coverage ($MtCO_2e$)	Year	Emission coverage (percentage)	Liable entities	Gas coverage	Sectors	Allocation	Offsets and credits
1	28 EU member states, Iceland, Liechtenstein and Norway	EUETS	In force	1939	2017	45%	11,000+	CO_2, N_20, PFCs	Phase one (2005–07): Power stations and other combustion installations with >20 MW thermal rated input, industry (various thresholds) including oil refineries, coke ovens, iron and steel plants and production of cement, glass, lime, bricks, ceramics, pulp, paper and board. Phase two (2008–12): Phase one sectors plus aviation from 2012. Phase three (2013–20): Phase two sectors plus CCS installations, production of petrochemicals, ammonia, non-ferrous and ferrous metals, gypsum, aluminum, nitric, adipic and glyoxylic acid. International Aviation: in 2012, the EU temporarily suspended enforcement for extra-EU flights operating from or to non European countries, while continuing to apply the legislation to flights within Europe. Exemptions for operators with low emissions	Auctioning and free allocation	International offsets
2	Schwitzerland	Swiss ETS	In force	S.2	2017	11[a]	55	CO_2, N_20, PFCs	Industries listed under Annex 6 of the revised CO_2 Ordinance (2S sub-sectors)	Auctioning and free allocation	International offsets
3	Kazakhstan	KAZETS	*Scheduled*	*153.7*	*2017*	*49–50%*	*140*	CO_2	*Energy sector (including oil and gas) mining and chemical industry (>20,000 tCO_2/year). Inclusion thresholds: for phase 1 (2013) and phase II (2014–2015), thresholds are based on 2010 and 2012 emission levels. For phase III, 2014 emission levels are used*	Free allocation	Domestic offsets
4	Ukraine		Scheduled								

(continued)

Table 1 (continued)

No.	Countries	Name of the scheme	Status	Emission coverage ($MtCO_2e$)	Year	Emission coverage (percentage)	Liable entities	Gas coverage	Sectors	Allocation	Offsets and credits
5	9 US States (Connecticut, Delaware, Maine, Maryland, Massachusetts, New Hampshire, New York, Rhode Island, Vermont)	Regional Greenhouse Gas Initiative (RGGI)	In force	89.1	2015	20%	164	CO_2	Fossil fuel electric generating units (equal to or greater than 25 MW)	Auctioning	Domestic offsets
6	California (USA)	California Cap-and Trade Program	In force	370.4	2017	85%	450	CO_2, CH_4, N_2O	First compliance period (2013–14): Large industrial facilities, including cement production, glass production, hydrogen production, iron and steel production, lead production, lime manufacturing, nitric acid production, petroleum and natural gas systems, petroleum refining, pulp and paper manufacturing, including cogeneration facilities co-owned/operated at any of these facilities; electricity generation, electricity Imports, other stationary combustion, CO_2 suppliers. Second compliance period (2015–17): First period sectors plus suppliers of natural gas, suppliers of reformulated blendstock for oxygenate blending and distillate fuel oil, suppliers of liquid petroleum gas in California an —suppliers of liquefied natural gas. Inclusion thresholds: Facilities 225,000 tCO_2e/year	Auctioning and free allocation	Domestic offsets
7	Quebec (Canada)	Quebec Cap-and Trade System	In force	61.1	2017	85%	132	CO_2, CH_4, N_2O, SF_6, HFC. PFC. NO_2 and fluorinated GHGs	First compliance period (2013–14); Electricity, industry (>25,000 tCO_2e/year), Second (2015–17) and third compliance period (2018–20): first compliance period sectors plus the distribution and importation of fuels used for consumption in the transport and building sectors and in small and medium- sized businesses. Inclusion Thresholds: >25,000 tCO_2e/year	Auctioning and free allocation	Domestic offsets

(continued)

Table 1 (continued)

No.	Countries	Name of the scheme	Status	Emission coverage ($MtCO_2e$)	Year	Emission coverage (percentage)	Liable entities	Gas coverage	Sectors	Allocation	Offsets and credits
8	Ontario (Canada)	Ontario Cap-and-Trade Program	In force	142	2017	80–85%	Unknown	CO_2. CH_4. N_2O, NF_3 and fluorinated GHGs	Phase 1 (2017–20): Industrial and large commercial operators including manufacturing, base metal processing, steel, pulp and paper, food processing, facilities, electricity with more than 25,000 tCO_2e/year. Transportation fuel distributors (including propane and fuel oil) over 200 L of fuel/year. Natural gas distributors with more than 25,000 tCO_2e/year	Free allocation	Domestic offsets
9	Tokyo (Giappone)	Tokyo Cap-and Trade Program	In force	13.5	2014	20%	1300	CO_2	Commercial and industrial sectors. Inclusion thresholds: facilities that consume energy more than 1500 kL of crude oil equivalent/year	Free allocation	Domestic offsets
10	Sartama (Giappone)	Target setting emissions trading system in Saitama	In force	6.9	2014	18%	568	CO_2	Commercial and Industnal sectors. Inclusion Thresholds: Facilities that consume energy more than 1,S00kl of crude oil equivalent/year	Free allocation	Domestic offsets
11	New Zealand	New Zealand Emissions Trading St heme (NZ ETS)	In force	41.4	2014	51%	2.364	CO_2, CH_4. N_2O, SF_6, HFCs, PFCs	2008: Forestry (mandatory: deforesting pre-1990 forest land, voluntary: post-1989 forest land). 2010: Stationary energy, industrial processing and liquid fossil fuels (with various thresholds). 2013: Waste and synthetic GHGs (various thresholds). Biological emissions from agriculture must be reported	Auctioning and free allocation	International until May 2015 and domestic
12	Korea	Korean Emissions Trading System (KETS)	In force	551	2017	67.7%	525	CO_2, CH_4, N_20, PFCs. HFCs. Sf6	Phase one (2015–17): 23 sub-sectors.among which steel, cement, petrochemistry. refinery, power, buildings, waste and aviation. Inclusion thresholds: company >125,000 tCO_2/year, facility >25,000 tCO_2/year	Free allocation	Domestic offsets

(continued)

Table 1 (continued)

No.	Countries	Name of the scheme	Status	Emission coverage ($MtCO_2e$)	Year	Emission coverage (percentage)	Liable entities	Gas coverage	Sectors	Allocation	Offsets and credits
13	China	China Emissions Trading System	Scheduled	N/A		N/A	N/A	CO_2	Petrochemicals, chemicals, building materials. Iron and steel, non-ferrous metals, paper making, power (including power generation and grid) and aviation. Inclusion Thresholds: entities with an annual energy consumption of more than 10,000 tons of standard coal equivalent (around 26,000 tCO_2) In any year over 2013–2015 have to report their historical emissions and expect to be covered	Free allocation at the beginning; it it foreseen the introduction of auctioning later	Domestic offsets
14	Beijing (China)	Beijing (Pilot) Emissions Trading System	In force	46	2016	40%	947	CO_2	Industrial and non-industrial companies and entities, including electricity providers, heating sector, cement, petrochemicals, other industrial enterprises, manufacturers, service sector, and public transport. Inclusion thresholds: 5000 tCO_2/year, considering both direct and indirect emissions. Mandatory reporting: 2000 tons of standard coal equivalent energy consumption/year	Free allocation	Domestic offsets
15	Chongqing (China)	Chongqing (Pilot) Emissions Trading Scheme	In force	100.4	2016	40%	230	CO_2, CH_4, N_20, HFCS, PFCs, SF_6	Power, electrolytic aluminum, ferroalloys, calcium carbide, cement, caustic soda, and iron and steel, Inclusion threshold: 20,000 tCO_2e/year	Free allocation	Domestic offsets
16	Fujian (China)	Fujian (Pilot) Emissions Trading	In force	200	2016	<60%	277	CO_2	Electricity, petrochemical, chemical, building materials, iron and steel, nonferrous metals, paper, aviation and ceramics. Inclusion thresholds: Energy consumption of 10,000 tons of coal equivalent (tce)/year for any year between 2013–2015	Auctioning and free allocation	Domestic offsets

(continued)

Table 1 (continued)

No.	Countries	Name of the scheme	Status	Emission coverage ($MtCO_2e$)	Year	Emission coverage (percentage)	Liable entities	Gas coverage	Sectors	Allocation	Offsets and credits
17	Guangdong (China)	Guangdong (Pilot) Emissions Trading	In force	422	2016	60%	280	CO_2	Power, iron and steel, cement and petrochemicals. From 2016: aviation, paper and white cement. Reporting sectors: ceramics, textiles, non-ferrous metals, and chemicals. Inclusion thresholds: 20,000 tCO_2/year or energy consumption 10,000 tons coal equivalent (tce)/year.	Auctioning and free allocation	Domestic offsets
18	Hubei (China)	Hubei (Pilot) Emissions Trading System	In force	253	2016	35%	236	CO_2	Power and heat supply, iron and steel, non-ferrous metals, petrochemicals, chemicals, chemical fiber, cement, glass and other building materials, pulp and paper, ceramics, automobile and general equipment manufacturing, food, beverage and medicine producers. Inclusion threshold: Annual energy consumption more than 10,000 tons coal equivalent (tee) in any year between 2013 and 2015 for the power, steel, non-ferrous, chemicals, petrochemicals, building matenals and pulp and paper sectors and 60,000 tee for other sectors	Free allocation	Domestic offsets
19	Shanghai (China)	Shanghai (Pilot) Emissions Trading	In force	155	2016	57%	368	CO_2	Airports, aviation, chemical fiber, chemicals, commercial, power and heat, water suppliers, commercial, hotels, financial, iron and steel, petrochemicals, ports, shipping, non-ferrous metals, building materials, paper, railways, rubber, and textiles. Inclusion thresholds: various, ranging from 5000 to 100,000 tCO_2e/year.	Auctioning and free allocation	Domestic offsets
20	Shenzhen (China)	Shenihen (Pilot) Emissions Trading	In force	31.45	2015	40%	824	CO_2	Power, water, gas, manufacturing sectors, buildings, port and subway sectors, public buses and other non-transport sectors. Inclusion thresholds: 30000 tCO_2e/year for enterprises; 20.000m2 for public buildings and 10,000m2 for government buildings	Auctioning and free allocation	Domestic offsets

(continued)

Table 1 (continued)

No.	Countries	Name of the scheme	Status	Emission coverage ($MtCO_2e$)	Year	Emission coverage (percentage)	Liable entities	Gas coverage	Sectors	Allocation	Offsets and credits
21	Tianjin (China)	Tianjin (Pilot) Emissions Trading System	In force	160–170	2015	55%	109	CO_2	Heat and electricity production, iron and steel, petrochemicals. chemicals, exploration of oil and gas. Inclusion threshold: 20,000 tCO_2/year considering both direct and indirect emissions	Free allocation	Domestic offsets

Source ICAP 2017, own elaboration

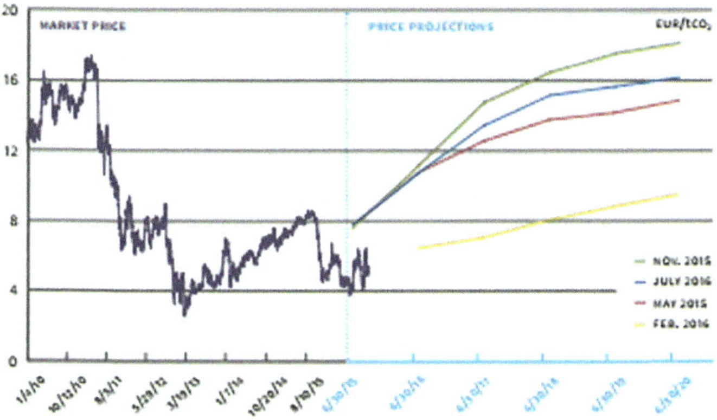

Fig. 1 EU ETS price projections have systematically been above realized prices. *Source* based on Thomson Reuters' consensus of analysts, ICAP, 2017

EU ETS carbon price, which is the international representation of European climate ambition, does not reflect the full cost of our policies and measures" (ICAP 2017a).

Several reforms have been studied to fix the problems within the EU ETS: in 2014 the "backloading reform" took 900 million emission out of the EU ETS 2014 (Caciagli 2013). In 2015, the 'Market Stability Reserve' (MSR) was introduced: the MSR will start operating in January 2019, aiming at addressing the current surplus of allowances and improving the system's resilience to major shocks by adjusting the supply of allowances to be auctioned (European Commission 2015).

The 900 million allowances that were back-loaded in 2014–2016 will be transferred to the reserve rather than auctioned in 2019–2020; unallocated allowances will also be transferred to the reserve. Market analysts estimate that around 550–700 million allowances could remain unallocated by 2020 (EU 2014).

The reserve will operate entirely according to pre-defined rules that leave no discretion to the Commission or Member States in its implementation. Each year, the Commission will publish the total number of allowances in circulation. This will serve as the indicator on whether allowances will be placed in the reserve, and if so how many, or whether allowances will be released from the reserve.

However, at the present time the price of an emission unit of the EU ETS remains low: still around EUR 5 at the end of June 2017.

Many experts suggest to insert a carbon price floor in the EU ETS, that would effectively provide the industries with a predictable price signal (Richstein et al. 2015; Wood and Jotzo 2011, Laing et al. 2013). This proposal was supported also by the new French President Emmanuel Macron, who tried to convince the German Chancellor Angela Merkel to set a common price floor (De Beaupuy and Fouquet 2017)—without succeeding so far.

Swiss Emissions trading System (Swiss ETS)

The Swiss Emission Trading System (Swiss ETS) begun just 3 years after the EU ETS, in 2008. In the first phase, covering 5 years, it was a voluntary system:

companies could alternatively pay a CO_2 levy or opt in for the voluntary emissions trading scheme. From 1st January 2013, the system became mandatory for bigger emitters, while smaller emitters can join on a voluntary basis. It covers now 11% of the Swiss emissions of CO_2, N_2O and PFCs (ICAP 2017a).

In January 2016, Switzerland and the EU concluded a bilateral agreement to link their emissions trading schemes: they will mutually recognize each other's emissions allowances (Vöhringer 2012).

Kazakhstan Emissions Trading System (KAZ ETS)
In 2013, the Kazakh government introduced the Kazakhstan Emissions Trading System (KAZ ETS): the system covered about 140 million tonnes of CO_2e in 2012, which is about 50% of national emissions. It covered the energy sector (including oil and gas,) mining and chemical industry, with a threshold of a minimum of 20,000 tCO_2/year. In February 2017 the system was suspended until 2018 to prepare for some reforms, with new allocation methods and trading procedures planned (ICAP 2017a).

Ukraine
Ukraine is planning to establish a national emissions trading scheme in line with its obligations under the Ukraine-EU Association Agreement, signed and ratified by the country in 2014. The first step will be the development of an MRV system (ICAP 2017a; Federal Office for the Environment 2017).

Regional Greenhouse Gas Initiative (RGGI)
The Regional Greenhouse Gas Initiative (RGGI) was the first compulsory US emissions trading scheme launched in 2009. It covers carbon dioxide emissions from power plants over 24 MW in 9 north-eastern states, including Maine, New York, Connecticut, for a total of about 90 million tonnes of CO_2e in 2012, accounting to about 20% of total emissions.

California Cap-and-Trade Program
The California Cap-and-Trade Program was set up in 2013, including not only CO_2 but also methane and other GHGs and covering 85% of California's emissions, for a total of about 390 million tonnes of CO_2e in 2014. The program has ambitious goals: by 2020, to return to the level of emissions of the year 1990; by 2040, reach a 40% cut compared to 1990; by 2050 a decrease of 80% compared to 1990. The Californian scheme has already been linked to other ETS from Canadian provinces, including Quebec, Manitoba, British Columbia and Ontario, forming the Western Climate Initiative (ICAP 2017a).

Québec Cap-and-Trade System
Québec's Cap-and-Trade system for GHG emissions was created in 2012, with a pilot year in which emitters could familiarize themselves with the initiative. It became mandatory from 1st January 2013, with a first compliance period ending on 31st December 2014. All of Québec's covered emitters could fulfil the requirement. Currently, 85% of the emissions of Quebec are covered. Québec has been a member

of the Western Climate Initiative (WCI) since 2008 and formally linked its system with that of California on 1st January 2014 (ICAP 2017a).

Ontario Cap-and-Trade Program

Ontario Cap-and-Trade Program was introduced for the first compliance period of 2017–2020. It is set to cover around 85% of the emissions: the industrial and large commercial operators including manufacturing, base metal processing, steel, pulp and paper, food processing and facilities, with annual emissions of more than 25,000 tCO_2e; electricity; transportation fuel distributors. Ontario has been a member of the Western Climate Initiative (WCI) since 2008 and intends to link its program with the Californian and Québec carbon market in 2018 (ICAP 2017a).

Tokyo Cap-and-Trade Program

In Japan, there are two systems linked to each other by local bodies: the Tokyo Cap-and-Trade Program and the Target Setting Emissions Trading System for the Saitama Prefecture. The Japanese government is considering adopting a national ETS, while companies across the country can already join a voluntary scheme, the Advanced Technologies Promotion Subsidy Scheme with Emission Reduction Targets (ASSET).

The Tokyo Metropolitan Government Cap-and-Trade Program (TMG ETS), launched in April 2010, is Japan's first mandatory ETS. It covers 20% of the emissions (ICAP 2017a).

Target Setting Emissions Trading System in Saitama

Target Setting Emissions Trading System in Saitama was created in April 2011 as part of the Saitama Prefecture Global Warming Strategy Promotion Ordinance. Saitama's ETS is bilaterally linked to the Tokyo Cap-and-Trade Program (ICAP 2017a).

New Zealand Emissions Trading Scheme (NZ ETS)

The New Zealand Emissions Trading Schemes was introduced in 2008. It covers 52% of emissions, including agricultural ones, for a total of 81 million tonnes of CO_2e in 2014. Despite the introduction of this scheme, New Zealand's emissions have increased by 23% over the period 1990–2014. The government is revising the scheme, to adapt it to the new commitments declared under the Paris Agreement, which foresee a 30% reduction in emissions by 2030 compared to 2005 (ICAP 2017a).

Korean Emissions Trading System (KETS)

In 2015, the Republic of Korea launched the Korean Emissions Trading System (KETS), covering more than 500 installations, corresponding to about 68% of the national GHGs. In addition to carbon dioxide, it considers direct emissions of six GHGs and indirect emissions from electricity consumption. It will play an important role in achieving Korea's target for 2030, to cut emissions by 37% compared to the business as usual (BAU) scenario (ICAP 2017a; Hawkins and Jegou 2014).

China Emissions Trading Systems
Among the various systems in the Asian continent, eyes are currently focused on China, that is foreseen to introduce a new national ETS in 2017.

In China there are already some pilot ETS operating. Since 2013, the province of Shenzhen has launched the first pilot scheme of ETS, covering about 40% of emissions. Subsequently, other programs were implemented in the provinces of Shanghai, Beijing, Guangdong, Hubei, Tianjin, Chongqing and Fujian (ICAP 2017a).

- The Beijing pilot ETS was launched on 28 November 2013. It covers about 45% of the city's total emissions, including both direct and indirect emissions from electricity providers, the heating sector, cement, petrochemicals, other industrial enterprises, manufacturers, the service sector and public transport.
- Chongqing begun its pilot ETS on 19 June 2014. The system covers 230 enterprises from seven sectors: power, electrolytic aluminium, ferroalloys, calcium carbide, cement, caustic soda, and iron and steel. It includes around 40% of the city's total emissions.
- The Fujian Province government introduced a one-year pilot ETS on 30th September 2016, to allow local enterprises to get familiar with an emissions trading scheme before the entrance into force of the national scheme. It covers 277 entities.
- On 19 December 2013, Guangdong started its pilot ETS: it is the largest of the Chinese ETS pilots. Covered sectors account for 60% of the province's emissions. The third compliance period was completed in June 2016 with 100% compliance rate. Guangdong and Shenzhen are the only two Chinese markets open to foreign investors.
- Hubei started it pilot ETS on 2nd April 2014. Til now, Hubei has been the most active market among the pilot ETSs in terms of trading.
- Shanghai was the second Chinese region, after Shenzhen, to start its pilot ETS on 26 November 2013. The pilot covers 57% of the city's emissions, including power, industrial, building, aviation and shipping.
- Shenzhen was the first of the Chinese pilot ETSs to start operations on 18th June 2013. It covers a total of 824 entities, 40% of total emissions. Shenzhen and Guangdong are the only two Chinese markets open to foreign investors.
- Tianjin started its pilot ETS on 26 December 2013. It covers 109 entities of five sectors: heat and electricity production, iron and steel, petrochemicals, chemicals, as well as oil and gas exploration, accounting for 50–60% of the city's total emissions.

The intention to develop a Chinese national ETS scheme was reiterated in the Intended Nationally Determined Contributions (INDC) Memorandum of Understanding. In September 2015, in a joint declaration with the United States on Climate Change, Chinese President Xi Jinping announced that the country will launch its national ETS in 2017.

In March 2017, the State Council approved the cap setting and allowance allocation framework plan of the new national emissions trading system. The first phase, covering 3 years, is planned to start by the end of 2017. It will form the world's largest carbon market, covering 3–5 billion tons of CO_2e. The expected trading volume is of 1.2–8 billion CNY, about EUR 161 million–1.08 billion (ICAP 2017d).

The plan contains some of the rules of the new market: the general principles and the allocation methodology: free allocation will be based on benchmarking and historical carbon intensity (ICAP 2017e; Peng et al. 2017).

Emissions Trading Systems Under Consideration

The following ETS systems are currently under consideration (ICAP 2017a).

- Brazil: in 2009, Brazil developed its National Climate Change Policy (PNMC), which promotes the introduction of a Brazilian market for emissions reductions. Brazil is currently assessing different carbon pricing instruments including an ETS and a carbon tax. Moreover, since 2013, a group of companies have been participating in a voluntary ETS simulation.
- Chile: Chile has already introduced a carbon tax for thermal power generation and is considering deepening the tax or transitioning to an emissions trading scheme. In 2009, Chile established also a local platform for trading voluntary GHG reductions (ICAP 2017a).
- Egypt: Egypt states in its NDC that "a national market for carbon trading may be established. This national market may further be developed into a regional market, which can attracting foreign direct investment in national carbon credit transactions, especially in the Arab and African region" (see "Chapter Understanding Climate Change Adaptation: The Role of Citizens' Perceptions and Appraisals About Extreme Weather Events").
- Japan: in 2016, an Expert Panel advising the Ministry of Environment Japan proposed the introduction of carbon pricing to achieve Japan's 2050 emissions reduction target. Currently Japanese companies can join a voluntary carbon market, called Advanced Technologies Promotion Subsidy Scheme with Emission Reduction Targets (ICAP 2017a).
- Mexico: Mexico intends to implement a national ETS by 2018. In 2014, Mexico introduced a USD 3.50 carbon tax on fossil fuel sales and imports (natural gas exempted). For compliance, companies can also choose to use offset credits from domestic projects. In the same year, a mandatory reporting system was introduced, both for direct and indirect GHG emissions, for facilities with more than 25,000 tCO_2e/year. In 2015, Mexico signed a Memorandum of Understanding with Québec that includes cooperation on ETS. In 2016 Mexico, Québec, and Ontario issued a joint declaration on carbon markets collaboration. In August 2016, the Ministry of Environment and Natural Resources, the Mexican stock exchange (Grupo BMV), and MÉXICO2 (the voluntary carbon platform at the BMV) signed a cooperation agreement to implement a voluntary pilot ETS for 60 entities in the power generation, manufacturing, and transport sector (ICAP 2017a).

- Nova Scotia (Canada): in 2016, Nova Scotia announced the implementation of an ETS from 2018, in line with Canada's federal carbon pricing policy. On 8 March 2017, Nova Scotia released a discussion paper on design options for the new scheme: it would cover combustion and fixed process emissions from industrial facilities that emit from 100,000 t/CO_2e/year and the electricity sector. Petroleum product suppliers and natural gas distributors, over a certain threshold and delivering into the Nova Scotia market will also be covered. The scheme will include less than 20 participants (ICAP 2017b).
- Panama: Panama states in its NDC that "plans to design a carbon market and will continue to participate in the international emissions trading" (see Chapter Understanding Climate Change Adaptation: The Role of Citizens' Perceptions and Appraisals About Extreme Weather Events).
- Russian Federation: in 2014, Russia designed a plan for the development and implementation of emissions reductions activities, introducing an MRV system at a company level. This may lead to the introduction of an ETS.
- Taiwan: in 2017, Taiwan approved its Climate Change Action Guidelines, which describes ten principles on how to achieve Taiwan's climate targets. The third principle is about the implementation of a cap-and-trade system (ICAP 2017c).
- Thailand: carbon markets are mentioned in various document of the Thai government, such as the 11th National Economic and Development Plan (2012–2016) and the National Climate Change Master Plan (2015–2050). In October 2014, the Thailand MRV system for the Thailand Voluntary ETS started its pilot phase, which will last until September 2017.
- Turkey: in 2012, Turkey introduced a mandatory MRV system. Monitoring started in 2015 (ICAP 2017a; PMR 2016). An ETS may be introduced later.
- Vietnam: Vietnam's planned MRV system and crediting NAMA could lead to a cap-and-trade system in the steel sector, which should start in 2020.
- Washington: in 2016, the Washington Department of Ecology published the Clean Air Rule, which regulates the businesses who are responsible for about two-thirds of carbon pollution in Washington, such as transportation, refining, and manufacturing. Under the proposed rule, regulated businesses would be able to comply by reducing their own emissions, buying or trading credits with other regulated parties, funding projects that reduce emissions or acquiring emissions reductions from external carbon markets (State of Washington 2017).

Carbon Markets in the National Domestic Contributions

The National Determined Contributions (NDCs) are the commitments that that the Parties of the UNFCCC intend to fulfil in the post-2020 climate regime.

At the 19° Conference of Parties in Warsaw (COP19), the Ad Hoc Working Group on the Durban Platform for Enhanced Action (ADP), by its decision 1/CP.19, invited all Parties to initiate or intensify domestic preparations for their

INDCs towards achieving the objective of the Convention as set out in its Article 2, which states the objective of the United Nations Framework Convention on Climate Change: "The ultimate objective of this Convention and any related legal instruments that the Conference of the Parties may adopt is to achieve, in accordance with the relevant provisions of the Convention, stabilization of GHG concentrations in the atmosphere at a level that would prevent dangerous anthropogenic interference with the climate system. Such a level should be achieved within a time frame sufficient to allow ecosystems to adapt naturally to climate change, to ensure that food production is not threatened and to enable economic development to proceed in a sustainable manner" (UNFCCC 2012).

In the request, the ADP specified that the invitation was "without prejudice to the legal nature of the contributions, in the context of adopting a protocol, another legal instrument or an agreed outcome with legal force under the Convention applicable to all Parties" (UNFCCC 2013).

One year later, reminding the Parties to communicate their INDCs well in advance of COP 21 (by the first quarter of 2015 by those Parties ready to do so), in decision 1/CP.20 the COP also invited all Parties to consider including their plans about adaptation or an adaptation component in their INDCs.

Moreover, in the decision 1/CP.20 it is further specified that, in order to facilitate clarity, transparency and understanding, the information to be provided by Parties communicating their intended nationally determined contributions may include, as appropriate, inter alia:

- quantifiable information on the reference point (including, as appropriate, a base year),
- time frames and/or periods for implementation,
- scope and coverage,
- planning processes,
- assumptions and methodological approaches including those for estimating and accounting for anthropogenic GHG emissions and, as appropriate, removals,
- how the Party considers that its intended nationally determined contribution is fair and ambitious, in light of its national circumstances,
- how it contributes towards achieving the objective of the Convention as set out in its Article 2 (UNFCCC 2014).

By decision 1/CP.21 the COP reiterated its invitation to all Parties that have not yet communicated their INDC, to do so as soon as possible and well in advance of COP 22 (UNFCCC 2015). The COP also invited Parties to communicate their first NDC no later than when the Party submits its respective instrument of ratification, acceptance, approval or accession of the Paris Agreement.

INDCs are to be renamed NDCs, Nationally Determined Contributions:

(a) after that the Paris Agreement has entered into force;
(b) if the document is submitted prior to the submission of an instrument of ratification, it will be named as INDC. If it submitted together with the instrument of ratification, then it will be named as NDC and thus becomes the first NDC of the country.

From this point, in this paper the term used will be NDC indifferently.

All communicated NDCs have been made available on the public NDC registry. At the time of writing (30 June 2017), only 2 countries have not signed the Paris Agreement: Syria and Nicaragua. 151 Parties have completed the process of ratification; the Paris Agreement entered into force on 4th November 2016, 30 days after the date on which at least 55 Parties to the Convention, accounting in total for at least an estimated 55% of the total global GHGs, had deposited their instruments of ratification, acceptance, approval or accession with the Depositary.

164 NDCs are present on the registry, representing 191 countries. Table 2 provide a comprehensive overview about what the countries include in their NDCs about national ETS and about carbon markets.

Table 2 National emission trading schemes in the NDCs

Country	GHG emissions including LUCF 2012 ($MtCO_2e$)	ETS or carbon mechanisms mentioned in the INDC
China	10,684.29	Promoting Carbon Emission Trading Market: To build on carbon emission trading pilots, steadily implementing a nationwide carbon emission trading system and gradually establishing the carbon emission trading mechanism so as to make the market play the decisive role in resource allocation;—To develop mechanisms for the reporting, verifying and certificating of carbon emissions and to improve rules and regulations for carbon emission trading to ensure openness, fairness and justice in the operation of the carbon ET market
Egypt	287.97	A national market for carbon trading may be established. This national market may further be developed into a regional market, which can attracting foreign direct investment in national carbon credit transactions, especially in the Arab and African region
European Union[a]	4122.64	Domestic legally-binding legislation already in place for the 2020 climate and energy package. The existing legislation for land use, land-use change and forestry (EU Decision 529/2013) is based on the existing accounting rules under the second commitment period of the Kyoto Protocol. Legislative proposals to implement the 2030 climate and energy framework, both in the ET sector and in the non-traded sector, to be submitted by the European Commission to the Council and European Parliament in 2015–2016 on the basis of the general political directions by the European Council, taking into account environmental integrity

(continued)

Table 2 (continued)

Country	GHG emissions including LUCF 2012 ($MtCO_2e$)	ETS or carbon mechanisms mentioned in the INDC
Iceland	2.64	Iceland is a party to the EU-ETS, due to membership of the European Economic Area, and over 40% of Iceland's emissions fall under the trading scheme. It is foreseen that this will continue after 2020. For a small economy like Iceland access to markets is important, as it provides flexibility that bigger economies do not need to the same extent; a single project can contribute a high ercentage of emissions in a small economy like Iceland's but hardly register in bigger economies
Liechtenstein		The decoupling of greenhouse gas emissions and economic growth is the consequence of the establishment of rigorous environmental standards and a comprehensive set of climate related policy measures such as the participation in the European Emissions Trading regime, levying of CO_2 emissions from the combustion of fossil fuels or the obligation to compensate emissions caused by motor fuel consumption
Norway	24.1	Norwegian emissions are covered by the EU ETS, and Norway will through our participation in the ETS contribute to the necessary emission reductions. The EU ETS ensures that no double counting occurs
Panama	22.05	Panama plans to design a carbon market and will continue to participate in the international emissions trading. The country will take voluntary action in the international maritime and air sectors to facilitate the fulfillment of these industries, within the framework of the Organization International Maritime Organization (IMO) and the Civil Aviation Organization (ICAO), through market-based measures
South Korea	661.39	In promoting cost-effective measures for mitigation, Korea adopted the Act on the Allocation and Trading of Greenhouse Gas Emission Permits in 2012, and launched a nation-wide Emissions Trading Scheme (ETS) in 2015. The ETS covers 525 business entities which account for 67.7% of national greenhouse gas emissions

Source CAIT-WRI and UNFCCC, own elaboration

[a]Belgium, Bulgaria, Croatia, Czech Republic, Denmark, Germany, Estonia, Ireland, Greece, Spain, France, Italy Cyprus, Latvia, Lithuania, Luxembourg, Hungary, Malta, Netherlands, Austria, Poland, Portugal, Romania, Slovenia, Slovakia, Finland, Sweden, United Kingdom acting jointly

National ETS and carbon markets in the NDCs

Evaluating the NDCs presents some difficulties. The first problem is formal, connected to the language: not all the NDCs are written in English. Some of them are only in the original languages, which are Spanish, French and Arab. The Chinese NDC reports an English unofficial translation.

The second issue is substantial, about the consistency of the contents: while some of them follow the requirements indicated by the decision 1/CP.20 (UNFCCC 2014) and some further guidance (see for further reference: WRI/UNDEP 2015; Holdaway and Dodwell 2015), some others do not. Therefore, the NDCs do not contain the same types of information.

Also, the inclusion of the specific measures to be followed to reduce GHGs is not required by UNFCCC, nor the country's position on carbon markets. They are left to the country's discretion. Moreover, in some cases the interpretation is not univocal: for example in the case of Cuba, it states that "priority should be given to public financing" but does not seem to completely role out carbon markets. The interpretation leaves space for doubts.

Tables 2 and 3 summarizes what the Parties wrote about emission trading schemes and carbon markets in their NDCs. Because of the above mentioned issues, the choice is to report exactly what the countries wrote in their NDC (with a translation into English where the original language was used).

Only 8 Parties mention explicitly emissions trading schemes on achieving their climate change mitigation targets: China, Egypt, European Union, Iceland, Liechtenstein, Norway, Panama and South Korea.

Kazakhstan, Switzerland, New Zealand, Canada, USA do not mention their national or state cap-and-trade systems. Of the countries were a legislation is under way to introduce a national carbon market, only China, Egypt and Panama mention it.

When it comes to a reference to international carbon markets, the number of Parties that mention them in their NDC grows: 110 Parties included a section, a paragraph or a phrase about the national position on the use of international carbon credits. This amounts to the 67% of the NDCs submitted.

Only 2 countries, Bolivia and Venezuela, declare that their strong opposition against carbon markets. Cuba notes its preference for public finance.

62 countries would like to sell carbon credits, while only 7 countries declare that they are willing to buy them (Canada, Costa Rica, Japan, New Zealand, South Korea, Switzerland, Turkey). Of them, 34 countries would like to sell REDD + credits.

The rest does not explicitly declare their intention about carbon markets, expressing a generic interest or support.

From the first round of presentation of the NDCs, it appears the possibility that the demand for international carbon assets for compliance would be declining in the nearest future. The huge discrepancy between intentions to buy and to sell opens to new questions about the chances for developing countries, eager to receive additional revenues to fulfil part of their climate mitigation commitments through the carbon markets, to obtain this kind of finance in the amount that they expect.

Table 3 INDCs, carbon mechanisms and carbon credits

	Country	GHG emissions including LUCF 2012 ($MtCO_2e$)	ETS or carbon mechanisms mentioned in the INDC	Carbon credits in the INDC	Willingness to buy internationa 1 carbon credits	Willingness to sell international carbon credits	Against carbon markets
1	Afghanistan	30.76		Among the "Technology and Capacity Building Needs": Carbon finance and project development skills	No	Yes	
2	Albania	7.22		Albania intends to sell carbon credits during the period until 2030 to contribute to cost-effective implementation of the low emission development pathway and its sustainable development. Albania foresees that for the utilization of international market mechanism is conditional on having effective accounting rules developed under the UNFCCC to ensure the environmental integrity of the mechanisms	No	Yes	
3	Algeria	189.08					
4	Andorra						
5	Angola	199.72		The Republic of Angola recognizes the roll that Carbon Market can play for the mobilization of resources and promotion of the development and transfers of climate friendly technology. Presently, the potential income from REDD+ projects in the country is considered to be substantial. Calculating the potential emission reductions from REDD + activities in Angola demonstrates that there is mitigation potential if deforestation is avoided completely. Assuming that the baseline is entirely based on historical emissions, avoided emissions are calculated by multiplying the annual deforestation in Angola, estimated to be 124,800 ha per year (based on numbers from the period 1990–2010), by 82 tC/	No	Yes	

(continued)

Table 3 (continued)

	Country	GHG emissions including LUCF 2012 ($MtCO_2e$)	ETS or carbon mechanisms mentioned in the INDC	Carbon credits in the INDC	Willingness to buy internationa 1 carbon credits	Willingness to sell international carbon credits	Against carbon markets
				ha, which is the approximate amount of tons of carbon stored per ha in the country"s forests. Avoiding deforestation, alone, in Angola has the potential to contribute to more than 35 million tons in CO_2 emission reductions every year. The budget required is above 500 million USD			
6	Antigua and Barbuda	1.14		Antigua and Barbuda considers the establishment of an international market mechanism as an important complementary option to reduce total costs associated with limiting GHG emissions and to assist global efforts limiting temperature to 1.5 °C above ore-industrial levels. Antigua and Barbuda acknowledges the potential for a renewed and reformed Clean Development Mechanism to fulfill this roll through its existing structure. The final mechanism should be a robust system that guarantees transparency and environmental integrity, and delivers real, permanent and verified emissions reductions and ensures that double counting is avoided	No	Yes	
7	Argentina	405.03					
8	Armenia	9.3		In case of non-exceeding its total emissions quota (633 million tons) set for the period of 20152050 Armenia can credit non-utilized reduction to 'carbon market', or transfer it to the balance of emissions limitation envisaged for the period of 2050–2100	No	Yes	

(continued)

Table 3 (continued)

	Country	GHG emissions including LUCF 2012 ($MtCO_2e$)	ETS or carbon mechanisms mentioned in the INDC	Carbon credits in the INDC	Willingness to buy internationa l carbon credits	Willingness to sell international carbon credits	Against carbon markets
9	Australia	685.05					
10	Austria						
11	Azerbaijan	69.34					
12	Bahrain	33.25					
13	Bangladesh	189.86		Bangladesh does not rule out the use of international market based mechanisms in line with agreed modalities and accounting rules	No mention	No mention	
14	Barbados	3.35		Intention to use market-based mechanisms to meet contribution: Clean Development Mechanism (COM), Nationally Appropriate Mitigation Action (NAMA)	No	Yes	
15	Belarus	75.35					
16	Belgium						
17	Belize	14.59		Belize is willing to explore the potential of market mechanisms, including COM and other mechanisms under the UNFCCC process, that demonstrate environmental integrity; result in real, long-term, verified mitigation outcomes; prevent double counting; and which are accessible	No	Yes	
18	Benin	21.76					
19	Bhutan	−7.97		Pursue sustainable and clean hydropower development with support from CDM or other climate market mechanisms to reduce emissions within Bhutan and the region by exporting surplus electricity. The forest monitoring and inventory system being developed in conjunction with a national forest monitoring system for REDD+ will	No	Yes	

(continued)

Table 3 (continued)

	Country	GHG emissions including LUCF 2012 ($MtCO_2e$)	ETS or carbon mechanisms mentioned in the INDC	Carbon credits in the INDC	Willingness to buy internationa l carbon credits	Willingness to sell international carbon credits	Against carbon markets
				enable monitoring and assessment of forest cover overtime			
20	Bolivia	136.47		Eradication of commodification of nature and carbon markets promoting business climate millionaires, which do not solve the problem of the climate crisis	No	No	Yes
21	Bosnia and Herzegovina	27.75		International Market Based Mechanisms: Conditional emission reduction is only possible with international support	No	Yes	
22	Botswana	23.9		Market Mechanisms: Botswana will use market mechanisms under the convention	No	Yes	
23	Brazil	1823.15		Brazil reserves its position in relation to the possible use of any market mechanisms that may be established under the Paris agreement. Brazil emphasizes that any transfer of units resulting from mitigation outcomes achieved in the Brazilian territory will be subject to prior and formal consent by the Federal Government. Brazil will not recognize the use by other Parties of any units resulting from mitigation outcomes achieved in the Brazilian territory that have been acquired through any mechanism, instrument or arrangement established outside the Convention, its Kyoto Protocol or its Paris agreement	No mention	Yes	

(continued)

Table 3 (continued)

	Country	GHG emissions including LUCF 2012 ($MtCO_2e$)	ETS or carbon mechanisms mentioned in the INDC	Carbon credits in the INDC	Willingness to buy internationa l carbon credits	Willingness to sell international carbon credits	Against carbon markets
24	Brunei	22.07		Expertise in climate mitigation and adaptation needs to be enhanced. This includes capacity to study and explore the potential adoption of international market mechanisms as a means of achieving its mitigation objectives	No	Yes	
25	Bulgaria						
26	Burkina Faso	33.01		Burkina Faso supports the use of market mechanisms such as the Clean Development Mechanism (CDM) as an effective monitoring, reporting and verification tool for mitigation activities and an instrument for results-oriented financing. Thus, Burkina Faso supports the use of certified emission reduction units (CER's) delivered by CDM project, programmes and activities to reach the pre-2020 mitigation goals. Hence, payments for carbon such as to make this economically viable in the specific context of the less advanced countries, the developing countries and the small developing island states is a priority. To do this, the establishment of new accounting rules within the framework of the UNFCCC is necessary to guarantee the environmental integrity of market mechanisms and avoid double counting. These accounting rules will also be introduced for the EIF in order to achieve the required financial transparency	No	Yes	

(continued)

Table 3 (continued)

	Country	GHG emissions including LUCF 2012 ($MtCO_2e$)	ETS or carbon mechanisms mentioned in the INDC	Carbon credits in the INDC	Willingness to buy internationa 1 carbon credits	Willingness to sell international carbon credits	Against carbon markets
27	Burundi	6.91		In terms of compensation for any loss of revenue or for the restriction of certain economic activities due to the implementation of the INDC programme, Burundi will rely on international greenhouse gas emissions compensation mechanisms and on current national legislation. In terms of forestry in particular, the plan is to promote the development of ecosystem services	No	Yes	
28	Cambodia	49.12		The support received will be channelled through bilateral and multilateral mechanisms, including market based mechanisms. […] Cambodia is already participating in REDD* mechanism with respect to forestry related actions	No	Yes	
29	Cameroon	191.2		Cameroon supports the inclusion of international carbon markets in a post-2020 climate agreement and proposes that such an instrument, coupled with an appropriate accounting regime, could be used to help finance certain investments in low-carbon infrastructure and for the resilient to climate change. Cameroon considers that some of the action in this INDC, or additional actions, could be financed in whole or in part through the international transfer of carbon assets by ensuring respect for the principles of environmental integrity and transparency	No	Yes	
30	Canada	856.28		Canada may use international mechanisms to achieve its 2030 target, subject to robust systems that deliver real and verified emissions reductions	Yes	No	

(continued)

Table 3 (continued)

	Country	GHG emissions including LUCF 2012 ($MtCO_2e$)	ETS or carbon mechanisms mentioned in the INDC	Carbon credits in the INDC	Willingness to buy internationa 1 carbon credits	Willingness to sell international carbon credits	Against carbon markets
31	Cape Verde	0.43		Cabo Verde supports the use of market-based mechanisms to implement and achieve the conditional portion of the contributions mentioned in this document. Several conditional measures envisaged may be financed through mechanisms and/ or carbon markets, including the Clean Development Mechanism, new market and nonmarket based mechanisms, and credited NAMAs. A proper GHG accounting system needs to be established to address the risk of, and ultimately avoid, doublecounting of outcomes, in accordance with UNFCCC guidance and technical specifications	No	Yes	
32	Central African Republic	60.52		Moreover, the Central African Republic, a non-Annex 1 country, supports the inclusion of international market instruments, such as the Clean Development Mechanism, in a post-2020 climate agreement. Such an instrument may be used to help finance certain investments in low-carbon infrastructure that is resilient to climate change	No	Yes	
33	Chad	43.27		The country wants to encourage investment in mitigation projects on its own soil thanks in particular to the Clean Development Mechanism (CDM) and the REDD+ program. These initiatives should place particular emphasis on co-benefits for local populations, such as (i) access to sustainable energy for all, (ii) creation of local jobs, (iii) reduction of environmental impacts health and the environment, (iv) reductions in inequality including gender, and (v) respect for human rights	No	Yes	

(continued)

Table 3 (continued)

	Country	GHG emissions including LUCF 2012 ($MtCO_2e$)	ETS or carbon mechanisms mentioned in the INDC	Carbon credits in the INDC	Willingness to buy internationa l carbon credits	Willingness to sell international carbon credits	Against carbon markets
34	Chile	93.74					
35	China	10684.29	II. Policies and measures to implement enhanced actions on climate change: […] L. Promoting Carbon Emission Trading Market to build on carbon emission trading pilots, steadily implementing a nationwide carbon emission trading system and gradually establishing the carbon emission trading mechanism so as to make the market play the decisive role in resource				
36	Colombia	199.68		With the objective of contributing to achieve the emissions reduction target, with a focus on cost-efficiency, Colombia will explore the use of market instruments (or other economic instruments) that guarantee the principles of transparency and environmental integrity, which result in real, permanent, additional, verified mitigation outcomes and prevent double counting	No mention	No mention	
37	Comoros	0.6					
38	Cook Islands	0.19					
39	Costa Rica	4.97		Costa Rica reserves its sovereign right to use international compensation units to accomplish its goals within the National Contribution or, as well, within its Domestic Compensation Market. Any compensation units traded abroad will be registered in the National Emissions Inventory to avoid double accounting	Yes	No mention	

(continued)

Table 3 (continued)

	Country	GHG emissions including LUCF 2012 ($MtCO_2e$)	ETS or carbon mechanisms mentioned in the INDC	Carbon credits in the INDC	Willingness to buy internationa 1 carbon credits	Willingness to sell international carbon credits	Against carbon markets
40	Croatia						
41	Cuba	34.79		While all sources, including public funds and private, bilateral and multilateral flows, among others, may contribute to the necessary financing required by developing countries, priority should be given to public financing. The role of private financing and investment flows can not be overestimated, given the instability and volatility of such flows, such as that which has prevailed since 2008. What happened in the global carbon markets during this period of economic contraction and deceleration reveals the above	No	May be	
42	Cyprus						
43	Czech Republic						
44	Democratic Republic of the Congo	201.53		1.3 […] to realize a series of activities: […]—The implementation of processes to reduce emissions from deforestation and forest degradation (REDD, 2009): (I) commitment to the Forest Sector Investment Program (2010); (II) adoption of the National Framework Strategy on REDD+ (2012); (III) the creation of the National REDD+ Fund in 2012	No	Yes	
45	Denmark						
46	Djibouti	2.58					
47	Dominica	0.32					

(continued)

Table 3 (continued)

	Country	GHG emissions including LUCF 2012 ($MtCO_2e$)	ETS or carbon mechanisms mentioned in the INDC	Carbon credits in the INDC	Willingness to buy internationa 1 carbon credits	Willingness to sell international carbon credits	Against carbon markets
48	Dominican Republic	31.18		As the NDS indicates a vision for the country to be competitively inserted into a global economy, there is potential to participate in carbon market mechanisms. The development of carbon markets must ensure environmental integrity both nationally and internationally	Maybe	Maybe	
49	Ecuador	138.16					

Conclusions

The fight against climate change is a top priority in the global agenda: the Paris Agreement outlines a clear path for climate and energy policies over the coming decades and the Sustainable Development Goals include, under goal n. 13, the commitment to take urgent action to combat climate change and its impacts.

The Paris Agreement is a milestone in the climate negotiations, but in order to achieve a maximum increase in temperature of +1.5°C, studies show that studies show that, to achieve a maximum increase in temperature of +1.5 °C, more rapid reductions of GHGs are needed. According to Rogelj et al. (2016), the world should reach zero net emissions already between 2045 and 2060. Additional and efficient climate policies are essential to accelerate the transition to a low carbon economy. The Agreement contains also a long-term plan, with a process of review of the nationally determined contributions (NDCs) every five years.

To the present time, 164 NDCs have been submitted to the UNFCCC, representing 191 countries. In this paper it was presented an overview of the existing ETS, of the incumbent ones and how about what the countries included in their NDCs about international markets.

The evaluation of the NDCs presents some difficulties, mostly because of the different content and language barriers.

The questions that it addresses is to what extend the countries are planning to make use of ETS and international carbon credits, through an in depth screening of the NDCs and whether demand and supply will be possibly match.

62 countries have the intention to sell carbon credits (34 of them would like to sell REDD+ credits): this shows that developing countries are striving for climate finance to achieve their mitigation climate change commitments. But this collide with the fact that only 7 countries declare that they are willing to buy international carbon credits (Canada, Costa Rica, Japan, New Zealand, South Korea, Switzerland, Turkey). This discrepancy may need further attention in terms of quantification.

While the role of national emissions trading system is expanding, with new cap-and-trade system to become operational in the coming months and years (and among them the Chinese one), from the first round of presentation of the NDCs, it appears the possibility that the demand for international compliance carbon assets would be declining in the nearest future. The huge discrepancy between intentions to buy and to sell opens to new questions about the chances for developing countries, eager to receive additional revenues to fulfil part of their climate mitigation commitments through the carbon markets, to obtain this kind of finance in the amount that they expect.

References

Caciagli, V. (2013). *Emission trading, del back-loading e di altre riforme, Qualenergia.it*. Available at http://www.qualenergia.it/articoli/20130220-emission-trading-del-back-loading-e-di-altre-riforme. Accessed June 21, 2017.

De Beaupuy, F., & Fouquet, H. (2017). *Macron seeking stiff carbon costs to avert climate change*. Bloomberg.com, June 9, 2017. Available at: https://www.bloomberg.com/news/articles/2017-06-09/macron-said-to-push-for-franco-german-c02-emissions-price-floor. Accessed June, 23, 2017.

De Perthuis, C., & Trotignon, R. (2014). Governance of CO_2 markets: Lessons from the EU ETS. *Energy Policy, 75*, 100–106.

Ellerman, A. D., Convery, F., et al. (2010). *Pricing carbon: The European Union emissions trading scheme*. Cambridge: Cambridge University Press.

European Commission. (2014). Impact assessment, 22/01/2014 – SWD (2014) 17.

European Commission. (2015). *EU ETS handbook*. Available at https://ec.europa.eu/clima/sites/clima/files/docs/ets_handbook_en.pdf. Accessed June 25, 2017.

Federal Office for the Environment. (2017). *Swiss emissions trading scheme (ETS)*. Available at: https://www.bafu.admin.ch/bafu/en/home/topics/climate/info-specialists/climate-policy/emissions-trading/swiss-emissions-trading-scheme–ets-.html. Accessed June 18, 2017.

Hawkins, S., & Jegou, I. (2014). *Linking emissions trading schemes—Considerations and recommendations for a joint EU-Korean carbon market*. Available at: http://www.ictsd.org/sites/default/files/research/linking-emissions-trading-schemes-considerations-and-recommendations-for-a-joint-eu-korean-carbon-market.pdf Accessed June 20, 2017.

Holdaway, E., & Dodwell, C. (2015). *A guide to INDCs, intended nationally determinated contributions*, CDKN. Available at https://www.transparency-partnership.net/sites/default/files/u2055/cdkn-ricardo-aea-guide-to-indcs_final_web.pdf. Accessed June 20, 2017.

ICAP. (2017a). *Emissions trading worldwide—Status Report 2017*. Available at https://icapcarbonaction.com/en/?option=com_attach&task=download&id=447. Accessed June 25, 2017.

ICAP. (2017b). *Nova Scotia releases cap-and-trade design options*. https://icapcarbonaction.com/en/news-archive/454-nova-scotia-releases-cap-and-trade-design-options. Accessed June 16, 2017.

ICAP. (2017c). *Taiwan to introduce emissions trading system*. https://icapcarbonaction.com/en/news-archive/456-taiwan-to-introduce-national-emissions-trading-system. Accessed June 16, 2017.

ICAP. (2017d). *ICAP Quarterly—Global trends in emissions trading*. Issue n. 13, 05 April 2017. Available at https://icapcarbonaction.com/en/newsletter-archive/mailing/view/listid-/mailingid-52/listtype-1#PRICES. Accessed June 16, 2017.

ICAP. (2017e). Rules on cap setting and allowance allocation for emissions trading in China approved. Available at https://icapcarbonaction.com/en/news-archive/457-rules-on-cap-setting-and-allowance-allocation-for-emissions-trading-in-china-approved. Accessed June 19, 2017.

IEA. (2016). *CO_2 emissions from fuel combustion*. Available at https://www.iea.org/publications/freepublications/publication/CO2EmissionsfromFuelCombustion_Highlights_2016.pdf. Accessed June 10, 2017.

Kitous, A., & Keramidas, K. (2015). *Analysis of scenarios integrating the INDCs*. Sevilla, Spain: JRC Policy Brief.

Koch, N., Fuss, S., Grosjean, G., & Edenhofer, O. (2014). Causes of the EU ETS price drop: Recession, CDM, renewable policies or a bit of everything? *Energy Policy, 73*, 676–685.

Laing, T., Sato, M., Grubb, M., Comberti, C. (2013). Assessing the effectiveness of the EU emissions trading scheme. Available at http://www.lse.ac.uk/GranthamInstitute/wp-content/uploads/2014/02/WP106-effectiveness-eu-emissions-trading-system.pdf. Accessed June 15, 2017.

OECD. (2016). *Effective carbon rates: Pricing CO_2 through taxes and emissions trading systems*. Paris: OECD Publishing. Available at http://dx.doi.org/10.1787/9789264260115-en. Accessed June 15, 2017.

Peng, N., Yuqing, L., & Guang, Y. (2017). The establishment of carbon trading market in People's Republic of China: A legislation and policy perspective. *International Journal of Climate Change Strategies and Management, 9*(2), 138–150.

PMR. (2016). *Project implementation status report*. Available at http://pmrturkiye.org/wp-content/uploads/2016/09/Turkey_PMR-Project-Implementation-Status-Report_2016.pdf. Accessed June 15, 2017.

Richstein, J. C., Chappin, E. J. L., et al. (2015). The market (in-)stability reserve for EU carbon emission trading: Why it might fail and how to improve it. *Utilities Policy, 35*(2015), 1–18.

Rogelj, J., den Elzen, M., Höhne, N., Fransen, T., Fekete, H., Winkler, H. (2016). Paris Agreement climate proposals need a boost to keep warming well below 2 °C. *Nature, 534*, 631–639. Available at http://www.nature.com/nature/journal/v534/n7609/abs/nature18307.html. Accessed June 05, 2017.

Sato, M., Laing, T., Cooper, S,, & Wang, L. (2016). Methods for evaluating the performance of emissions trading schemes. Available at http://climatestrategies.org/wp-content/uploads/2015/11/CS-ChinaETS-Evaluation-paper-final1.pdf. Accessed June 25, 2017.

State of Washington—Department of Ecology. (2017). *Clean air rule*. Available at: http://www.ecy.wa.gov/programs/air/permit_register/Clean_Air_Rule/car.htm. Accessed June 16, 2017.

UNFCCC. (2012). *United Nations framework convention on climate change*. Available at https://unfccc.int/resource/docs/convkp/conveng.pdf. Accessed June 19, 2017.

UNFCCC. (2013). Report of the conference of the parties on its nineteenth session, held in Warsaw from 11 to 23 November 2013. Addendum. Part two: Action taken by the Conference of the Parties at its nineteenth session, FCCC/CP/2013/10/Add.1. Available at http://unfccc.int/documentation/documents/advanced_search/items/6911.php?priref=600007788. Accessed June 18, 2017.

UNFCCC. (2014). *Decision 1/CP.20, Lima call for climate action*. http://unfccc.int/files/meetings/lima_dec_2014/application/pdf/auv_cop20_lima_call_for_climate_action.pdf. Accessed June 18, 2017.

UNFCCC. (2015). *Decision 1/CP.21, adoption of the Paris Agreement*. Available at http://unfccc.int/resource/docs/2015/cop21/eng/10a01.pdf#page=2. Accessed June 18, 2017.

Vöhringer, F. (2012). Linking the Swiss emissions trading system with the EU ETS: Economic effects of regulatory design alternatives. *Swiss Journal of Economics and Statistics, 148*(2), 167–196.

Wood, P. J., & Jotzo, F. (2011). Price floors for emissions trading. *Energy Policy, 39*(3), 1746–1753.

WRI/UNDEP. (2015). *Designing and preparing intended nationally determinated contributions (INDCs)*. Available at https://www.wri.org/sites/default/files/designing-preparing-indcs-report.pdf. Accessed June 20, 2017.

Adaptation Technologies in Water Sector Demanded by Developing Countries and the Potential of Technology Transfer of SMEs in South Korea

Ho-Sik Chon, Huncheol Im and SeJin An

Abstract According to the Paris Agreement, the global including both developed and developing countries has responsibility not only to mitigate greenhouse gas emission but also to enhance capacity of climate change adaptation. For the response to climate change, developing countries have requested technical and financial support to developed countries. Water resource management is one of the crucial issues on adapting to climate change. A number of developing countries have suffered from water shortage for agricultural and domestic withdrawal, asking for technology assistance to improve water quality and water supply at local level. In this paper, adaptation technologies in water sector demanded by developing countries were provided by reviewing Technology Needs Assessment (TNA) reports that 79 developing countries submitted to the UNFCCC. Considering national indicators such as GDP growth rate and the Human Development Index (HDI), technology needs from developing countries were analysed. Moreover, this paper identified technology groups in water sector developed by small and medium enterprises (SMEs) in South Korea that could be transferred and implemented to developing countries. By comparing technology needs from developing countries with the number of associated SMEs and carrying out expert survey, promising technologies in water sector that SMEs in South Korea have relatively high competitiveness in technology transfer to developing countries were determined.

Keywords Adaptation technology · Water sector · Technology Needs Assessment (TNA) · Technology transfer · Developing country · Small and medium enterprises (SMEs)

H.-S. Chon (✉) · H. Im · S. An
Green Technology Center, 17th Floor Namsan Square Bldg, 173, Toegye-ro, Jung-gu, Seoul 04554, South Korea
e-mail: hschon@gtck.re.kr

F. Alves et al. (eds.), *Theory and Practice of Climate Adaptation*, Climate Change Management, https://doi.org/10.1007/978-3-319-72874-2_32

Introduction

As a new climate change response framework, Paris Agreement was announced through COP21 in 2015. The Kyoto Protocol concentrated on the greenhouse gas reduction from developed countries; however, this new regime requires the efforts of both developed and developing countries to conduct actions against climate change (UNFCCC 2015). Under the Paris Agreement, it is expected that the world has the same goal and responsibilities for climate change response.

Awareness of the adaptation to enhance climate resilience has been spreading in most countries around the world (APAN 2016; EC 2017). Based on *the EU strategy on adaptation to climate change* (EC 2013), a number of member states in European Union have established and implemented adaptation policies with a wide range of adaptive measures[1]. Developing countries, however, request support to the international community for reducing their vulnerability to climate impacts and increasing their capacity and knowledge (CTCN 2016). In particular, as climate risks such as depletion of water resources, food shortage, and damage from natural disasters steadily increase, technical assistances have been implemented for developing countries. Due to increasing demands of adaptation technologies from developing countries, it is necessary to facilitate technology transfer at regional or country levels based on the analysis of their capacity and economic situation.

South Korea used to be a developing country that received technical and financial support; however, this country becomes a country with high level of technology development and has a potential for continuously conducting south-south cooperation under the UNFCCC. Although private and public sectors in South Korea have a capacity for technical support and knowledge sharing to developing countries, they have paid their attention to domestic market. Likewise, the South Korean Government has been more interested in developing level of national technology R&D rather than contributing to bilateral and multilateral cooperation with developing countries for climate change response. Under the Paris Agreement, it is required that the Government and companies in South Korea communicate with developing countries in terms of their technology needs not only for conducting national responsibility as a donor of climate technologies in the international community but also for expanding their business opportunity.

Based on reviewing Technology Needs Assessment (TNA) reports that developing countries submitted to the UNFCCC, this paper evaluated their technology demands in water sector. With this, promising water technologies in South Korea in association with technology transfer to developing countries were identified by considering their demands and supply capacity of small and medium enterprises (SMEs) in water sector.

---

[1]For more information, visit the following website: http://climate-adapt.eea.europa.eu/.

Methodology

This study comprised four different stages. As the first stage, water technologies demanded from developing countries were assessed based on literature review. TNA reports that 79 developing countries submitted to the UNFCCC by April 2016 (UNEP-DTU Partnership 2016; UNFCCC 2016) were reviewed to identify countries with their water technology demands at technology class level[2]. Differences of technology demands at regional and country group levels were provided by using socio-economic and climate change-related indicators.

At the second stage, classification (taxonomy) of water technologies in relation to technology demands from developing countries was developed. Technology classes demanded from developing countries were classified into different technology groups in consultation with experts in water sector. Following this stage, SMEs in South Korea were evaluated to ascertain their capacity as a provider of water technologies to developing countries. All of the SMEs with green technology certification[3] in water sector were investigated, and those with water technologies in relation to demands from developing countries were screened. Technologies of screened SMEs were then classified into technology groups defined in the previous stage. Moreover, an additional list of SMEs in water sector was made, using Korean company information service database (KIS-VALUE)[4]. All of the SMEs in each sector of water industry with the top 10% sales, listed in the KIS-VALUE, were collected. Products of the collected SMEs, such as patents and technologies, were then evaluated to secure SMEs with water technologies in relation to demands from developing countries. Information of the SMEs and their products were also classified into technology groups defined in this paper.

At the last stage, promising water technologies were determined, using two different approaches. In the first approach, technology groups with higher competency of technology transfer to developing countries than others were identified through a linkage between technology demands from developing countries and the number of associated SMEs in South Korea. Expert survey was conducted, as the second approach, using four evaluation indicators including technical effect, economic effect, greenhouse gas reduction effect and the necessity of government support. Eleven experts in water sector evaluated each technology class in consideration with the cases of its transfer to developing countries. An average of scores for each evaluation indicator of the technology class was used to represent the score for that of the technology group. Results from these approaches were

---

[2]In the TNA process, a climate technology demanded by a developing country is classified into a technology class. For more information, visit the following website: http://www.database.tech-action.org/.

[3]This certification is approved by South Korean Government. A list of SMEs with the green technology certification in South Korea can be found in the follow website (in Korean): https://www.greencertif.or.kr.

[4]For more information, visit the following website (in Korean): http://www.kisvalue.com.

taken into account together to decide promising water technologies in South Korea with regards to technology transfer to developing countries.

Technology Demands in Water Sector from Developing Countries

Among the 79 developing countries submitting TNA reports by April 2016, 56 countries were identified to have technology demands in water sector (Table 1). The water technologies demanded from developing countries were categorised into 37 technology classes which are presented in Table 2. The top 5 technology classes with the highest number of demands from developing countries were 'rainwater harvesting', 'wastewater treatment and reuse', 'desalination', 'planning for safe water supply', 'water infrastructure operation'.

The technology demands from different regions (Asia-Pacific, Latin America and Caribbean (LAC), Africa and Europe) were compared at technology class level, as shown in Fig. 1. 'Rainwater harvesting' was the highest demanded technology from all regions. Asia-Pacific, LAC and Africa showed similar pattern of technology demands with slightly different technology priority. Technology demands from European countries were related to water storage and supply (construction and maintenance of dams and reservoirs, irrigation—surface self-flow) and leakage water management (leakage reduction and loss management in water supply/Water loss reduction).

By applying various socio-economic and climate change-related indicators, developing countries were intended to be grouped and compared to identify the difference of technology demands depending on the social, economic and environmental status of the countries. In Fig. 2, developing countries were categorised into two different groups by using GDP growth rate (World Bank 2016) and Human Development Index (HDI) (UNDP 2016). In general, water technologies were less

Table 1 A list of developing countries demanding technologies in water sector

Region	Developing country
Asia-Pacific (17 countries)	Armenia, Bhutan, Cambodia, Indonesia, Iran (Islamic Republic of), Jordan, Kazakhstan, Laos, Lebanon, Mongolia, Niue, Philippines, Samoa, Sri Lanka, Tajikistan, Thailand, Vietnam
Latin America and Caribbean (12 countries)	Antigua and Barbuda, Costa Rica, Cuba, Dominica, Dominican Republic, Ecuador, Guyana, Haiti, Jamaica, Peru, Saint Kitts and Nevis, Saint Lucia
Africa (22 countries)	Botswana, Burkina Faso, Cape Verde, Chad, Comoros, Cote d'Ivoire, Egypt, Ghana, Guinea, Kenya, Madagascar, Mali, Mauritius, Morocco, Namibia, Rwanda, Senegal, South Africa, Sudan, Tunisia, Zambia, Zimbabwe
Europe (5 countries)	Albania, Azerbaijan, Georgia, Malta, Republic of Moldova

Table 2 Technology demands from developing countries at technology class level

TNA technology class	Number of demands from developing countries
Atmospheric water generation	1
Artificial recharge of aquifers	5
Boreholes for water supply	4
Community-based water management	1
Construction and maintenance of dams and reservoirs	1
Deep wells for water supply in dry season	1
Desalination	15
Fog harvesting	1
Groundwater extraction—renewable energy	2
Groundwater assessment and monitoring	1
Integrated river basin management (IRBM)	1
Integrated urban water resource management	1
Irrigation information system and best practice dissemination	1
Irrigation—surface self-flow	1
Leakage reduction and loss management in water supply/ Water loss reduction	4
Localised irrigation	1
Micro hydropower—adaptation	2
Planning for safe water supply	12
Rain and snow water harvesting—herder groups	1
Rainwater harvesting	31
Small dams for continuous water supply	2
Water harvesting—earth dam	2
Water infrastructure operation	9
Water reclaim and reuse	4
Water resource assessment and prediction	3
Water safety plan	3
Water saving and reuse—production system change	1
Water saving at taps	2
Water saving irrigation	1
Watershed management	2
Water user associations	2
Water treatment and storage—household	2
Water treatment—filtration	3
Water treatment—solar distillation	1
Wastewater treatment and reuse	18
Weather modification—artificial raining	1
Wells for groundwater extraction	3

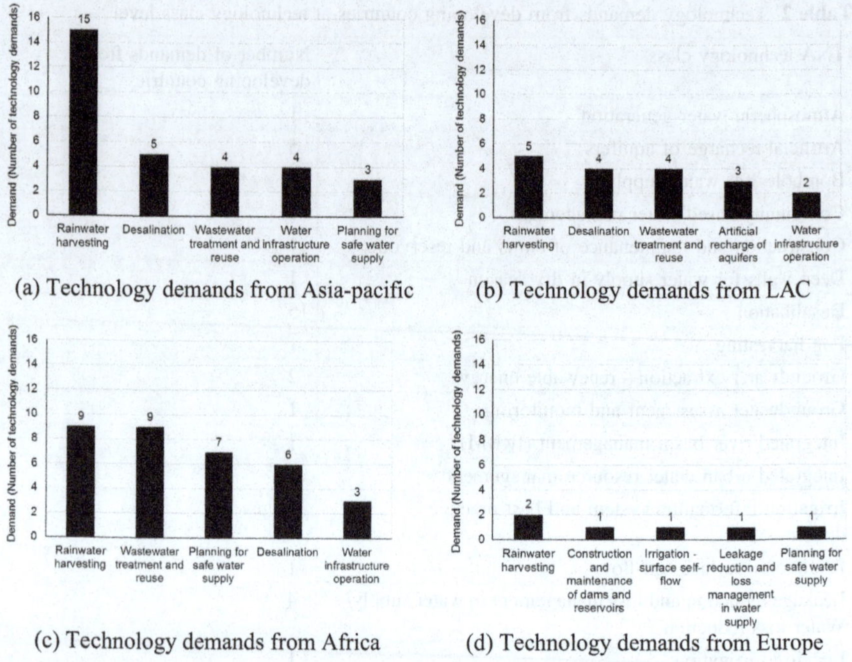

Fig. 1 Technology demands from different regions

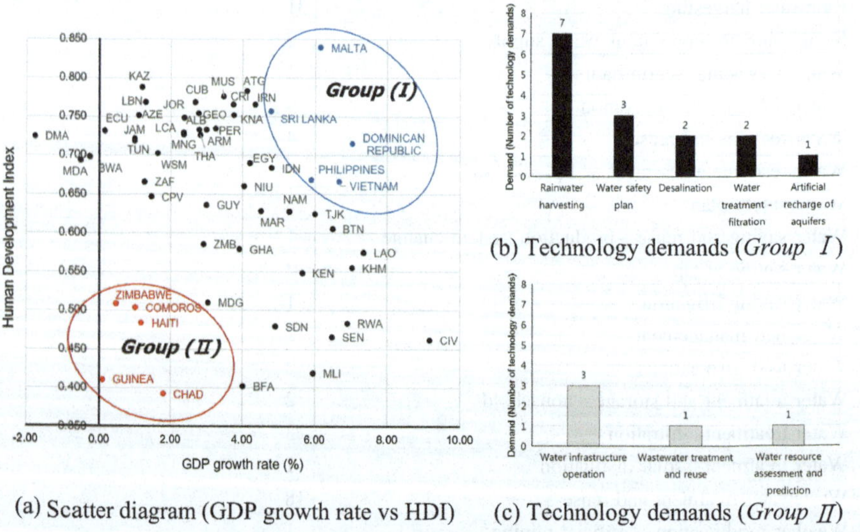

Fig. 2 Technology demands from different country groups categorised by using socio-economic indicators

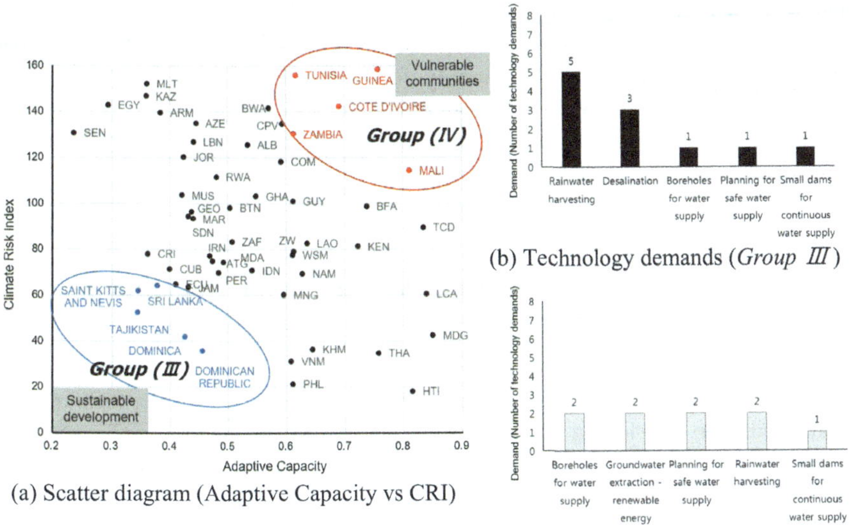

(a) Scatter diagram (Adaptive Capacity vs CRI)

(b) Technology demands (*Group Ⅲ*)

(c) Technology demands (*Group Ⅳ*)

Fig. 3 Technology demands from different country groups categorised by using climate change-related indicators

requested from developing countries in *group II* compared to those in *group I*. In developing countries of *group I*, which were Malta, Sri Lanka, Dominican Republic, Philippines and Vietnam, 'rainwater harvesting' was the most highly demanded technology followed by 'water safety plan', 'desalination' and 'water treatment-filtration'. Developing countries in *group II* with low GDP growth rate and low HDI selected 'water infrastructure operation', 'wastewater treatment and reuse' and 'water resource assessment and prediction' as technologies needed for adaptation in water sector.

As Downing and Patwardhan (2005) suggested, adaptation-related indicators, Adaptive Capacity[5] (ND-GAIN 2016) and Climate Risk Index (CRI)[6] (Germanwatch 2016), were applied to categorise developing countries into two different groups with regards to climate vulnerability. As presented in Fig. 3, St. Kitts and Navis, Sri Lanka, Tajikistan, Dominica and Dominican Republic were classified as 'sustainable development' group (*group III*), whereas Tunisia, Guinea, Cote d'Ivoire, Zambia and Mali were 'vulnerable communities' group (*group IV*). It was identified that developing countries in *group IV* had relatively high exposure to unsafe water, sanitation and hygiene (WASH) and low population density (World Bank 2016). Developing countries in *group III* prioritised 'rainwater harvesting'

[5]An average of Adaptive Capacity values of a developing country between 2011 and 2015 was represented as a value for Adaptive Capacity of that country.

[6]An average of CRIs of a developing country between 1996 and 2015 was represented as a value for CRI of that country.

and 'desalination' as technologies for adaptation to climate change in water sector. However, diverse technologies such as 'boreholes for water supply', 'groundwater extraction-renewable energy', 'planning for safe water supply' and 'rainwater harvesting' were requested by developing countries in *group IV*. It was estimated that implementation of technologies for groundwater supply was a major concern in developing countries of *group IV*.

Classification of Technologies in Water Sector Demanded by Developing Countries

Based on a review of TNA reports, classification (taxonomy) of water technologies demanded from developing countries was provided in consultation with experts in water sector (Fig. 4). Technology demands in water sector from developing countries were classified into two sub-sectors, which were 'water quantity' and 'water quality'. The sub-sector 'water quantity' consisted of 4 categories including 'groundwater utilisation', 'surface water utilisation', 'other water resource utilisation' and 'system operation and management', and ten different technology groups were distributed in these categories. Whereas, the sub-sector 'water quality' had one category 'water treatment', and it consisted of one technology group

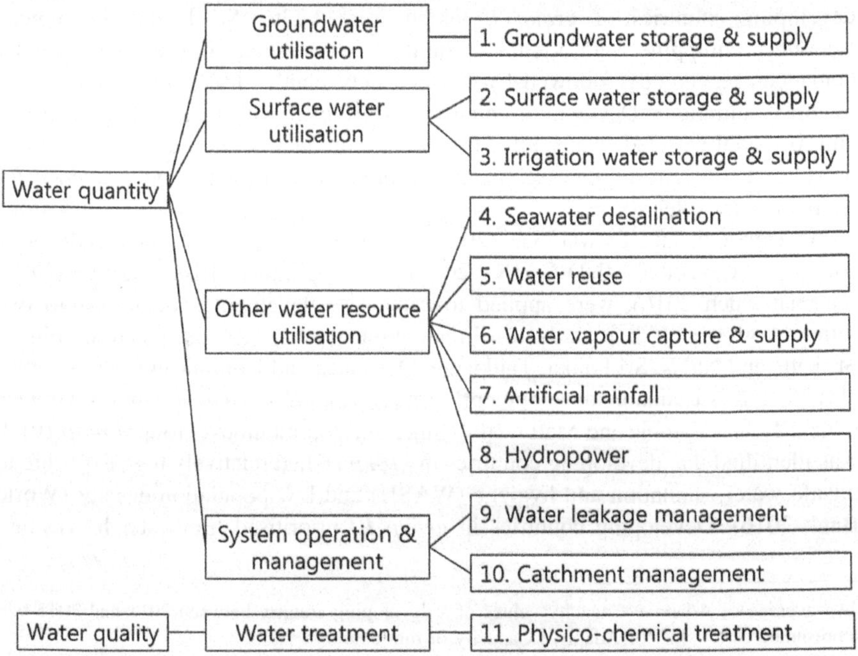

Fig. 4 Classification of technologies in water sector demanded from developing countries

Table 3 A list of technology groups and associated technologies demanded from developing countries

Technology group	TNA technology class
1. Groundwater storage and supply	– Wells for groundwater extraction – Artificial recharge of aquifers – Boreholes for water supply – Groundwater extraction—renewable energy
2. Surface water storage and supply	– Construction and maintenance of dams and reservoirs – Rainwater harvesting – Small dams for continuous water supply – Water harvesting—earth dam – Rain and snow water harvesting—hearder groups
3. Irrigation water storage and supply	– Irrigation information system and best practice dessemination – Localised irrigation – Water saving irrigation – Irrigation—surface self-flow
4. Seawater desalination	– Desalination
5. Water reuse	– Water reclaim and reuse – Wastewater treatment and reuse – Water treatment and storage—household – Water saving and reuse—production system change
6. Water vapour capture and supply	– Atmospheric water generation – Fog harvesting
7. Artificial rainfall	– Weather modification—artificial raining
8. Hydropower	– Micro hydropower—adaptation
9. Water leakage management	– Leakage reduction and loss management in water supply/ Water loss reduction – Water saving at taps
10. Catchment management	– Integrated river basin management (IRBM) – Integrated urban water resource management – Groundwater assessment and monitoring – Water infrastructure operation Watershed management – Community-based water management – Water resource assessment and prediction – Deep wells for water supply in dry season – Planning for safe water supply – Water user associations – Water safety plan
11. Physico-chemical treatment	– Water treatment—filtration – Water treatment—solar distillation

'physico-chemical treatment'. A list of technology groups and associated technologies demanded from developing countries in water sector is presented in Table 3.

A Potential of SMEs in South Korea for Technology Transfer in Water Sector to Developing Countries

The statistical results of SMEs in South Korea with water technologies (patent, technology) demanded from developing countries were presented in Table 4. It was identified that 209 SMEs[7] had water technologies in 8 technology groups. 'Water reuse' was the technology group with the highest number of SMEs in South Korea, followed by 'catchment management', 'physico-chemical treatment' and 'surface water storage and supply'. The highest number of patents and technologies SMEs owned were related to 'water reuse'. SMEs with water technologies in 'physico-chemical treatment' and 'catchment management' were assessed to have 71 and 47 patents respectively. 'Catchment management' and 'water leakage management' were the second and third technology groups with the high number of technologies produced by SMEs in South Korea.

The number of SMEs in each technology group was compared with the number of technology demands from developing countries (Fig. 5). In 'surface water storage and supply' and 'seawater desalination', the number of SMEs seemed to be not enough to respond to technology demands from developing countries. On the other hand, South Korea had the high number of SMEs in 'water reuse', 'water leakage management', 'catchment management' and 'physico-chemical treatment', which could provide their products (patents and technologies) to developing countries. In particular, SMEs in South Korea seemed to have high competitiveness in wastewater treatment technology.

Identification of Promising Water Technologies in South Korea with High Competitiveness in Technology Transfer to Developing Countries

A linkage between technology demands and supply capacity in South Korea was carried out to identify technology groups that could be considered as high priority areas for technical support to developing countries. As shown in Fig. 6, GDP of developing countries from World Bank (2016) was used to define the degree of their demands in water sector (X-axis). A supply capacity (Y-axis) for each technology group was calculated by dividing the number of SMEs with a technology of concern by the total number of SMEs with water technologies identified in this

[7]Duplicates were made when calculating the number of SMEs at technology class level.

Table 4 Statistics of SMEs and their technology ownership in South Korea

Technology group	TNA technology class	SME		Patent			Technology			Technology demand	
		No.	%	No.	No. of SMEs	%	No.	No. of SMEs	%	No.	%
Groundwater storage and supply	Wells for groundwater extraction	2	1.0	3	2	0.8	0	0	0	3	2.8
Surface water storage and supply	Construction and maintenance of dams and reservoirs	4(1[a])	1.9	4	4(1[a])	6.2	0	0	3.1	1	34.0
	Rainwater harvesting	9(6[a])	4.3	14	7(5[a])		2	2(1[a])		31	
	Small dams for continuous water supply	1	0.5	3	1		0	0		2	
	Water harvesting—earth dam	2(2[a])	1.0	2	2(2[a])		0	0		2	
Seawater desalination	Desalination	4(1[a])	1.9	5	3	1.3	1	1(1[a])	1.5	15	14.1
Water reuse	Water reclaim and reuse	15(8[a])	7.2	21	14(7[a])	57.6	5	1(1[a])	43.1	4	22.6
	Wastewater treatment and reuse	75 (35[a])	35.8	193	61 (34[a])		23	14(1[a])		18	
	Water treatment and storage—household	1(1[a])	0.5	1	1(1[a])		0	0		2	
Hydropower	Micro hydropower—adaptation	1(1[a])	0.5	4	1(1[a])	1.1	0	0	0	2	1.9
Water leakage management	Leakage reduction and loss management in water supply/ Water loss reduction	15(4[a])	7.2	5	4(3[a])	1.3	12	11(1[a])	18.5	4	3.8

(continued)

Table 4 (continued)

Technology group	TNA technology class	SME		Patent			Technology			Technology demand	
		No.	%	No.	No. of SMEs	%	No.	No. of SMEs	%	No.	%
Catchment management	Integrated urban water resource management	6(1[a])	2.9	4	4	12.6	2	2(1[a])	21.5	1	17.0
	Groundwater assessment and monitoring	10(6[a])	4.8	21	10(6[a])		0	0		1	
	Integrated river basin management (IRBM)	6(2[a])	2.9	2	1(1[a])		5	5(1[a])		1	
	Water infrastructure operation	7(2[a])	3.3	9	7(2[a])		0	0		9	
	Watershed management	12(7[a])	5.7	10	7(7[a])		3	5		2	
	Community-based water management	2	1.0	0	0		2	2		1	
	Water resource assessment and prediction	3	1.4	1	1		2	2		3	
Physico-chemical treatment	Water treatment—filtration	33 (19[a])	15.7	71	28 (18[a])	19.1	7	5(1[a])	12.3	3	3.8
	Water treatment—solar distillation	1(1[a])	0.5	0	0		1	1(1[a])		1	
Total		209	100	373	158	100	65	51	100	106	100
Total (without duplication of companies)		131			85			46			

[a]The number of SMEs with two or more technologies

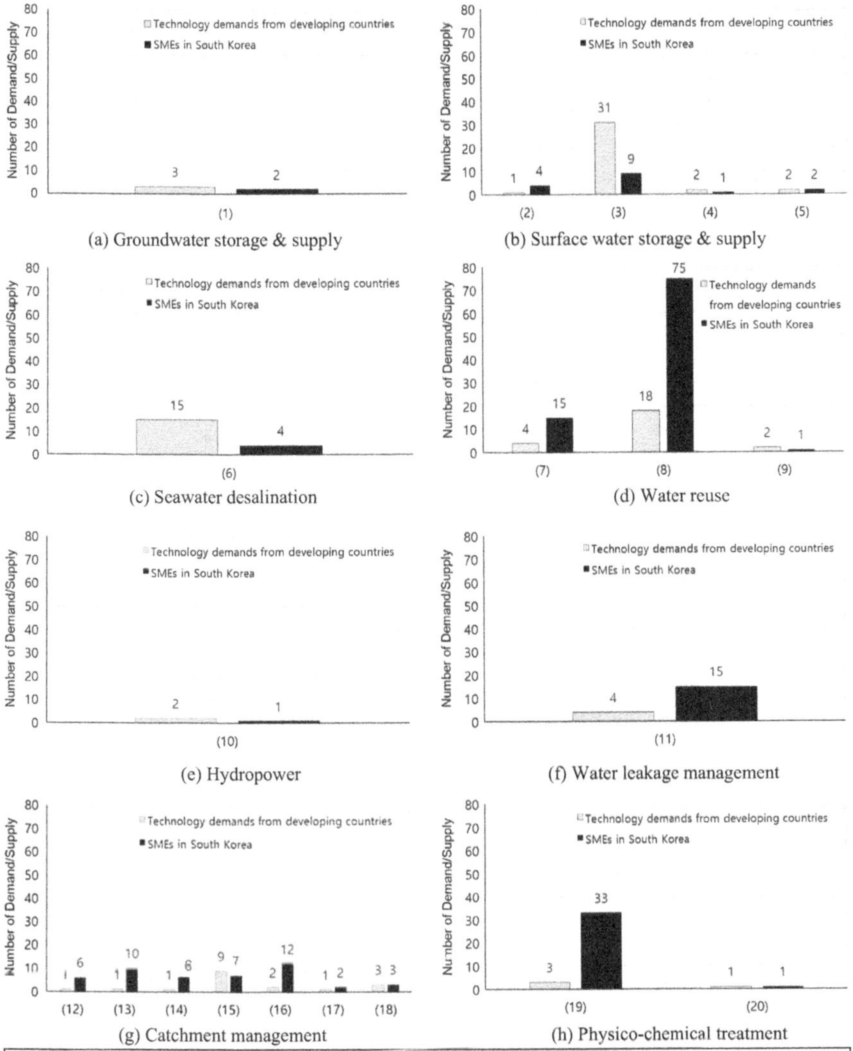

(1) Wells for groundwater extraction	(11) Leakage reduction and loss management in water supply
(2) Construction and maintenance of dams and reservoirs	(12) Integrated urban water resource management
(3) Rainwater harvesting	(13) Groundwater assessment and monitoring
(4) Small dams for continuous water supply	(14) Integrated river basin management(IRBM)
(5) Water harvesting - earth dam	(15) Water infrastructure operation
(6) Desalination	(16) Watershed management
(7) Water reclaim and reuse	(17) Community-based water management
(8) Wastewater treatment and reuse	(18) Water resource assessment and prediction
(9) Water treatment and storage – household	(19) Water treatment – filtration
(10) Micro hydropower – Adaptation	(20) Water treatment - solar distillation

Fig. 5 Comparison of the number of SMEs in South Korea with the number of technology demands from developing countries

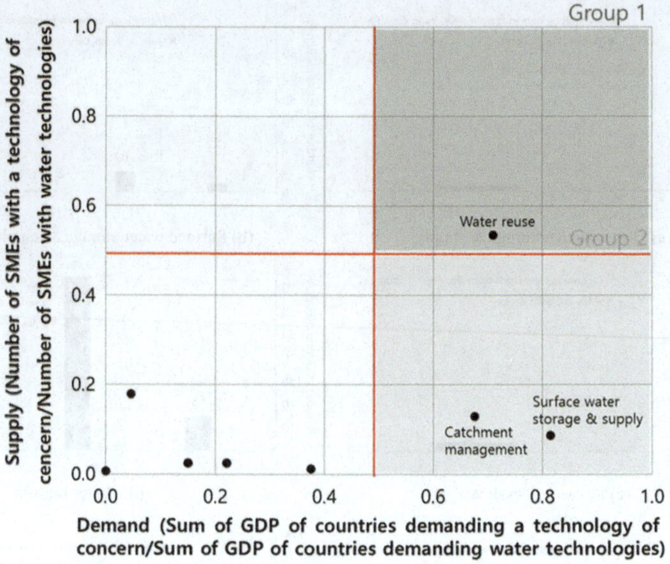

Fig. 6 Results of the linkage between technology demands and supply capacity in South Korea

paper. At the quadrant matrix, 'water reuse' was identified as a technology group with relatively high demands and high supply capacity (*Group 1*). 'Surface water storage and supply' and 'catchment management' were technology groups that had higher demands but relatively lower supply capacity in South Korea (*Group 2*).

With this, expert survey was conducted to estimate benefits that SMEs could have from transferring their water technologies to developing countries. Four evaluation indicators including technical effect, economic effect, greenhouse gas reduction effect and necessity of government support were answered by eleven experts in private and public companies, academia and research institutions. The experts scored each indicator based on 5-point Likert scale ranging from 1 (indicating the lowest effect) to 5 (indicating the highest effect). Results of expert survey are presented in Fig. 7. 'Seawater desalination', 'hydropower', 'water reuse', 'catchment management' and 'water leakage management' were evaluated as technology groups providing higher technical and economic benefits, greenhouse gas emission reduction and governmental support to SMEs in South Korea. In particular, 'seawater desalination' was assessed to make the highest technical and economic effects among the technology groups.

Findings from the linkage between technology demands and supply capacity and expert survey were considered together to identify promising water technologies in South Korea with regards to technology transfer to developing countries. Based on the findings, 'water reuse' and 'catchment management' were technology groups

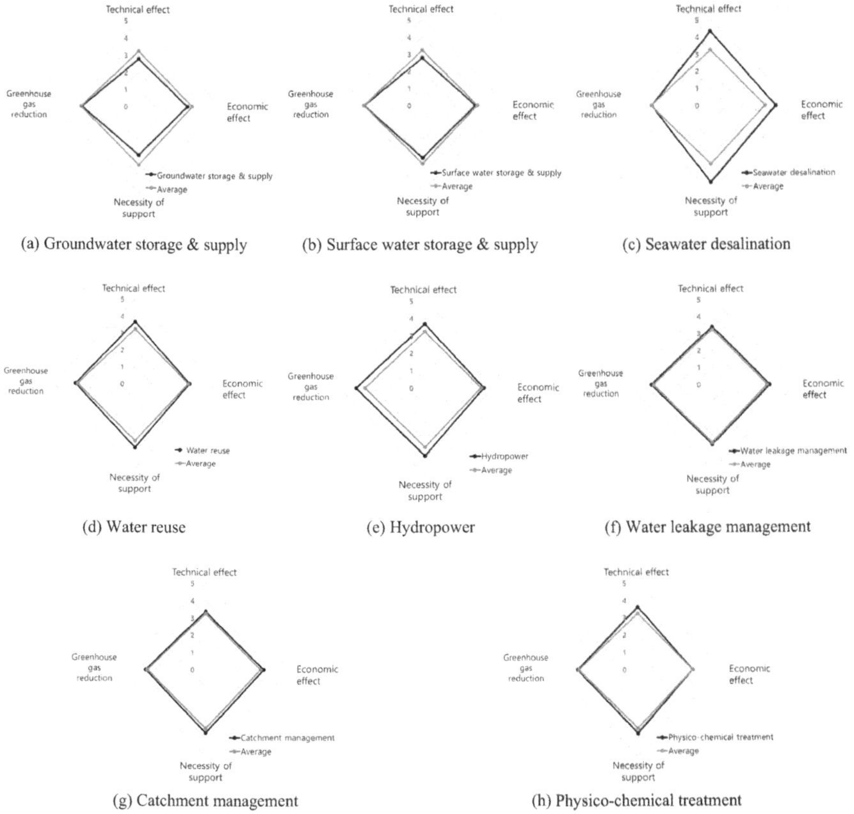

(a) Groundwater storage & supply (b) Surface water storage & supply (c) Seawater desalination

(d) Water reuse (e) Hydropower (f) Water leakage management

(g) Catchment management (h) Physico-chemical treatment

Fig. 7 Results of expert survey

determined as promising technologies in water sector. Although 'seawater desalination' and 'hydropower' had higher evaluation scores from the expert survey compared to other technology groups, they were not included as promising water technologies due to the limited number of SMEs in South Korea.

Discussion and Conclusions

Results from this study showed that SMEs in South Korea could not contribute to all of the technical assistances in water sector demanded from developing countries (54% and 73% of technology demands at technology class level and technology group level respectively could be assisted and implemented by SMEs in South Korea). In South Korea, water industry and associated technologies have been developed mainly based on domestic market. In general, the level of water technologies used in the domestic market in South Korea is higher than that demanded from developing countries. Some technologies requested from developing countries are not developed and/or produced by SMEs in South Korea due to the fact that there are no demands, and it is difficult to expect the economic benefits in the domestic market.

Under the Paris Agreement, it is required that the South Korean Government develops a plan or a strategy for the R&D and implementation of water technologies in consideration with technical support to developing countries. With regards to technology transfer, programmes for enhancing capacity of SMEs will also need to be provided in order to expand their market externally and contribute to climate change response at global level. Moreover, it is necessary that SMEs in South Korea carry out proactive approach to identify technology needs and evaluate the relevance of their technologies to the needs in collaboration with the government and stakeholders of developing countries. For this, the South Korean Government needs to develop good bilateral relationship with developing countries, provide financial support to SMEs and incorporate with international organisation to facilitate the transfer of water technologies to developing countries.

Acknowledgements This paper partially used the data from the research project, entitled "Promising adaptation technologies and evaluation framework in response to demand from developing countries (No. 2016-011)", which was supported by the Green Technology Center (GTC) and Ministry of Science and ICT of South Korea (MSIT).

References

APAN. (2016). Adapting and living below 2°C: Outcomes of the 5th Asia Pacific Climate Change Adaptation Forum. Asia-Pacific Adaptation Network.

CTCN. (2016). Global climate commitments in action: 2016 Progress report. Climate Technology Centre & Network.

Downing, T. E. & Patwardhan, A. (2005). Assessing vulnerability for climate adaptation. In: B. Lim, E. Spanger, I. Burton, E. L. Malone & S. Huq (Eds.), *Adaptation policy frameworks for climate change: Developing strategies, policies and measures* (pp. 67–89). UNDP.

EC. (2013). Communication from the Commission to the European Parliament, the Council, the European Economic and Social Committee and the Committee of the Regions: An EU strategy on adaptation to climate change. European Commission.

EC. (2017). *Adaptation to climate change*. [Online] Available from https://ec.europa.eu/clima/policies/adaptation_en. Accessed September 29, 2017.

Germanwatch. (2016). *Global climate risk index 2017*. [Online] Available from https://germanwatch.org/en/download/16411.pdf. Accessed September 4, 2017.

ND-GAIN. (2016). *Notre Dame global adaptation index*. [Online] Available from http://index.gain.org. Accessed December 23, 2016.

UNDP. (2016). *Human development report 2015*. [Online] Available from http://report.hdr.undp.org. Accessed December 23, 2016.

UNFCCC. (2015). *Paris Agreement*. [Online] Available from http://unfccc.int/meetings/paris_nov_2015/items/9445.php . Accessed September 28, 2017.

UNFCCC. (2016). *Technology needs assessment reports (TNA): Country reports*. [Online] Available from http://unfccc.int/ttclear/templates/render_cms_page?TNR_cre. Accessed November 15, 2016.

UNEP-DTU Partnership. (2016). *TNA fact sheet.* [Online] Available from http://www.database.tech-action.org . Accessed November 15, 2016.
World Bank. (2016). *World development indicators.* [Online] Available from http://data.worldbank.org/data-catalog/world-development-indicators. Accessed December 23, 2016.

Author Biographies

Dr. Ho-Sik Chon is a Senior Researcher at the Green Technology Center (GTC) which is a technology-based policy research institute funded by the South Korean Government. He obtained Ph.D. in water management at the Centre for Environmental Policy, Imperial College London in 2012. Since July 2013, he has joined the GTC, supporting the Government to establish national R&D plans and strategies for climate technologies in South Korea. Dr. Chon is carrying out research on the forecasting of promising climate technologies in response to demands from developing countries. He uses Technology Needs Assessment (TNA) reports submitted to the UNFCCC as key reference to identify technology demands from developing countries and to evaluate the potential of SMEs in South Korea as providers of adaptation and mitigation technologies.

Dr. Huncheol Im is a Researcher at the GTC. He received Ph.D. in wastewater treatment at the School of Earth Science and Environmental Engineering, Gwangju Institute of Science and Technology (GIST) in 2016. Since September 2016, he has worked at the GTC, researching climate technology R&D policy and international cooperation for technology transfer. Dr. Im is currently carrying out research on developing a methodology for international cooperation of climate technology with developing countries.

Mr. SeJin An is a Researcher at the GTC as a specialist in international cooperation and financial linkage of climate technology. He obtained a master's degree at the Department of International Development Studies, Hankuk University of Foreign Studies. He has experienced technological development cooperation in south-east Asia for several years. He worked for the Korean Water Resources Corporation and Korea International Cooperation Agency as a mechanical engineer in domestic and international environment and has participated in several research projects related to technology transfer at the GTC.

Erratum to: Promoting Private Sector Engagement in Climate Change Adaptation and Flood Resilience—A Case Study of Innovative Approaches Applied by MSMEs in Mumbai, India

Caroline Schaer and Archana Pantakar

Erratum to:
Chapter "Promoting Private Sector Engagement in Climate Change Adaptation and Flood Resilience—A Case Study of Innovative Approaches Applied by MSMEs in Mumbai, India" in: F. Alves et al. (eds.), *Theory and Practice of Climate Adaptation*, Climate Change Management, https://doi.org/10.1007/978-3-319-72874-2_10

The original version of the book was inadvertently published with misplaced Tables 2 and 3 in Chapter 10, which have been now corrected. The erratum chapter and the book have been updated with the changes.

The updated online version of this chapter can be found at
https://doi.org/10.1007/978-3-319-72874-2_10

F. Alves et al. (eds.), *Theory and Practice of Climate Adaptation*,
Climate Change Management, https://doi.org/10.1007/978-3-319-72874-2_33

The manufacturer's authorised representative in the EU is Springer Nature Customer Service Centre GmbH, Europaplatz 3, 69115 Heidelberg, Germany. If you have any concerns regarding our products, please contact ProductSafety@springernature.com

Printed and bound by CPI Group (UK) Ltd, Croydon, CR0 4YY
30/09/2026
02225658-0016